Lecture Notes in Civil Engineering

Volume 103

Series Editors

Marco di Prisco, Politecnico di Milano, Milano, Italy

Sheng-Hong Chen, School of Water Resources and Hydropower Engineering, Wuhan University, Wuhan, China

Ioannis Vayas, Institute of Steel Structures, National Technical University of Athens, Athens, Greece

Sanjay Kumar Shukla, School of Engineering, Edith Cowan University, Joondalup, WA, Australia

Anuj Sharma, Iowa State University, Ames, IA, USA

Nagesh Kumar, Department of Civil Engineering, Indian Institute of Science Bangalore, Bengaluru, Karnataka, India

Chien Ming Wang, School of Civil Engineering, The University of Queensland, Brisbane, QLD, Australia

Lecture Notes in Civil Engineering (LNCE) publishes the latest developments in Civil Engineering - quickly, informally and in top quality. Though original research reported in proceedings and post-proceedings represents the core of LNCE, edited volumes of exceptionally high quality and interest may also be considered for publication. Volumes published in LNCE embrace all aspects and subfields of, as well as new challenges in, Civil Engineering. Topics in the series include:

- Construction and Structural Mechanics
- Building Materials
- Concrete, Steel and Timber Structures
- Geotechnical Engineering
- Earthquake Engineering
- Coastal Engineering
- Ocean and Offshore Engineering; Ships and Floating Structures
- Hydraulics, Hydrology and Water Resources Engineering
- Environmental Engineering and Sustainability
- Structural Health and Monitoring
- Surveying and Geographical Information Systems
- Indoor Environments
- Transportation and Traffic
- Risk Analysis
- Safety and Security

To submit a proposal or request further information, please contact the appropriate Springer Editor:

– Mr. Pierpaolo Riva at pierpaolo.riva@springer.com (Europe and Americas);
– Ms. Swati Meherishi at swati.meherishi@springer.com (Asia - except China, and Australia, New Zealand);
– Dr. Mengchu Huang at mengchu.huang@springer.com (China).

All books in the series now indexed by Scopus and EI Compendex database!

More information about this series at http://www.springer.com/series/15087

Sandip Kumar Saha · Mousumi Mukherjee
Editors

Recent Advances in Computational Mechanics and Simulations

Volume-I: Materials to Structures

Editors
Sandip Kumar Saha
School of Engineering
Indian Institute of Technology Mandi
Mandi, Himachal Pradesh, India

Mousumi Mukherjee
School of Engineering
Indian Institute of Technology Mandi
Mandi, Himachal Pradesh, India

ISSN 2366-2557　　　　　　ISSN 2366-2565　(electronic)
Lecture Notes in Civil Engineering
ISBN 978-981-15-8137-3　　　ISBN 978-981-15-8138-0　(eBook)
https://doi.org/10.1007/978-981-15-8138-0

© The Editor(s) (if applicable) and The Author(s), under exclusive license to Springer Nature Singapore Pte Ltd. 2021
This work is subject to copyright. All rights are solely and exclusively licensed by the Publisher, whether the whole or part of the material is concerned, specifically the rights of translation, reprinting, reuse of illustrations, recitation, broadcasting, reproduction on microfilms or in any other physical way, and transmission or information storage and retrieval, electronic adaptation, computer software, or by similar or dissimilar methodology now known or hereafter developed.
The use of general descriptive names, registered names, trademarks, service marks, etc. in this publication does not imply, even in the absence of a specific statement, that such names are exempt from the relevant protective laws and regulations and therefore free for general use.
The publisher, the authors and the editors are safe to assume that the advice and information in this book are believed to be true and accurate at the date of publication. Neither the publisher nor the authors or the editors give a warranty, expressed or implied, with respect to the material contained herein or for any errors or omissions that may have been made. The publisher remains neutral with regard to jurisdictional claims in published maps and institutional affiliations.

This Springer imprint is published by the registered company Springer Nature Singapore Pte Ltd.
The registered company address is: 152 Beach Road, #21-01/04 Gateway East, Singapore 189721, Singapore

Foreword

Indian Institute of Technology Mandi (IIT Mandi) organized the 7th International Congress on Computational Mechanics and Simulation (ICCMS 2019), which witnessed presentations of exceptionally high-quality research in various domains of computation mechanics and allied areas. Selected papers, presented in the congress, are compiled into two book volumes. The first volume primarily reports the advances made in recent years in mechanics of materials and structures, and the second volume reports the advances made in nano- to macro-mechanics and application of computational techniques in several emerging areas in engineering and technology.

This volume is comprised of 50 chapters categorized into seven sections. The first section, having eight chapters, is dedicated to the mechanics of solids and structures. These chapters report new developments in the understanding of wave propagation in solids, their deformation behavior, and mechanics of structural systems. There are two chapters that focus on the mechanics of steel-concrete composite elements. Another two chapters in this section highlight the mechanics of offshore structures. This section provides some readily useful information on the behavior of steel-concrete composite sections, which are becoming common nowadays in modern constructions. Further, some observations reported in this section help to uncover the mechanics of innovative offshore structures and their components. Behavior of these structures needs more attention as they are associated with priority services such as renewable energy and potable water supply to be provided uninterruptedly.

Specific focus on special structural elements, namely plates and shells, is notable in the second section. The six chapters in this section are devoted to the advanced mechanics of plates and shells. Specially, the studies related to plates with functionally graded material, having several futuristic applications, are included herein. Use of artificial neural network (ANN) in the failure analysis of the laminated composite plates is also kept in this section. Considering the developing era of data analytics and artificial intelligence, increased use of high-end computer-aided techniques is required to complement the computational mechanics research fraternity in solving the twenty-first-century problems.

The emphasis of the third section, containing six chapters, is on the geomechanics ranging from micromechanical modeling of materials to modeling of deep foundations. The problems involving multi-phase materials such as avalanches, debris flow need better a understanding of the micromechanical behavior of the materials, and this section provides a step forward in this direction. This section also highlights the simulation of rock fragmentation, stability of tunnels, and modeling of seismic action on earth retaining structures. Future research activities will be required to address the design issues related to the interaction of buried and foundation structures with their surrounding soil having site-specific varied inhomogeneities subjected to dynamic loading of different characteristics.

The fourth section deals with recent developments in the domain of computational structural dynamics. There are four chapters elaborating various computational techniques for damage detection and structural health monitoring (SHM), whereas one chapter each is dedicated toward elastodynamics and vibration control. The considered topics are extremely relevant toward restoring and safeguarding deteriorating infrastructure. More research is required to make use of damage detection and SHM techniques in real time for post-disaster rescue and recovery. The need is highlighted to continuously monitor important infrastructure for its health and condition assessment in real time and analyze the sensor data employing advanced computational techniques, to facilitate taking appropriate decisions on the requirements of structural strengthening, upgradation, and/or rehabilitation based on the cost-benefit and risk analyses.

Further, the behavior of structures under extreme and accidental loading has been presented in the fifth and sixth sections. The nine chapters in the fifth section cover recent trends in the use of computational mechanics concepts in modeling earthquake-induced shaking to simulate the structural response. Fracture-mechanics-based modeling of fault rupture and improved understanding on the propagation of seismic waves may help us to predict impending earthquakes in future. This section also discusses seismic response control using advanced dynamic response mitigation/abatement devices. The seven chapters in the sixth section encompasses research developments in modeling and simulation for blast-resistant structures and their designs. This include modeling of blast pressure, numerical and simulation techniques in blast response analysis. Considering the increased threat of man-made and accidental explosions, it is important to enhance our understanding of the structural mechanics under such extreme loading. Understandably, all the future design codes may consider blast resistance as one of the objectives in the performance-based design framework for strategically important infrastructures.

The last section is devoted to the chapters discussing uncertainty quantification and reliability analyses. There are three chapters reporting uncertainty quantification in materials and structural response. Next two chapters discuss the reliability assessment of composite plates and tall structure, respectively. Last three chapters present stochastic analyses for design optimization under blast load and for vibration of plates and shells made with composite materials. In the past few decades, we have improved our understanding of the constitutive properties of traditional and new materials and their behavior under various loading scenarios. However, with progress in the computational techniques, it is high time to pay attention toward the propagation of parametric uncertainties from the material level to the structural level and study their effects on the structural response under scenario-specific conditions.

Overall, the readers can expect, from this book volume, a wholesome package of state-of-the-art knowledge in computational mechanics and simulations for materials and structures. Indeed, it was a great pleasure to write this foreword for such a unique book volume enriched with a wealth of knowledge in cutting-edge research areas in the broader domain of computational mechanics, and I must congratulate the editors of this text for their sincere efforts in making this contribution.

Prof. Vasant Matsagar
Dogra Chair Professor
Department of Civil Engineering
Indian Institute of Technology (IIT) Delhi
New Delhi, India

Preface

Indian Institute of Technology Mandi (IIT Mandi) successfully hosted the 7th International Congress on Computational Mechanics and Simulation (ICCMS 2019). This prestigious biennial event was attended by over 250 delegates from all over the world. This unique congress has an exceptional reputation for showcasing the latest developments in various fronts of computational mechanics—from theories to applications. In the modern era, scientific exploration has gained a new height with the advancements in the mathematical and computational methods. Its significant impact can further be realized through the possibilities of performing engineering analysis and design tasks, which could have not been managed by manual efforts earlier. Computational Mechanics emphasizes the development of mathematical models representing physical phenomena and applies modern computing methods to analyze these phenomena. Essentially, being interdisciplinary in nature, it thrives on the fields related to physics, mechanics, mathematics, and computer science, and encompasses applying numerical methods to various problems in science and engineering. Recent advances in the field of Computational Mechanics have generated considerable interest among researchers and practicing engineers to gain more knowledge and insight into the various aspects of modeling, analysis, and design. The principles of Computational Mechanics can be effectively applied to the rational design of engineering components under various extreme and complex loading conditions. Within the aforementioned general framework, Computational Mechanics is currently being used in a broad range of applications including civil, offshore, aerospace, automotive, naval, biomechanics, and nuclear structures. Keeping such multi-disciplinary aspects in mind, this book volume presents the recent advancements which took place in the field of Computational Mechanics by compiling selected papers presented in the ICCMS 2019.

About 215 technical papers were presented during the 3 days' congress by researchers and practitioners from a wide spectrum of research background. In addition, there were 15 plenary speeches, 10 keynote speeches, and 4 invited talks delivered by eminent researches in respective areas. All the submitted articles went through rigorous a three-level review process. The review comments were shared with the authors to comply and improve. The articles were accepted for presentation only after receiving positive recommendations from the reviewers. It was indeed an enthralling experience to assimilate all the outstanding research papers from the multidisciplinary domains. However, based on the reviewers' feedback and recommendations, only 105 papers were selected for compilation as book volumes. We are delighted that the state-of-the-art technical articles are finally coming up in the form of two exclusive book volumes by Springer titled "Recent Advances in Computational Mechanics and Simulations". The technical articles are categorically presented as 105 chapters distributed in two volumes. Volume-I is broadly dedicated to the use of computational mechanics and simulations at the level of *Materials to Structures*; Volume-II broadly focuses on computational mechanics at *Nano to Macro* levels and its applications in emerging areas. Nevertheless, there are several chapters that encompass multiple frontiers of computational mechanics. Therefore, drawing a fine boundary while distributing the articles in two volumes was really challenging.

The wholehearted support received from the organizing committee members and colleagues at IIT Mandi in organizing the ICCMS 2019 is truly appreciated. We are grateful to Prof. Tarun Kant, Founding President of Indian Association of Computational Mechanics (IndACM), and Prof. Vasant Matsagar, Executive Secretary of IndACM, for their mentoring in the overall organization of the congress. We would like to thank the authors, who contributed the exceptional technical articles for the congress. We express our sincere gratitude to all the reviewers for their valuable time and painstaking effort in reviewing the articles. We are indebted to Prof. Vasant Matsagar for writing the foreword of this volume. The list of acknowledgements will be incomplete without mentioning the passionate efforts put in by our postgraduate students at various levels of the organization. Without their active support and devotion, it would not have been possible to organize an event of this stature. We thank Springer for accepting our proposal and Ms Swati Meherishi, Executive Editor and her team for continuous support in successfully bringing out these book volumes.

Mandi, Himachal Pradesh, India
May 2020

Sandip Kumar Saha
Mousumi Mukherjee

7th International Congress on Computational Mechanics and Simulation (ICCMS 2019)

Under the auspices of the Indian Association for Computational Mechanics (IndACM), International Congress on Computational Mechanics and Simulation (ICCMS) is organized biennially. The Association was founded on January 1, 2000 and has about 200 life members. The previous congresses (ICCMS) were held at IIT Bombay (2015, 2009), CSIR-SERC Chennai (2014), IIT Hyderabad (2012), IIT Guwahati (2006), and IIT Kanpur (2004). Computational Mechanics, being an interdisciplinary domain of mechanics, involving mathematical models of physical real-life problems and their solutions, the congress is typically attended by delegates with a background in civil engineering, mechanical engineering, aerospace engineering, materials engineering/ science, physics, mathematics, and other disciplines.

The 7th edition of ICCMS was organized by Indian Institute of Technology Mandi (IIT Mandi), during December 11–13, 2019, at the serene Kamand Valley located in the lap of the great Himalaya. The congress hosted 15 plenary speakers, 10 keynote speakers, and 4 invited speakers from, Australia, India, Japan, UK and USA. More than 215 technical papers were presented by the delegates from across the world in addition to the plenary, keynote, and invited presentations. The technical papers were invited under 14 different subdomains, such as biomechanics; computational fluid dynamics and transport phenomena; computational geomechanics and geotechnics; natural materials; computational structural dynamics; constitutive modelling of materials; composites and multifunctional materials; fracture and failure mechanics; interfaces, contacts and interactions; multiscale and multiphysics problems, simulation; numerical methods and algorithms in engineering and science; simulation and analysis under accidental and extreme loadings; structural health monitoring; vibration control; structural mechanics, materials and engineering; uncertainty quantification, reliability analysis; and application of computational techniques in other areas. The technical presentations were conducted in six parallel sessions during the 3 days of the congress.

Along with the regular technical sessions, International Society for Soil Mechanics and Geotechnical Engineering (ISSMGE), in association with ICCMS 2019, organized a day-long Mini-Symposium of ISSMGE's Technical committee TC-105 on GeoMechanics from Micro to Macro. This comprised of seven invited lectures and four regular presentations by young researchers. The objective of this mini-symposium was to promote the micromechanics-based modeling and research within the geomechanics community of India.

After seven successful occasions, the 8th edition of ICCMS will be organized by Indian Institute of Technology Indore in 2021.

Organizing Committee (ICCMS 2019)

Patron

Prof. Timothy A. Gonsalves, Director, IIT Mandi

Mentor

Prof. Tarun Kant, Founding President, IndACM, Professor Emeritus, IIT Bombay, Visiting Distinguished Professor, IIT Mandi

Core Organizing Committee

Dr. Sandip Kumar Saha, Assistant Professor, School of Engineering, IIT Mandi (Convenor)
Dr. Mousumi Mukherjee, Assistant Professor, School of Engineering, IIT Mandi
Dr. Maheshreddy Gade, Assistant Professor, School of Engineering, IIT Mandi
Dr. Venkata Uday Kala, Assistant Professor, School of Engineering, IIT Mandi
Dr. Rajesh Ghosh, Assistant Professor, School of Engineering, IIT Mandi
Dr. Subhamoy Sen, Assistant Professor, School of Engineering, IIT Mandi
Dr. Gaurav Bhutani, Assistant Professor, School of Engineering, IIT Mandi
Dr. Rajneesh Sharma, Assistant Professor, School of Engineering, IIT Mandi
Dr. Kaustav Sarkar, Assistant Professor, School of Engineering, IIT Mandi

Contents

Mechanics of Solids and Structures

Propagation of Elastic Waves in Nonlocal Bars and Beams 3
V. S. Mutnuri and S. Gopalakrishnan

Mechanics of Damage at Steel-Concrete Interface in RC Structures 23
Saiwal Krishna, Pritam Chakraborty, and S. K. Chakrabarti

A Discussion on Locking and Nonlocking Gradient-Enhancement Formulations for Concrete Behavior 33
Ajmal Hasan Monnamitheen Abdul Gafoor and Dieter Dinkler

On Flexure of Shear Deformable Isotropic Rectangular Propped Cantilever Beams 43
Kedar S. Pakhare, P. J. Guruprasad, and Rameshchandra P. Shimpi

Progressive Collapse Potential of Steel Frames Sustaining Post-Hazard Support-Yielding 53
Anil Kumar, Niranjan Muley, Pranesh Murnal, and Vasant A. Matsagar

Influence of Concrete Fill on the Buckling Characteristics of Slender Circular Steel Tubes 73
Rebecca Mary Paul, M. Madhu Karthik, and M. V. Anil Kumar

Vibration of Flexible Member in Offshore Structures 85
Madagala Sravani and Kiran Vijayan

Nonlinear Analysis of Mooring System for an Offshore Desalination Platform 95
Ashwani Vishwanath and Purnima Jalihal

Advanced Mechanics of Plates and Shells

Finite Element Studies and Dynamic Analysis of Initially Stressed Functionally Graded Plates Using a Refined Higher Order Shear Deformation Theory 109
S. Jayaraman, Tarun Kant, and N. Shanmuga Sundaram

Vibration Analysis of Functionally Graded Material Plate 119
N. I. Narayanan, Sauvik Banerjee, Akshay Prakash Kalgutkar, and T. Rajanna

ANN-Based Random First-Ply Failure Analyses of Laminated Composite Plates 131
Subrata Kushari, A. Chakraborty, T. Mukhopadhyay, S. R. Maity, and S. Dey

Interaction of Higher Buckling Modes in Uniformly Compressed Simply Supported Unstiffened Plates 143
K. C. Kalam Aswathy and M. V. Anil Kumar

Natural Frequency of Higher-Order Shear Deformable FGM Plates with Initial Geometric Imperfection Resting on Elastic Foundation 153
Mohammed Shakir and Mohammad Talha

A Geometrically Inspired Model for Kirchhoff Shells with Cartan's Moving Frames 163
Bensingh Dhas and Debasish Roy

Geomechanics and Geotechnics

3-Dimensional Analysis of Fixed-Headed Single Pile and 2 × 2 Pile Group in Multilayered Soil 179
Arindam Dey and Somenath Mukherjee

Significance of Interface Modeling in the Analysis of Laterally Loaded Deep Foundations 199
Ramyasri Rachamadugu and Gyan Vikash

Micromechanical Modeling of Material Cross-anisotropy 211
Geethesh Naiyyalga and Mousumi Mukherjee

Stability of an Unsupported Elliptical Tunnel Subjected to Surcharge Loading in Cohesive-Frictional Soil 225
Puja Dutta and Paramita Bhattacharya

PFC3D Modeling of Rock Fragmentation by Pressure Pulse 237
Jason Furtney and Jay Aglawe

Modeling of Seismic Actions on Earth Retaining Structures 247
Rohit Tiwari, Nelson Lam, and Elisa Lumantarna

Computational Structural Dynamics

Detection of Damages in Structures Using Changes in Stiffness and Damping 259
Uday Sinha and Sushanta Chakraborty

A Data-Based Technique for Damage Detection Handling Environmental Variability During Online Structural Health Monitoring 273
K. Lakshmi and Junia Blessy

Damage Detection in Presence of Varying Temperature Using Mode Shape and a Two-Step Neural Network 285
Smriti Sharma and Subhamoy Sen

Breathing Crack Localization Using Nonlinear Intermodulation Based Exponential Weighting Function Augmented Spatial Curvature Approach 301
J. Prawin

Error in Constitutive Equation based Approach for Isotropic Material Parameter Estimation in Frequency-Domain Elastodynamics 311
Shyamal Guchhait and Biswanath Banerjee

A Novel Sloshing Damper for Vibration Control of Short Period Structures 323
Anuja Roy, Atanu Sahu, and Debasish Bandyopadhyay

Modeling and Simulation in Earthquake Engineering

Fracture Mechanics Based Unilateral and Bilateral Earthquake Simulations: Application to Cable-Stayed Bridge Response 333
K. S. K. Karthik Reddy and Surendra Nadh Somala

Broadband Ground Motion in Indo-Gangetic Basin for Hypothetical Earthquakes in Himalaya 351
J. Dhanya, S. Jayalakshmi, and S. T. G. Raghukanth

Earthquake Engineering in Areas Away from Tectonic Plate Boundaries 367
Nelson Lam

Floor Response Spectra Generation Considering Nonlinearity of Reinforced Concrete Shear Walls 381
Paresh Kothari, Y. M. Parulekar, G. V. Ramarao, and G. V. Shenai

Efficient Arrangement of Friction Damped Bracing System (FDBS) for Multi-storey Steel Frame 397
Saikat Bagchi, Avirup Sarkar, and Ashutosh Bagchi

Bidirectional Pushover Analysis considering the Effect of Angle of Seismic Incidence 413
Prabakaran Kesavan and Arun Menon

Seismic Energy Loss in Semi-rigid Steel Frames Under Near-Field Earthquakes 431
Vijay Sharma, M. K. Shrimali, S. D. Bharti, and T. K. Datta

Seismic Behavior of Baffled Liquid Storage Tank Under Far-Field and Near-Field Earthquake 445
Sourabh Vern, M. K. Shrimali, S. D. Bharti, and T. K. Datta

Seismic Response of Asymmetric Structure with Soil Structure Interaction Using Semi-Active MR Damper 457
Shuvadeep Panchanan, Praveen Kumar, Swagata Basu, and R. S. Jangid

Modelling and Simulation in Blast Resistant Design

A Critical Review of TNT Equivalence Factors for Various Explosives 471
P. A. Shirbhate and M. D. Goel

Limitations of Simplified Analysis Procedures Used for Calculation of Blast Response of Structures 479
K. K. Anjani and Manish Kumar

A Comparative Analysis of Computation of Peak Overpressure in Near Field Scenario During High Energetic Material's Detonation 497
Praveen K. Verma, Rohit Sankrityayan, Devendra K. Dubey, and Anoop Chawla

Development of Performance-Based Design Guidelines for Reinforced Concrete Columns Subject to Blast Loads 509
Vishal Kochar, K. K. Anjani, and Manish Kumar

Numerical Damage Modelling of RC Slabs Under Blast Loading Using K&C Concrete Model 529
K. Akshaya Gomathi and A. Rajagopal

Numerical Modeling of Tunnel Subjected to Surface Blast Loading 543
Jagriti Mandal, M. D. Goel, and Ajay Kumar Agarwal

Dynamic Response of Tunnel Under Blast Loading and Its Blast Mitigation Using CFRP as Protective Barrier 555
V. S. Phulari and M. D. Goel

Uncertainty Quantification and Reliability Analysis

Uncertainty Quantification of Random Heterogeneous Media Using XFEM 565
Ashutosh Rawat and Suparno Mukhopadhyay

Uncertainty Propagation in Estimated Structural Parameters Owing to Univariate Uncertain Parameter Using RSM and PDEM 575
Kumar Anjneya, Divya Grover, and Koushik Roy

Seismic Response of Liquid Storage Tank Considering Uncertain Soil Parameters 589
Hitesh Kumar and Sandip Kumar Saha

Reliability Assessment of CFRP Composite Laminate Subjected to Low Velocity Impact Damage 605
Shivdayal Patel, Akshay Sontakke, and Suhail Ahmad

Passive Vibration Control of Tall Structures with Uncertain Parameters—A Reliability Analysis 613
Said Elias, Deepika Gill, Rajesh Rupakhety, and Simon Olafsson

Dual Polynomial Response Surface-Based Robust Design Optimization of Structure Under Stochastic Blast Load 623
Gaurav Datta and Soumya Bhattacharjya

Stochastic Modal Damping Analysis of Stiffened Laminated Composite Plate 635
Sourav Chandra, Kheirollah Sepahvand, Vasant Matsagar, and Steffen Marburg

Support Vector Model Based Thermal Uncertainty on Stochastic Natural Frequency of Functionally Graded Cylindrical Shells 651
Vaishali and S. Dey

Editors and Contributors

About the Editors

Sandip Kumar Saha is an Assistant Professor in the School of Engineering at the Indian Institute of Technology (IIT) Mandi. He obtained his Ph.D. from IIT Delhi, India, and postdoctoral work in the University of Canterbury, New Zealand. His research interests include computational structural dynamics, performance-based earthquake engineering, uncertainty modeling in dynamical systems and multi-hazard (earthquake, wind, fire, blast, etc.) protection of structures.

Mousumi Mukherjee is an Assistant Professor in the School of Engineering at the Indian Institute of Technology (IIT) Mandi, India. She received her Ph.D. in civil engineering from Indian Institute of Technology Kanpur, India and pursued postdoctoral work in the School of Civil, Environmental and Mining Engineering at the University of Adelaide, Australia. Her research interests include geomechanics, constitutive modeling of frictional material, soil instability analysis, modeling of debris flow, numerical modelling of large deformation problems pertinent to geotechnical engineering.

Contributors

Ajay Kumar Agarwal Department of Applied Mechanics, Visvesvaraya National Institute of Technology, Nagpur, India

Jay Aglawe Itasca India Consulting Private Limited, Nagpur, India

Suhail Ahmad Indian Institute of Technology, Delhi, India

Ashutosh Bagchi Concordia University, Montreal, QC, Canada

Saikat Bagchi Concordia University, Montreal, QC, Canada

Debasish Bandyopadhyay Jadavpur University, Kolkata, India

Sauvik Banerjee Department of Civil Engineering, Indian Institute of Technology Bombay, Mumbai, India

Swagata Basu Indian Institute of Technology Bombay, Mumbai, India

S. D. Bharti Malaviya National Institute of Technology Jaipur, Jaipur, India

Soumya Bhattacharjya Indian Institute of Engineering Science and Technology, Shibpur, Howrah, India

Paramita Bhattacharya Indian Institute of Technology Kharagpur, Kharagpur, India

Junia Blessy CSIR-SERC, CSIR Complex, Taramani, Chennai, India

A. Chakraborty Indian Institute of Technology, Guwahati, India

Sushanta Chakraborty IIT Kharagpur, Kharagpur, India

Anoop Chawla Department of Mechanical Engineering, Indian Institute of Technology, New Delhi, India

Gaurav Datta Indian Institute of Engineering Science and Technology, Shibpur, Howrah, India

T. K. Datta Malaviya National Institute of Technology Jaipur, Jaipur, India

Arindam Dey Indian Institute of Technology Guwahati, Guwahati, India

S. Dey National Institute of Technology Silchar, Assam, India

Devendra K. Dubey Department of Mechanical Engineering, Indian Institute of Technology, New Delhi, India

Puja Dutta Indian Institute of Technology Kharagpur, Kharagpur, India

Said Elias Earthquake Engineering Research Centre, University of Iceland, Selfoss, Iceland

Jason Furtney Itasca Consulting Group, Minneapolis, MN, USA

Deepika Gill Indian Institute of Technology (IIT) Delhi, New Delhi, India

M. D. Goel Department of Applied Mechanics, Visvesvaraya National Institute of Technology, Nagpur, India

Purnima Jalihal National Institute of Ocean Technology, Chennai, India

R. S. Jangid Indian Institute of Technology Bombay, Mumbai, India

S. Jayaraman U R Rao Satellite Centre, ISRO, Bengaluru, India

Akshay Prakash Kalgutkar Department of Civil Engineering, Indian Institute of Technology Bombay, Mumbai, India

Tarun Kant IIT-Bombay, Mumbai, India

K. S. K. Karthik Reddy Indian Institute of Technology, Hyderabad, India

Vishal Kochar Indian Institute of Technology Bombay, Mumbai, India

Paresh Kothari Bhabha Atomic Research Centre, Mumbai, India

Anil Kumar Indian Institute of Technology (IIT) Delhi, New Delhi, India

Hitesh Kumar Indian Institute of Technology Mandi, Mandi, India

Manish Kumar Indian Institute of Technology Bombay, Mumbai, India

Praveen Kumar Bhabha Atomic Research Centre, Mumbai, India

Subrata Kushari National Institute of Technology, Silchar, India

K. Lakshmi CSIR-SERC, CSIR Complex, Taramani, Chennai, India

Nelson Lam The Department of Infrastructure Engineering, The University of Melbourne, Parkville, VIC, Australia

Elisa Lumantarna The Department of Infrastructure Engineering, The University of Melbourne, Parkville, VIC, Australia

S. R. Maity National Institute of Technology, Silchar, India

Jagriti Mandal Department of Mining Engineering, Visvesvaraya National Institute of Technology, Nagpur, India

Vasant A. Matsagar Indian Institute of Technology (IIT) Delhi, New Delhi, India

Mousumi Mukherjee School of Engineering, IIT Mandi, Mandi, India

Somenath Mukherjee C. E. Testing Co. Pvt. Ltd., Kolkata, India

T. Mukhopadhyay University of Oxford, Oxford, UK

Niranjan Muley Government College of Engineering, Aurangabad, India

Pranesh Murnal Government College of Engineering, Aurangabad, India

Geethesh Naiyyalga School of Engineering, IIT Mandi, Mandi, India

Simon Olafsson Earthquake Engineering Research Centre, University of Iceland, Selfoss, Iceland

Shuvadeep Panchanan Bhabha Atomic Research Centre, Mumbai, India

Y. M. Parulekar Bhabha Atomic Research Centre, Mumbai, India

Shivdayal Patel Indian Institute of Information Technology, Design and Manufacturing, Jabalpur, India

Rebecca Mary Paul Indian Institute of Technology Palakkad, Palakkad, India

V. S. Phulari Department of Applied Mechanics, Visvesvaraya National Institute of Technology, Nagpur, India

J. Prawin CSIR-Structural Engineering Research Centre, CSIR Campus, Chennai, TN, India

Ramyasri Rachamadugu Shiv Nadar University, Greater Noida, India

A. Rajagopal Indian Institute of Technology Hyderabad, Hyderabad, India

T. Rajanna B. M. S. College of Engineering, Bengaluru, Karnataka, India

G. V. Ramarao CSIR-Structural Engineering Research Centre, Chennai, India

Anuja Roy Jadavpur University, Kolkata, India

Rajesh Rupakhety Earthquake Engineering Research Centre, University of Iceland, Selfoss, Iceland

Sandip Kumar Saha Indian Institute of Technology Mandi, Mandi, India

Atanu Sahu National Institute of Technology, Cachar, Assam, India

Rohit Sankrityayan Department of Mechanical Engineering, Indian Institute of Technology, New Delhi, India

Avirup Sarkar Concordia University, Montreal, QC, Canada

Mohammed Shakir Indian Institute of Technology, Mandi, India

G. V. Shenai Bhabha Atomic Research Centre, Mumbai, India

P. A. Shirbhate Department of Applied Mechanics, Visvesvaraya National Institute of Technology, Nagpur, India

M. K. Shrimali Malaviya National Institute of Technology Jaipur, Jaipur, India

Uday Sinha IIT Kharagpur, Kharagpur, India

Surendra Nadh Somala Indian Institute of Technology, Hyderabad, India

Akshay Sontakke Indian Institute of Information Technology, Design and Manufacturing, Jabalpur, India

N. Shanmuga Sundaram U R Rao Satellite Centre, ISRO, Bengaluru, India

Mohammad Talha Indian Institute of Technology, Mandi, India

Rohit Tiwari The Department of Infrastructure Engineering, The University of Melbourne, Parkville, VIC, Australia

Vaishali National Institute of Technology Silchar, Assam, India

Praveen K. Verma Department of Mechanical Engineering, Indian Institute of Technology, New Delhi, India

Sourabh Vern Malaviya National Institute of Technology, Jaipur, India

Gyan Vikash Shiv Nadar University, Greater Noida, India

Ashwani Vishwanath National Institute of Ocean Technology, Chennai, India

Mechanics of Solids and Structures

Propagation of Elastic Waves in Nonlocal Bars and Beams

V. S. Mutnuri and S. Gopalakrishnan

Abstract In this paper, wave dispersion properties in nonlocal theories of elasticity models are critically examined. Both gradient type as well as integral-type non-locality within a setting of the rod and beam are considered. The mathematical framework here involves the Fourier frequency analysis that leads to the frequency spectrum relation (FSR) and system transfer function. Utilizing the FSR, *wave modes* and group speeds are examined. One main difference that arises between the two nonlocal model types is in the number of wave modes that are possible from the FSR. In gradient-type non-locality, the number of modes is finite and equal to the order of the displacement gradient in the governing equation of motion. However, in an integral-type non-locality, wave modes are infinite in number. Further, in contrast to classical theories there exist nonclassical wave modes showing the existence of nonphysical features, such as either exponential instabilities or undefined wavenumbers, negative group speeds, infinitesimally small, and infinite group speed values. Existence of such features, then, naturally raises the aspect of physically realizable wave motion and *causality* in these models. In literature, there exist dispersion relations or Kramers–Kronig (K-K) relations as an aid to examine wave motion in a linear, passive, and *causal* system. In this paper, K-K relations are utilized in order to further examine the wave dispersion properties. Discrepancies are seen between FSR predictions and K-K predictions, especially at frequencies with nonphysical features. An example is also presented that contradicts the general concept that all time-domain formulations agree well with the K-K relations.

Keywords Nonlocal elasticity · Kramers–Kronig relations · Causality · Wave dispersion

---

V. S. Mutnuri · S. Gopalakrishnan (✉)
Indian Institute of Science, Bangalore 560012, India
e-mail: krishnan@iisc.ac.in

1 Introduction

Nonlocal theories of elasticity are generalizations of the classical elasticity theories in order to account for physics in solids with microstructure. From wave motion point of view, solids with microstructure bring in various wave phenomena including wave scattering, wave diffraction, wave mode conversion, etc., upon interaction with boundaries [1]. Therefore, wave propagation in these solids is then associated with not only dispersion but also with wave dissipation as waves propagate in solid. Nonlocal elasticity models bring this heterogeneous microscale wave physics into the macroscopic scale through the introduction of length-scale parameters into the governing equation of motion of an effective homogenized medium. It is of interest of this paper to examine the above aspect of dispersion as well as dissipation in the nonlocal models, without considering any damping physics, such as viscous damping.

A closely related concept to the above aspect, which has been extensively applied to the literature in the fields of electromagnetism and acoustics, is the idea of *causality*. Causality dictates that cause precedes the effect but not the vice versa. Mathematical expression, due to a Fourier frequency analysis, of the causality is known in the literature as dispersion relations or Kramers–Kronig (K-K) relations [2]. It essentially relates real and imaginary parts of an analytic complex function over an entire real frequency range as a pair of Hilbert transforms. These relations were also derived explicitly in terms of wave parameters such as phase speeds and attenuation as well as real and imaginary parts of wave numbers [3]. Therefore, from wave motion point of view, if waves in a causal medium are dispersive in nature then waves are dissipative in nature as well. As wave motion is known to be highly dispersive in nonlocal models, and therefore, it is of interest here to examine the dissipative part of waves in these structures utilizing K-K relations.

This paper is organized as follows. In Sect. 2, Fourier frequency analysis of governing equations of motion of gradient and integral-type nonlocal rod models and gradient elasticity Euler–Bernoulli beam (EBB) model, leading to frequency spectrum relations and/or system transfer function, is presented. Wave modes and group speeds behaviors are examined in Sect. 3. The forms of K-K relations utilized here is briefly discussed in Sect. 4. Application of these K-K relations to the considered models is examined in Sect. 5. In Sect. 6, a pre-stressed EBB is investigated in its agreement/disagreement with K-K relations. Finally, in Sect. 7 observations are summarized.

2 Fourier Frequency Analysis

The governing equations of motion (EOM) of the considered nonlocal models are of the two types: partial differential equation (PDE) and integro-differential equation (IDE). As a first step in the Fourier frequency analysis [4], displacement is assumed in the form of a discrete Fourier transform (DFT) as

$$u(x,t) = \sum_n \hat{u}(x,\omega_n)e^{i\omega_n t} \quad (1)$$

where u is the displacement of a particle at x and at time t, $\hat{u}$ is displacement of the entire structure at a angular frequency ω_n, and n is the number of DFT points. Upon substitution of Eq. (1) into PDE and IDE results in a set of ordinary differential equations (ODE) and integral equations (IE) at each ω_n. This procedure is applied for various nonlocal elasticity models below.

Notes and Comments. It is worth mentioning here some contrasting features that Fourier frequency analysis brings into wave dispersion analysis. In practice, a plane wave solution is substituted into the governing equation of motion obtaining the frequency spectrum relation (FSR). Then, a set of *real values* of wavenumbers are given as input to FSR with the output being the frequency. In contrast to this practice, Fourier frequency analysis considers frequency as the independent variable obtaining wavenumber as part of the solution of the ODE/IE. By this procedure, *real, imaginary, and complex values* to wavenumbers within the model can be obtained. This dependency of wavenumbers as a function of frequency is defined as *wave modes* here. Also, as detailed later, wave modes are the necessary ingredients for the K-K relations.

In the following, ρ is the mass density, E is the Young's modulus, A is the area of cross section, $l's$ are the length-scale parameters, and F is the external force field per unit length.

2.1 Nonlocal Rods

2.1.1 Gradient Model

This nonlocal gradient elasticity model is due to [5]. Its equation of motion is given as

$$\rho A \frac{\partial^2 u(x,t)}{\partial t^2} - EA\frac{\partial^2 u(x,t)}{\partial x^2} + \frac{\rho^2 l_1^2 A}{E}\frac{\partial^4 u(x,t)}{\partial t^4} - \rho A l_2^2 \frac{\partial^4 u(x,t)}{\partial t^2 \partial x^2} = F(x,t) \quad (2)$$

There are two length-scale parameters l_1 and l_2 due to higher order time and mixed space–time derivatives. Motivation for using higher order time derivatives is from the aspect of causality [6]. Mixed space–time derivatives appears in various models including Love rod [7] and stress-gradient model [8]. Motivation in the current model was to resolve the problem of secularity in the wave propagation studies of periodic structures [9]. Upon substitution of Eq. (1) into Eq. (2) gives

$$-\rho A\omega^2 \hat{u} - EA\frac{d^2\hat{u}}{dx^2} + \frac{\rho^2 l_1^2 A}{E}\omega^4 \hat{u} + \rho A l_2^2 \omega^2 \frac{d^2\hat{u}}{dx^2} = \hat{F} \quad (3)$$

Here, subscript n is dropped. Equation (3) has to be satisfied at each ω. For a given ω, Eq. (3) is a constant coefficient equation. Then, $\hat{u}$ is assumed as the following form:

$$\hat{u}(x, \omega) = ae^{-ikx} \tag{4}$$

Upon substitution of Eq. (4) into Eq. (3), under absence of external forces, gives the FSR as

$$-\rho A \omega^2 + EAk^2 + \frac{\rho^2 l_1^2 A}{E}\omega^4 - \rho A l_2^2 \omega^2 k^2 = 0 \tag{5}$$

which can be further simplified as

$$k = \pm \sqrt{\frac{-\frac{\rho^2 l_1^2 A}{E}\omega^4 + \rho A \omega^2}{EA - \rho A l_2^2 \omega^2}} \tag{6}$$

2.1.2 Integral Model

The considered integral nonlocal model is the bond-based peridynamics (BBPD) model due to [10]. Its equation of motion is given as

$$\rho A \frac{\partial^2 u(x,t)}{\partial t^2} - \int_{-\infty}^{\infty} C(|\xi|) A[u(x+\xi, t) - u(x,t)]d\xi = F(x,t) \tag{7}$$

Here, C is the micro-modulus function that defines strength of interaction between particles at $x + \xi$ and x in the reference configuration. Upon substitution of Eq. (1) into Eq. (7) gives

$$-\rho A \omega^2 \hat{u} - \int_{-\infty}^{\infty} C(|\xi|) A[\hat{u}(x+\xi, \omega) - \hat{u}(x, \omega)]d\xi = \hat{F}(x, \omega) \tag{8}$$

Taylor series expansion of $\hat{u}(x + \xi)$, in terms of $\hat{u}(x)$, is utilized here as a further step of study of Eq. (8). Carrying out these calculations results in the following series equation:

$$-\rho A \omega^2 \hat{u} - \sum_{m=1}^{\infty} \frac{\int_{-\infty}^{\infty} C(|\xi|) A \xi^m d\xi}{m!} \frac{d^m \hat{u}}{dx^m} = \hat{F}(x, \omega) \tag{9}$$

Notes and Comments. Motivation to transform the integral equation into an infinite series equation is twofold. First, being an integral equation, boundary conditions do not emerge out naturally [11] and hence series form can fill this need. Second, one can conclude from Eq. (9) that the original integral equation, for well-behaved functions of C, actually represents a constant coefficient equation, and thus its solution can take the form of Eq. (4) and allows performing a wave motion study.

Substitution of Eq. (4) into Eq. (9), under absence of body forces, results in the FSR in a series form as

$$\rho\omega^2 = \sum_{m=1}^{\infty}(-1)^{m+1}k^{2m}I_m \qquad (10)$$

where $I_m = \int_{-\infty}^{\infty} \frac{C(|\xi|)\xi^m}{m!}d\xi$. From Eq. (10), it can be inferred that for a given ω there exist infinite number of roots to k.

Notes and Comments. Existence of an infinite number of wave modes is a feature not only unique to BBPD formulation but also to any integral-type nonlocal formulation proposed in the literature. For details on various models, see [12] and references therein. This feature arises due to the fundamental assumption in all of these models that there exists a direct influence between long-range particles in the medium. Depending on the constitutive behavior, there may arise problematic aspects with respect to wave dispersion properties. For instance, the existence of infinite values to k, implies

$$\hat{u}(x,\omega) = \sum_{n=1}^{\infty} a_n e^{-ik_n x} \qquad (11)$$

Then, one needs an infinite number of conditions to obtain wave coefficients ($a'_n s$). Therefore, constructing a boundary value problem for wave motion analysis seems to be an ill-posed problem as there is a need for infinite number of conditions.

2.2 Nonlocal Beams

A class of stress gradient and combined strain/inertia gradients models for Euler–Bernoulli and Timoshenko beams have been formulated and examined with respect to wave dispersion in [13]. The scope of examination there was limited to real wavenumbers. The motivation behind the utilization of stress and combined strain/inertia gradients was to remove the unstable wave dispersion properties due to strain gradients. For discussion on this, please refer to [13] and references therein. In this section *wave modes* of one of the models considered in [13] are examined.

2.2.1 Combined Strain/Inertia Gradients Model—Euler–Bernoulli Beam

The governing equation of motion of this model is given as

$$\rho A \frac{\partial^2 u(x,t)}{\partial t^2} = -EI\left(\frac{\partial^4 u(x,t)}{\partial x^4} - l_s^2 \frac{\partial^6 u(x,t)}{\partial x^6}\right) - \rho I l_m^2 \frac{\partial^6 u(x,t)}{\partial x^4 \partial t^2} + F(x,t) \qquad (12)$$

where I is the moment of inertia and u is the transverse displacement. Upon substitution of Eq. (1) into the governing equation of motion gives

$$-\rho A \omega^2 \hat{u} = -EI\left(\frac{d^4\hat{u}}{dx^4} - l_s^2\frac{d^6\hat{u}}{dx^6}\right) + \rho I l_m^2 \omega^2 \frac{d^4\hat{u}}{dx^4} + \hat{F}(x,\omega) \quad (13)$$

Its frequency spectrum relation, under absence of body forces, is given as

$$EIl_s^2 k^6 + (EI - \rho I l_m^2 \omega^2)k^4 - \rho A \omega^2 = 0 \quad (14)$$

Its a cubic polynomial in k^2. Therefore, there exist six roots to k at a given ω and if $+k$ is a root then $-k$ is also a root. Out of six values to k, three values represent forward-moving wave components and the other three represent backward or reflected wave components. In a semi-infinite beam setting, then solution $\hat{u}$ can be given as

$$\hat{u}(x,\omega) = a_1 e^{-ik_1 x} + a_2 e^{-ik_2 x} + a_3 e^{-ik_3 x} \quad (15)$$

As part of the derivation procedure of weak form from Eq. (13), boundary conditions in a semi-infinite beam can be obtained as

$$\hat{u}|_{x=0} = 0 \text{ or } \hat{F}|_{x=0} = [(\rho I l_m^2 \omega^2 - EI)\frac{d^3\hat{u}}{dx^3} + EIl_s^2\frac{d^5\hat{u}}{dx^5}]|_{x=0} \quad (16)$$

$$\frac{d\hat{u}}{dx}|_{x=0} = 0 \text{ or } [(\rho I l_m^2 \omega^2 - EI)\frac{d^2\hat{u}}{dx^2} + EIl_s^2\frac{d^4\hat{u}}{dx^4}]|_{x=0} = 0 \quad (17)$$

$$\frac{d^2\hat{u}}{dx^2}|_{x=0} = 0 \text{ or } -EIl_s^2\frac{d^3\hat{u}}{dx^3}|_{x=0} = 0 \quad (18)$$

At the free end $x = 0$, in a semi-infinite beam, $\hat{u} \neq 0$ and $\frac{d\hat{u}}{dx} \neq 0$. There exist, however, an ambiguity in the third boundary condition whether, curvature to be taken as zero or higher order moment has to be taken as zero. Following arguments put forward in [5] within the context of a nonlocal rod, higher order stress is assumed here to be zero.

Utilizing the boundary conditions and Eq. (15), the system transfer function (G) can be expressed in terms of wavenumbers as

$$G = T_1 a e^{-ik_1 x} - T_2 a e^{-ik_2 x} + a e^{-ik_3 x} \quad (19)$$

where

$$a = \frac{1}{-ib_2(T_1 k_1^5 - T_2 k_2^5 + k_3^5)}$$

$$T_1 = \frac{B_3 k_2^3 - B_2 k_3^3}{-B_1 k_2^3 + B_2 k_1^3}$$

$$T_2 = \frac{-B_1 k_3^3 + B_3 k_1^3}{-B_1 k_2^3 + B_2 k_1^3}$$
$$B_1 = b_2 k_1^4 - b_1 k_1^2$$
$$B_2 = b_2 k_2^4 - b_1 k_2^2$$
$$B_3 = b_2 k_3^4 - b_1 k_3^2$$
$$b_1 = \rho I l_m^2 \omega^2 - EI$$
$$b_2 = EI l_s^2$$

3 Wave Modes and Group Speeds

In this section, roots of the FSR expressions obtained in the previous section, for a set of given ω are presented. Further, group speeds are also examined. Definition of group speed used here is

$$C_g = Real\left(\frac{d\omega}{dk}\right) \tag{20}$$

3.1 Nonlocal Rods

3.1.1 Gradient Model

Expression of wavenumber is given in Eq. (6). It is a single wave mode model and it is shown in Fig. 1 along with group speed behavior. Numerical values used here are: $E = 70$ GPa, $\rho = 2700$ kg/m^3, $A = 0.001$ m^2, $l_2 = l\sqrt{\frac{2}{15}}$, $l_1 = \frac{l}{\sqrt{20}}$ and $l = 0.01$ m.

There exist two characteristic frequencies, escape (ω_e) and cut-off frequencies (ω_c) in this model such that $\omega_c > \omega_e$. At $\omega_e, k \to \infty$ and at $\omega_c, k = 0$. For $\omega < \omega_e$ and $\omega > \omega_c$, waves are homogeneous. However, for $\omega_e < \omega < \omega_c$, waves are evanescent. Thus, this model filters out a bandwidth of ω as evanescent waves and hence can be defined as band no-pass filter. Wave dispersion takes place for $\omega < \omega_e$ and $\omega > \omega_c$ where as wave dissipation takes place for $\omega_e < \omega < \omega_c$.

3.1.2 Integral Model

Expression of wavenumber is given in Eq. (10). As noted earlier, there exists infinite number of wave modes in the rod. For brevity, only first three modes are shown here and the discussion extends to all higher modes without any loss of generality. Wave modes and group speeds are shown in Fig. 2. Computer algebra software, *Maxima*, is utilized to obtain the roots. Numerical values used here are, $E = 70$ GPa, $\rho = 2700$ kg/m^3, and an uniform micro-modulus function is given by [14]

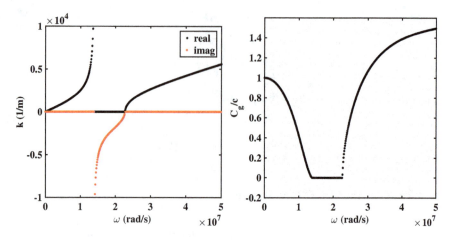

Fig. 1 Wave mode and group speed in a gradient rod model. Here, $c = \sqrt{\frac{E}{\rho}}$

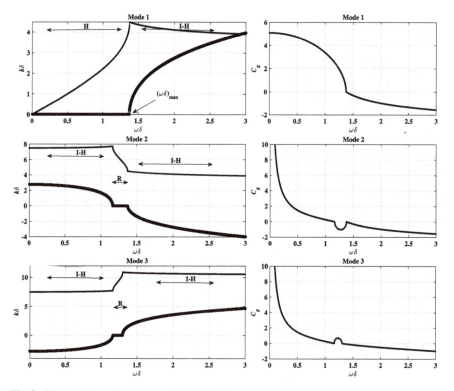

Fig. 2 Wave modes and group speeds in BBPD bar

$$C(\xi) = \frac{3E}{\delta^3} \ for \ 0 \le |\xi| \le \delta \ otherwise \ 0$$

is used. Here, δ is the horizon, a length-scale parameter.

Realizing that C is an even function and using Descartes rule of signs, Eq. (10) can be shown to have no pure imaginary roots, that is, evanescent modes are absent in a BBPD bar. Therefore, values to k are either real (H or R) or complex (I-H) as shown in Fig. 2. Mode 1 has a characteristic frequency $(\omega\delta)_{max}$, at which real k changes to complex k. That is, all higher frequencies damp in the rod and hence behaves as a low-pass filter [15]. However, the existence of a bandwidth of frequencies with real k in all of the higher modes suggests, they behave as a bandpass filter. From the group speed plots, one can notice infinite values as well as infinitesimally small and zero as well as negative speeds. Zero group speed frequencies are the locations where complex k originates from/culminates to real k in a wave mode. This is a very peculiar feature of BBPD. At these zero C_g frequencies, essentially, two distinct real roots k coalesce to form a case of repeated roots. At higher frequencies, complex k occur as a pair of conjugates between the modes. This is a problematic issue because one of the conjugate pairs leads to exponential instability of the system transfer function. If that particular pair is avoided, then, it leads to a case of discontinuity in k. Discontinuity in k is allowed typically at $k \to \infty$, as in stress-gradient model [8]. However, here it happens at a finite k. Negative group speeds (NGS) are seen in real k as well as in complex k. Existence of NGS in complex k may not be a problem because wave damps with propagation. However, NGS in real k leads to violation of causality because NGS wave packet overtakes the signal front [16].

3.2 Nonlocal Beams

3.2.1 Combined Strain/Inertia Gradient Model–EBB

Expression for FSR is given in Eq. (14). Its a sixth-order polynomial in k and cubic in k^2. If k is a root then $-k$ is also a root. In general, k represents a forward-moving wave and $-k$ represents backward propagating wave. All the six wave modes are shown in Fig. 3. Numerical values used here are: $E = 70\,\text{GPa}$, $\rho = 2700\,\text{kg/m}^3$, $A = 2\pi rt$, $I = \pi r^3 t$, $r = 0.01\,\text{m}$, $t = 0.001\,\text{m}$, $l_s = 0.001\,\text{m}$, $l_m = 0.001\,\text{m}$. Roots are obtained utilizing MATLAB in-built function *roots*.

For $\omega < \omega_c$, there exist two real ($\pm k_1$) and four imaginary roots ($\pm ik_2$ and $\pm ik_3$). For $\omega > \omega_c$, there exist two real roots ($\pm k_1$) and four complex roots $\pm(a \pm ib)$. Therefore, in general, solution to $\hat{u}$ is

$$\hat{u} = a_1 e^{-ik_1 x} + a_2 e^{-ik_2 x} + a_3 e^{-ik_3 x} + a_4 e^{-ik_4 x} + a_5 e^{-ik_5 x} + a_6 e^{-ik_6 x} \qquad (21)$$

Now, consider a semi-infinite beam under impact type of transverse loading applied at free end. Then, solution in the domain $x > 0$ reads

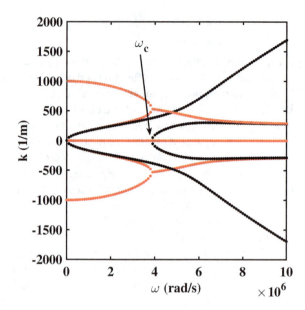

Fig. 3 All six wave modes in the nonlocal EBB model. Here, real part of k is black and imaginary part of k is red

$$\hat{u} = a_1 e^{-ik_1 x} + a_2 e^{-ik_2 x} + a_3 e^{-ik_3 x} \qquad (22)$$

This is because, out of the six wave coefficients (a_i for $i = 1, 2, ..6$) in $\hat{u}$, three represents forward moving and the other three represents reflected components. As, in a semi-infinite medium there cannot be any reflections possible, hence, $\hat{u}$ simplifies as the above.

For $\omega < \omega_c$, $k_1 = A_1$, $k_2 = -iA_2$ and $k_3 = -iA_3$, for some real $A_i > 0$ and for $i = 1, 2, 3$. Then, solution further simplifies as

$$\hat{u} = a_1 e^{-iA_1 x} + a_2 e^{-A_2 x} + a_3 e^{-A_3 x} \qquad (23)$$

For $\omega > \omega_c$, there exist one real and two complex roots. In the case of real in the region $x > 0$, $k_1 = A_1$. However, in the case of complex roots there exist two possibilities for each of k_2 and k_3:

- $A_2 + iB_2$
- $A_2 - iB_2$
- $-A_2 + iB_2$
- $-A_2 - iB_2$

for some positive real numbers A_2 and B_2. In the beam, if the choice of imaginary part of wavenumbers is $+iB_2$, then, although waves propagate in the beam, however, response suffers from exponential instability and thus violates K-K relations. If $-iB_2$ is choice, then, one wave corresponds to anomalous dispersion or with negative group speeds.

Therefore, from the exponential stability point of view, the $\hat{u}$ for $\omega > \omega_c$ reads

$$\hat{u} = a_1 e^{-iA_1 x} + a_2 e^{-iA_2 x} e^{-B_2 x} + a_3 e^{iA_2 x} e^{-B_2 x} \tag{24}$$

These wave modes and their group speeds are shown in Fig. 4. It can be observed that group speeds become unbounded with ω in mode 2 and mode 3.

For $\omega = \omega_c$, wave modes coalesce giving repeated roots. That is, the six wavenumbers are $\pm A_1$, iB_2, iB_2, $-iB_2$, and $-iB_2$. Then, solution $\hat{u}$ in a semi-infinite beam reads

$$\hat{u} = a_1 e^{-iA_1 x} + (a_2 + a_3 x) e^{-B_2 x} \tag{25}$$

The exponential damping overtakes the linear rise in the wave coefficient and hence, a well-behaved response. However, this is not the case with, for example, in the peridynamics rod wherein, coalescence of wave modes happens with real k. This is also the same case of coalescence in real k with unstable strain gradient model [13]. This feature leads to a linear rise of wave amplitude with x in the response and hence an ill-behaved and unstable solution.

4 Kramers–Kronig Relations

In literature, there exist dispersion relations or Kramers–Kronig (K-K) relations as an aid to examine wave motion behavior in any linear, passive, and causal medium [2]. These relations have been applied to various branches of natural sciences and various forms have been proposed in the literature [17]. In this paper, the form of interest, due to [3], relating real (Re) and imaginary (Im) parts of k is considered. These relations are given as

$$Re\, k(\omega') = B\omega' + \frac{2\omega'}{\pi} \int_0^\infty \frac{Im\, k(\omega) - Im\, k(\omega')}{\omega^2 - \omega'^2} d\omega + Re\, k(0) \tag{26}$$

and

$$Im\, k(\omega') = \frac{-2\omega'^2}{\pi} \int_0^\infty [\frac{Re\, k(\omega)}{\omega} - \frac{Re\, k(\omega')}{\omega'}] \frac{d\omega}{\omega^2 - \omega'^2} + Im\, k(0) \tag{27}$$

where B is the real limit of $\frac{k(\omega)}{\omega}$ as $\omega \to \infty$. Further, a second form due to Titchmarsh [2], which is the classical form, relating real and imaginary parts of system transfer function (G) is also considered. They are given as

$$Re\, G(\omega') = \frac{2}{\pi} \int_0^\infty \frac{\omega Im\, G(\omega)}{\omega^2 - \omega'^2} d\omega \tag{28}$$

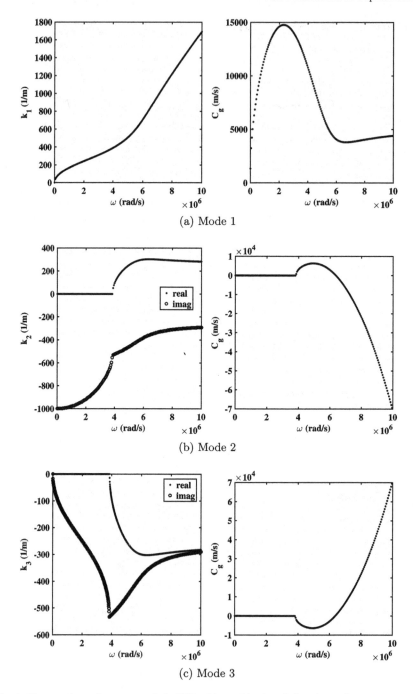

Fig. 4 Wave modes and group speeds in EBB with combined strain/inertia gradients

and

$$Im\ G(\omega') = \frac{-2\omega'}{\pi} \int_0^\infty \frac{Re\ G(\omega)}{\omega^2 - \omega'^2} d\omega \tag{29}$$

The G form of K-K relations need to be extended to account for long-wavelength limit of models as well as the existence of characteristic frequencies. Without going into details of involved derivation procedure, these extended K-K relations will be just stated later. In this paper, real part of k or G is given as input and imaginary part of k or G is predicted as output. Prediction by K-K relations are compared to the corresponding values as obtained from governing equation of motion. This is the topic of the next section.

5 Comparisons

Predictions from K-K relations and the governing equation of motion are compared in this section. For the gradient rod model, these results are shown in Fig. 5. In general, K-K relations introduce a nonzero imaginary component as the real part of k in a nonlocal model is nonlinear. However, the imaginary component is seen to be significant around the frequencies that show infinitesimal or zero group speeds, as compared to other frequencies.

The considered integral rod model is the bond-based perdiynamics. As noted earlier, there exist infinite wave modes in the rod. There exist some interesting features that need some discussion. Firstly, at $\omega = 0$, there exist zero and non-zero values to k. These nonzero k are known to define a Saint Venant's effect in the peridynamic rod. However, as noted before, these wavenumbers correspond to quasi-static load-

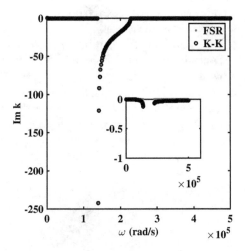

Fig. 5 K-K predictions in strain/inertia gradient rod

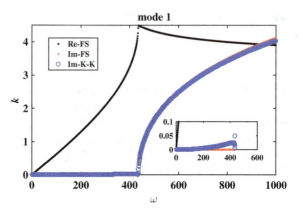

Fig. 6 K-K predictions in peridynamics rod. Here, FS means predictions from the frequency spectrum relation

ing frequencies that show infinite value to groups speeds. Second, one can show that peridynamic rod does not show any evanescent waves and hence only either real or complex k can exist in the rod. Complex k exist along with its conjugates, which is a well-known fact. That is, if $a + ib$ is a root then $a - ib$ is also a root. Typically, complex conjugate represents in wave propagation terminologies, a wave propagating in $x < 0$ region or a reflected wave in $x > 0$ region. However, in the case of peridynamic rod, due to coalescence of wave modes at certain discrete values of frequencies, both a complex k as well as its conjugate should exist in a region. The values to these frequencies are at those where group speeds go to zero. This is a problematic feature because, system transfer function, then, suffers from exponential instability. If complex values are so chosen that, exponential instability is avoided, even then, wave mode suffers discontinuities in wavenumbers, that is, there exists a frequency where wavenumber is undefined. These coalescences of wavemodes are also seen in certain strain gradient models. In peridynamic rod, this feature of coalescence is seen in all linear-type micro-modulus functions. Existence of exponential instability implies violation of K-K relations and hence causality. Existence of undefined k at a frequency is unphysical.

Upon avoiding the wave modes contributing to infinite speeds, negative speeds, exponential instabilities, and undefined k, then results in an existence of a single wave mode which being Mode 1. Then, Eq. (27) is applied to this wave mode and the comparison is shown in Fig. 6. It can be seen that there exist a good agreement. Also, one can see a significant imaginary k, at the frequency bandwidth where group speed goes to zero, as compared to lower frequencies. Being a low-pass filter behavior of the rod, dispersion and dissipation are built-in within a single wave mode.

For the case of nonlocal EBB beam model, K-K relations can be derived as

$$Im\ G(\omega') = \frac{-2}{\pi \omega'} \int_0^\infty \frac{Re\ G(\omega)\omega^2}{\omega^2 - \omega'^2} d\omega \qquad (30)$$

Fig. 7 K-K prediction in a nonlocal EBB

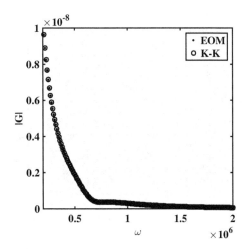

Utilizing this relation, comparison is shown in Fig. 7. Numerical values used here are different for $l_s = 0.01$ m and $l_m = 0.01$ m, while others being the same as before. There seems to be a very good agreement between both the predictions. Existence of very high group speeds at higher frequencies seems to be not a problematic feature from the K-K relations' point of view. This corroborates with [5, 6], wherein a new causality condition is defined as Einstein causality, being, speeds cannot be more than the speed of the light in the vacuum, has to be separately invoked while studying causality properties of a model.

6 An Anomaly

In this section, a classical Euler–Bernoulli beam theory which is under a pre-stress condition is revisited. In [5], it is discussed that all time-domain formulations satisfy K-K relations. However, in contradiction to this idea, classical EBB under pre-compression is shown here that it violates K-K relations.

The governing equation of motion here reads

$$EI\frac{\partial^4 v}{\partial x^4} + \rho A \frac{\partial^2 v}{\partial t^2} + T\frac{\partial^2 v}{\partial x^2} = P(x,t) \tag{31}$$

Upon substitution of Eq. (1), equation transforms to

$$EI\frac{d^4 \hat{v}}{dx^4} - \rho A \omega^2 \hat{v} + T\frac{d^2 \hat{v}}{dx^2} = \hat{P}(x,\omega) \tag{32}$$

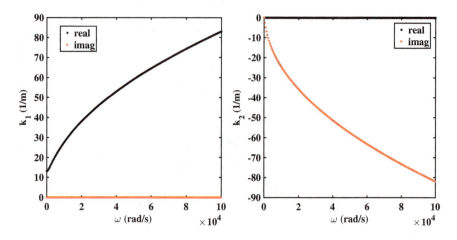

Fig. 8 Wave modes in an EBB under pre-compression

Being a constant coefficient equation and assuming solution as complex exponential in k, above equation then, under absence of body forces, reads

$$EIk^4 - \rho A\omega^2 - Tk^2 = 0 \tag{33}$$

Then the wave modes are

$$k = \pm\sqrt{\frac{T \pm \sqrt{T^2 + 4EI\rho A\omega^2}}{2EI}} \tag{34}$$

The four wave modes are due to $\pm$ inside and outside of the first square root. In general, $+$ and $-$ signs outside square root corresponds to forward and backward or reflected waves in a beam. However, in the case of a semi-infinite beam the two wave modes are given as

$$k_1 = \sqrt{\frac{T + \sqrt{T^2 + 4EI\rho A\omega^2}}{2EI}} \quad k_2 = -\sqrt{\frac{T - \sqrt{T^2 + 4EI\rho A\omega^2}}{2EI}} \tag{35}$$

and are shown in Fig. 8. Numerical values used here are: $E = 70\,\text{GPa}$, $\rho = 2700\,\text{kg/m}^3$, width($b$) = 0.01 m, depth($h$) = 0.01 m, $I = \frac{bh^3}{12}$, $T = 10000\,\text{N}$.

One wave mode is purely real and other is pure imaginary. However, instead of $k_1 = k_2 = 0$ at $\omega = 0$, as in classical EBB, $k_1 \neq 0$. That is, a wave is set up in this EBB under a static load, which is unphysical. In order to further examine this behavior, the STF form of K-K relations is utilized. To this end, the STF has to be derived and it is achieved as follows.

Fig. 9 System transfer function in EBB under pre-compression

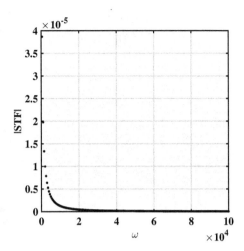

Again, weak form of Eq. (32) is derived, leading to the following boundary conditions:

$$\left[-EI\frac{d^3\hat{v}}{dx^3} - T\frac{d\hat{v}}{dx} - \hat{P}\right]|_{x=0} = 0 \quad \frac{d^2\hat{v}}{dx^2}|_{x=0} = 0 \quad (36)$$

where

$$\hat{v} = a_1 e^{-ik_1 x} + a_2 e^{-ik_2 x} \quad (37)$$

Utilizing the boundary conditions and $\hat{v}$, STF (G) can be obtained as

$$G = \frac{-k_2^2 e^{-ik_1 x} + k_1^2 e^{-ik_2 x}}{i * term} \quad term = EI(k_1^3 k_2^2 - k_1^2 k_2^3) - T(k_1 k_2^2 - k_1^2 k_2) \quad (38)$$

Magnitude of STF is shown in Fig. 9 at $x = 0.005$ m. It can be seen from STF that $G \to \infty$ as $\omega \to 0$. Without going into details, STF form of K-K relations in this case reads

$$Im\ G(\omega') = \frac{-2\omega'}{\pi} \int_0^\infty \frac{Re\ G(\omega)}{\omega^2 - \omega'^2} d\omega \quad (39)$$

Real part of G is given as input to Eq. (39) to predict the imaginary part of G. This prediction is compared to corresponding value from Eq. (38). This is shown in Fig. 10. It is clear that system transfer function as derived from the equation of motion and boundary conditions violates the K-K relations.

Fig. 10 Imaginary part of $\hat{G}$ from Eq. (38) (EOM) and Eq. (39) (K-K)

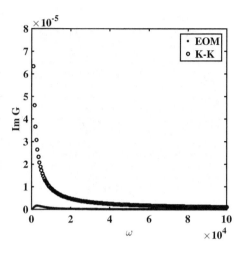

7 Summary

In this paper, gradient-type and integral-type nonlocal models are examined with respect to their respective wave dispersion properties. In general, it is observed that, there exist finite and infinite number of wave modes in gradient and integral nonlocal models, respectively. Examination of these wave modes showed the existence of unphysical attributes such as exponentially unstable responses, undefined wavenumbers at certain frequencies, very large or very small group speeds and negative group speeds.

In order to understand the causality aspect in these models, K-K relations have been applied. It is seen that certain models straight away violates K-K relations due to exponential instability. Existence of very high group speeds seems not to cause any violation of K-K relations. This observation can be seen as a limitation on the applicability of K-K relations. In particular, K-K relations checks the causality with respect to time of arrival of a signal away from the source, rather, the response behavior of the system. Existence of negative group speeds also seems not to cause violation of K-K relations as the involved waves are in-homogeneous. Existence of infinitesimally small speeds violates K-K relations by the introduction of a small imaginary k at these frequencies.

Finally, an example is presented that contradicts a general perception that time-domain formulations do not violate K-K relations. The considered example is a classical Euler–Bernoulli beam under compression. Wave mode shows that even under quasi-static loading a wave is set up in the structure with infinitesimally small or zero phase speeds. Upon application of K-K relations, a disagreement between K-K predictions and equation of motion predictions of the system transfer function is seen.

References

1. Achenbach JD (1973) Wave propagation in elastic solids. Elsevier, North Holland
2. Nussenzveig HM (1972) Causality and dispersion relations. Elsevier, Academic Press, New York
3. Weaver RL, Pao Y-H (1981) Dispersion relations for linear wave propagation in homogeneous and inhomogeneous media. J Math Phys 22(9):1909–1918. doi.org/10.1063/1.525164
4. Chakraborty A, Gopalakrishnan S (2003) A spectrally formulated finite element for wave propagation analysis in functionally graded beams. Int J Solids Struct 40:2421–2448. https://doi.org/10.1016/S0020-7683(03)00029-5
5. Askes H, Metrikine AV, Pichugin AV, Bennett T (2008) Four simplified gradient elasticity models for the simulation of dispersive wave propagation. Philos Mag 88(28–29):3415–3443. https://doi.org/10.1080/14786430802524108
6. Metrikine AV (2006) On causality of the gradient elasticity models. J Sound Vib 297:727–742. https://doi.org/10.1016/j.jsv.2006.04.017
7. Graff KF (1975) Wave motion in elastic solids. Dover, New York
8. Eringen AC (1983) On differential equations of nonlocal elasticity and solutions of screw dislocation and surface waves. J Appl Phys 54:4703–4710. https://doi.org/10.1063/1.332803
9. Chen W, Fish J (2001) A dispersive model for wave propagation in periodic heterogeneous media based on homogenization with multiple spatial and temporal scales. ASME J Appl Mech 68:153–161. https://doi.org/10.1115/1.357165
10. Silling SA (2000) Reformulation of the elasticity theory for discontinuities and long-range forces. J Mech Phys Solids 48(1):175–209
11. Emmrich E, Weckner O (2007) On the well-posedness of the linear peridynamic model and its convergence towards Navier equation of linear elasticity. Commun Math Sci 5(4):851–864
12. Zingales M (2011) Wave propagation in 1D elastic solids in presence of long-range central interactions. J Sound Vib 330:3973–3989. https://doi.org/10.1016/j.jsv.2010.10.027
13. Askes H, Aifantis EC (2009) Gradient elasticity and flexural wave dispersion in carbon nanotubes. Phys Rev B 330:195412-1–195412-8. https://doi.org/10.1103/PhysRevB.80.195412
14. Mikata Y (2012) Analytical solutions of peristatic and peridynamic problems in a 1D infinite rod. Int J Solids Struct 49:2887–2897. https://doi.org/10.1016/j.ijsolstr.2012.02.012
15. Bazant ZP, Luo W, Chau VT, Bessa MA (2016) Wave dispersion and basic concepts of peridynamics compared to classical nonlocal damaged models. ASME J Appl Mech 83:111004-1–111004-16
16. Mobley J (2007) The time-domain signature of negative acoustic group velocity in microsphere suspensions. J Acoust Soc Am 122(1):EL8–EL14 (2007)
17. Waters KR, Mobley J, Miller JG (2005) Causality-imposed (Kramers-Kronig) relationships between attenuation and dispersion. IEEE Trans Ultrason Ferroelectr Freq Control 52(5):822–833

Mechanics of Damage at Steel-Concrete Interface in RC Structures

Saiwal Krishna, Pritam Chakraborty, and S. K. Chakrabarti

Abstract One of the major problems affecting present infrastructure is shortening of service life due to durability issues. A large number of steel-concrete composite structures fail not from overloading but because of durability issues. One of the main causes of short service life is corrosion of steel in steel-concrete composite structures. The current state of practice in design codes deals with this issue by assuming a perfect bond between steel and concrete for design considerations and provision of minimum cover depth for prevention against corrosion. The degradation at the interface that can have long-term deteriorative effects on strength and durability cannot be addressed with the present framework. Thus a need for more exact design and analysis recommendations is needed to address and deal with the issue of durability caused by corrosion. The objective of this study is to investigate the effect of corrosion at the steel-concrete interface on the structural behaviour of steel-concrete composite structures through numerical methods. The mechanics of degradation of the interface have been studied using finite element method. Effects of corrosion at the steel-concrete interface such as weakening of the interface, reduction of structural steel area and de-lamination at the interface have been studied. These phenomenon can be modelled using finite element approaches. Modelling of corrosion using solid elements presented with problems of mesh refinement, coupling damage with a constitutive response. To address the above, a UEL has been implemented in ABAQUS using a traction separation law based on the Mohr-Coulomb failure criteria, which will be used to model corrosion at the interface. Experimental data on corrosion samples were used to obtain the material properties. The developed model shall pro-

---

S. Krishna (✉)
Department of Civil Engineering, Indian Institute of Technology Kanpur, Kanpur 208 016, India
e-mail: saiwal@iitk.ac.in

P. Chakraborty
Department of Aerospace Engineering, Indian Institute of Technology Kanpur, Kanpur, India

S. K. Chakrabarti
Department of Civil Engineering, Budge Budge Institute of Technology, Kolkata, India

S. Krishna · P. Chakraborty · S. K. Chakrabarti
Indian Institute of Technology Kanpur, Kanpur, Uttar Pradesh, India

© The Editor(s) (if applicable) and The Author(s), under exclusive license to Springer Nature Singapore Pte Ltd. 2021
S. K. Saha and M. Mukherjee (eds.), *Recent Advances in Computational Mechanics and Simulations*, Lecture Notes in Civil Engineering 103,
https://doi.org/10.1007/978-981-15-8138-0_2

vide a way to study the effect of the presence of varying amounts of corrosion on the performance of a structure. The present study aims to provide a robust numerical scheme to analyse composite structures with varying degrees of corrosion. On the basis of obtained results, recommendations for addressing deterioration due to corrosion would be recommended.

Keywords Steel-concrete interface · Corrosion · Finite element method

1 Introduction

Steel-concrete structures which make up the majority of modern infrastructure face deterioration and shortened service life because of early corrosion. Early studies on corrosion were focused on the problem of corrosion in embedded steel reinforcing bars which were investigated by experiments involving pullout tests on corroded rebars embedded in concrete [1–5]. A number of studies have studied the phenomenon of corrosion specifically involving rebars in concrete for which analytical models have been devised, largely involving a bond-slip relationship which is calibrated from experimental studies performed earlier [6, 7]. Others have studied the effect of corroded reinforcement on the behaviour of structures [8–10]. However, these models are confined to the case of corroded rebars embedded in concrete. Studies specific to the modelling of the interface, which has undergone corrosion are lacking. Few studies have investigated effects of corrosion on friction characteristics of the interface [11], which provides an insight into some aspects of corrosion at the interface but do not address other aspects such as the effect on the strength of the interface.

A review of the major design codes [12–15] shows the present methods of ensuring adequate durability requirements is limited to cover thickness according to exposure conditions and monitoring cover crack width caused by corrosion of embedded steel. However, an approach to address loss in strength and performance of a structure due to corrosion hasn't been adopted in the design approach. The aim of this study is to provide a numerical framework for the analysis of damage/deterioration in composite structures due to corrosion. The developed model has been used to demonstrate the effect of corrosion on embedded reinforcing bars. For this, a pullout test commonly used to estimate bond stress for embedded reinforcement bars is used.

2 Model of Corroded Layer as Thick Interface

To study the effect of the presence of corrosion products on the structural behaviour of steel embedded in concrete, a finitely thick corroded steel-concrete interface was considered and discretized using solid elements. The pullout test was considered to investigate the workability of the approach and the corresponding Finite Element

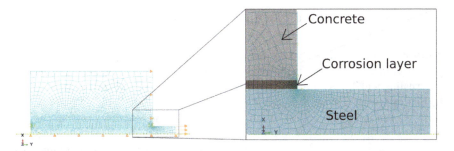

Fig. 1 Details of the FEM model used for simulating pullout test

Method (FEM) model is shown in Fig. 1. The specimen dimensions have been chosen to be comparable with the experimental setup utilized in [2, 16]. In the FEM model, a thickness of 1 mm of corrosion products for a bar of 10 mm diameter corresponding to a corrosion level of 10% by area of steel was considered [2]. Since the steel bar has a cylindrical cross section and was assumed to corrode uniformly, hence axisymmetric analysis was performed in this study. The behaviour of concrete was assumed as a linear elastic. An elastic modulus of 300 GPa and a Poisson's ratio of 0.27 was used corresponding to M40 concrete mix. An elastic perfectly plastic behaviour was chosen for the steel bar with a yield stress of 250 MPa and Poisson's ratio of 0.17 [17]. The Mohr-Coulomb elasto-plastic model was adopted to represent the deformation behaviour of the corroded layer. The model is suitable for granular materials like soil, which have a lower strength in tension as compared to compression. The elastic properties were obtained from the tests performed by Zhao et al. [18]. As reported [18], the Young's Modulus for corrosion product varied between 95 and 610 MPa depending on the physical properties of the test specimen which in turn varied according to the exposure conditions, source of sample, and location of steel. The parameters required for Mohr-Coulomb model are the yield stress (σ_y), dilation angle (ϕ) and cohesion (c). The yield stress under compression was obtained from the uniaxial compression tests performed on corroded samples [18]. However, due to the lack of experimental data on corroded material, the other parameters, i.e. dilation angle and cohesion, were assigned typical values of granular material such as soil and sand [19]. The values of the parameters used in this study are, $\phi = 27°$, $\sigma_y = 10$ MPa and $c = 10$ MPa. Pullout simulations were performed in ABAQUS, a commercial FEM software, utilizing the boundary conditions shown in Fig. 1. As can be seen from the figure, the axis of the bar was restrained to displace in the x-direction, while pulled on the right face along the y-direction at a constant velocity. The right face of the concrete was restrained to displace along the y-direction. These boundary conditions are representative of the pullout tests.

2.1 Limitations of the Thick Interface Representation

The use of finitely thick corroded layer presented a few challenges. One of the issues was of erroneous stresses in elements at corners showing a checkerboard pattern (alternating tensile and compressive stress) as shown in Figs. 2 and 3. The issue may be due to a very constrained corner region resulting in large gradients difficult to capture even with very refined mesh. Also, these corner elements undergo severe deformation leading to excessive mesh distortion. These numerical issues can be addressed by considering a cohesive zone representation of the corroded interface and is discussed in Sect. 3. Also, the thickness dimension of the corroded layer is significantly smaller than the concrete or steel members in a realistic structure. Hence, the assumption of a cohesive zone is much more appropriate under all practical situations.

Numerical issues at the corner elements of the corroded layer.

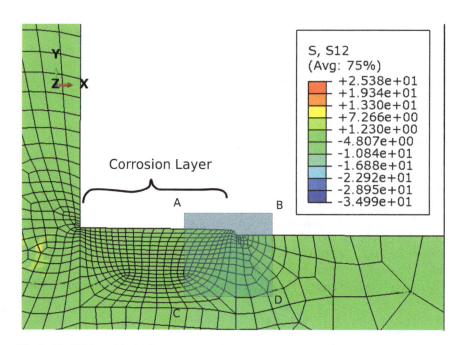

Fig. 2 The FEM model of pullout specimen showing the corroded interface layer

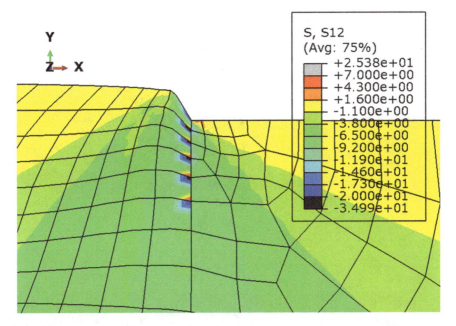

Fig. 3 Magnified view of A-B-C-D shown in Fig. 2 highlighting the anamolous stress distribution at the corner of the corroded layer

3 Cohesive Zone Model of the Corroded Interface

In the cohesive zone model developed in this work, the traction separation law is based on the Mohr-Coulomb model provided in [20]. The yield surface for the Mohr-Coulomb model in the cohesive zone is defined as

$$F = |\tau| + (\sigma - \sigma_{max})tan(\phi) \quad (1)$$

where τ is the shear traction, σ is the normal traction, σ_{max} is the cohesion obtained from uniaxial tests and ϕ is the friction angle. A schematic of the yield surface is shown in Fig. 4. A non-associative flow rule is used to describe the inelastic separations. The potential function, ψ, thus used is described by

$$\psi = \begin{cases} \frac{\sigma^2}{\sigma_{max}^2} + \frac{\tau^2}{\tau_{max}^2} - 1 & \text{if } \sigma > 0 \\ sgn(\tau)\tau_{max} & \text{otherwise} \end{cases} \quad (2)$$

The potential surface is schematically shown in Fig. 4. It is presently assumed that post yield the corroded material shows a perfectly plastic response and accumulates inelastic separations in the normal (w_i) and shear (v_i) directions. Thus, the tractions can be given by

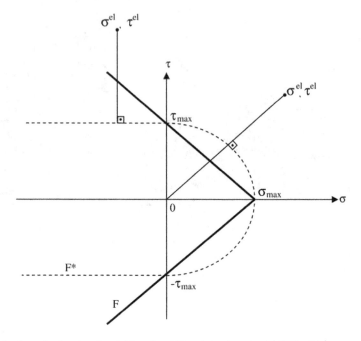

Fig. 4 A schematic showing the yield surface (F) and plastic potential (F^*) of the non-associative Mohr-Columb-based traction separation law of the cohesive zone

$$\sigma = K_n(w - w_i) \quad (3)$$

$$\tau = K_t(v - v_i) \quad (4)$$

where w and v correspond to the total normal and shear separations, respectively. K_n and K_t are the stiffness parameters in the normal and shear mode, respectively.

The plastic separation rates are obtained from

$$\dot{w}_i = \dot{\lambda}\frac{\partial \psi}{\partial \sigma} \quad \text{and} \quad \dot{v}_i = \dot{\lambda}\frac{\partial \psi}{\partial \tau} \quad \text{if } F = 0 \quad (5)$$

$$\dot{w}_i = \dot{v}_i = 0 \quad \text{otherwise} \quad (6)$$

where λ is the plastic multiplier which can be obtained from the consistency condition. The model is implemented as user element (UEL) subroutine in ABAQUS. The return mapping algorithm is used to integrate the constitutive behaviour [21]. Further, large deformation is taken into account by considering the rotation of the mid-plane of cohesive zone following [22].

4 Verification of the Cohesive Zone Model Using Pullout Tests

The cohesive zone model was verified against experimental results from pullout tests [2, 16]. An axisymmetric 2D analysis was carried out to obtain bond-slip curves and bond stress distribution along the bar. The FEM model used is shown in Fig. 5. The dimensions and boundary conditions of the FEM model were chosen the same as described in Sect. 2. From the simulation, the corresponding bond stress distribution along the bar length is shown in Fig. 6 and compared with the experiments performed on uncorroded rebars [16] shown in Fig. 7. The bond-slip relationship for corroded bars was also compared with [2] and is shown in Fig. 8.

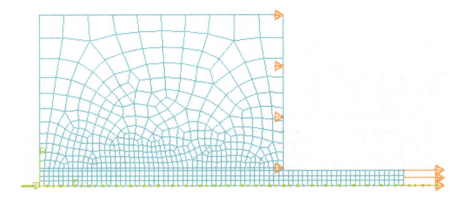

Fig. 5 FEM model used for pullout simulation using cohesive zone model

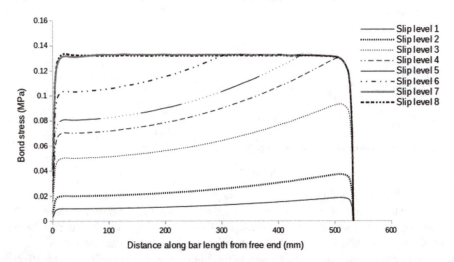

Fig. 6 FEM results where level refers to the extent of bar slip, 1 being the minimum and 8 is maximum

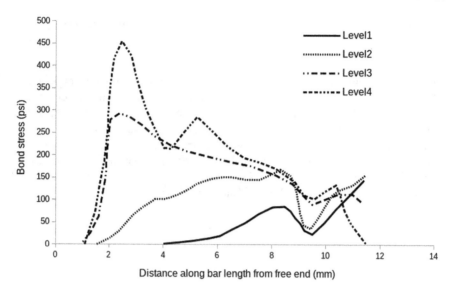

Fig. 7 Experimental results [16]

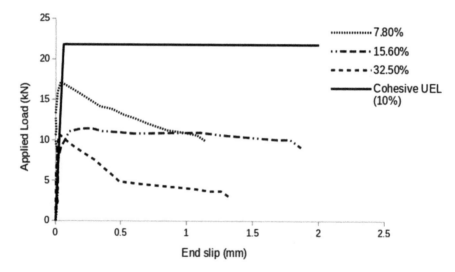

Fig. 8 Bond versus slip relationship for pullout specimens

Distribution of bond stress along the length of the embedded bar.

From the figures, it can be observed that the bond stress distribution matches well with the experimental observations from Mains [16] shown in Fig. 7. The simulated bond-slip curve shows a perfectly plastic response which was expected, however in order to effectively simulate damage at the interface a post-yield softening effect needs to be incorporated in the model which can be seen in the experimental results shown in Fig. 8. The few aberrations at the loading end can be accounted for by the fact that the bond stress obtained from experiments was obtained at few points along the bar and the condition of the interface after attaching strain gauges gets disturbed and is prone to faulty measurement. The development of bond stress with progressive slip is seen to be similar to experimental observations, where after the failure of the interface at the loading surface the interface stresses increase at the end of the bar (slip level 7,8). The bond stresses have been extracted for different end slip levels referred to as level 1–8 in Fig. 6.

5 Conclusions

In this work, a cohesive zone model to capture the effect of corrosion at the steel-concrete interface on the response of RC structures has been developed. The cohesive representation of the interface ignores the thickness dimension of the interface, which is few orders smaller than the structural dimension. Such a treatment has also been shown in this study to alleviate the numerical difficulties experienced in the thick interface model. The traction separation law of the cohesive zone is derived from Mohr-Columb elasto-plastic model for soils. The model parameters have been calibrated and FEM simulation of the pullout test has been performed to demonstrate the appropriateness of the model. From the simulation, it has been observed that the model captures the trend in bond stress distribution along the bar at different levels of pullout. However, since the model behaves in a perfectly plastic manner without damage, thus the peaks and drops as observed in the experiments are missing. Hence, a post-yield softening behaviour needs to incorporated in the model to effectively capture damage of the interface during pullout. The proposed cohesive zone model can be extended to a wide variety of structures with inhomogeneous and progressive corrosion, thus enabling realistic simulation of RC structures and their optimal design.

References

1. Al-Sulaimani GJ, Kaleemullah M, Rasheeduzzafar Basunbul IA (1990) Influence of corrosion and cracking on bond behavior and strength of reinforced concrete members. ACI Struct J 87(2):220–231
2. Almusallam AA, Al-Gahtani AS, Aziz AR, Dakhil FH, Rasheeduzzafar (2002) Effect of reinforcement corrosion on flexural behavior of concrete slabs. J Mater Civil Eng 8(3):123–127. https://doi.org/10.1061/(asce)0899-1561(1996)8:3(123)
3. Fang C, Lundgren K, Plos M, Gylltoft K (2006) Bond behaviour of corroded reinforcing steel bars in concrete. Cem Concr Res 36(10):1931–1938. https://doi.org/10.1016/j.cemconres.2006.05.008
4. Cairns J, Du Y, Law D (2006) Residual bond strength of corroded plain round bars. Mag Concr Res 58(4):221–231. https://doi.org/10.1680/macr.2006.58.4.221

5. Bilcik J, Holly I (2013) Effect of reinforcement corrosion on bond behaviour. Procedia Eng 65:248–253. https://doi.org/10.1016/j.proeng.2013.09.038
6. Coronelli D (2002) Corrosion cracking and bond strength modeling for corroded bars in reinforced concrete. ACI Struct J 99(3):267–276
7. Blomfors M, Zandi K, Lundgren K, Coronelli D (2018) Engineering bond model for corroded reinforcement. Eng Struct 156:394–410. https://doi.org/10.1016/J.ENGSTRUCT.2017.11.030
8. Lundgren K (2001) Bond between corroded reinforcement and concrete. Department of Structural Engineering, Chalmers University of Technology, Sweden. Report No. 00, 3
9. Lee H, Noguchi T, Tomosawa F (2002) Evaluation of the bond properties between concrete and reinforcement as a function of the degree of reinforcement corrosion. Cem Concr Res 32:1313–1318
10. Du YG, Clark LA, Chan AHC (2005) Residual capacity of corroded reinforcing bars. Mag Concr Res 57(3):135–147. https://doi.org/10.1680/macr.2005.57.3.135
11. Cairns J, Du Y, Law D (2007) Influence of corrosion on the friction characteristics of the steel/concrete interface. Constr Build Mater 21(1):190–197. https://doi.org/10.1016/j.conbuildmat.2005.06.054
12. American Concrete Institute (ACI) (2011) Building code requirements for structural concrete (ACI 318-11) and commentary
13. Standard I (2000) IS-456. 2000 Plain and reinforced concrete-code of practice. Bureau of Indian Standards, Manak Bhawan, p 9
14. EN 1992-1-1 (2004) Eurocode 2: design of concrete structures–part 1-1: general rules and rules for buildings. European Committee for Standardization
15. Japan Society of Civil Engineers Concrete Committee (2007) Standard specifications for concrete structures: design
16. Mains (1951) Measurement of the distribution of tensile and bond stresses along reinforcing bars. ACI J Proc 48(11). https://doi.org/10.14359/11882
17. IS-2062 (1999) Steel for general structural purposes-specification (Fifth Revision). Bureau of Indian Standards, New Delhi
18. Zhao Y, Dai H, Ren H, Jin W (2012) Experimental study of the modulus of steel corrosion in a concrete port. Corros Sci 56:17–25
19. Dusko Hadzijanev Ardiaca (2009) Mohr-Coulomb parameters for modelling of concrete structures. Plaxis Bull 25:12–15
20. Lens LN, Bittencourt E, d'Avila VMR (2009) Constitutive models for cohesive zones in mixed-mode fracture of plain concrete. Eng Fract Mech 76(14):2281–2297. https://doi.org/10.1016/j.engfracmech.2009.07.020
21. Belytschko T, Liu WK, Moran B, Elkhodary K (2013) Nonlinear finite elements for continua and structures. Wiley, New Jersey
22. Park K, Paulino GH (2012) Computational implementation of the PPR potential-based cohesive model in ABAQUS: educational perspective. Eng Fract Mech 93:239–262. https://doi.org/10.1016/j.engfracmech.2012.02.007

A Discussion on Locking and Nonlocking Gradient-Enhancement Formulations for Concrete Behavior

Ajmal Hasan Monnamitheen Abdul Gafoor and Dieter Dinkler

Abstract Modeling of strain-softening behavior exhibited by quasi-brittle materials such as concrete and rocks leads to mesh-sensitive and physically objectionable solutions upon mesh refinement and thereby causing ill-posedness of the boundary value problem. These numerical difficulties can be avoided by employing regularization methods that include an internal length scale. Although integral-type nonlocal regularization and gradient-type methods among those are widely accepted in the scientific community, gradient methods remain as a convenient approach due to its straightforwardness in linearization. The present paper provides a comparison of two different types of implicit-gradient formulations that consider an equivalent strain measure and the damage variable for gradient-enhancement. It discusses the locking and nonlocking effects on softening behavior. An elasticity-based isotropic damage model is adopted for the comparison in gradient-enhanced modeling.

Keywords Damage · Localization · Gradient · Locking · Nonlocking

1 Introduction

Many material models within the context of continuum mechanics have been developed to study the nonlinear deformation behavior and fracture mechanisms of quasi-brittle materials such as concrete and rocks. These include the models based on plasticity theories, fracture, and continuum damage mechanics. As the continuum damage models overcome the drawbacks of fracture-based models such as the needs of node-splitting and predefined crack, etc., they remain as efficient tools in engineering applications. The distributed micro-cracking pattern exhibited by materials like concrete upon the increase in loading can be modeled as strain-softening in the structural analyses. Strain-softening is nothing but the decrease of stress in the post-peak region upon the increase in strains [2]. This kind of softening response is

A. H. Monnamitheen Abdul Gafoor (✉) · D. Dinkler
Institute of Structural Analysis, Technische Universität Braunschweig, Braunschweig, Germany
e-mail: a.abdul-gafoor@tu-braunschweig.de

always accompanied by a reduction of the unloading stiffness of the material, and irreversible (permanent) deformations, which are localized in narrow zones often called cracks or shear bands.

When these local damage models are implemented into finite element programs, the localization of deformation occurs into a size of vanishing volume and subsequently the dissipation of energy ceases to zero due to strain-softening. Thus, the local approach eventually leads to spurious and physically objectionable results [4]. Moreover, as soon as strain-softening begins, any material loses stability when the matrix of its tangent moduli ceases to be positive definite. As the tangent modulus is negative on the strain-softening (declining) branch, the matrix of the tangent moduli is not positive-definite anymore. Thus, the characterization of strain-softening in terms of stress, strain, and state variables of continuum causes material instability. Subsequently, material instability leads to structural instability. Consequently, these instabilities become the reason for the ill-posedness of boundary value problems [1, 3, 7].

Thus, the local continuum approach that is still a mathematically meaningful concept does not yield a proper representation of strain-softening materials in which a finite energy dissipation and a finite size of localization zone are witnessed by experiments. Therefore, the micro-cracking pattern observed in the strain-softening materials must be described in a nonlocal sense instead of simply a local sense. A search of some regularization methods that could overcome the numerical difficulties rendered method of introducing internal length scales such as viscous regularization [7], Cosserat continua [7], nonlocal integral method [6, 15] or implicit/explicit gradient methods [14]. Nonetheless, the advancements from local models to nonlocal models (either integral type or gradient type) make continuum damage mechanics a promising approach to study the deformation behavior of any material.

However, the approach of introducing an internal length scale using the implicit-gradient method is widely accepted for regularizing the boundary problems, as its linearization is straightforward. A comparison of nonlocal integral formulations of damage or plasticity [5, 9] shows that models based on certain nonlocal formulations are intrinsically incapable of describing the entire fracture processes of materials up to complete failure and modeling macroscopic crack. The theoretical and numerical analyses have revealed that such models exhibit a special type of stress locking effect during the localization process. On the other hand, a comparison of damage formulation enhanced by implicit-gradient method [8] has been presented using separate examples with different materials for each model. Nevertheless, an in-depth study of damage models that adopt a constant length scale parameter and their influence on the localization behavior has not been compared previously. This paper concerns two types of implicit-gradient formulations that consider an equivalent strain measure and the damage variable for gradient-enhancement. The elasticity-based damage model proposed by the authors [11, 12] will be presented here for monotonic behavior for the sake of comparison. Attention is focused only on isotropic damage evolution and the internal length scale is treated as constant. The numerical results of a benchmark problem are discussed.

2 Local Modeling of Damage

According to classical continuum damage theory, the nonlinear deformation behavior of the damaged quasi-brittle material can be characterized by the constitutive relation that adopts energy equivalence principle as follows:

$$\boldsymbol{\sigma} = (1 - D)^2 \mathbb{C} : \boldsymbol{\varepsilon}, \tag{1}$$

where $\boldsymbol{\sigma}$ and $\boldsymbol{\varepsilon}$ are Cauchy stress and strain tensors; $\mathbb{C}$ is the fourth-order elasticity tensor of the undamaged material. The isotropic scalar damage D is defined explicitly as a function of a history variable κ such that $D = D(\kappa)$ [13]. The history variable κ registers the severe most deformation that the material experienced during the entire load history. The damage variable D ranges between $0 \leq D \leq 1$.

The local state of deformation is identified by a variable that accelerates the growth of damage. Therefore, a unified local equivalent strain ϵ that describes both tensile as well as compression behavior of softening materials like concrete as presented:

$$\epsilon = \left(\alpha I_1 + \sqrt{3 J_2} + \beta H(\sigma_{\max})\sigma_{\max}\right)/(1-\alpha)E, \tag{2}$$

is adopted as a local deformation measure. Herein, I_1 and J_2 are the first invariant of stress tensor and the second invariant of deviatoric stress tensor, respectively. $\sigma_{\max}$ refers to the maximum principal stress of predicted stress. H is a Heaviside function, which equals to 1 for all positive maximum principal stresses, otherwise equals to 0. α and β are dimensionless constants as defined in [10]. The expression (2) is inspired from the extended Lubliner–Lee failure criterion [10] and slightly transformed to express a damage criterion f_d conveniently in strain space. The relation between κ and ϵ is postulated by the function f_d as follows:

$$f_d = \epsilon - \kappa. \tag{3}$$

Thus, the function f_d decides the possibility of damage evolution at a point of continuum. During the damage process, the evolution of the history parameter κ must always satisfy the Kuhn–Tucker loading/unloading conditions, which are mathematically expressed as

$$f_d \leq 0; \qquad \dot{\kappa} \geq 0; \qquad f_d \dot{\kappa} = 0, \tag{4}$$

where $(\dot{\ })$ represents the derivative of a variable with respect to time t. In addition, to characterize the unilateral/cyclic behaviors of concrete, two different history parameters κ_t and κ_c are introduced corresponding to tension and compression such that

$$\kappa_t = \operatorname{Sup}\left[\kappa_{0t}, \max \epsilon\right]; \; \kappa_c = \operatorname{Sup}\left[\kappa_{0c}, \max \epsilon\right], \tag{5}$$

$$\kappa = \kappa_t H + \kappa_c (1 - H), \tag{6}$$

where κ_{0t} and κ_{0c} are the initial thresholds under tension and compression. κ_t and κ_c describe the most severe deformation that the material experienced during the tensile load history and the compression load history, respectively.

3 Gradient Modeling

Modeling of local damage theories in a nonlocal sense has become popular by incorporation of implicit-gradient-enhancement method. Thus, the nonlocal averaging procedure is applied to the local variable denoted by Φ. Consequently, the nonlocal counterpart of Φ, i.e., $\hat{\Phi}$ can be approximated by the partial differential equation called implicit-gradient equation [14] as follows:

$$\hat{\Phi} - l_c^2 \nabla^2 \hat{\Phi} = \Phi, \qquad (7)$$

where ∇^2 is the Laplacian operator and l_c is the characteristic internal length scale. To describe the microstructural interactions occurring in the fracture process zone, the nonlocal variable $\hat{\Phi}$ is considered. It represents the average of its local counterpart Φ within the volume of domain considered. Equation (7) must always be supplemented by an additional boundary condition. Hither, we adopt a natural boundary condition

$$\nabla \hat{\Phi} \cdot \boldsymbol{n} = 0, \qquad (8)$$

at every point of the boundary, as it ensures that the average of nonlocal variable $\hat{\Phi}$ over the entire domain equals that of its local counterpart Φ [14]. The nonlocal averaging can be applied to various types of variables involved in the fracture process of the materials. These variables include a measure of an equivalent strain ϵ, history deformation variable κ, damage energy release rate Y, and damage variable D itself [5, 9]. Nevertheless, the authors' attention is focused on the two gradient-enhanced damage formulations using a constant internal length scale. One considers the damage variable D [6] as a local deformation variable and the other adopts the equivalent strain ϵ [14] as local deformation variable for nonlocal averaging.

3.1 Gradient-Enhancement of Equivalent Strain

Firstly, the averaging procedure is applied to the local equivalent strain ϵ, i.e., $\Phi = \epsilon$, as the local variable ϵ drives the damage growth. Therefore, its nonlocal counterpart $\hat{\epsilon}$ is approximated by a partial differential equation such that

$$\hat{\epsilon} - l_c^2 \nabla^2 \hat{\epsilon} = \epsilon, \qquad (9)$$

along with a natural boundary condition at every point of the boundary as

$$\nabla \hat{\epsilon} \cdot \boldsymbol{n} = 0. \tag{10}$$

As a consequence, κ is later related to the distributed nonlocal equivalent strain $\hat{\epsilon}$. Thus, $\hat{\epsilon}$ enters into the damage criterion (3) instead of its local counterpart and subsequently the damage criterion can be rewritten as follows:

$$f_d = \hat{\epsilon} - \kappa. \tag{11}$$

The Kuhn–Tucker loading/unloading relations (4) remain valid. Assuming the length scale $l_c \to 0$, the local damage model is resumed as $\hat{\epsilon} = \epsilon$.

3.2 Gradient-Enhancement of Damage

On the other hand, as the presence of damage variable helps in modeling the strain-softening behavior of any material, this quantity can also be considered for the gradient enrichment. Thus, the nonlocal quantity $\hat{D}$ is approximated by the partial differential equation such that

$$\hat{D} - l_c^2 \nabla^2 \hat{D} = D, \tag{12}$$

along with a natural boundary condition at every point of the boundary as

$$\nabla \hat{D} \cdot \boldsymbol{n} = 0. \tag{13}$$

Consequently, the nonlocal damage variable $\hat{D}$ that enters into the constitutive law (1) instead of D yields the stress–strain relation as follows:

$$\boldsymbol{\sigma} = (1 - \hat{D})^2 \mathbb{H} : \boldsymbol{\varepsilon}. \tag{14}$$

The damage criterion (3) and the Kuhn–Tucker loading/unloading relations (4) still remain valid.

4 Localization Analysis

To visualize the localization phenomena, a one-dimensional bar, as depicted in Fig. 1a, is investigated under tension. Tensile loading is applied under displacement control by means of an imposed displacement, as depicted in Fig. 1b. Three different FE meshes (discretized with $n = 40, 80, 160$-equal-sized solid elements along the bar) are considered. An imperfection of $10\,[\text{mm}]$ wide at the center of the bar is

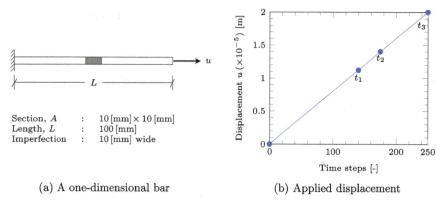

(a) A one-dimensional bar   (b) Applied displacement

Fig. 1 A one-dimensional bar with a central imperfection subjected to tension

Table 1 Material and model parameters adopted

Material Parameters	f_c (MPa)	f_t (MPa)	E (GPa)	ν (–)		
	27.6	3.48	31.7	0.2		
Model Parameters	β_{1t}	β_{2t}	β_{1c}	β_{2c}	κ_{0t}	κ_{0c}
	0.85	0.18	0.0	0.095	f_c/E	$10f_t/3E$

introduced to trigger the damage localization in the middle of the bar by reducing the initial damage threshold κ_0 by approximately 13%. The other material and model parameters used in the analysis are provided in Table 1. The nonlocal enhancement of local quantities, namely, ϵ and D based on the implicit-gradient method has been analyzed. Linear shape functions have been used for the nonlocal variables. Two different internal length values $l_c = \{5, 10\}$ [mm] are used in the numerical analysis. Mesh-independent solutions are achievable with convergence. The converged stress responses are shown in Fig. 2. The corresponding local and nonlocal strains and damage distributions obtained from nonlocal models are illustrated along the bar in Figs. 3 and 4.

4.1 Nonlocking Formulation

In the case of the gradient-enhanced equivalent strain, i.e., $\hat{\Phi} = \hat{\epsilon}$, the stress responses obtained for the three meshes and for the different internal length scales are depicted in Fig. 2a. It is clearly observed that the softening curves of every mesh are practically identical and converging faster. A finite energy dissipation apparently has been observed. The higher values of l_c lead to the increase in peak stress, as noticed in Fig. 2a as well as the broader width of strain/damage localization, as seen in Fig. 3a. The model responses become brittle; the model loses its stiffness and the peak stress

A Discussion on Locking and Nonlocking Gradient-Enhancement ...

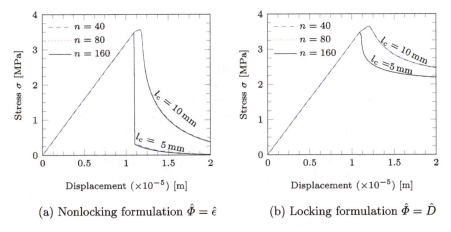

Fig. 2 A one-dimensional bar with a central imperfection subjected to tension and stress responses for nonlocking and locking formulations

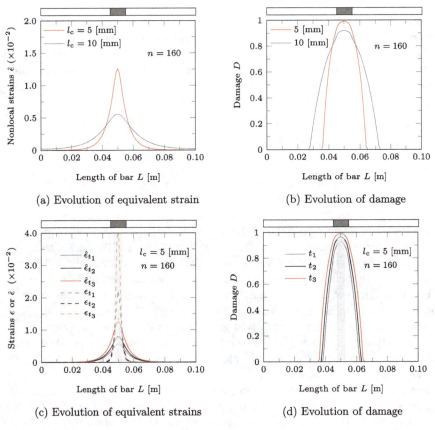

Fig. 3 Nonlocal solutions of bar for nonlocking formulations $\hat{\Phi} = \hat{\epsilon}$

value reduces for decreasing internal lengths and vice versa. An increase in the internal length scale makes the localization zone wider. Moreover, the softening responses are also physically meaningful and are free from any locking effects.

Figure 3a and b display the distributions of nonlocal equivalent strains and resulting damage along the bar obtained for the fine mesh ($n = 160$) at the loading step t_3 using the different length scales, respectively. A larger value of l_c allows wider spreading of the nonlocal equivalent strain. Furthermore, Fig. 3c compares the evolution of local and nonlocal equivalent strains corresponding to the three different loading steps (t_1, t_2, t_3) in the post-peak regime for the fine mesh ($n = 160$) with $l_c = 5$ [mm]. As realized, the finite width of localization of deformation or damage occurs and subsequent loading causes broadening of the localization band. As observed, the local equivalent strains (dashed lines) are concentrated near the vicinity of the triggered region and further loading increases the narrowing of the zone leading to propagation of macro-crack. The nonlocal equivalent strains (thick lines) are spread over the localization zone, physically meaning the occurrence of microstructural interaction in the neighborhood.

The respective evolution of damage is plotted over the length of the bar in Fig. 3d. The continued loading leads to the subsequent development of damage. The damage is formed in a broader area. Nonetheless, the narrow region of strains is obtained. Therefore, the gradient-enhancement of the present model describes the physical meaning of crack initiation, growth of micro-cracks, and coalescence of these into a macro-crack, as observed by Peerlings et al. [14].

4.2 Locking Formulation

In the case of the gradient-enhanced damage variable, i.e., $\hat{\Phi} = \hat{D}$, the stress responses obtained for the three meshes and for varying internal length scales are depicted in Fig. 2b. It is observed that the softening curves are identical, and therefore show convergence upon the mesh refinement. As depicted in Fig. 2b, the model responses show brittleness due to the loss of its stiffness and the peak stress value reduces for decreasing internal lengths and vice versa. It is anticipated that the larger internal length makes the localization zone wider. But the present formulation, on the whole, does not yield a reasonable behavior at later stages of softening upon the increase in the applied displacement, as observed in Fig. 2b. Consequently, this leads to locking effects and therefore fails to predict the material behavior correctly.

Furthermore, Fig. 4a and b illustrates the distributions of local equivalent strains and nonlocal damage along the bar obtained for the fine mesh ($n = 160$) at the loading step t_3 using the different length scales, respectively. A larger value of l_c allows wider spreading of the equivalent strain, as the gradient-enhanced damage (nonlocal damage) expands wider. In other words, a smaller value of l_c yields a strong (narrowed) localization of damage at the center of the triggered zone. At the same instant, the widening of the damage zone or softening zone occurs across the whole bar, as noted in Fig. 4b.

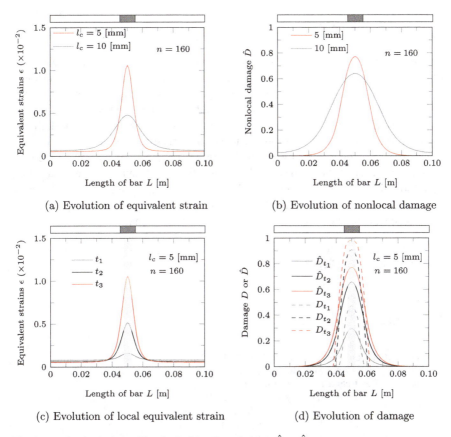

Fig. 4 Nonlocal solutions of bar for locking formulations $\hat{\Phi} = \hat{D}$

Figure 4c shows the evolution of local equivalent strains corresponding to the three different loading steps (t_1, t_2, t_3) in the post-peak regime obtained for the fine mesh ($n = 160$) with $l_c = 5$ [mm]. As can be observed in Fig. 4d, the local damage (dashed lines) is concentrated near the vicinity of the triggered region, whereas the nonlocal damage (thick lines) are spread over the localization zone.

Nevertheless, the gradient-enhancement of D leads to locking effects in the softening behavior. It merely represents the model failing to produce a realistic behavior of brittle materials like concrete. As demonstrated, it typically displays a ductile nature beyond a certain level of considerable damage even though strains increase. Because the averaging of damage variable leads to the gradual expansion of the softening zone across the whole bar. It is worth prominently mentioning this kind of locking effect in stress is also observed in the use of nonlocal integral models [5, 9].

5 Concluding Remarks

The local variables such as equivalent strain and damage variable are considered for nonlocal or gradient-enhancements in the localization analysis. It is demonstrated that the gradient-enhancement on the local equivalent strain yields physically meaningful mesh-independent responses, as the width of localization becomes finite. The maximum values of local strains and the smooth distribution of nonlocal strains replicate the macro-crack and the existence of microstructural interactions in the fracture process zone, respectively. On the other hand, the gradient-enhancement on the damage variable shows locking effects in softening behavior, as the localization of deformation expands towards the ends of the bar. In this manner, the exact behavior of the material cannot be produced, as the gradient-enhanced damage variable never reaches unity because of the restrictions on local damage growth.

Acknowledgements Financial support by the German Research Foundation (DFG) in the framework of the Graduate College 2075 is acknowledged.

References

1. Bažant Z (1976) J Eng Mech Divis 102(2):331
2. Bažant Z (1986) Appl Mech Rev 39(5):675
3. Bažant Z, Belytschko T (1985) J Eng Mech 111(3):381
4. Bažant Z, Belytschko T, Chang T (1984) J Eng Mech 110(12):1666
5. Bažant Z, Jirásek M (2002) J Eng Mech 128(11):1119
6. Bažant Z, Pijaudier-Cabot G (1988) J Appl Mech 55(2):287
7. De Borst R, Sluys L, Mühlhaus H, Pamin J (1993) Eng Comput 10(2):99
8. Geers M, Peerlings R, Brekelmans W, De Borst R (2000) Acta Mechanica 144(1–2):1
9. Jirásek M (1998) Int J Solids Struct 35(31):4133
10. Lee J, Fenves G (1998) J Eng Mech 124
11. Monnamitheen Abdul Gafoor AH, Dinkler D (2017) A simplified damage model for unilateral behavior of concrete. In: Scheven Mv, Keip, M-A, Karajan, N (eds) Conference proceedings of 7th GACM colloquium on computational mechanics for young scientists from academia and industry 2017, pp 312–315. Institute for Structural Mechanics, University of Stuttgart, Stuttgart. https://doi.org/10.18419/opus-9334
12. Monnamitheen Abdul Gafoor AH, Dinkler D (2017) PAMM 17(1):241. https://doi.org/10.1002/pamm.201710090. https://onlinelibrary.wiley.com/doi/abs/10.1002/pamm.201710090
13. Monnamitheen Abdul Gafoor, AH, Dinkler, D (2016) PAMM 16(1):155. https://doi.org/10.1002/pamm.201610066. https://onlinelibrary.wiley.com/doi/abs/10.1002/pamm.201610066
14. Peerlings R, De Borst R, Brekelmans W, De Vree J (1996) Int J Numer Methods Eng 39:3391
15. Pijaudier-Cabot G, Bažant Z (1987) J Eng Mech 113(10):1512

On Flexure of Shear Deformable Isotropic Rectangular Propped Cantilever Beams

Kedar S. Pakhare, P. J. Guruprasad, and Rameshchandra P. Shimpi

Abstract This paper presents a comparison study on the flexure of shear deformable isotropic rectangular propped cantilever beams by utilizing newly developed variationally consistent two-variable-refined beam theory, variationally inconsistent single-variable refined beam theory, and variationally inconsistent Levinson beam theory. The beam is assumed to be under the action of uniformly distributed transverse load. It should be noted that governing differential equations of two-variable-refined beam theory are derived by utilizing Hamilton's principle. Whereas, beam gross equilibrium equations are utilized in the case of single-variable-refined beam theory as well as Levinson beam theory to derive their respective governing differential equations. All three beam theories take into account a parabolic variation of the beam transverse shear strain and hence the beam transverse shear stress through the beam thickness. These theories satisfy transverse shear stress-free beam surface conditions. Hence, these theories do not require a shear correction factor. Effects of the beam thickness-to-length ratio on the location of maximum beam transverse displacement and values of maximum non-dimensional beam transverse displacement, non-dimensional beam axial stress, and non-dimensional beam transverse shear stress are presented. Profiles of the non-dimensional beam transverse displacement, non-dimensional beam axial stress, and non-dimensional beam transverse shear stress for various values of the beam thickness-to-length ratio are also presented.

Keywords Shear deformable beams · Propped cantilever · Flexure

1 Introduction

The Bernoulli–Euler beam theory (*BEBT*) is the simplest displacement-based beam theory, which neglects transverse shear deformation effects present through the beam thickness. Although effects of the transverse shear with regard to the beam deforma-

---

K. S. Pakhare (✉) · P. J. Guruprasad · R. P. Shimpi
Department of Aerospace Engineering, IIT Bombay, Mumbai 400 076, India
e-mail: kedar188200@gmail.com

© The Editor(s) (if applicable) and The Author(s), under exclusive license to Springer Nature Singapore Pte Ltd. 2021
S. K. Saha and M. Mukherjee (eds.), *Recent Advances in Computational Mechanics and Simulations*, Lecture Notes in Civil Engineering 103,
https://doi.org/10.1007/978-981-15-8138-0_4

tion generally remain insignificant as compared to effects of the flexural deformation for slender beams, these effects become prominent in case of thick beams as well as slender beams vibrating at higher modes. Hence, the *BEBT* can provide reasonably accurate results only for slender beams. In order to incorporate these effects, various displacement-based first-order (*FSDT*) and higher-order shear deformation beam theories (*HSDT*) have been proposed in the literature. Although *FSDT* predict the mechanical behavior of shear deformable beams with reasonable accuracy, their assumed displacement fields give rise to an unrealistic constant transverse shear strain (and hence the constant transverse shear stress) through the beam thickness. In case of *FSDT*, the resulting transverse shear stress does not satisfy transverse shear-stress-free beam surface conditions. *HSDT* address the just-mentioned drawback of *FSDT*. *HSDT* assume displacement fields such that they give rise to a more realistic nonlinear variation of transverse shear strain (and hence the transverse shear stress) through the beam thickness. Resulting transverse shear stress in case of *HSDT* satisfy transverse shear-stress-free beam surface conditions (Ghugal and Shimpi [1]).

The aim of this paper is to present a comparative study on the flexure of shear deformable isotropic rectangular propped cantilever beams by utilizing newly developed variationally consistent two-variable-refined beam theory (Shimpi et al. [2]), variationally inconsistent single-variable-refined beam theory (Shimpi et al. [3]), and variationally inconsistent Levinson beam theory (references [4, 5]), which are all *HSDT*. Effects of the beam thickness-to-length ratio on the location of maximum beam transverse displacement and values of maximum non-dimensional beam transverse displacement, non-dimensional beam axial stress, and non-dimensional beam transverse shear stress are presented. Profiles of the non-dimensional beam transverse displacement, non-dimensional beam axial stress, and non-dimensional beam transverse shear stress for various values of the beam thickness-to-length ratio are also presented.

It should be noted that the literature presents various displacement-based *HSDT* for analyzing the mechanical behavior of shear deformable beams. However, to the best of authors' knowledge, these theories have not been implemented for analyzing the deformation behavior of shear deformable isotropic rectangular propped cantilever beams. The present study aims at discussing the effects of the shear deformation on the maximum beam transverse displacement and its location, beam axial, and transverse shear stress for such beams.

2 Theoretical Formulation

For the static flexure, governing differential equations of the two-variable-refined beam theory (*TVRBT*), single-variable-refined beam theory (*SVRBT*), and Levinson beam theory (*LBT*) are as follows:

1. The *TVRBT* (Shimpi et al. [2]):

$$E I \frac{d^4 w_b}{dx^4} = q \tag{1}$$

$$\frac{E I}{84} \frac{d^4 w_s}{dx^4} - \frac{5 E A}{12(1+\mu)} \frac{d^2 w_s}{dx^2} = q \tag{2}$$

with

$$w = w_b + w_s \tag{3}$$

2. The *SVRBT* (Shimpi et al. [3]):

$$E I \frac{d^4 w_b}{dx^4} = q \tag{4}$$

with

$$w = w_b - \frac{h^2 (1+\mu)}{5} \frac{d^2 w_b}{dx^2} \tag{5}$$

3. The *LBT* (references [4, 5]):

$$\frac{2}{3} A G \left(\frac{d\psi}{dx} + \frac{d^2 w}{dx^2} \right) = -q \tag{6}$$

$$\frac{2}{3} A G \left(\psi + \frac{dw}{dx} \right) + \frac{E I}{5} \left(\frac{d^3 w}{dx^3} - 4 \frac{d^2 \psi}{dx^2} \right) = 0 \tag{7}$$

Symbols appearing in Eqs. (1) through (7) have the same meanings as those of the corresponding symbols appearing in respective cited references.

The beam geometry and co-ordinate system considered here are the same as those of Shimpi et al. [2]. The beam is under the influence of uniformly distributed transverse load for which $q = q_o$, where q_o is the amplitude of applied transverse load. For the propped cantilever beam under consideration, the beam end $x = 0$ is simply supported and $x = L$ is clamped. Boundary conditions utilized to solve governing differential equations of the *TVRBT* (Eqs. (1) and (2)), *SVRBT* (Eq. (4)) and *LBT* (Eqs. (6), and (7)) are as follows:

2.1 Beam Boundary Conditions

1. For the *TVRBT*:

   a. At the beam end $x = 0$:

   $$[w_b]_{x=0} = 0 \tag{8}$$

   $$\left[\frac{d^2 w_b}{dx^2}\right]_{x=0} = 0 \tag{9}$$

   $$[w_s]_{x=0} = 0 \tag{10}$$

   $$\left[\frac{d^2 w_s}{dx^2}\right]_{x=0} = 0 \tag{11}$$

   b. At the beam end $x = L$:

   $$[w_b]_{x=L} = 0 \tag{12}$$

   $$\left[\frac{d w_b}{dx}\right]_{x=L} = 0 \tag{13}$$

   $$[w_s]_{x=L} = 0 \tag{14}$$

   $$\left[\frac{d^2 w_s}{dx^2}\right]_{x=L} = 0 \tag{15}$$

2. For the *SVRBT*:

   a. At the beam end $x = 0$:

   $$\left[w_b - \frac{h^2(1+\mu)}{5}\frac{d^2 w_b}{dx^2}\right]_{x=0} = 0 \tag{16}$$

   $$\left[\frac{d^2 w_b}{dx^2}\right]_{x=0} = 0 \tag{17}$$

   b. At the beam end $x = L$:

   $$\left[w_b - \frac{h^2(1+\mu)}{5}\frac{d^2 w_b}{dx^2}\right]_{x=L} = 0 \tag{18}$$

   $$\left[\frac{d w_b}{dx} + \frac{h^2(1+\mu)}{20}\frac{d^3 w_b}{dx^3}\right]_{x=L} = 0 \tag{19}$$

3. For the *LBT*:

   a. At the beam end $x = 0$:

$$[w]_{x=0} = 0 \tag{20}$$

$$\left[4\left(\frac{d\psi}{dx}\right) - \frac{d^2w}{dx^2}\right]_{x=0} = 0 \tag{21}$$

b. At the beam end $x = L$:

$$[w]_{x=L} = 0 \tag{22}$$

$$[\psi]_{x=L} = 0 \tag{23}$$

Boundary conditions (Eqs. (8) through (23)) are utilized to evaluate primary unknown quantities of respective theories (w_b and w_s for the *TVRBT*, w_b for the *SVRBT*, w and ψ for the *LBT*) and thereby secondary unknown quantities for the shear deformable propped cantilever beam.

3 Numerical Results and Discussion

In Tables 1 through 4, comparison of the following quantities is obtained by utilizing the *TVRBT*, *SVRBT* and *LBT* are presented for various values of the beam thickness-to-length ratio (h/L):

Table 1: Maximum non-dimensional beam transverse displacement ($\overline{w}_{max}$).
Table 2: Non-dimensional location ($\overline{x}_{max}$) of maximum beam transverse displacement from simply supported beam end ($x = 0$).
Table 3: Non-dimensional beam axial stress ($\overline{\sigma}_x$) at the location $x = L, z = h/2$.
Table 4: Non-dimensional beam transverse shear stress ($\overline{\tau}_{zx}$) at the location $x = L$, $z = 0$.

In Tables 1 through 4, values written in parenthesis indicate percentage difference involved in predicting various quantities with respect to well-established *LBT*.

Figures 1 through 3 present profiles of the non-dimensional beam transverse displacement ($\overline{w}$), beam along-the-thickness profiles of non-dimensional beam axial

Table 1 $\overline{w}_{max}$ for propped cantilever beam, Poisson's ratio $\mu = 0.3$

Theory	$\overline{w}_{max} = (100\, w_{max}\, E\, I)/(q_o\, L^4)$			
	$h/L = 0.01$	$h/L = 0.05$	$h/L = 0.10$	$h/L = 0.15$
TVRBT [2]	0.541929	0.549538	0.573333	0.613036
	(0.02 %)	(0.57 %)	(2.16 %)	(4.41 %)
SVRBT [3]	0.542056	0.552704	0.585959	0.641293
	(0.00 %)	(0.00 %)	(0.00 %)	(0.00 %)
LBT [4, 5]	0.542056	0.552704	0.585959	0.641293

Table 2 $\overline{x}_{max}$ of maximum beam transverse displacement from simply supported beam end ($x = 0$), Poisson's ratio $\mu = 0.3$

Theory	$\overline{x}_{max} = x_{max}/L$			
	$h/L = 0.01$	$h/L = 0.05$	$h/L = 0.10$	$h/L = 0.15$
TVRBT [2]	0.421564	0.422265	0.424378	0.427661
	(0.01 %)	(0.14 %)	(0.54 %)	(1.10 %)
SVRBT [3]	0.421589	0.422870	0.426672	0.432396
	(0.00 %)	(0.00 %)	(0.00 %)	(0.00 %)
LBT [4, 5]	0.421589	0.422870	0.426672	0.432396

Table 3 $\overline{\sigma}_x$ for propped cantilever beam at location $x = L, z = h/2$, Poisson's ratio $\mu = 0.3$

Theory	$\overline{\sigma}_x = (\sigma_x b)/q_o$			
	$h/L = 0.01$	$h/L = 0.05$	$h/L = 0.10$	$h/L = 0.15$
TVRBT [2]	−7500.00	−300.00	−75.00	−33.33
	(−0.02 %)	(−0.53 %)	(−2.12 %)	(−4.84 %)
SVRBT [3]	−7498.42	−298.43	−73.44	−31.79
	(0.00 %)	(0.00 %)	(0.00 %)	(0.00 %)
LBT [4, 5]	−7498.42	−298.43	−73.44	−31.79

Table 4 $\overline{\tau}_{zx}$ for propped cantilever beam at location $x = L, z = 0$, Poisson's ratio $\mu = 0.3$

Theory	$\overline{\tau}_{zx} = (\tau_{zx} b)/q_o$			
	$h/L = 0.01$	$h/L = 0.05$	$h/L = 0.10$	$h/L = 0.15$
TVRBT [2]	−74.92	−14.92	−7.42	−4.92
	(20.09 %)	(20.34 %)	(20.56 %)	(20.65 %)
SVRBT [3]	−93.75	−18.73	−9.34	−6.20
	(0.00 %)	(0.00 %)	(0.00 %)	(0.00 %)
LBT [4, 5]	−93.75	−18.73	−9.34	−6.20

stress ($\overline{\sigma}_x$) and non-dimensional beam transverse shear stress ($\overline{\tau}_{zx}$), respectively, for various values of h/L.

The following points should be noted with regard to Tables 1 through 4:

1. Table 1: The maximum non-dimensional beam transverse displacement ($\overline{w}_{max}$) obtained by utilizing the *SVRBT* matches exactly with the corresponding values obtained by utilizing the *LBT*. Whereas, $\overline{w}_{max}$ obtained by utilizing the *TVRBT* has a quantitative agreement with corresponding values of the *LBT* (maximum % difference of 4.41% for $h/L = 0.15$).

2. Table 2: The non-dimensional location ($\overline{x}_{max}$) of the maximum beam transverse displacement from the simply supported beam end obtained by utilizing the *SVRBT* matches exactly with corresponding values obtained by utilizing the *LBT*. Whereas, $\overline{x}_{max}$ obtained by utilizing the *TVRBT* has a quantitative agree-

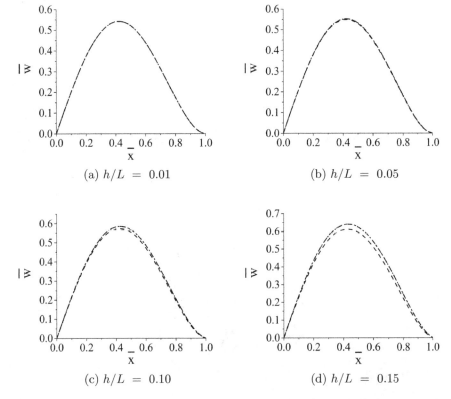

Fig. 1 For the propped cantilever beam, the non-dimensional beam transverse displacement ($\bar{w} = (100\, w\, E\, I)/(q_o\, L^4)$) vs non-dimensional beam axial location ($\bar{x} = x/L$) for various values of the beam thickness-to-length ratio (h/L), - - - for the *TVRBT*, · · · for the *SVRBT* and - · - for the *LBT*

ment with corresponding values of the *LBT* (maximum % difference of 1.10% for $h/L = 0.15$).

3. Table 3: The non-dimensional beam axial stress ($\bar{\sigma}_x$) obtained by utilizing the *SVRBT* matches exactly with corresponding values obtained by utilizing the *LBT*. Whereas, $\bar{\sigma}_x$ obtained by utilizing the *TVRBT* has a quantitative agreement with corresponding values of the *LBT* (maximum % difference of -4.84% for $h/L = 0.15$).

4. Table 4: The non-dimensional beam transverse shear stress ($\bar{\tau}_{zx}$) obtained by utilizing the *SVRBT* matches exactly with corresponding values obtained by utilizing the *LBT*. Whereas, $\bar{\tau}_{zx}$ obtained by utilizing the *TVRBT* has an appreciable quantitative disagreement with corresponding values of the *LBT* (maximum % difference of 20.65% for $h/L = 0.15$).

The following points should be noted with regard to Figs. 1 through 3:

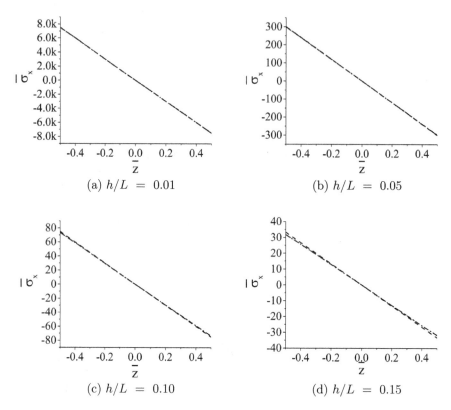

Fig. 2 For the propped cantilever beam, the non-dimensional beam axial stress ($\overline{\sigma}_x$) versus non-dimensional beam thickness location ($\overline{z} = z/h$) at $x = L$ for various values of the beam thickness-to-length ratio (h/L), - - - for the *TVRBT*, · · · for the *SVRBT* and - · - for the *LBT*

1. Profiles of the non-dimensional beam transverse displacement ($\overline{w}$) and beam along-the-thickness profiles of non-dimensional beam axial stress ($\overline{\sigma}_x$) for various values of h/L obtained by utilizing the *TVRBT*, *SVRBT* and *LBT* (Figs. 1 and 2, respectively) closely match each other.
2. Beam along-the-thickness profiles of the non-dimensional beam transverse shear stress ($\overline{\tau}_{zx}$) for various values of h/L obtained by utilizing that the *SVRBT* and *LBT* closely match each other (Fig. 3). Profiles of $\overline{\tau}_{zx}$ obtained by utilizing the *TVRBT* have certain deviations with respect to corresponding profiles of $\overline{\tau}_{zx}$ obtained by utilizing the *SVRBT* and *LBT*.

Hence, the flexural analysis of shear deformable isotropic rectangular propped cantilever beams carried out by utilizing the variationally inconsistent *SVRBT* and *LBT*, variationally consistent *TVRBT* has a quantitative agreement with each other. The above-mentioned observations also point out that the location of the maximum beam transverse displacement for just-mentioned beams is a function of the beam thickness-to-length ratio.

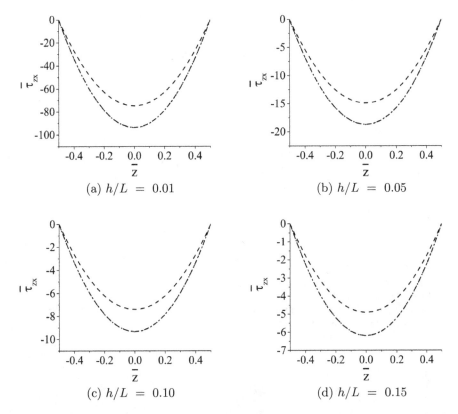

Fig. 3 For the propped cantilever beam, the non-dimensional beam transverse shear stress ($\bar{\tau}_{zx}$) versus non-dimensional beam thickness location ($\bar{z} = z/h$) at $x = L$ for various values of the beam thickness-to-length ratio (h/L), - - - for the *TVRBT*, · · · for the *SVRBT* and - · - for the *LBT*

4 Concluding Remarks

In this paper, a comparison study on the flexure of shear deformable isotropic rectangular propped cantilever beams by utilizing the two-variable-refined beam theory, single-variable-refined beam theory, and Levinson beam theory is presented. Effects of the beam thickness-to-length ratio on the maximum non-dimensional beam transverse displacement and its location, non-dimensional beam axial stress, and non-dimensional beam transverse shear stress are presented. In addition, effects of the beam thickness-to-length ratio on profiles of the non-dimensional beam transverse displacement, non-dimensional beam axial stress, and non-dimensional beam transverse shear stress are presented. These results and comparisons prove that the flexure of shear deformable isotropic rectangular propped cantilever beams carried out using these three theories is comparable. To the best of authors' knowledge, displacement-based higher-order theories have not been utilized to analyze the effects of shear deformation on the mechanical behavior of isotropic rectangular propped cantilever beams in the literature. The present study aims at doing so. One of the major outcomes of this study is that the location of the maximum beam transverse displacement for shear deformable isotropic rectangular propped cantilever beams is a function of the beam thickness-to-length ratio.

References

1. Ghugal YM, Shimpi RP (2001) A review of refined shear deformation theories for isotropic and anisotropic laminated beams. J Reinf Plast Comp 20:255–272. https://doi.org/10.1177/073168401772678283
2. Shimpi RP, Guruprasad PJ, Pakhare KS (2020) Simple two variable refined theory for shear deformable isotropic rectangular beams. J Appl Comp Mech 6:394–415. https://doi.org/10.22055/jacm.2019.29555.1615
3. Shimpi RP, Shetty RA, Guha A (2017) A simple single variable shear deformation theory for a rectangular beam. P I Mech Eng C-J Mec 231:4576–4591. https://doi.org/10.1177/0954406216670682
4. Levinson M (1981) A new rectangular beam theory. J Sound Vib 74:81–87. https://doi.org/10.1016/0022-460X(81)90493-4
5. Levinson M (1981) Further results of a new beam theory. J Sound Vib 77:440–444. https://doi.org/10.1016/S0022-460X(81)80180-0

Progressive Collapse Potential of Steel Frames Sustaining Post-Hazard Support-Yielding

Anil Kumar, Niranjan Muley, Pranesh Murnal, and Vasant A. Matsagar

Abstract Hazards such as fire, blast or severe seismic event may cause a local failure of columns leading to disproportionate or progressive collapse of a structure. Progressive collapse is a complex dynamic process greatly influenced by several factors including the integrity and tie-strength of beam-and-slab system over the damaged column, load-redistribution mechanism and post-hazard-exposure condition of the columns' bases. In the present study, simple steel space frames designed conventionally as per Indian standard codes of practice are considered for numerical parametric study. Column removal locations as recommended by Unified Facilities Criteria (UFC 4-023-03) are considered. Progressive collapse potential is assessed for different column sections subjected to loss of major load-bearing element while adjoining columns have partially lost their fixity with the base. This is modelled by varying support conditions as fixed, hinged or roller; whilst joints adjacent to the lost column are changed to pinned connections. The effect of material properties degradation is also evaluated considering variation in steel grade. The progressive collapse potential is seen to be higher for frames with yielded supports emphasizing that the static indeterminacy plays an important role in progressive collapse potential of frames. Amongst the steel sections considered herein, square-section columns are found to be the most effective against progressive collapse.

Keywords Abnormal loading · Column removal · Nonlinear dynamic analysis · Progressive collapse

---

A. Kumar (✉) · V. A. Matsagar
Indian Institute of Technology (IIT) Delhi, New Delhi 110016, India
e-mail: anilkumar@ccet.ac.in

N. Muley · P. Murnal
Government College of Engineering, Aurangabad 431005, India

1 Introduction

Progressive collapse is the spread of an initial local failure from element to element eventually resulting in the collapse of a significant part of a structure or sometimes of the entire structure [1]. This extent of collapse is generally disproportionate to the magnitude of initial event. Such events might be manmade, natural, intentional or unintentional generating extreme or abnormal loading such as explosion, heavy fire or severe seismic action. Progressive collapse is a complex dynamic process greatly influenced by several factors including the integrity and tie-strength of beam-and-slab system spanning over the damaged column, load-redistribution mechanism and post-hazard-exposure condition of the columns' bases. The collapsing system redistributes the loads enhanced due to the failure of major load-bearing element for which the beams, columns, and frame connections are designed to handle the potential redistribution of the large loads [2].

The buildings are structurally designed to withstand the loads safely as per the needs of consumer, which include the aesthetic as well as functional requirements. However, after the partial collapse of 22-storey Ronan Point apartment, London, in 1968 which was caused due to gas explosion in a kitchen at 18th floor provoked the structural designers to consider these abnormal loadings while designing the structures [3]. For the general structural design of buildings, the loads or forces generated due to extreme loading events are normally not taken into account which may cause progressive collapse. Nowadays, progressive collapse has become an important consideration in designing important buildings; especially after the progressive collapse of the World Trade Center in 2001. Recently, various government building norms, design codes and standards have addressed the progressive collapse criterion in the design guidelines. The American Society of Civil Engineers standard 'Minimum design loads for buildings and other structures' [1] has a section on general structural integrity which states that 'buildings and the other structures designed in such way that they should sustain the local damage with the structural system as a whole remaining stable and not being damaged to an extent disproportionate to the original event'. This standard focuses on redundancy and alternate load path for redistribution of loads to avoid susceptibility to disproportionate collapse. The American Concrete Institute's code on 'Requirements for structural concrete' [4] includes the general provisions to achieve building integrity and continuity. This shall be ensured in such a way that the partial damage of structure does not result in the total collapse of the structure. To reduce the potential of progressive collapse, the General Services Administration [5, 6] and Department of Defense, USA presented guidelines for the new and existing buildings [7]. These guidelines recommend the use of an alternate path method in which, firstly the susceptible major load-bearing element is removed from the structure and analysis is performed; then the limit states of members are evaluated. If a limit state is exceeded for any member, it is removed, and load is redistributed to the adjacent members. This process is repeated until no remaining elements exceed the limit state. For progressive collapse analysis, the alternate path method can be executed by using four procedures: linear static analysis, nonlinear static analysis, linear dynamic analysis, and nonlinear dynamic analysis.

Various studies are available on progressive collapse including the investigations and preventive measures for improving the resistance of the structure to the progressive collapse. Ellingwood and Leyendecker [8] discussed three approaches for the prevention of progressive collapse: event control, indirect design method, and direct design method. Nair in 2004 [9] highlighted three alternative approaches to reduce the potential of progressive collapse whilst designing structures, which are: (1) redundancy or alternate load paths, (b) local resistance, and (c) interconnection or continuity. Marjanishivili [10] presented four sophisticated analysis procedures for assessing the hazard of progressive collapse. In progressive analysis, response of a structure is evaluated by initial static analysis and then proceeding to the more complex analysis method as essential, until the low possibility of progressive collapse is determined or until available engineering analysis methodologies are exhausted. Powell [11] discussed the principles of progressive collapse analysis for the alternate path method and compared the static and dynamic analyses methods. The author suggested that the use of static analysis for progressive collapse analysis with present guidelines may lead to over-conservative design, whereas, the use of nonlinear dynamic analysis is realistic, more accurate and not difficult to carry out. Kaewkulchai and Williamson [12] proposed a beam element formulation for dynamic progressive collapse analysis. Grierson et al. [13] discussed simplified methods for progressive collapse analysis of buildings. Marjanishivili and Angew [14] analysed a nine-storey steel moment-resisting frame using four different analysis methods. The authors have recommended that the nonlinear static analysis method should be used in combination with nonlinear dynamic analysis method as a supplemental analysis to determine the first yield and ultimate capacity limits. Khandelwal and El-Tawil [15] presented a study on steel moment connections behaviour followed by a new analysis technique pushdown analysis [16], which can be used to determine the collapse modes and ultimate load-carrying capacity of structure with the lost or damaged critical load-bearing member.

Kim and Hwang [17] in 2008 investigated the effect of catenary action on collapse of steel frames. The results showed that considering the catenary action, the nonlinear static pushdown curve formed upper bound of the curve than that obtained without considering catenary action. Helmy et al. [18] in 2013 carried out the assessment of ten-storey reinforced concrete structure for progressive collapse using applied element method. The numerical studies have shown that for economical design, the structure should be analysed considering the slabs in the analysis. Neglecting slab elements in progressive collapse analysis, for both 2D frame and 3D frame, may lead to incorrect structural behaviour and uneconomical design.

During an extreme loading event such as heavy fire or explosion, etc., not only are the columns lost but the supports of adjacent columns are also compromised. Beam-column connections may also get affected to a certain extent. Progressive collapse studies considering such partial fixity losses of column supports and connections, and material degradation adjacent to column lost are very limited. Fu [19] conducted 3D finite element analysis of a 20-story building for progressive analysis with variation in steel strength and found it affecting progressive collapse response.

In the present study, the progressive collapse assessment of simple steel structures subjected to abnormal loading due to the failure of major load-bearing element is carried out using Unified Facilities Criteria 4-023-03 [7] guidelines, whilst adjoining columns have partially lost their fixity with the base. This is modelled simply by varying support conditions as fixed, hinged or roller; whilst joints adjacent to the lost column are changed to pinned connections. The effect of material properties degradation is also evaluated considering variation in steel grade. The structure is conventionally designed as per Indian standard code of practice. Column sections are selected as I, circular and square sections. The column-removal sequence is decided as per GSA (2013) [6] and UFC 4-023-03 (2009) [7] guidelines. Once the column failure scenario is decided, the column-removal is performed using nonlinear dynamic analysis procedure incorporating material nonlinearity [11].

2 Modelling and Analysis

The location of the column to be removed whilst assessing the potential of progressive collapse is decided as follows [6, 7]:

1. The column located nearest to the middle of the exterior side in the long direction,
2. The column located nearest to the middle of the exterior side in the short direction,
3. The corner column, and
4. An interior column in case of irregular public access or parking.

The load combination for the dynamic analysis is recommended to be 1.2DL+0.5LL [6, 7], in which DL is the dead load and LL is the imposed load. In this study, the modelling and analysis are conducted by adopting the general structural analysis program SAP2000 [20]. Nonlinear rotational plastic hinges are defined including both the positive and negative yield; ultimate moment capacities and associated rotations. Figure 1 shows a typical plastic hinge definition used in the modelling. Beams and columns are modelled as given in Table 1.

2.1 Nonlinear Dynamic Progressive Collapse Analysis

The nonlinear dynamic analysis is carried out considering the sudden failure of load-bearing elements. In the dynamic analysis, the column-removal simulation is performed by suddenly stepping down the reaction forces to zero applied at the joint above the removed column using an appropriate time history function. These reaction forces are equal to the internal forces developed in the column to be removed. In this step, the gravity loads are kept acting continuously and the column reaction forces are released suddenly. The load-removal time depends on the column-failure time

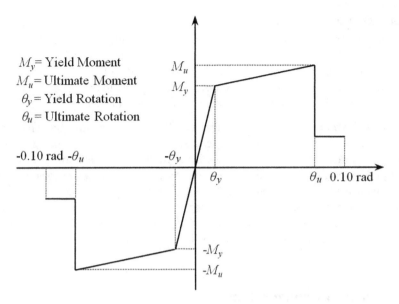

Fig. 1 Plastic hinge definition for members

Table 1 Member details of the structure

Section	Column	Main beam	Secondary beam
I	ISHB 200 @ 37.3 kg/m	ISMB 100 @11.5 kg/m	ISMB 100 @11.5 kg/m
Circular	ISNB 125 M	SHS 91.5 × 91.5 × 5.4 mm	SHS 49.5 × 49.5 × 3.6 mm
Square	SHS 113.5 × 113.5 × 5.4 mm	SHS 91.5 × 91.5 × 5.4 mm	SHS 49.5 × 49.5 × 3.6 mm

and is generally taken as equal to or less than 1/10th the time period of vibration mode involving vertical motion the structure at column-removal point. The column-removal time is an important parameter in progressive collapse analysis to achieve reliable results since it influences dynamic behaviour of the structure. The dynamic analysis procedure for progressive collapse assessment consists of three stages of analysis that include the static analysis, modal analysis, and nonlinear dynamics analysis.

Static Analysis. In static analysis, the structure is statically analysed under the effect of gravity loads and without removing the critical column. The load combination used for the analysis is taken as given by [7] for the dynamic analysis without magnification factor, i.e. 1.2DL+0.5LL. Then, the reaction forces (axial force, shear force and bending moment) at the upper end of the critical column to be removed are calculated so that they can be used in further analysis.

Modal Analysis. In modal analysis, the natural frequencies and their associated mode shapes are determined by solving undamped vibration equation of motion of building model having removed the critical column. The time period of the structure is obtained corresponding to the mode shape associated with the vertical motion of the portion where column has been removed. It is required to determine the value of the time step so that it can be used in the nonlinear dynamic analysis.

Nonlinear Dynamic Analysis. Progressive collapse is a dynamic phenomenon in which the structure is exposed to impact energy resulted from the sudden transfer of residual forces that were otherwise carried by the failing member. The impact energy excites the structure, which continues to vibrate until it is damped and reaches to the final steady deformation state. Nonlinear dynamic analysis is carried out for evaluating the performance of structure in terms of several parameters such as number and state of plastic hinges formed, progression of vertical deflection of the column-removal point, and changes in axial forces in the adjacent columns.

3 Structural Configuration

The structure selected for the present study is a simple two-storey steel frame office building, with two 3-m wide bays in each direction. The floor height of 3.0 m is considered for both the floors. The floor diaphragms are made of composite metal deck with two-way slab having a thickness of 150 mm. The beams (main girders and secondary girders) and columns comprise conventional Indian steel sections (given in Table 1). Figure 2 shows the typical plan and side view of the structure.

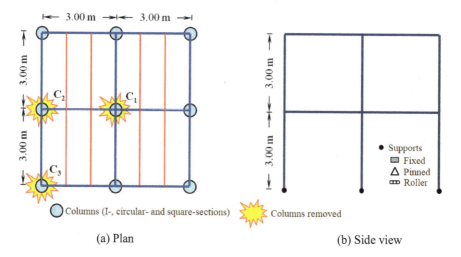

Fig. 2 Steel-framed structure

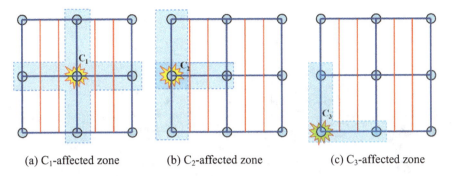

(a) C_1-affected zone (b) C_2-affected zone (c) C_3-affected zone

Fig. 3 Supports and joints considered yielded adjacent to the failed column

Table 2 Denotation for support and connection types

Support condition/Joint type	Fixed support	Hinged support	Roller support
Rigid connection	F1	H1	R1
Pinned connection	F2	H2	R2

The steel-framed structure is designed as per Indian Standard Code of Practice IS 800:2007 [21], to resist the estimated gravity loads as well as seismic lateral loads estimated as per IS 1893:2016 (Part I) [22]. The beam-column joints adjacent to the removed column (as shown in Fig. 3) are considered rigid and pinned simulating stiffness loss. The members are designed to satisfy the limit state of collapse and checked for limit state of serviceability. Table 2 gives the boundary conditions and connections considered for supports and joints of the structure adjacent to the lost column.

The load combination used for the dynamic analysis whilst evaluating progressive collapse potential is 1.2DL+0.5LL [7]. Live load is assumed to be 3.5 kN/m^2 distributed uniformly across the whole floor area including roof. In addition to this, a dead load of 8.55 kN/m^2 coming from perimeter and internal walls at first floor is considered on beams.

4 Column Failure Scenarios

Several researchers studied and concluded that the nonlinear analysis based on the moment plastic hinge concept is acceptable for progressive collapse study. Herein, the moment plastic hinge concept is used [23]. Figure 4 shows the generalized force-deformation relationship for steel elements or components. In this investigation, three independent failure scenarios are considered after performing the pushover analysis and then, the potential of the building to sustain progressive collapse is assessed.

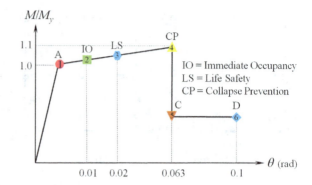

Fig. 4 Generalized moment-rotation relationship for steel elements

For progressive collapse analysis, UFC 4-023-03 [7] mandates several loss scenarios; however, for each analysis, only one element removal is required. The locations of column removal taken into account for performing the progressive collapse analysis are already given in Fig. 3. It also shows the extent of damage considered spread adjoining the failing column. The supports and beam-column connections at first floor are considered vulnerable and altered to hinged or roller. Whilst performing analysis, the cause of member failure is not considered; this is referred to as threat-independent approach. The basis for this fact is that the element is suddenly incapable of carrying load. The column failure scenario is decided by using the pushover analysis. The sequence of plastic hinge formation and state of hinges at various levels of building performance were obtained from SAP2000 [20] and shown in Fig. 5. This gives clues about the weakest member, which can be removed to perform the progressive collapse analysis.

5 Progressive Collapse Analysis

The basic procedure for performing a progressive analysis using nonlinear dynamic analysis in any of the structural analysis program is through a time history analysis (transient analysis). In this study, the forces are kept unchanged which represent the stable condition of the structure (Fig. 6), and the applied reverse loads are suddenly ramped down within a short time step. This time step is taken equal to one-tenth of the modal period associated with the structural response mode for the element removal. This mode involves vertical motion the frame around the column-removal point. In case of column C_1 failure, this time period is 0.08 s, and therefore, the reverse load applied is removed within 0.008 s. Its time history function is shown in Fig. 7.

The determination of the subsequent free dynamic response under the column-removal scenario includes geometric nonlinearity and inserting the plastic hinges. The structural response of the frames is observed in terms of number and state of plastic hinges formed, progression of vertical deflection of the column-removal point and changes in axial forces in the columns adjacent to the removed column.

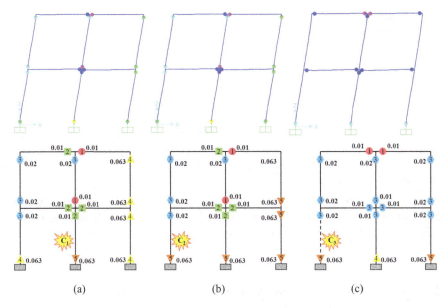

Fig. 5 Formation of plastic hinges in pushover analysis for **a** column C_1, **b** column C_2 and **c** column C_3 (values written at member ends show rotations in radian)

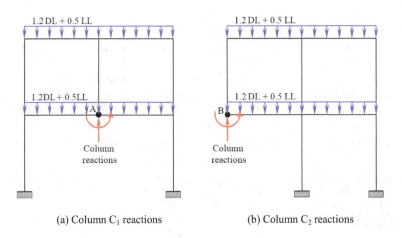

Fig. 6 Reactions applied at column removal points

6 Analysis and Discussion of Results

The response of structural frames with fixed support as well as yielded supports (modelled as hinged and roller) having rigid and pinned joint (modelled for damaged joints) connections is investigated after the sudden failure of columns. Also, the

Fig. 7 Column-load removal time history

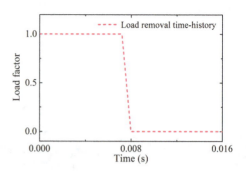

vertical deflection of the column-removal point is investigated for steel strength degradation from Fe345 to Fe300 and Fe250 after the failure of column. The factors by which axial forces in adjoining columns get escalated are also evaluated.

6.1 Vertical Deflection of the Column-Removal Joint

Figure 8 shows the vertical deflection time history of the joint in the floor above the failed column C_2 for rigid joint as well as pinned joint connections for I-, circular- and square-section columns. In both the cases, the frame with I section shows less vertical deflection and deformation as compared to that of circular and square sections. For the frame with rigid joint connection, the maximum vertical deformation is observed in case of circular-section frame; whereas, in case of pinned joint connection the vertical deflection behaviour and deformation for both circular and square-section frame is almost identical. It has also been observed that vertical deflection is less following the failure of column C_1 as compared to the other two columns. It is evident, since in this case, more structural members participate in redistribution of loads and resisting the progressive collapse. In case of C_2 column failure, even

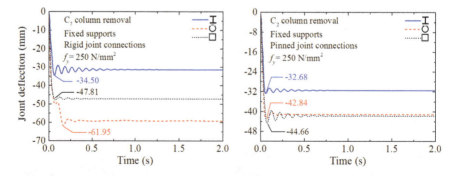

Fig. 8 Vertical deflection of the upper joint of the column C_2 after its removal for fixed support frames with rigid and pinned connections

though moments of inertia are same, circular-section frame shows higher vertical deflections and deformations than that of square-section frame, whereas, in case of C_3 column failure, square-section frame shows higher vertical deflections than that of circular-section frame. This is due to the difference of axial forces in the beams; higher the axial force in the beam member, lesser will be the vertical deflection and deformations. It was also evident from the case of frame with I-section columns, in which, after the failure of column C_3, the vertical joint deflection for pinned joint connections are higher than that of rigid joint connection, as the axial force in the beam member in case of rigid joint frame (2170 N) is more than that of pinned joint frame (1520 N). For frames with roller supports and rigid joint connections, it was seen that the downward deflection at the joints above the column failed increased drastically, indicating the failure of bays which in turn initiates the progressive failure of the structure.

Figure 9 shows the vertical deflection time history for column C_2 removal for rigid joint and pinned joint connections in I-section for variation in steel grade. It is seen that the joint vertical deflection increases with a decrease in steel strength. This increase is not very significant and is of the order of almost 10–12%. This might be

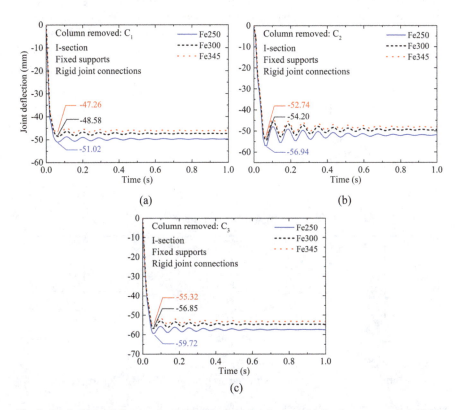

Fig. 9 Vertical joint deflection time history for I-section frames with fixed supports and rigid joint connections for change in steel strength and removal of column **a** C_1, **b** C_2 and **c** C_3

attributed to the fact that due to the membrane action of the slabs gives the structure an ability to act as one unit.

6.2 Number and State of Plastic Hinges

Figure 10 shows the location of plastic hinges and their rotation angles in radians in the framed structure subjected to the gravity load after the removal of column C_2. In case of frame with rigid joint connections, after the removal of column C_2 the plastic hinges are formed in beams as well as in columns. The plastic rotations in a frame with I-, circular and square section are within the limits specified, whereas, in case of the frame with pinned joint connection, the plastic hinges are seen to form in columns only. The plastic rotations in these frames have been observed to be relatively higher in case of I-, and square-section frames than those obtained for the frames with rigid joint connections. These rotations exceeded the acceptance criteria specified forming the collapse mechanism which is evident in this case.

The results show that the excessive plastic hinges rotations developed in the columns demonstrated that the bays above the failed columns experience the progressive collapse after the failure of certain columns, as is clear from Fig. 11. For the frames with circular-section columns and pinned joint connections, it was seen that

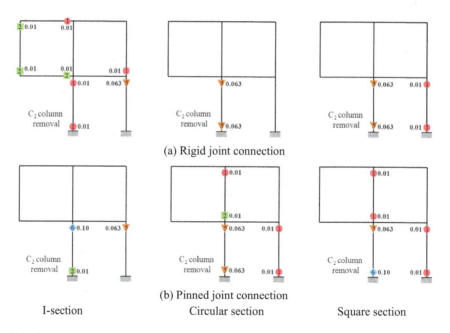

Fig. 10 Location of hinges in a frame with fixed supports for column C_2 removal

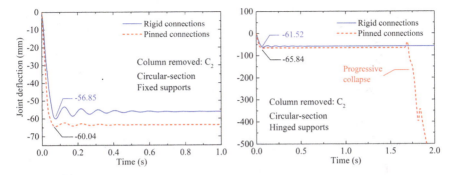

Fig. 11 Vertical joint deflection time history for circular-section frame with rigid and pinned joint connections for fixed and hinged supports followed by C_2 column removal

the downward deflection at the joint above the column failed have increased drastically, which indicates the failure of bays, which in turn initiates the progressive failure of the structure.

Figure 12 shows the number and states of hinges formed in various frames considered herein. In case of C_1 column failure in I-section frame (Fig. 12a), the ultimate rotations associated with the members are reached up to the effective yield point (point A in Fig. 4); whereas, in some members it has reached up to the ultimate strength (C, 0.063 rad) limit. In case of column C_2 and column C_3 failures, the ultimate rotations associated with the column members exceeded the collapse (E, 0.105 rad) limit specified by the guidelines. The results show that after the failure of column C_2, and C_3, excessive rotations in the plastic hinges at the ends of members are developed in above the failed columns.

Accordingly, it is demonstrated that the bays above and adjacent to the failed columns experience the progressive collapse following the failure of these columns and thereby initiating global failure of the building. Hence, I-section frame with fixed end support having rigid joint connection is ineffective in resisting the progressive collapse. In all column failure scenarios for pinned connections I-section frames, hinges are formed in columns only. In case of C_1 and C_3 column failures, the ultimate rotations in maximum number of members are reached up to the ultimate strength (C, 0.063 rad) limit. In case of column C_2 failure, the ultimate rotations associated with the column members exceeded the collapse (E, 0.105 rad) limit.

From Fig. 12d–f, it is demonstrated that the failure of any of the columns will not lead to progressive collapse of the building, and circular-section frame with rigid joint connection is effective in absorbing the failure of major load-bearing element. Hence, circular-section columns with fixed end supports having rigid joint connection have potential to reduce the progressive collapse vulnerability. For pin-jointed circular-section frame having fixed end supports, hinges are formed in columns only. In case of C_1, and C_2 column failure, the ultimate rotations associated with the members are reached up to the effective yield point (A); whereas, in some members it has reached up to the immediate occupancy (IO, 0.01 rad) and ultimate strength (C,

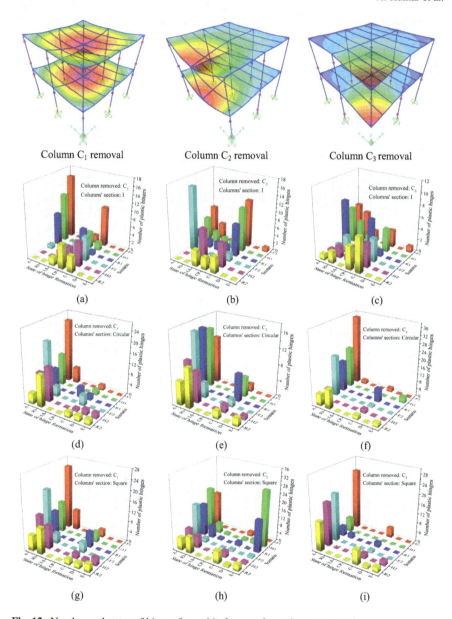

Fig. 12 Number and states of hinges formed in frames after columns removal

0.063 rad) limits. In case of C_3 column failure, the ultimate rotations associated with the members are reached up to the effective yield point (A); whereas, in some members it has reached up to the ultimate strength (C, 0.063 rad) limit. The ultimate rotations of beams as well as of columns are within the limits specified.

Figure 12g–i shows hinges formed in case of square-section frames. From these results, it is observed that that the failure of any of the columns will not lead to the progressive collapse of the building, and square-section frame with rigid joint connection is more effective against progressive collapse than the circular-section frames. However, the square-section frames with pinned connections are found to be ineffective to resist the progressive collapse because the ultimate rotations associated with the adjacent column member exceeded the collapse (E, 0.105 rad) limit specified. Also, the results indicate that after the failure of column C_2 excessive rotations in the plastic hinges are developed in beams above the failed column.

It is further evident from Fig. 12 that in case of hinged and roller supports with rigid connections (H1 and R1), or pinned connections (H2 and R2), none of the frames come close to progressive collapse resistance of frames with fixed supports-rigid connections (F1).

6.3 Increase in Axial Forces in Columns

Tables 3 and 4 give the variation in axial forces in the columns adjacent to the column lost. The column which carries maximum axial force has been selected for displaying these results. Figures 13, 14 and 15 show axial force increase factor for all the column-removal scenarios in different frames considered. For any column, this factor is the increase in axial force expressed as a fraction of original axial force

Table 3 Variation in adjacent column-axial-forces (kN) for rigid joint connections

Support conditions		Fixed supports		Hinged supports		Roller supports	
Type of section	Column removal location	With column	After column removal	With column	After column removal	With column	After column removal
I section	C1	109.21	334.40	109.22	334.90	109.23	335.40
	C2	219.50	596.60	219.81	599.30	220.12	602.01
	C3	106.31	305.80	106.21	306.70	106.11	310.50
Circular section	C1	111.14	328.10	111.18	319.00	111.22	310.15
	C2	205.92	481.90	206.59	374.20	207.26	368.33
	C3	111.27	314.30	111.31	315.40	111.35	300.25
Square section	C1	111.05	320.30	111.09	318.50	111.13	307.21
	C2	208.52	480.20	209.17	311.09	209.82	310.25
	C3	111.19	311.80	111.22	310.90	111.25	310.00

Table 4 Variation in adjacent column-axial-forces (kN) for pinned joint connections

Support conditions		Fixed supports		Hinged supports		Roller supports	
Type of section	Column removal location	With column	After column removal	With column	After column removal	With column	After column removal
I section	C1	108.43	380.14	108.30	314.70	108.17	260.52
	C2	204.83	540.40	205.40	514.70	205.97	490.22
	C3	114.94	312.40	114.97	324.00	115.00	336.03
Circular section	C1	110.99	343.70	111.01	344.50	111.03	345.30
	C2	205.36	540.10	206.02	501.20	206.68	465.10
	C3	111.11	318.00	111.13	318.00	111.15	318.00
Square section	C1	110.93	346.30	110.95	349.00	110.97	351.72
	C2	207.99	558.60	208.64	514.10	209.29	473.14
	C3	111.05	317.70	111.07	317.80	111.09	317.90

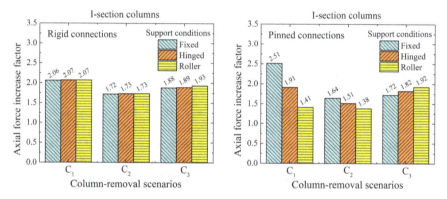

Fig. 13 Axial force increase factors in columns of I-section frames

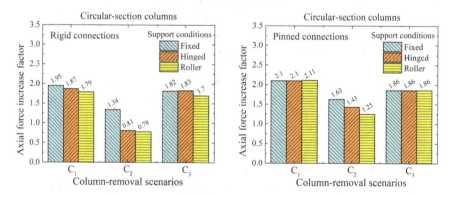

Fig. 14 Axial force increase factors in columns of circular-section frames

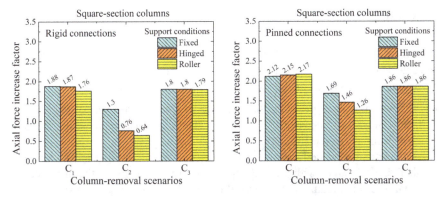

Fig. 15 Axial force increase factors in columns of square-section frames

(i.e., before column removal). It is seen from here that after the failure of major load-bearing element, the axial forces in the adjacent elements/members almost increase by three times. The increase in the axial force in the column is due to dynamic effect generated by the failure of column in a very short period of time, i.e. 8 ms, 5 ms and 3 ms. It is evident here that it is necessary to consider the axial forces to be higher under the dynamic load combination specified by the guidelines. If the designed columns fail to accommodate these extreme loads generated due to the abnormal events, it will lead to the progressive collapse.

6.4 Influence of Support Conditions, Connections and Column Sections

From the results obtained, it has been observed that the support conditions, joint connection types, and section properties play an important role in reducing the potential of progressive collapse. The potential of progressive collapse is relatively higher in case of yielded supports as compared to the fixed supports. Joint connections, when altered from rigid to pinned connections, the potential of progressive collapse increases significantly since the number of hinges required to form the collapse mechanism is less in the latter case. The frames with roller supports and pinned joint connections are found to be vulnerable to progressive collapse. The joint conditions are also seen to affect beams' behaviour in axial tension (tie-forces), which play a significant role in progressive collapse resistance.

From the results reported herein, it can be interpreted that the use of square-section columns with fixed end support and rigid joint connections is effective in resisting or reducing the vulnerability to progressive collapse. It is also re-established that the redundancy of a structure plays an important role in progressive collapse resistance.

7 Conclusions

The potential of progressive collapse for simple steel-framed structures was investigated using nonlinear dynamic analysis. From this numerical study and interpretation of results, the following conclusions are drawn.

1. The potential of progressive collapse increases after the fixed supports are considered yielded to hinged or roller supports. This is due to an increase in kinematic indeterminacy of the structure.
2. The vertical joint deflections for rigid-jointed frames are observed to be less than that of frames with yielded connections (pinned), due to the difference of axial forces in members. As the axial force in beam decreases, the vertical joint deflection increases.
3. Number of plastic hinges required to form collapse mechanism is more in case of rigid connections than that of pinned connections due to higher degree of redundancy. The structure with supports yielded to roller, and pinned connections, is more vulnerable to progressive collapse.
4. Decrease in steel strength is seen to have a moderate effect on the joint deflection pattern of the structure after the failure of the critical column.
5. Amongst the column sections used herein, square-section columns with fixed end support and rigid joint connections are found to be effective in reducing the vulnerability to progressive collapse. It is also established that the redundancy of a structure plays an important role in progressive collapse resistance. Supports and joints vulnerable to damage due to abnormal loading events should be designed and protected by suitable structural mechanisms.

As an extension to the study presented here, supports may be modelled as partially fixed with springs of varying stiffness depending upon footing conditions, and post-hazard residual capacities of adjacent beams and columns and their degraded connections may also be taken into account. Finite element modelling tools may be used for simulating collapse scenarios to a more realistic extent.

References

1. ASCE 7-05 (2005) Minimum design loads for buildings and other structures. American Society of Civil Engineers, Reston, Virginia, USA
2. Starossek U (2009) Progressive collapse of structures, 1st edn. Thomas Telford, Reston, Virginia, USA
3. Ellingwood BR, Smilowitz R, Dusenberry DO, Duthinh D, Lew HS, Carino NJ (2007) Best practices for reducing the potential for progressive collapse in buildings. National Institute of Standards and Technology, Gaithersburg, Maryland, USA
4. ACI 318-02 (2002) Building code requirements for structural concrete. American Concrete Institute, Farmington Hills, Michigan, USA
5. GSA 2003 (2003) Progressive collapse analysis and design guidelines for new federal office buildings and major modernization projects. General Services Administration, Washington, DC, USA

6. GSA 2013 (2013) Alternate path analysis and design guidelines for progressive collapse resistance. General Services Administration, Washington, DC, USA
7. UFC 4-023-03 (2009) Design of buildings to resist progressive collapse. Unified Facilities Criteria, Department of Defense, Washington, DC, USA
8. Ellingwood BR, Leyendecker EV (1978) Approaches for design against progressive collapse. J Struct Div 104(3):413–423
9. Nair RS (2004) Progressive collapse basics. Modern Steel Construction. AISC, Chicago, USA, 44(3):37–44
10. Marjanishvili SM (2004) Progressive analysis procedure for progressive collapse. J Perform Constr Facil 18(2):79–85
11. Powell G (2005) Progressive collapse: case studies using nonlinear analysis. In: Structures congress 2005: metropolis and beyond. ASCE, New York, USA, pp 1–14
12. Kaewkulchai G, Williamson EB (2004) Beam element formulation and solution procedures for dynamic progressive collapse analysis. Comput Struct 82(7–8):639–651
13. Grierson D, Safi M, Xu L, Liu Y (2005) Simplified methods for progressive collapse analysis of buildings. In: Structures congress 2005: metropolis and beyond. ASCE, New York, USA, pp 1–8
14. Marjanishvili S, Agnew E (2006) Comparison of various procedures for progressive collapse analysis. J Perform Constr Facil 20(4):365–374
15. Khandelwal K, El-Tawil S (2007) Collapse behavior of steel special moment resisting frame connections. J Perform Constr Facil 113(5):646–665
16. Khandelwal K, El-Tawil S (2008) Assessment of progressive collapse residual capacity using pushdown analysis. In: Structures congress 2008: crossing borders. ASCE, Vancouver, British Columbia, Canada, pp 1–8
17. Kim J, Hwang J (2008) Effect of catenary action on collapse of steel frames. In: Ofner R, Beg D, Fink, J, Greiner R, Unterwegger H (eds) Proceedings of the 5th European conference on steel and composite structures, EUROSTEEL 2008, Graz, Austria
18. Helmy H, Salem H, Mourad S (2013) Computer-aided assessment of progressive collapse of reinforced concrete structures according to GSA code. J Perform Constr Facil 27(5):529–539
19. Fu F (2010) 3-D nonlinear dynamic progressive collapse analysis of multi-storey steel composite frame buildings—Parametric study. Eng Struct 32:3974–3980
20. SAP2000 (2016) Structural analysis program. Computers and structures Inc., Berkeley, California, USA
21. IS 800:2007 (2007) General construction in steel-code of practice. Bureau of Indian Standards, New Delhi, India
22. IS 1893:2016 (2016) (Part I): Criteria for earthquake resistant design of structures. Bureau of Indian Standards, New Delhi, India
23. FEMA 356 (2000) Prestandard and commentary for the seismic rehabilitation of buildings. Federal Emergency Management Agency, ASCE, Washington, DC, USA

Influence of Concrete Fill on the Buckling Characteristics of Slender Circular Steel Tubes

Rebecca Mary Paul, M. Madhu Karthik, and M. V. Anil Kumar

Abstract The strength, stiffness and buckling characteristics of hollow structural steel tubes can be significantly enhanced by filling it with concrete. AISC 360-16 classifies circular concrete-filled steel tubes (CFSTs) as compact, non-compact and slender based on the section slenderness ratio $\lambda = D/t$ (outer diameter/thickness of the steel tube). The steel in compact and non-compact sections yield before buckling, and slender sections are assumed to buckle elastically. However, a comparison of the design provisions suggested by AISC 360-16 with the existing experimental results shows that the predicted strength is highly conservative for slender sections. To investigate the influence of concrete core on the buckling characteristics of slender circular CFSTs, a finite element (FE) analysis is performed using ABAQUS. A parametric study is conducted to study the influence of the input parameters of the concrete damage plasticity model. The post-peak branch of the stress-strain relation of unconfined concrete is also modified to account for the increase in the peak-axial strain of confined concrete. The results from the FE analysis of a slender CFST short column is in good agreement with the experimental observations. The analysis shows that the steel tube did not elastically buckle, and crushing of concrete led to the failure of the specimen. This is contrary to AISC 360-16 provisions, which assumes that slender CFSTs fail by elastic local buckling. This could probably be the reason for the overly conservative prediction of the ultimate load capacity of slender CFSTs by AISC 360-16, and needs further investigation.

Keywords Concrete-filled steel tube · Finite element modelling · Steel tube · Buckling · Slender

---

R. M. Paul (✉) · M. M. Karthik · M. V. Anil Kumar
Indian Institute of Technology Palakkad, Palakkad 678557, India
e-mail: 101803102@smail.iitpkd.ac.in

© The Editor(s) (if applicable) and The Author(s), under exclusive license to Springer Nature Singapore Pte Ltd. 2021
S. K. Saha and M. Mukherjee (eds.), *Recent Advances in Computational Mechanics and Simulations*, Lecture Notes in Civil Engineering 103,
https://doi.org/10.1007/978-981-15-8138-0_6

1 Introduction

Concrete-filled steel tubes (CFSTs) are composite members where the hollow steel tubes are filled with concrete. The introduction of concrete in steel tubes not only enhances the load capacity of the sections, but also results in improved performance of concrete and steel, when compared to their individual performances. CFSTs are predominantly used as compression members, where both the concrete and steel contribute to the axial load carrying capacity of the composite member. They are primarily used as columns for the lower storeys of high-rise buildings due to their low cross-sectional area-to-load carrying capacity ratio. The steel tube also provides lateral confinement, which in turn confines the concrete, thereby increasing its strength and ductility. Besides providing confinement, the steel tube also acts like a formwork which further provides speedy and economical construction. Due to its inherent advantages, CFSTs are mainly adopted for construction of buildings situated in seismic zones, bridge piers (with internal reinforcements), transmission towers and for retrofitting purposes.

When CFSTs are axially loaded, both the steel and concrete contribute to their axial load carrying capacity. Due to the difference in Poisson's ratio of steel ($\mu_s \sim 0.3$) and concrete ($\mu_c \sim 0.15$), they do not influence the mechanical behaviour of each other during the initial loading stages and share the load independently [1]. As the applied load increases, micro-cracks start to develop in the concrete core at approximately 0.4–0.7 f'_c (f'_c = concrete compressive strength) [1, 2], which causes concrete to expand laterally. This increases the Poisson's ratio of concrete to about 0.6 [3], thereby resulting in the concrete applying lateral pressure on the steel tube. The tensile hoop stress causes a biaxial state of stress in the steel tube, and results in a reduction of the peak axial stress in the steel tube. Even though there is a reduction in the axial capacity of the steel tube, there is an increase in the overall capacity of the CFST because of the enhancement of strength of concrete due to confinement.

2 Classification of Circular CFSTs and Comparison with Experimental Database

Based on the ratio of overall diameter to the steel tube thickness of concrete-filled steel tubes (D/t), termed as slenderness ratio λ, AISC 360-16 [4], classifies CFSTs as compact, non-compact and slender. Compact sections are those in which yielding happens much before buckling, thus providing adequate confinement to the concrete core. In non-compact sections, the buckling and yielding of steel happen simultaneously, which may offer marginal confinement to concrete. In slender sections, the steel is assumed to elastically buckle before yielding, resulting in ineffective confinement of the core concrete. Depending on the section classification, AISC 360-16 proposes different expressions for calculating the ultimate capacity of circular CFSTs.

Compact CFSTs, $\lambda \leq \lambda_p = 0.15\, E_s/F_y$

$$P = F_y A_s + 0.95 f'_c A_c \tag{1}$$

Non-compact CFSTs $\lambda_p < \lambda \leq \lambda_r = 0.19\, E_s/F_y$

$$P = P_p - \frac{(P_p - P_y)(\lambda - \lambda_p)^2}{(\lambda_r - \lambda_p)^2} \tag{2}$$

Slender CFSTs, $\lambda > \lambda_r \leq 0.31\, E_s/F_y$

$$P = F_{cr} A_s + 0.7 f'_c A_c \tag{3}$$

in which, F_y = yield strength of steel; f'_c = compressive strength of concrete; A_s, A_c = area of steel and concrete; E_s = modulus of elasticity of steel; P_p = P as defined in Eq. (1); and the yield load and the critical buckling stress F_{cr} are given as

$$P_y = F_y A_s + 0.7 f'_c A_c \tag{4}$$

$$F_{cr} = \frac{0.72 F_y}{\left(\frac{D}{t}\frac{F_y}{E_s}\right)^{0.2}} \tag{5}$$

These design expressions are valid for steel strength less than 525 MPa and concrete strength within the range of 21–69 MPa.

Figure 1 shows a comparison of the axial load capacity of circular CFSTs predicted by AISC 360-16, N_{AISC} with the experimental database, N_{exp} [3, 5]. It is evident from Fig. 1, that the expressions defined in AISC 360-16 underestimate the axial load carrying capacity of circular CFSTs. This is particularly true for slender sections where the mean and standard deviation are 1.345 and 0.131, respectively. This warrants further investigation on the behaviour of slender CFSTs.

To understand the effect of the concrete fill in slender CFSTs, a numerical model using the finite element software, ABAQUS [6], is conducted. A parametric study on the concrete model is undertaken to obtain the parameters required for defining

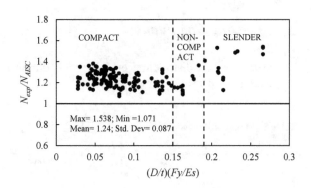

Fig. 1 Comparison of axial load carrying capacity of experimental database [3, 5] with AISC 360-16 [4]

the behaviour of concrete. A hollow steel tube with geometric imperfections is also analysed to validate the modelling procedure, before extending the modelling to a slender CFST column.

3 Finite Element Modelling

In order to investigate the effect of concrete on the buckling characteristics of steel tube in slender CFSTs, a finite element analysis using the finite element software, ABAQUS [6], was conducted. The details of the model are presented in this section.

3.1 Mesh Details

The steel tube was modelled using four-noded linear shell elements (S4R) with reduced integration. For the concrete core, eight-noded linear brick elements with reduced integration (C3D8R) was adopted. To determine the optimal mesh size, an eigenvalue buckling analysis was carried out in ABAQUS [6], for a hollow steel tube with diameter $D = 300$ mm, thickness $t = 2$ mm, modulus of elasticity $E_s = 200$ GPa and Poisson's ratio $\mu = 0.3$. The theoretical buckling load $P_{Th} = 3021.9$ kN was determined using the following equation:

$$P_{Th} = \frac{2E_s}{\sqrt{3(1-\mu^2)}} \frac{t}{D} \qquad (6)$$

Table 1 presents the details regarding the mesh convergence study. Based on the ratio of P_{Th} to the buckling load generated from ABAQUS P_{FEM}, and considering the tradeoff between accuracy and computational time, a mesh size of 15 mm was adopted for further analysis. In order to ensure compatibility at the interface of steel and concrete in CFST, the same mesh size was adopted for the concrete core as well.

3.2 Loading Conditions

To ensure uniform distribution of axial loads during experimental tests, most of the experimental tests on hollow steel tubes and CFSTs applied loads via steel

Table 1 Mesh convergence study

Mesh size (mm)	25	20	15	10	5
P_{Th}/P_{FEM}	0.986	0.992	0.995	0.997	1.006

bearing plates welded to their ends. To accurately represent the experimental loading conditions in the analysis, loads were evenly applied to the top nodes of steel and/or concrete. Both the ends of the CFST were restrained against all translational movements except the axial deformation of the top nodes.

3.3 Constitutive Relation for Steel

Any appropriate stress-strain relation that represents the behaviour of steel can be adopted in the analysis. For simplicity, the stress-strain relationship of steel shown in Fig. 2, was adopted in this study. The initial modulus of elasticity E_s and Poisson's ratio μ were taken as 200 GPa and 0.3 respectively. The post-yield hardening modulus was taken as 3% of E_s.

3.4 Constitutive Relation for Concrete

In CFSTs, the steel tube confines the core concrete, thereby subjecting the concrete to a triaxial state of compression. Concrete subjected to triaxial stresses can be modelled using the concrete damage plasticity (CDP) model, which accounts for the multiaxial behaviour of concrete. The parameters required to define the plasticity of concrete are the dilation angle ψ, flow potential eccentricity ratio e, ratio of the second stress invariant on tensile meridian to compressive meridian K_c, ratio of biaxial compressive strength to uniaxial compressive strength f_{bo}/f_{co} and viscosity v. Together with these parameters, the model demands for the compressive and tensile

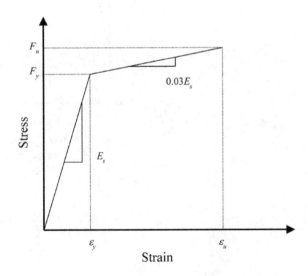

Fig. 2 Stress-strain behaviour of steel

behaviour of concrete. A parametric study was conducted on a 150 mm × 300 mm concrete cylinder with a compressive strength of 20 MPa, to determine the optimal parameters for the CDP model. The results of the parametric study are presented herein.

Dilation angle, ψ: This is defined as the angle of inclination of the asymptote of the failure surface of concrete with respect to hydrostatic axis [7]. Figure 3a shows the results of the parametric study conducted on the concrete cylinder by varying the dilation angle from 5° to 30°. Figure 3a shows a comparison of the defined stress-strain relation of unconfined concrete with the stress-strain relation of concrete extracted from the FE analysis with various dilation angles. There is a slight increase in the compressive strength with an increase in the dilation angle. From the FE results a dilation angle of 5° was adopted for further analysis.

Flow potential eccentricity, e: This is the ratio of tensile to compressive strength of concrete [7]. The variation of the flow potential eccentricity parameter from 0.05 to 0.2 did not have any effect on the stress-strain behaviour of concrete. Therefore, the default value of 0.1 specified in ABAQUS manual [8], was adopted.

Ratio of biaxial compressive strength to uniaxial compressive strength, f_{bo}/f_{co}: The ratio represents the point at which the concrete fails in a biaxial state of stress. ABAQUS manual [8] recommends a default value of 1.16. To study the effect of f_{bo}/f_{co} on the behaviour of modelled concrete, a parametric study was conducted

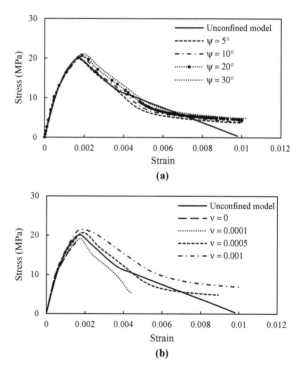

Fig. 3 Parametric study on the influence of **a** dilation angle ψ, and **b** viscosity parameter v on unconfined concrete

by varying the parameter from 1 to 1.5. The default value of 1.16 was found to model the behaviour of concrete quite accurately, and therefore, was adopted for further analysis.

Ratio of the second stress invariant on tensile meridian to compressive meridian, K_c: Physically it is the ratio of the distances from hydrostatic axis to compression and tension meridians in the deviatoric cross-section [7]. The ratio ranges between 0.5 and 1, where $K_c = 1$ represents a circular failure surface in the deviatoric cross-section. From the studies reported by Śledziewski [7] $K_c = 0.67$ was adopted, as the shape of the concrete failure surface represented William Warnke strength criterion.

Viscosity, v: While modelling the nonlinear behaviour of concrete in ABAQUS, difficulties in convergence arise due to the softening and stiffness degradation behaviour of concrete [9]. The viscosity parameter helps to overcome these convergence issues. Figure 3b shows the parametric study conducted on a standard concrete cylinder by varying the viscosity parameter from 0 to 0.001. Based on the parametric study, a viscosity parameter of 0.0005 was found to model the behaviour of unconfined concrete under uniaxial compression quite well.

Table 2 presents the plastic parameters that were adopted for modelling the behaviour of concrete under compression, based on the parametric study. The same parameters were adopted for modelling concrete in the slender CFST specimen [10], presented later.

The compressive stress-strain behaviour of concrete suggested by Karthik and Mander [11], was adopted in this study. Figure 4a shows the unconfined and confined concrete models. From the preliminary FE analysis of CFST with the unconfined

Table 2 Plasticity parameters for CDP model

f'_c (MPa)	Dilation angle	Eccentricity	f_{bo}/f_{co}	K_c	Viscosity
20	5°	0.1	1.16	0.67	0.0005

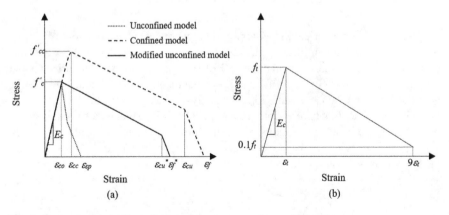

Fig. 4 Stress-strain behaviour of concrete in **a** compression, and **b** tension

model, it was found that the interaction between steel and concrete accounts for the increase in compressive strength of concrete due to confinement. However, the increase in ductility of confined concrete (ε_{cc}) was not being accounted for. To account for the increase in ductility of confined concrete, as shown in Fig. 4a, the post-peak behaviour of the unconfined concrete was modified by replacing it with the post-peak behaviour of the confined model. Figure 4b shows the tensile behaviour of concrete that was adopted in this study, where the tensile strength is assumed to be 0.1 f_c' [12].

3.5 Modelling Contact

To ensure that the interaction between concrete and steel in the CFST are modelled appropriately, general contact was employed. A hard contact in the normal direction was specified for the interface between concrete and steel, to ensure compatible deformation of concrete and steel during compression. For the tangential contact, the coefficient of friction between concrete and steel was taken as 0.55 as reported in Lai and Varma [5].

3.6 Geometric Imperfections

The buckling mode shapes of the hollow steel tube were determined from the eigenvalue buckling analysis. The mode shape for the first eigenvalue was used to define the geometric imperfection in the steel tube for further nonlinear analysis.

3.7 Validation of Modelled Hollow Steel Tube

To ensure that the numerical model appropriately simulates the behaviour of hollow steel tubes without any infill, a slender tube with $F_y = 210.7$ MPa and $D/t = 220$ was modelled using the finite element method (FEM). Figure 5 shows a comparison of the average stress-strain behaviour of the hollow steel tube member obtained from FEM with the experimental test result [13]. The analysis clearly shows that the steel tube failed by local buckling before yielding at a stress of 175.3 MPa, which is comparable with the experimental observations. Since the finite element model could reasonably simulate the behaviour of the slender steel tube, the model was further used to study the behaviour of slender CFSTs.

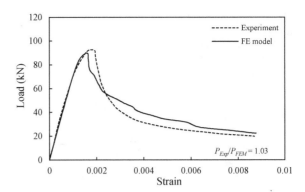

Fig. 5 Comparison of stress-strain behaviour of slender hollow steel tube from FEM with test result [13]

4 Modelling of Concrete-Filled Steel Tubes

A finite element model of a slender CFST was developed to simulate the behaviour of an experimental specimen [10]. Table 3 presents the details of the experimental specimen. The slenderness ratio of the specimen was 150 which was greater than $\lambda_r = 111$, and based on AISC 360-16 [4], classifications the section can be classified as slender.

The parameters for the CDP model that were determined in Sect. 3.4, was used to model concrete in the CFST. From a preliminary FE analysis of the CFST using the unconfined concrete model presented in Fig. 4a, the confinement in concrete was determined to be 1.2. To account for the improvement in ductility of the CFST specimen, as shown in Fig. 4a and discussed in Sect. 3.4, the post-peak stress-strain behaviour of unconfined concrete was modified with that of confined concrete with a confinement factor $K = 1.2$. To simulate the load versus axial strain behaviour of the CFST column, a displacement controlled analysis with Newton Raphson method was adopted.

Figure 6 presents a comparison of load versus axial strain behaviour of the slender CFST obtained from the finite element analysis with the experimental results [10]. The initial stiffness, ultimate load and the post-peak behaviour obtained from the analysis compare well with the experimental results. Points 'a' and 'b' indicated on the load versus axial strain behaviour in Fig. 6, represents the key events during the loading of the CFST. Point 'a' corresponds to yielding of steel tube (Von Mises stress equal to yield stress over full cross-section) whereas point 'b' corresponds to crushing of the concrete core when the CFST attained its ultimate load. The occurrence of these events shows that even though the CFST specimen is classified as slender

Table 3 Specimen details for slender CFST specimen CU-150 [10]

Specimen	Diameter (mm)	Thickness (mm)	Length (mm)	F_y (MPa)	f_c' (MPa)	N_{Exp}/N_{AISC}	N_{Exp}/N_{FEM}
CU-150	300	2	840	341.7	27.2	1.36	0.98

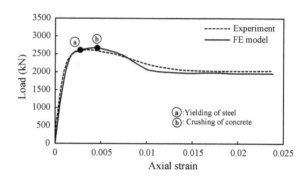

Fig. 6 Comparison of FEM results with experimental load versus axial strain for slender CFST specimen CU-150 [10]

according to AISC 360-16 [4], the column failed due to crushing of concrete after the steel tube yielded rather than the elastic buckling of the steel tube.

These observations are indicative of the influence of concrete fill on the buckling characteristics of slender CFSTs. Therefore, it can be concluded that not all sections classified as slender according to AISC 360-16 [4], undergo elastic buckling. This is a plausible reason for the conservative prediction of the ultimate load carrying capacity of slender CFSTs by AISC 360-16 (Fig. 1), when compared to experimental results. Additional analysis may be required for an in-depth understanding of the failure mechanism, and is currently under consideration.

5 Conclusion

To investigate the influence of concrete on the buckling characteristics of slender CFSTs, a numerical model was developed using the finite element application ABAQUS. A parametric study was conducted to determine the appropriate parameters for the concrete damage plasticity model of concrete. The post-peak branch of the stress-strain relation of unconfined concrete was modified to account for the increase in the peak-strain of confined core concrete. From the finite element analysis of slender CFST, the following can be concluded.

- The concrete fill influences the behaviour of slender CFSTs.
- Slender CFSTs can fail by crushing of core concrete that occurs after yielding of the steel tube.
- Not all CFSTs classified as slender according to AISC 360-16, fail by elastic local buckling of the steel tube.
- This change in failure mode could possibly be the reason for the conservative prediction of ultimate load capacity of slender CFSTs by AISC 360-16 provisions.
- The behaviour of slender CFSTs needs to be investigated further to have a better understanding of their failure mechanism.

References

1. Lin S, Zhao Y-G, He L (2018) Stress paths of confined concrete in axially loaded circular concrete-filled steel tube stub columns. Eng Struct 173:1019–1028
2. Chen WF, Han DJ (2007) Plasticity for structural engineers. J. Ross Publishing (2007)
3. Lai M, Ho J (2016) A theoretical axial stress-strain model for circular concrete-filled-steel tube columns. Eng Struct 125:124–143
4. AISC 360-16 (2016) Load and resistance factor design (LRFD) specification for structural steel buildings. American Institute of Steel Construction, Chicago (IL, USA)
5. Lai Z, Varma AH (2015) Noncompact and slender circular CFT members: experimental database, analysis, and design. J Constr Steel Res 106:220–233
6. ABAQUS: Version 6.8 [Computer software]. Dassault Systemes Simulia, Providence, RI
7. Śledziewski K (2017) Selection of appropriate concrete model in numerical calculation. In: ITM Web of conferences, vol 15. EDP Sciences, p 07012
8. ABAQUS Analysis User's Guide: Version 6.14. Dassault Systèmes Corp, Providence, RI (USA)
9. Demir A, Ozturk H, Edip K, Stojmanovska M, Bogdanovic A (2018) Effect of viscosity parameter on the numerical simulation of reinforced concrete deep beam behaviour. Online J Sci Technol 8(3):50–56
10. Huang CS, Yeh YK, Liu GY, Hu HT, Tsai KC, Weng YT, Wu MH (2002) Axial load behaviour of stiffened concrete-filled steel columns. J Struct Eng 128(9):1222–1230
11. Karthik MM, Mander JB (2011) Stress-block parameters for unconfined and confined concrete based on a unified stress-strain model. J Struct Eng 137(2):270–273
12. Wahalathantri BL, Thambiratnam DP, Chan THT, Fawzia S (2011) A material model for flexural crack simulation in reinforced concrete elements using ABAQUS. In: Proceedings of the first international conference on engineering, designing and developing the built environment for sustainable wellbeing. Queensland University of Technology, pp 260–264 (2011)
13. O'Shea MD, Bridge RQ (1997) Local buckling of thin-walled circular steel sections with or without internal restraint. J Constr Steel Res 41(2–3):137–157

Vibration of Flexible Member in Offshore Structures

Madagala Sravani and Kiran Vijayan

Abstract There are many flexible vertical members in offshore platforms like mooring lines and risers. These are exposed to wave, vessel motion and drilling loads during operation. For the smooth operation of these structures, the lateral excursion of these members should be minimal. This paper explains about the vibration analysis of a flexible vertical member subjected to base excitation. Initially, the flexible member was analysed as a SDOF non-linear spring-mass system. A qualitative understanding of the system was determined by solving the system using continuation method. Further, a numerical and experimental analysis was carried out on this member. Numerically, the member was analysed as an Euler–Bernoulli beam in MATLAB using the finite element method with three degrees of freedom at each node. The buckling load was determined using the Eigen value analysis. The minimum length required for buckling under gravity load was determined. The system was subjected to base excitation and the variation in the maximum response with respect to base excitation frequency was noted in the axial and lateral direction. The maximum response was obtained near the resonant frequency. Experimentally, the flexible member with a force transducer at its base was connected to the modal exciter through a stinger. The material of the member chosen was acrylic polymer as it buckles easily and can be used for testing. Forced excitation was provided using the modal exciter. On varying the frequency, the response of the member was observed. The length of the member was also a parameter that was varied such that the excursion was within controllable limits. Also, the area of the cross section in the upper half of the section was increased as an additional parameter for controlling the excursion. It was observed that the excursion was reduced as obtained from the theoretical analysis.

M. Sravani · K. Vijayan (✉)
Department of Ocean Engineering and Naval Architecture, IIT Kharagpur,
Kharagpur 721302, India
e-mail: kiran.vijayan@naval.iitkgp.ac.in

Keywords Buckling analysis · Non-linear dynamics · Continuation method

1 Introduction

Floating structures are generally attached to flexible members such as riser, mooring lines and data cables, etc. Each of these members serves a different purpose. Risers are used to contain fluids for well control (drilling risers) and to convey hydrocarbons from the seabed to the platform (production risers). The mooring system consists of freely hanging lines connecting the surface platform to anchors at the seabed mainly used for the station-keeping function. The data cables are generally attached to buoy for transmitting data. These structures are similar to a reinforced concrete chimneys on a flexible foundation [1, 2] and offshore wind turbines [3]. Datta et al. [1, 2] carried out an analysis on response reduction of these structures when subjected to wind loading using tuned mass damper. One of the key aspects for the proper functioning of these members is to maintain the structural integrity. A critical instability which can interfere with the structural integrity is the buckling. The buckled member can undergo a large lateral excursion. This can interfere with the structural integrity and curtail the intended purpose of the flexible member [4]. The purpose of this study is to understand the lateral dynamics of a flexible member when subjected to excitation. An experimental and theoretical study would be carried out on the system. The influence of system parameter on the lateral excursion would be determined.

2 Conceptual Model

To analyse the excursion of these flexible vertical members, a suitable member made of acrylic polymer of an appropriate length was chosen. The member was selected in such a way that there is minimum lateral excursion under gravity for the vertical member under gravity. Acrylic polymer was chosen as the material. The length and cross section are details are provided in Table 1. The length and cross section area was chosen taking into account the buckling load.

2.1 Theoretical Model

A theoretical model for the conceptual model was developed. The flexible member was modelled as an Euler–Bernoulli beam. At each node, there are three degrees of freedom, i.e. axial, bending and rotation, respectively. The system was modelled using Finite element analysis and the elemental mass and stiffness matrix [5–8] was obtained as

Vibration of Flexible Member in Offshore Structures

Table 1 Properties of the flexible vertical member

Property	Value
Material	Acrylic polymer
Length	12.6 cm
Density	1180 kg/m^3
Cross sectional area	2.9 * 2.3 mm^2
Mass	9.1 g

$$M = \frac{\rho AL}{420} \begin{bmatrix} 140 & 0 & 0 & 70 & 0 & 0 \\ 0 & 156 & 22L & 0 & 54 & -13L \\ 0 & 22L & 4L^2 & 0 & 13L & -3L^2 \\ 70 & 0 & 0 & 140 & 0 & 0 \\ 0 & 54 & 13L & 0 & 156 & -22L \\ 0 & -13L & -3L^2 & 0 & -22L & 4L^2 \end{bmatrix} \quad (1)$$

$$K = \begin{bmatrix} \frac{AE}{L} & 0 & 0 & -\frac{AE}{L} & 0 & 0 \\ 0 & 12\frac{EI}{L^3} & 6\frac{EI}{L^2} & 0 & -12\frac{EI}{L^3} & 6\frac{EI}{L^2} \\ 0 & 6\frac{EI}{L^2} & 4\frac{EI}{L} & 0 & -6\frac{EI}{L^2} & 2\frac{EI}{L} \\ -\frac{AE}{L} & 0 & 0 & \frac{AE}{L} & 0 & 0 \\ 0 & -12\frac{EI}{L^3} & -6\frac{EI}{L^2} & 0 & 12\frac{EI}{L^3} & -6\frac{EI}{L^2} \\ 0 & 6\frac{EI}{L^2} & 2\frac{EI}{L} & 0 & -6\frac{EI}{L^2} & 4\frac{EI}{L} \end{bmatrix} \quad (2)$$

Before proceeding with the natural frequency determination an optimal length for the member has to be determined. Geometric stiffness matrix is obtained as [6]

$$\left| (K - \lambda K_g) - M\omega^2 \right| X = 0 \quad (3)$$

$$K_g = \frac{1}{L} \begin{bmatrix} 0 & 0 & 0 & 0 & 0 & 0 \\ & \frac{6}{5} & \frac{L}{10} & 0 & -\frac{6}{5} & \frac{L}{10} \\ & & 2\frac{L^2}{15} & 0 & -\frac{L}{10} & -\frac{L^2}{30} \\ & & & 0 & 0 & 0 \\ Sym. & & & & \frac{6}{5} & -\frac{L}{10} \\ & & & & & \frac{2L^2}{15} \end{bmatrix} \quad (4)$$

The eigenvalue problem

$$\left| (K - \lambda K_g) \right| X = 0 \quad (5)$$

is solved to determined the buckling load values. The length of the member was chosen in such a manner that the minimum buckling load was self weight.

2.2 Experimental Model

Experimentally, the member was tested to understanding the lateral excursion (Fig. 1). The specimen was base excited using a modal shaker. The shaker is an electromechanical device that provides a mechanical motion due to the input signal sent to its coil. Despite being widely used, it is well known that the shaker interacts with the structure under test. Generally, a stinger is attached between Structure Under Test (SUT) and shaker, and the input force is measured. Flexible stinger which is used to transmit the excitation signals to the SUT in a single direction, reducing secondary forms of excitation (e.g. bending moments) due to possible misalignments [9]. A power amplifier was used to vary the gain so that the power can be varied in current mode or voltage mode. The OROS analyzer was an FFT analyzer that takes the input signal from the force transducer and displays the signal for which the NVgate software platform was used. We can vary the frequency through this software.

Force transducer is placed between the stinger and the flexible member. The member here was fixed to a hollow brass stud fixed on the forced transducer. The threads on this brass bolt get fit into that of force transducer which was fixed to the stinger rod [10]. This stinger rod was fitted at its base to the modal exciter. The length of the member was 9.11 cm at which it was nearly straight under static load.

The specimen was base excited for a range of frequencies. The shaker was operated in current mode and the gain was fixed to a particular value at each frequency. The

Fig. 1 Experimental set up showing the specimen attached with a stinger to the modal exciter

shaker was operated in current mode for sufficient excitation force [9]. In addition to the frequency, the cross sectional area and length of the member was varied as parameters. The area of the cross section was varied using two approaches. Firstly, the area of the cross section was increased in the upper portion of the member and was excited to check for the minimum excursion on varying the frequency. Secondly, the experiment was repeated on increasing the area of the cross section at the mid-portion of the member. It was found that the dynamics of the member was affected for certain frequencies, the lateral excursion is minimised and the flexible member was nearly straight. Experimentally, the natural frequency at which the flexible member tries to attain minimum excursion was at 25 Hz. This indicates a possibility of occurrence of parametric resonance. Therefore, a non-linear analysis on the system was carried out [11].

2.3 Non-linear Analysis

For the non-linear analysis, an equivalent SDOF system was designed with the natural frequency near to the second natural frequency of the MDOF system. Under the presumption that the system is buckled, a non-linearity in stiffness was assumed. Ben et al. [11] carried out a study on a cable with a suspended mass using a piecewise model. The piecewise model was smoothed using a non-linear exponential function model. However, for this study, the non-linear model is assumed to be a duffing model.

A hardening cubic non-linear stiffness was assumed. The non-dimensional Duffing equation [12], with damping and external forcing has the form as follows:

$$\ddot{y} + 2\beta\dot{y} + y + \gamma y^3 = q \sin(\omega t) \tag{6}$$

The cubic non-linearity and the time-dependent forcing are responsible for the rich dynamics exhibited by the Duffing equation [13]. Non-linear behaviour occurs such as the jump phenomenon where the steady state behaviour changes dramatically due to a transition from one stable solution to another stable solution as a control parameter such as the excitation frequency or the excitation amplitude is varied.

A numerical bifurcation analysis is carried out using continuation method [14]. Numerical continuation is a technique to compute a consecutive sequence of points which approximate the desired branch. Most continuation algorithms implement a predictor-corrector method. The system was modelled in Matcont as a non-autonomous system. Depending on the solution of floquet multiplier, the stability of limit cycle can be determined. The excitation frequency of the non-autonomous force was varied to determine the limit cycle and branching point.

3 Results

The natural frequency and mode shapes were determined using

$$\left|(K - \lambda K_g) - M\omega^2\right| X = 0 \quad (7)$$

Here, the λ value corresponds to the lowest buckling load value. The first five natural frequencies are given in Table 2. The first frequency is above zero which indicates absence of buckling. This is obtained by choosing the length of the member above the buckling load.

The mode shapes are shown in Fig. 2, which clearly indicates that most of the initial modes are predominant in the lateral direction as expected. The axial modes are observed for higher modes wherein the lateral degrees of freedom are having minimal modal amplitude change.

Next, the equation was formulated in state space and a base excitation was provided to the system.

Table 2 First five natural frequency of the beam

Mode	ω_n (Hz)
1	0.002
2	31.02
3	92.74
4	184.3
5	306.5

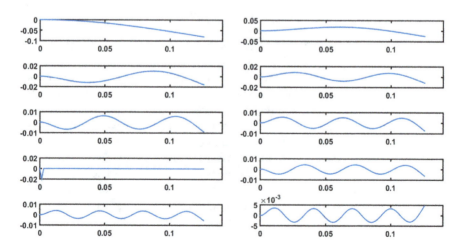

Fig. 2 Bending mode shapes

Fig. 3 Temporal variation of axial displacement

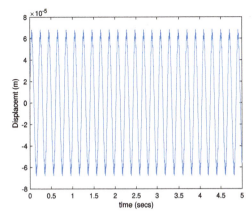

Fig. 4 Temporal variation of lateral displacement

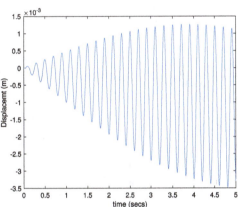

$$[M][\ddot{X}] + [[K] - \lambda[K_g]][X] = -\omega_e^2 \alpha[M] \tag{8}$$

The temporal variation of the axial and transverse response are shown in Figs. 3 and 4. Maximum lateral excursion value of 35.5 mm is observed. The system was excited near second natural frequency which is the bending mode. Hence as expected, the amplitude of the response in the lateral direction is higher than the axial direction. The frequency response of the system was obtained as shown in Fig. 5, by varying the excitation frequency and noting the steady state lateral oscillation magnitude.

Based on the parameters of MDOF system a SDOF duffing model was developed. The mass of the lumped model was taken as 9.1 g and the stiffness was adjusted to match the second natural frequency of the MDOF system. A force of 0.05 N, at a cubic stiffness parameter value of 7500 was analysed using continuation method and Fig. 6, was obtained. The corresponding displacement was found to be 39.9 mm. On varying this parameter, the branch point cycles were observed and the arching of the curve

Fig. 5 Variation of maximum lateral response with excitation frequency

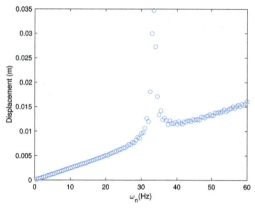

Fig. 6 Frequency response curve using continuation method

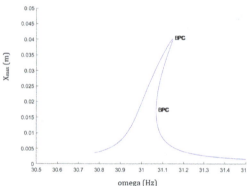

will be slightly higher for a large variation in the value of cubic stiffness parameter. Also, the maximum lateral displacement was found to have a slight decrease from the original value of 35.5 mm.

4 Conclusion

A conceptual model was developed to study the lateral excursion on vertical flexible members. The natural frequencies and mode shapes were determined for the vertical cantilever. The length and cross section were manipulated to prevent buckling under self-weight. Experimental results indicated the possibility of reduced lateral excursion due to non-linear effect. The effect of lateral excursion due to non-linearity was verified by continuation method.

References

1. Elias S, Matsagar V, Datta TK (2017) Distributed multiple tuned mass dampers for wind response control of chimney with flexible foundation. Procedia Eng 199:1641–1646
2. Elias S, Matsagar V, Datta TK (2019) Along-wind response control of chimneys with distributed multiple tuned mass dampers. Struct Control Health Monit 26(1):e2275
3. Banerjee A, Chakraborty T, Matsagar V, Achmus M (2019) Dynamic analysis of an offshore wind turbine under random wind and wave excitation with soil-structure interaction and blade tower coupling. Soil Dyn Earthq Eng 125:105699
4. Goss VGA, Van der Heijden GHM, Thompson JMT, Neukirch S (2005) Experiments on snap buckling, hysteresis and loop formation in twisted rods. Exp Mech 45(2):101–111. https://doi.org/10.1007/BF02428182
5. Cook RD (2007) Concepts and applications of finite element analysis. Wiley, New York
6. Menon D (2009) Advanced structural analysis. Alpha Science International, Oxford
7. Menon D (2008) Structural analysis. Alpha Science International, Oxford
8. Rao SS, Yap FF (2011) Mechanical vibrations, vol 4. Prentice Hall, Upper Saddle River
9. Varoto PS, de Oliveira LP (2002) On the force drop off phenomenon in shaker testing in experimental modal analysis. Shock Vib 9(4–5):165–175
10. Ewins DJ (1984) Modal testing: theory and practice. Research Studies Press, Letchworth
11. Ben-Gal N, Moore KS (2005) Bifurcation and stability properties of periodic solutions to two nonlinear spring-mass systems. Nonlinear Anal: Theory Methods Appl 61(6):1015–1030
12. Kalmár-Nagy T, Balachandran B (2011) Forced harmonic vibration of a Duffing oscillator with linear viscous damping. The Duffing equation: nonlinear oscillators and their behaviour. Wiley, New York, pp 139–173
13. Strogatz SH (2018) Nonlinear dynamics and chaos with student solutions manual: with applications to physics, biology, chemistry, and engineering. CRC Press, Boca Raton
14. Dhooge A, Govaerts W, Kuznetsov YA (2003) MATCONT: a MATLAB package for numerical bifurcation analysis of ODEs. ACM Trans Math Softw (TOMS) 29(2):141–164

Nonlinear Analysis of Mooring System for an Offshore Desalination Platform

Ashwani Vishwanath and Purnima Jalihal

Abstract The article describes the analysis and design of an optimum mooring system for the given spar configuration. National Institute of Ocean Technology (NIOT), Chennai has studied various configurations of floating platform and has designed and tested scaled down models of Semi-submersible and Spar based floating platform for accommodating Low Temperature Thermal Desalination (LTTD) plant. The Spar platform is to be permanently moored at 1000 m of water depth. The mooring configuration is very critical in the overall operation of the system. Hence a detailed study on this was carried out taking care of the nonlinearities of external forces induced in the system. Low frequency motions in surge, sway and yaw are excited by second order, nonlinear coupled effects between the wave and the Spar. The forces induced in the moorings are as well nonlinear due to nonlinear characteristics of the mooring material. The nonlinear mooring behavior characteristics and its effect on the platform motion discussed in the article, were carried out in potential/diffraction commercial simulation tool MOSES. The crucial parameters for the design including length, material, and mooring pretension were arrived from the simulation.

Keywords Spar · Mooring · LTTD · Diffraction

1 Introduction

Ocean is a vast source of energy and potable water both of which, need to be harnessed in a clean and green manner. Among different sources of energy from the ocean, ocean thermal gradient perhaps has the highest potential for both extraction of energy and fresh water since we are in the tropics. The process of extraction of potable water from the ocean has already reached its advanced state. Sea water desalination is attaining increasing attention of present day policy makers, especially with the

A. Vishwanath (✉) · P. Jalihal
National Institute of Ocean Technology, Chennai 600100, India
e-mail: ashwani.niot@gov.in

growing demands that urbanization, population explosion, irregular rainfall, and ground water contamination place on the fragile natural resources. LTTD is one process that uses the availability of a temperature gradient between two water bodies or flows to evaporate the warm water at low pressure and condense the resultant vapor with cold water to obtain fresh water. The temperature difference that exists between the warm surface sea water (28–30 °C) and deep sea cold water (7–15 °C) could be effectively utilized to produce potable water. The deep sea cold water is available at a distance of 30–40 km from the mainland shore. Because of the nonavailability of the deeper depths near shore, the plant needs to be installed at deep waters offshore at around 1000 m water depth. Therefore, it is essential that a suitable floating offshore structural system with good station keeping characteristics has to be designed and installed for installing the desalination plant, such as SPAR platform.

National Institute of Ocean Technology (NIOT) under the Ministry of Earth Sciences, strives to develop reliable indigenous technology to solve the various engineering problems associated with harvesting of living and nonliving resources in the Indian Exclusive Economic Zone (EEZ), which is about two-thirds of the land area of India. In the past, NIOT has installed LTTD plants at various places in Lakshadweep Islands. For mainland application, NIOT has proposed an offshore platform to be installed at 40 km off Chennai. NIOT has studied various configurations of floating platform and has designed and tested scaled down models of Semi-submersible and Spar based floating platforms for accommodating LTTD plant. The Spar platform is to be permanently moored at 1000 m of water depth. The mooring configuration is very critical in the overall operation of the platform system. Hence, a detailed study on this was carried out taking care of the nonlinearities of external forces induced in the system. Selection of Spar platform configuration and global analysis including risers were carried out in the article [1].

For a moored platform, low frequency motions occur in surge, sway, and yaw. Low frequency motions are resonance oscillations exited by second order, nonlinear coupled effects between the wave and the spar [2]. For moored large structures as the Spar, the natural periods in the horizontal degrees of freedom are much larger than the wave periods with considerable energy. The horizontal low frequency excitation is in general larger than the linear wave frequency motions, despite the fact that second order difference frequency forces are generally an order of magnitude smaller than linear wave frequency forces. This effect is, therefore, important in relation to the design of the mooring system [3].

A study had investigated the dependence of response of SPAR platforms on the connection point of mooring line and the length of mooring line. For reducing the surge response, the fairlead point should be placed near to the center of buoyancy, whereas to arrest the pitch motion, the fairlead point should be shifted near to the sea surface, to provide the restoring moment [4]. Effect of dynamic coupling between a Spar and its mooring lines was studied, in which the importance of doing a coupled dynamic analysis was stressed. Ordinary quasi-static analysis actually, under predicts the maximum force which comes in the mooring lines [5]. The time domain wind analysis based on Newmark's method of average acceleration was employed for the non-classically damped system for studying the multi-mode response [6]. Dynamic interaction was studied between the monopile and the underlying soil subjected to realistic random wind and wave loading [7]. Offshore random loading using Kaimal and the Pierson Moskowitz spectrum was taken into account for analyzing a monopile for offshore wind turbines [8].

Quantifying mooring pretensions and its nonlinear behavior on the platform motion require computer simulation incorporating all the coupled dynamic behavior of mooring and platform for arriving at the design of this complex offshore system. For a platform of the size accommodating LTTD plant (~23,000 tons displacement) with a riser system for drawing deep cold water, it is imperative to carry out an extensive global analysis for the design of the mooring component and assessing its behavior in operating and survival conditions.

2 Theoretical Background

2.1 First Order Wave Forces

Numerical models in fluid structure interaction problems in ocean engineering have been very useful for the design related parametric studies where several iterations have to be carried out prior to implementation. The interaction of a structure with regular waves can be solved by diffraction–radiation technique which is in use for several years. The equation of motion of a floating body can be described by the following equation in six degrees of freedom.

$$[M + A]\{\ddot{x}\} + [B]\{\dot{x}\} + [C]\{x\} = F(t) \tag{1}$$

where M is displaced mass of floating structure, A = added mass, B = damping force, C = restoring force, $F(t)$ = external force, $\{\ddot{x}\}$ = acceleration of vector, $\{\dot{x}\}$ = velocity vector and $\{x\}$ = displacement vector.

The external applied force $F(t)$ is due to incoming waves and these forces results in motion of the structure. Due to these motions of the floating structure, waves radiate from the structure boundary. The computation of external force on the structure can be divided into a linear diffraction problem governed by Laplace equation with appropriate boundary conditions and to solve for velocity potential for the radiated waves generated by body motion due to incident and diffracted waves. Hence, the total potential can be written as

$$\Phi^T = \Phi^I + \Phi^S + \sum_{j=1}^{6} \Phi_j^R \tag{2}$$

where Φ^T = total velocity potential, Φ^I = incident wave velocity potential, Φ^S = scattered wave velocity potential and Φ_j^R = radiated velocity potential, corresponding to the motion in jth degree of freedom.

Once the solution to the velocity potential Φ is obtained, the pressure along the body can be obtained using the linear form of Bernoulli's equation, the hydrodynamic pressure field due to wave motion is given by

$$P = -\rho \frac{d\Phi^T}{dt} \qquad (3)$$

2.2 Second Order Forces for the Design of Moorings

Floating structures in irregular waves are subjected to large first order wave forces and moments which are linearly proportional to the wave height and contain the same frequencies as the waves. They are also subjected to small, nonlinear second order wave forces and moments which are proportional to the square of the wave height. Even though small in magnitude, they are of importance in structures with natural periods of their movements are in regions of high and low frequency outside the normally encountered wave frequencies. Second order wave forces and moments are composed of three components, namely a steady (mean) force, a difference frequency (low frequency) component, and a sum frequency (high frequency) component. Due to the nonlinear and dynamic behavior of mooring lines, Time Domain simulations provide the most accurate results, but are considerably more time-consuming.

The hydrodynamic loads, motion response and the mooring line tensions have been calculated by using a commercial software package MOSES (Multi-Operational Structural Engineering Simulator). This program was developed by Ultramarine Inc. in Houston Texas, and is an integrated hydrostatic, hydrodynamic and structural analysis package. It is a diffraction radiation program based on linear potential theory. It works on the principle of panel methods.

3 Global Analysis of Spar

3.1 Platform Description

Various configurations of Spar were studied for sizing the platform to support the LTTD plant equipments. After carrying out detailed hydrostatics and motion response studies a stepped Spar configuration as shown in Fig. 1 was finalized [1]. Table 1 lists the hydrostatics parameters of Spar.

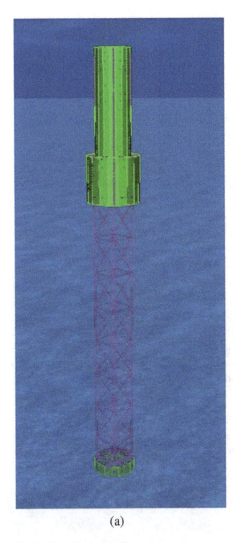 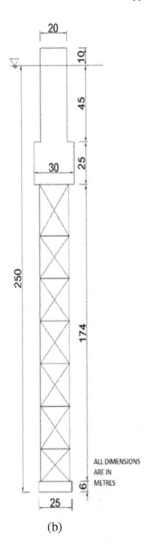

Fig. 1 Stepped spar configuration

3.2 Mooring Line Description

The floating platforms accommodating the LTTD plants are required to sustain extreme weather conditions as disconnecting it with the non-retrievable appendages attached would not prove to be a realistic proposition. The station keeping of the all weather offshore platform(s) is achieved by securing the permanent moorings to the seabed through specialized anchoring systems.

The forces induced in the moorings are as well nonlinear due to nonlinear characteristics of the mooring material. The nonlinear mooring behavior and its effect on

Table 1 Hydrostatic parameters of spar

Free board (m)	10
Displacement of structure (t)	23,278
Hard tank top diameter (m)	20
Hard tank bottom diameter (m)	30
Draft of structure (m)	250
Position of COG from keel KG (m)	163.77
Position of COB from keel KB (m)	190.35
Metacentric height GM (m)	12.91
Natural time period in heave motion (s)	34.8
Natural time period in pitch motion (s)	74

the platform's motion discussed in the article were carried out in potential/diffraction commercial tool MOSES. The mooring lines used for restraining the platform motions have three segments. The portion of the mooring line used for connecting it with the sea bed, as well as with the Spar is made of chain, whereas the middle portion in between the chains is made up of polyester rope. The properties of the mooring sections are given in Table 2. The fairlead is fixed near to the center of buoyancy. Nonlinear variation of horizontal force from the mooring line at the fairlead, with respect to anchor position can be studied in the Fig. 2. The nonlinearity in the forces is exhibited due to its material properties. It is observed that for the anchor position of 2000 m with respect to platform, the horizontal force for 2300 m length mooring

Table 2 Properties of mooring lines

Section	Diameter (mm)	Modulus of elasticity (MPa)	Breaking tension (t)	Weight per unit length (t/m)
Chain (fairlead)	137	112,000	1574	0.375
Polyester	223	70,000	1500	0.0349
Chain (touchdown)	147	112,000	1583	0.379

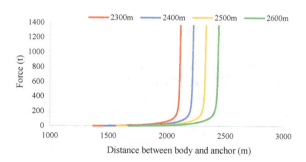

Fig. 2 Variation of horizontal force with anchor position

line is approximately 60t, whereas that for 2600 m mooring line is about 10t. Chains at fairlead and touchdown point are of 150 m in length, while length of polyester rope is varied to get required total length.

3.3 Forces in Mooring Line

After several iterations, spread mooring line configuration as shown in Fig. 3, was used for further analysis. Table 3 lists the environmental parameters used for operating condition in the time domain analysis. The static force acting on the structure (due to current and slow varying wave drift) is predicted around 150t. The variation in force excursion characteristics of the system is studied in Fig. 4, with respect to various mooring line length for pretension of 150t.

From Fig. 4, it is clear that, the horizontal excursion of the system will reduce by decreasing mooring line length for given pretension. However, at the static offset equilibrium position, the stiffness of the system for lesser mooring lengths is higher, which indirectly means that the tensions in the mooring lines will be much higher due to the platform motions, under the given environment. Therefore, the offset, as well as the line tensions have to kept optimum. The tensions in the line can be reduced by

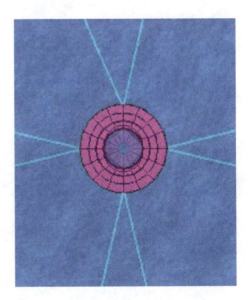

Fig. 3 Spread mooring configuration for Spar

Table 3 Operating environment conditions

Parameters	Values
Wave	Jonswap spectrum, Hs = 5.9 m, T_p = 11.1 s
Current	2.34 m/s at surface, 0.3 m/s near sea bed

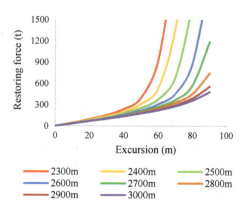

Fig. 4 Variation of restoring force with excursion for different line lengths

increasing the line length but keeping the static offset value under limit. Therefore, it was decided to use line length of 2800 m, with a pretension of 150t for restraining the structure.

In order to analyze the platform motions for different wave and current directions, few cases as shown in Fig. 5, are considered. Figures 6, 7, and 8 shows the variation of connector force ratio (mooring tension/breaking strength) for these three cases.

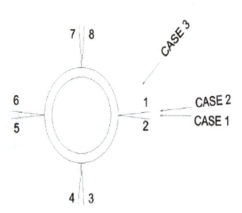

Fig. 5 Loading cases

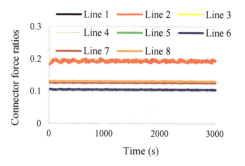

Fig. 6 Connector force ratios—case 1

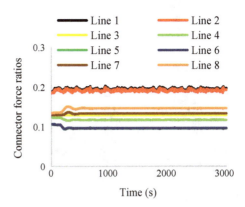

Fig. 7 Connector force ratios—case 2

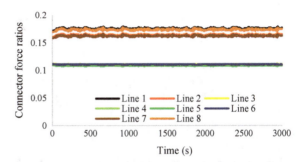

Fig. 8 Connector force ratios—case 3

As expected, in-line wave and current generate higher tensions in mooring line as compared to when they are at 45° to each other.

Figures 9 and 10 show the excursion and mooring tension behavior in damaged condition. It is observed that when the maximum loaded mooring line (line 1) is broken, the tension in the next adjacent line (line 2) increases and reaches a maximum of 60% of the ultimate breaking strength which is less than the maximum stipulated value of 80% of ultimate breaking strength as laid out in API RP 2 SK [10]. Also, the maximum offset of the platform after damaging the maximum loaded mooring line is approximately 52 m, which is less than the maximum recommended value of

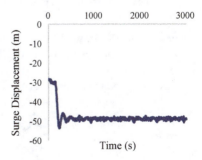

Fig. 9 Surge displacement in damaged condition

Fig. 10 Connector force ratios in damaged condition

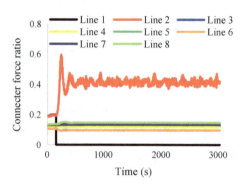

Fig. 11 Connector force ratios of different lines during survival condition

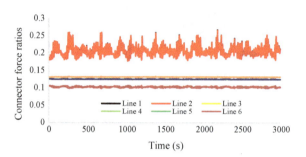

70 m (7% of water depth) as laid out in API RP 2 SK. Thus, tensions in mooring line and excursion of the platform are within the limit as prescribed in normal and damaged condition.

Mooring tensions are also checked in survival condition of the platform. JONSWAP spectrum with Hs = 17.3 m and Tp = 16.1 s, is used for representing the 100 year return period waves in survival condition. Same current profile is used as in operating condition. The maximum mooring line tension (Fig. 11), is approximately 26% of the ultimate breaking strength of the mooring line, which is less than the maximum recommended value of 60% as per API RP 2 SK.

3.4 Motions of Structure

Figures 12 and 13 show the time series variation of crucial heave and pitch motion of the platform in survival condition. Maximum heave and pitch motion is found to be less than 3 m and 3°, respectively. Resulting accelerations (maximum 0.5 m/s^2) were found to be well within the satisfactory limit of human tolerance.

Fig. 12 Platform heave motion

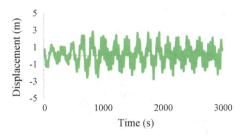

Fig. 13 Platform pitch motion

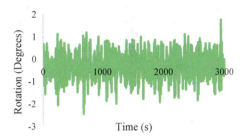

4 Conclusions

The analysis was carried out to design a suitable mooring system for the station keeping of a Spar platform housing a LTTD plant in operating and survival conditions. The numerical study was carried in a commercial computer code MOSES which is based on potential/diffraction theory. Several simulation iterations were carried out on given Spar configuration to arrive at an optimum mooring configuration. Spread mooring configuration of 8 mooring lines with each line consisting of a combination of steel chains and polyester ropes were arrived at. The platform motions and mooring line tensions were analyzed and found to be within the recommended limits as per API RP 2SK standard. Maximum tension in the mooring line was found to 20 and 26% of breaking tension in operating and survival condition. Platform static offset and tension in most heavily loaded mooring line in the damaged condition was found to be 52 m and 60% of breaking tension, respectively. The designed mooring lines consist of eight 2800 m long mooring lines, each having a pretension of 150t.

References

1. Sooraj (2015) Response analysis of SPAR platforms for a deep sea desalination plant. M. Tech Thesis, IIT Madras (2015)
2. Faltinsen O (1990) Sea Loads on Ships and Offshore Structure. Cambridge, Ocean Technology Series, Cambridge University Press, ISBN 0 521 45870 (paperback), UK. 1990
3. Haslum HA (2000) Simplified methods applied to nonlinear motion of spar platforms. Ph.D. thesis, Norwegian University of Science and Technology, Trondheim, Norway

4. Chen X, Zhang J, Ma W (2001) On dynamic coupling effects between SPAR and mooring lines. Ocean Eng 28:863–887
5. Jeon SH, Cho YU, Sea MW, Cho JR, Jeong WB (2013) Dynamic response of floating substructure of SPAR-type offshore wind turbine with catenary mooring cables. Ocean Eng 72:356–364
6. Elias S, Matsagar V, Datta TK (2017) Distributed multiple tuned mass dampers for wind response control of chimney with flexible foundation. Proc Eng 199:1641–1646
7. Banerjee A, Chakraborty T, Matsagar V (2019) Dynamic analysis of an offshore wind turbine under random wind and wave excitation with soil-structure interaction and blade tower coupling. Soil Dyn Earthquake Eng 125:105699
8. Banerjee A, Chakraborty T, Matsagar V (2019) Dynamic analysis of an offshore monopile foundation used as heat exchanger for energy extraction. Renewable Energy 131:518–548
9. Chakrabarti SK (1987) Hydrodynamics of offshore structures. Computational Mechanics Publication, GB, pp 168–232
10. API RP 2 SK (2005) Design and analysis of station keeping systems for floating structures

Advanced Mechanics of Plates and Shells

Finite Element Studies and Dynamic Analysis of Initially Stressed Functionally Graded Plates Using a Refined Higher Order Shear Deformation Theory

S. Jayaraman, Tarun Kant, and N. Shanmuga Sundaram

Abstract A C^0 isoparametric finite element formulation is presented to calculate the required number of lowest natural frequencies of functionally graded plates (FGPs) subjected to in-plane prestress. The material properties of the FGPs are assumed to vary continuously from one surface to another, according to a simple power law distribution in terms of the constituent volume fractions. The present theory is based on a higher-order displacement model and the three-dimensional Hooke's laws for plate material. The theory represents a more realistic quadratic variation of the transverse shearing and normal strains through the thickness of the plate. Nine-noded Lagrangian elements have been used for the purpose of discretization using a refined Higher Order Shear deformation Theory (HOST12) that includes the effects of transverse shear deformations, transverse normal deformation and rotary inertia. The plate structure is idealized into an assemblage of nine-noded isoparametric quadrilateral elements with twelve degrees of freedom per node. Hamilton's principle is used for the formulation. The effect of in-plane pre-stress is taken care by calculating the geometric stiffness matrix. The same shape functions are used to calculate the elastic stiffness matrix, geometric stiffness matrix and the element mass matrix. Subspace Iteration technique is applied to extract the natural frequencies. Numerical results for first seven natural frequencies are presented for rectangular plates under various boundary conditions. The results show good agreement with three-dimensional analytical formulation.

Keywords Finite elements · Functionally graded materials · Higher order theory

---

S. Jayaraman (✉) · N. S. Sundaram
U R Rao Satellite Centre, ISRO, Bengaluru 560 0017, India
e-mail: sjayaram@ursc.gov.in

T. Kant
IIT-Bombay, Mumbai 400 076, India

© The Editor(s) (if applicable) and The Author(s), under exclusive license to Springer Nature Singapore Pte Ltd. 2021
S. K. Saha and M. Mukherjee (eds.), *Recent Advances in Computational Mechanics and Simulations*, Lecture Notes in Civil Engineering 103,
https://doi.org/10.1007/978-981-15-8138-0_9

1 Introduction

Plates are structural elements of very small thickness as compared to other two dimensions. In general, plates are loaded in transverse direction. Plates fabricated with Functionally Graded Material (FGM) have a specified variation in material properties within the plate at any location. Generally, the variation of the properties in plates is only across the thickness. The gradient in mechanical properties is caused by the variation in chemical composition at different locations. An accurate prediction of dynamic behaviour of plates not only improves the safety but also reduce the material cost. Also, these plates will be experiencing several types of time varying loads of different intervals causing structures to vibrate. Hence it is required to extract the fundamental frequencies of vibration as well as response to the dynamic loads. Most of the structures referred above are subjected to the stresses due to static loads before being subjected to the dynamic loads. These stresses change the fundamental frequencies and structural response to the dynamic loads. It is essential to account for the pre-stress in the vibration analysis. The determination of the response of a plate structure or shell subjected to dynamic loads is a problem of immense practical significance in structural design. The evaluation of its free vibration response characteristics is a first step as a prelude to the prediction of the system's response to any excitations. Hence the evaluation of free vibration characteristics of pre-stressed plates has been presented in this paper. The effect of pre-stress by in-plane forces is to increase or decrease the stiffness for out of plane deformation depending on whether the stress is tensile or compressive.

FGMs possess smooth gradual variation in their mechanical properties such as elastic modulus, Poisson's ratio, coefficient of thermal expansion, tensile strength etc. In 1984, FGMs was proposed by material scientists in Japan for manufacturing thermal barrier materials [4]. FGMs composites prepared by using two different materials, heat-resistant ceramics to withstand the high-temperatures and highly conductive tough metals on the low-temperature side, with variation in composition being gradual from ceramic to metal. Therefore, FGMs are considered as advanced composites with inhomogeneous character at lower scales. Usually, the non metal will be ceramics like Zirconia, Silicon Carbide, Silicon Nitride and the metal will be Aluminium, Stainless Steel etc. Figure 1 shows that the variation of volume fraction of constituents from one surface to the other is gradual and smooth.

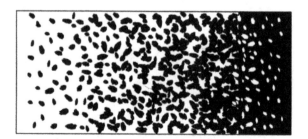

Fig. 1 Structure of functionally graded material

Plates with functionally graded materials are widely used because of their special property of resisting temperature gradient. Since there are no layers in between and no usage of adhesives, delaminating is not possible. Generally, they are used as thermal barriers. They are used in spacecraft and re-entry vehicles like space shuttle. Generally, the external surface of spacecraft will be experiencing the temperature of the order of 2100 K and hence, materials required at the surface will be designed so as to withstand temperatures at and around 2100 K and temperature gradients of about 1600 K. The other uses are circuit boards, solar panels etc. [4]. For power transmission purpose, gears are widely used to control the running speed and experience high stress concentration on the tooth surfaces in contact while in operation. Al-Qrimli et al. [1] studied and proved that the gears made out of FGMs give better hardness between two tooth surfaces owing to their microstructure augmentation and improved mechanical properties. Pre-stressing occurs in casting, stamping, forging, rolling, ball-peening, and welding operations of fabrication. It arises in buttressing, or any force fitting during assembly or erection. It may be caused by heating or cooling during service. Stress Stiffening can be produced by accumulated deformations under load, or by initial or residual stresses present before external load is applied. The term "stress stiffening" refers to a coupling between membrane stress and out of plane displacements associated with bending. The bending stiffness of a plate or shell is increased with the presence of tensile membrane while compressive membrane stress will decrease bending stiffness.

FGMs are classified depending on the direction and grading of the function. By varying the constituents continuously in a predetermined profile FGMs can be produced. Type of FGMs will be defined by the variation in constituents volume fractions. In most of the literature the following types of gradation functions are applied for the material properties of FGMs: Power-Functionally Graded Material, Sigmoidal Functionally Graded Material and Exponential Functionally Graded Material. Most of the researchers used the exponential function for describing the material properties:

$$E(z) = E_b e^{\lambda(z+h/2)} \text{ where } \lambda = \frac{1}{h} \ln\left[\frac{E_t}{E_b}\right] \qquad (1)$$

Hiroyuki [3] analyzed the fundamental frequencies, their mode shapes and related stresses for cross-ply composites subjected to in-plane pre-stresses of the plates using a global higher order theory accounting the effects of shear and normal deformations and also the influence of rotary inertia. The buckling stresses were obtained by increasing the compressive stresses values up to the zero natural frequency.

Based on Reddy's higher-order shear deformation plate theory, Yang et al. [14] performed large vibration analyses of Functionally Graded laminate plates subjected to pre-stress. Praveen and Reddy [6] studied the effects of transverse shear strains, rotary inertia and moderately large rotations in the von-Karman theory on the behaviour of FG plate. Static and dynamic responses of the FG plate are investigated using plate finite element method by varying the constituent volume fraction of the FGMs.

3-D elastic solutions of rectangular plate made out of FGMs under both mechanical and thermal loads on either/both sides of plate are studied by Reddy and Cheng [8] and Vel and Batra [10] Vel and Batra [11]. An exact solution for the cylindrical bending vibration of FG plates are suggested by them. Using first-order shear deformation theory (FOST) by Mindlin, Reddy et al. [7] examined axisymmetric bending and elastic deformations of circular FG plates. Bending vibration problems solved by both the classical theory (CPT) and the FOST and the relationships between the solutions are exact. Cheng and Batra [2] analyzed the steady state and buckling vibrations of a FG polygonal plate under simply supported conditions experiencing uniform in-plane hydrostatic loadings with Reddy's third order shear deformation theory (TSDT).

Yang et al. [13] predicted the responses of rectangular functionally graded thin plate under dynamic loads using the small deflection plate theory. Venkat [12] have worked on FGM beams and plates with CPT and HOST. Static analysis of FGM beams and plates and their free vibration characteristics have been studied using both HOST and FOST. Response prediction of moderately thick and thick composite plates or composite plates of anisotropic in nature is not possible using the Classical Plate Theories that are based on Kirchhoff hypothesis. For dynamic analysis of plates, shear correction factors are essential to correct the transverse shear stiffness and then obtain the gross responses from FOST by Mindlin (the shear strains in thickness direction are not varying). Hence the shear correction factors will influence the accuracy of solution. Reddy proposed a higher order plate theory, Third order Shear Deformation Theory (TSDT), the shear stresses are having quadratic variation through the transverse direction and accordingly, there is no requirement of any shear correction factors. Since the displacement model that neglects the transverse normal stress effect and satisfying the zero transverse shear stress on the bounding planes of the plate results in the second order derivatives of transverse displacement in the energy terms and thus, finite element formulation demands C1 continuous basis functions which are not preferable computationally as compared to the isoparametric formulation which is simpler and easily programmable.

The higher order shear deformation plate theory, **HOST12** proposed by Kant et.al, is based on the displacement model that contains cubic variation of displacements in each co-ordinate direction and the 3-D Hooke's law for material models. The theory represents variation of the transverse shear in quadratic form and normal strains in the thickness direction. The theory requires no shear correction factor and ultimately resulting in good estimation of shear strain energy. Vibration problems of both thin and thick plates can be solved using this formulation. Vibration analysis carried out using a **HOST12** that includes transverse shear deformations, transverse normal deformation and also the rotary inertia. Nine-noded Lagrangian elements have been chosen for the simulation case studies. Hence, in this present work, the normal modes analysis of pre-stressed FG plates using **HOST12** is considered, which is a gap found in the literature for vibration analyses. Here the aim is to formulate an element which is based on **HOST12**. Analytical solution has been developed for this purpose. Analysis results are validated by comparing with that of exact solutions. Also the results will be compared with other theories such as FOST, TSDT and 3D elasticity solutions to find the difference in error.

2 Displacement Fields for Higher Order Theories

Different number of terms is used for displacement field comprising all three directions by various researchers as per the conditions of problem and accuracy required. Figure 2 shows the positive set of displacement components in each direction and the following equations represent the displacement field associated with HOST12.

2.1 Higher Order Shear Deformation Theory (HOST12)

$$u(x, y, z) = u_0(x, y) + z\theta_x(x, y) + z^2 u_0^*(x, y) + z^3 \theta_x^*(x, y)$$
$$v(x, y, z) = v_0(x, y) + z\theta_y(x, y) + z^2 v_0^*(x, y) + z^3 \theta_y^*(x, y)$$
$$w(x, y, z) = w_0(x, y) + z\theta_z(x, y) + z^2 w_0^*(x, y) + z^3 \theta_z^*(x, y) \quad (2)$$

2.2 Finite Element Formulation

The region of interest is subdivided into sub regions known as finite elements. Within each sub region, the variables are assumed in the form of functions. Here, the derivations for the stiffness matrix, geometric stiffness matrix due to pre-stress and mass matrix are essential for the vibration analysis. Using these three matrices, an eigen

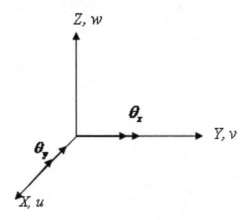

Fig. 2 Positive set of displacement components

value problem is formulated and then solved. Total stiffness matrix consists of three components namely:

$$[\underline{K}_T] = [\underline{K}_b] + [\underline{K}_s] \pm [\underline{K}_g] \qquad (3)$$

where $\underline{K}_e = \int_{V_e} \underline{B}^T . \underline{D}_e . \underline{B} \, dV; \underline{K}_s = \int_{V_s} \underline{B}^T . \underline{D}_s . \underline{B} \, dV; \underline{K}_g = \int_{V_g} \underline{N}^T . \sigma^0 . \underline{N}_g \, dV$

and $\underline{K}_e$ is stiffness matrix due to bending, $\underline{K}_s$ is Element elastic stiffness matrix due to shear and $\underline{K}_g$ is Geometric stiffness matrix.

2.3 Nine Node Iso-Parametric Quadrilateral Element

The displacement within an element is discretized in such a way that

$$\delta = \sum_{i=1}^{NN} N_i \delta_i \qquad (4)$$

where, N is the number of nodes in each element; N_i is the iso-parametric basis function related to node i in terms of ξ and; δ_i is generalized displacement.

The nine noded isoparametric quadrilateral element is as shown in Fig. 3. In order to solve the fundamental equations presented earlier a finite element displacement formulation can be conveniently used. The element properties are derived from the principle of minimum potential energy by assuming a displacement field which satisfies the criteria of completeness within the element and compatibility across the

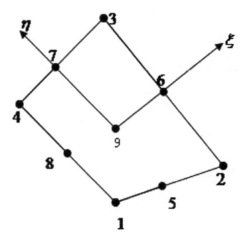

Fig. 3 Quadrilateral element with nine nodes

element boundaries. There are twelve degrees of freedom for each node of the finite element: [u_o, v_o, $w_o, \theta_x, \theta_y, \theta_z, u_o^*, v_o^*, w_o^*, \theta_x^*, \theta_y^*, \theta_z^*$]

2.4 Vibration Analysis

The vibration problem can be formulated by using Hamilton's principle [5] as follows:

$\delta(U - T) = 0$ where U is the Strain Energy and T is the Kinetic Energy

The governing equation for the natural vibration analysis (undamped) thus becomes,

$$[K_T]\{d\} - \omega^2[M]\{d\} = 0 \tag{5}$$

Knowing $[K_T]$ and $[M]$, the natural frequency and the associated mode shapes will be found by Subspace iteration technique.

3 Numerical Investigation and Results

Finite Element programming was carried out using MATLAB [15], since no commercial software package available supports the modelling of this special type of materials. The geometric properties are:

Length, a: 4 m, Width, b: 4 m and Thickness, h: 1 m. Table 1 gives the various material properties used for the analysis of FGPs. The exponential variation of Young's modulus is given as, $E = E_b e^{\lambda(z+h/2)}$.

The non-dimensional natural frequency is given by the formula [9], $\bar{\omega} = \omega * a^2 * \sqrt{(\rho * h)/D}$. Figure 4 shows the convergence of the first fundamental frequency of an FGM plate made out of Ceramic (Zirconia) and aluminium with increase in

Table 1 Material properties

S.No.	Material	Young's Modulus, E (GPa)	Poisson's ratio, ν	Density, ρ (kg/m^3)
1	Aluminium	70	0.3	2707
2	Ceramic (Zirconia)	151	0.3	3000
3	FG plate (Al + Ceramic)	70/151	0.3	2707/ 3000

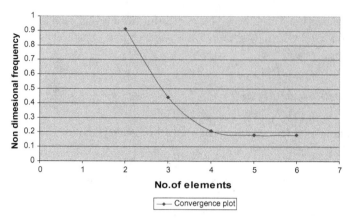

Fig. 4 Convergence plot for first natural frequency

the number of elements. From the plot, it can be concluded that it is an upper bound solution.

Table 2 summarises the comparison of results from all plate theories with that of 3D analytical solution and it can be concluded that HOST12 values are much closer to the analytical solution. Tables 3 and 4 give the first seven fundamental frequencies for different boundary and aspect ratio.

Table 2 Comparison of non-dimensional natural frequencies from various theories

Mode	3D Elasticity solution	HOST12	TSDT	FOST
1	0.0932	0.0932(0.0)*	0.0930(−0.21)	0.0929(−0.25)
2	0.2226	0.2226(0.0)	0.2220(−0.27)	0.2218(−0.33)
3	0.3421	0.3421(0.0)	0.3406(−0.44)	0.3403(−0.52)
4	0.4171	0.4172(0.02)	0.4151(−0.48)	0.4145(−0.63)
5	0.5239	0.5240(0.02)	0.5208(−0.59)	0.5202(−0.71)

*gives % error w.r.to 3D Elasticity solution

Table 3 Variation of Non-dimensional frequencies with 'n' for CCCC FGM square plates with biaxial tensile load ($N_x = 0.5$, $N_y = 0.5$)

	ω_1	ω_2	ω_3	ω_4	ω_5	ω_6	ω_7
Si3N4	53.26	100.89	100.89	144.25	172.44	172.44	213.66
n = 0.2	50.27	95.23	95.23	136.16	162.77	162.77	201.68
n = 3.0	47.14	89.29	89.29	127.67	152.62	152.62	189.10
n = 10	46.36	87.82	87.82	125.57	150.10	150.10	185.99
Steel	43.78	82.94	82.94	118.59	141.76	141.76	175.65

Table 4 Variation of Non-dimensional frequencies with 'n' for CSCS FGM rectangular plates with aspect ratio, 0.5

	ω_1	ω_2	ω_3	ω_4	ω_5	ω_6	ω_7
n = 0.2	15.71	27.16	44.42	48.91	59.38	67.35	76.17
n = 3.0	14.73	25.46	41.65	45.86	55.68	63.15	71.42
n = 10	14.49	25.04	40.96	45.11	54.76	62.11	70.24

4 Conclusions

A refined higher order displacement model (**HOST12**) using the simple C^0 isoparametric finite element formulation is more accurate than CPT, FOST and TSDT in predicting the natural frequencies of any FGPs. The results obtained by HOST12 are almost equal to 3-D Elasticity solutions unlike other theories. In contrast to the FOST, the present HOST needs no shear correction coefficient because of representation of the transverse shear deformation in quadratic variation. Good match is observed between the predicted simulation results with analytical results for isotropic plates, hence it can be concluded that the HOST12 is accurate enough to extract the natural frequencies of the FGPs. The error in predicting lower modes is lower as compared to higher modes. The volume fraction index, initial stresses and the boundary conditions have considerable influence on the dynamic characteristics of the FGPs. The natural frequencies of the FGPs lie in between the values of natural frequencies of the corresponding isotropic materials used in the manufacturing of the FGPs. The fundamental frequencies of the FGPs keep decreasing as the volume fraction index increases.

References

1. Al-Qrimli H, Oshkour A, Ismail F, Mahdi F (2015) Material design consideration for gear component using functionally graded materials. Int J Mater Eng Innov 6(4):243–256
2. Cheng ZQ, Batra RC (2000) Three-dimensional thermo-elastic deformations of a functionally graded elliptic plates. Compos B 31:97–106
3. Hiroyuki M (2001) Vibration of cross-ply laminated composite plates subjected to initial in-plane stresses. Thin-walled Struct 40:557–571
4. Koizumi M (1997) FGM activities in Japan. Compos B 28B:1–4
5. Petyt M (1990) Introduction to finite element vibration analysis. Cambridge University Press
6. Praveen GN, Reddy JN (1998) Nonlinear transient thermoelastic analysis of functionally graded ceramic-metal plates, International Journal of Solids and Structures, 35(33):4457-4476
7. Reddy JN, Wang CM, Kitipornchai S (1999) Axisymmetric bending of functionally graded circular and annular plates. Eur J Mech A Solids 18(1):185-199
8. Reddy JN, Cheng ZQ (2001) Three-dimensional thermo-mechanical deformations of functionally graded rectangular plates. Eur J Mech Solids 20:841–855
9. Cook, RD (1995) Finite element modeling for stress analysis. John Wiley & Sons, Inc
10. Vel SS, Batra RC (2002) Exact solutions for thermo-elastic deformations of functionally graded thick rectangular plates. AIAA J 40(7):1421–1433

11. Vel SS, Batra RC (2003) Three-dimensional analysis of transient thermal stresses in functionally graded rectangular plates. Int J Solids Struct 40(25):7181-7196
12. Venkat S (2005) Mechanics of functionally graded beams and plates M.Tech Dissertation, Deptt. of Civil Engg., IIT-Bombay
13. Yang J, Shen HS (2001) Dynamic response of initially stresses functionally graded rec-tangular thin plates. Compos Struct 54:497-508
14. Yang J, Kitipornchai S, Liew KM (2003) Large amplitude vibration of thermo-electro-mechanically stresses FGM laminated plates. Comput Method Appl Mech Eng 192:3681–3685
15. Young Kwon W, Bang H (1997) Finite element method using MATLAB. CRC Press

Vibration Analysis of Functionally Graded Material Plate

N. I. Narayanan, Sauvik Banerjee, Akshay Prakash Kalgutkar, and T. Rajanna

Abstract Functionally graded materials (FGMs) are the advanced materials in the family of engineering composites made of two or more constituent phases with continuous and smoothly varying composition. Nine noded heterosis plate element is used to formulate the elastic stiffness matrix and mass matrix. The results are also extracted from Abaqus CAE by using S8R5 shell elements. Free vibration analysis is done to obtain the different modes as well as the frequencies. Harmonic sine load is applied at the centre of the FGM plate to obtain a forced vibration response. Impulse forces of rectangular, half-cycle sine, triangular shapes are applied on the top of the plate at the centre and the shock spectra of C-Si C FGM plate is plotted.

Keywords FGMs · Heterosis plate element · Shock spectra

1 Introduction

The diverse and potential applications of FGMs in aerospace, medicine, defence, energy, and other industries have attracted a lot of attention recently. The application of these advanced materials was first visualized during a spaceplane project in 1984 in National Aerospace Laboratory of Japan to avoid the stress peaks at interfaces in coated panels for the space shuttle. Combination of materials used here served the purpose of a thermal barrier system capable of withstanding a surface temperature of 2000 K with a temperature gradient of 1000 K across a 10 mm thick section [1]. Later on, its applications have been expanded to also the components of chemical

---

N. I. Narayanan (✉) · S. Banerjee · A. P. Kalgutkar
Department of Civil Engineering, Indian Institute of Technology Bombay, Mumbai 400076, India
e-mail: 174043001@iitb.ac.in

N. I. Narayanan
Government College of Engineering, Kannur 670563, Kerala, India

T. Rajanna
B. M. S. College of Engineering, Bengaluru 560019, Karnataka, India

© The Editor(s) (if applicable) and The Author(s), under exclusive license to Springer Nature Singapore Pte Ltd. 2021
S. K. Saha and M. Mukherjee (eds.), *Recent Advances in Computational Mechanics and Simulations*, Lecture Notes in Civil Engineering 103,
https://doi.org/10.1007/978-981-15-8138-0_10

plants, solar energy generators, heat exchangers, nuclear reactors and high-efficiency combustion systems. The concept of FGMs has been successfully applied in thermal barrier coatings where requirements are aimed to improve thermal, oxidation and corrosion resistance. FGMs can also find application in communication and information techniques. Abrasive tools for metal and stone cutting are other important examples where the gradation of the surface layer has improved performance.

The variations of the volume fractions through the thickness are assumed to follow a power-law function. The Reissener-Mindlin first-order shear deformation theory is very much appropriate for thick plates [2]. It was taken to analyze the behaviour of the plate subjected to free and forced vibration. It has been found from the literature that not many studies are done to the vibration analysis of functionally graded plates. Reddy et al. [3] carried out the free vibration analysis of functionally graded plates. They have developed analytical formulations and solutions for the free vibration analysis of functionally graded plates using higher-order shear deformation theory (HSDT). The principle of virtual work was used to derive the equations of equilibrium and boundary conditions. Navier's technique was used to obtain the solutions for FGM plates. Vimal et al. [4] have studied the free vibration analysis of functionally graded skew plates using the finite element method. The first-order shear deformation plate theory is used to consider the transverse shear effect and rotary inertia. The properties of functionally graded skew plates are assumed to vary through the thickness according to a power law. It is found that when the length to thickness ratio of functionally graded skew plates increases beyond 25, the variation in the frequency parameter is very negligible and also found that a volume fraction exponent that ranges between 0 and 5 has a significant influence on the frequency. Gulshan et al. [5] carried out a free vibration analysis of functionally graded material (FGM) skew plates subjected to the thermal environment. It was concluded that the volume fraction index and skew angle plays an important role in predicting the vibration of FGM skew plate subjected to thermal load.

Reddy [6] have studied theoretical formulation and FEM model based on TSDT for FGM plate. The formulation accounted for thermo-mechanical effects combining change with time and geometric nonlinearity. In this higher-order theory, transverse shear stress was expressed as a quadratic function along with the depth. Hence this theory requires no shear correction factor. The plate was considered as the homogenous and material composition was varied along with the thickness. The Young's modulus was assumed to vary as per rule of the mixture in terms of the volume fractions of the material constituents. Hughes and Cohen [7] developed the heterosis element and elemental equation. They derived lumped positive definite mass matrix, element matrix and load vector and method for finding critical time step. High-accuracy finite element for thick and thin plate bending is developed, based upon Mindlin plate theory.

It has been found from the literature survey that not many researchers attempted to the vibration analysis of functionally graded plates. Further, we observed that many authors could model such problems with a stepped variation in material properties instead of continuous variation. This would have happened because of the limitations of the commercial software available. In this context, we felt that MATLAB code could be used for tailoring the continuous variation in material properties in FE Modelling. Hence MATLAB code was developed for vibration analysis of FG plate. The analysis was carried out for C-Si C FGM plate with different volume fraction indices. The results are compared with Abaqus CAE by using S8R5 shell elements.

2 Problem Formulation

First-order shear deformation theory is used for plate formulation. Displacement variation is linear, across the plate thickness. But there is no change in plate thickness during deformation. A further assumption is that the normal stress across the thickness is neglected. Properties are graded through the thickness direction which follows a volume fraction power-law distribution. The different elements of the plate are expected to undergo translational and rotational displacement. In the present work 9- noded heterosis element is used to discretize the plate.

2.1 Strain-Displacement Relations

The displacement field at any arbitrary distance z from the midplane based on the first-order shear deformation plate theory is given by

$$[\bar{u}_p(x, y, z), \bar{v}_p(x, y, z), \bar{w}_p(x,y,z)] = [u_0(x, y), v_0(x, y), w_0(x, y)] + z[\theta_x(x, y), \theta_y(x, y), 0] \quad (1)$$

where, $\bar{u}_p$, $\bar{v}_p$, $\bar{w}_p$ are displacements in x, y and z directions respectively, u_0, v_0 and w_0 are the associated midplane displacements along x, y and z axes respectively and θ_x and θ_y are the rotations about y and x-axes respectively.

The linear strain displacement relations are given by

$$\begin{aligned} \varepsilon_{xl} &= u_{0,x} + z\chi_x \\ \varepsilon_{yl} &= v_{0,y} + z\chi_y \\ \gamma_{xyl} &= u_{0,y} + v_{0,x} + z\chi_{xy} \\ \gamma_{xzl} &= w_{0,x} + \theta_x \\ \gamma_{yzl} &= w_{0,y} + \theta_y \end{aligned} \quad (2)$$

where, ε_{xl}, ε_{yl} and γ_{xyl} are the linear in-plane normal and shear strains, γ_{xzl} and γ_{xzl} are transverse shear strains, z is the distance of any layer from the middle plane of the plate and χ are the curvatures.

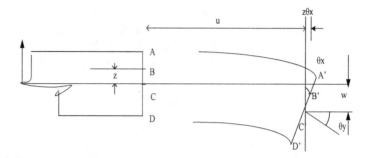

Fig. 1 Deformed and un-deformed beam

$$\{\chi\} = \begin{Bmatrix} \chi_x \\ \chi_y \\ \chi_{xy} \end{Bmatrix} = \begin{Bmatrix} \frac{\partial \theta_x}{\partial x} \\ \frac{\partial \theta_y}{\partial y} \\ \frac{\partial \theta_x}{\partial y} + \frac{\partial \theta_y}{\partial x} \end{Bmatrix} \quad (3)$$

The strain-displacement field at any distance z as shown in Fig. 1.

2.2 Finite Element Formulation

In the current work, the FGM plate has been discretized using 9-noded heterosis element with 5-degree of freedom (dofs) at all the edge nodes and 4 dofs at the internal node as shown in the Fig. 2. The serendipity shape functions have been used for the transverse dofs, w, and Lagrange shape function are used in the remaining dofs, u, v, θ_x, and θ_y

Resultant Forces and moments. The analysis of FGM plate is carried out to establish the relation between the forces and strains by considering transverse shear terms.

Constitutive matrix of the isotropic plate is

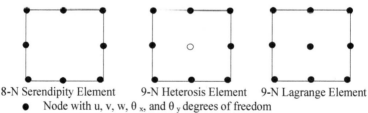

8-N Serendipity Element   9-N Heterosis Element   9-N Lagrange Element

●   Node with u, v, w, θ_x, and θ_y degrees of freedom
○   Node with u, v, θ_x, and θ_y degrees of freedom

Fig. 2 Nodal configuration of the plate element

$$Q = \begin{bmatrix} Q_{11} & Q_{12} & 0 \\ Q_{12} & Q_{11} & 0 \\ 0 & 0 & Q_{66} \end{bmatrix} \qquad (4)$$

where,

$$Q_{11} = \frac{E}{1-\nu^2}, \quad Q_{12} = \frac{\nu E}{1-\nu^2}, \quad Q_{66} = \frac{E}{2(1+\nu)}$$

The material properties P_z (Elastic constants E,ν, density) at distance, z from the middle surface of the plate is

$$P_z = P_b + (P_t - P_b)\left[\frac{(2z+h)}{2h}\right]^k = P_b + (P_t - P_b)Vf \qquad (5)$$

where, h is the plate thickness, t and b denotes the top and the bottom surface ($\pm z/2$), k is material volume fraction index, V_f is volume fraction.

Stress-strain relationship is

$$\{\sigma\} = [Q]\{\varepsilon\} \qquad (6)$$

where, $\{\sigma\} = \{\sigma_x, \sigma_y, \tau_{xy}\}^T, \varepsilon = \varepsilon_0 + z\chi$.

The in-plane resultant forces and moments in the kth layer are evaluated as

$$\{N, M\} = \sum_{k=1}^{N} \int_{z_{k-1}}^{z_k} \{\sigma\}(1, z)\, dz \qquad (7)$$

Resultant Transverse Shear Force on the kth layer is given by

$$\begin{bmatrix} Q_{xz} \\ Q_{yz} \end{bmatrix} = \sum_{k=1}^{N} \int_{z_{k-1}}^{z_k} \begin{Bmatrix} \tau_{xz} \\ \tau_{yz} \end{Bmatrix} dz = \sum_{k=1}^{N} \int_{z_{k-1}}^{z_k} \begin{bmatrix} Q_{44} & Q_{45} \\ Q_{45} & Q_{55} \end{bmatrix} \begin{Bmatrix} \gamma_{xz} \\ \gamma_{yz} \end{Bmatrix} dz$$

$$(8)$$

$$Q_{44} = (G_{13t} - G_{13b})\left[\frac{(2z+h)}{2h}\right]^k + G_{13b}$$

$$Q_{55} = (G_{23t} - G_{23b})\left[\frac{(2z+h)}{2h}\right]^k + G_{23b}$$

$$Q_{45} = 0 \qquad (9)$$

The constitutive relation for FGM plate is given by

$$\{N\} = [C]\{\varepsilon\} \tag{10}$$

where, $\{N\} = \begin{bmatrix} Nx, Ny, Nxy, Mx, My, Mxy, Qxz, Qyz \end{bmatrix}^T$ represents the in-plane stress resultants (N), out of plane bending moments (M) and shear resultants (Q). Here, [C] is the constitutive matrix [8] of the FGM plate. To compensate for the parabolic shear stress variation across the thickness of the plate, a correction factor of 5/6 is used in the shear-shear coupling components of the constitutive matrix [9]. Using Green-Lagrange's strain-displacement expression [10], the linear strain-displacement matrix[B] have been worked out.

The different participating element-level matrices such as elastic stiffness matrix [ke], and consistent mass matrix [me] have been derived using corresponding energy expression.

The element elastic stiffness matrix and element mass matrix are derived using the following relations

$$[ke] = \int_{-1}^{1} \int_{-1}^{1} [B]^T [C][B] |J| d\varepsilon d\eta \tag{11}$$

$$[me] = \int_{-1}^{1} \int_{-1}^{1} [\bar{N}]^T [I][\bar{N}] |J| d\varepsilon d\eta \tag{12}$$

In which, [I] is the inertia matrix

Computer coding and Implementation. To perform all the computations, a computer program is developed using MATLAB to implement the finite element formulation and include all the necessary parameters to investigate the vibration behaviour of the FGM plate.

In the present code, selective integration scheme is incorporated for the generation of the element stiffness matrix. The 3×3 Gauss quadrature rule is adopted to get the bending terms and 2×2 Gauss rule is used to solve shear terms in order to avoid possible shear locking. The mass matrix is evaluated by using 3×3 Gauss rule [11].

2.3 Formulation of Dynamic Problems

Validation of assembled stiffness matrix is done through bending problems and that of mass matrix is done through vibration problems. In order to validate the formulation of mass matrix, one has to solve a free vibration problem by incorporating the validated elastic stiffness matrix. The standard governing equation in matrix form for the deflection problem is

$$[K_e]\{q\} = \{P\} \tag{13}$$

{P} is the nodal load vector, $[K_e]$ is the system elastic stiffness matrix. For a given set of loads, the displacement $\{q\}$ can be determined using the above equation. If the displacement vector is validated, it ensures the correctness of formulation and coding of the stiffness matrix.

The standard governing equation in matrix form for the free vibration problem is

$$[M]\{\ddot{q}\} + +[K_e]\{q\} = \{P\} \tag{14}$$

The standard governing equation in matrix form for the force vibration problem is

$$[M]\{\ddot{q}\} + [C]\{\dot{q}\} + [K_e]\{q\} = \{P\} \tag{15}$$

$[M]$, $[K_e]$ and $[C]$ represents global mass matrix, global stiffness matrix and damping matrix respectively.

$$[C] = \alpha[M] + \beta[K_e] \tag{16}$$

where, α β and are the Rayleigh damping coefficients. From this, we can solve the forced vibration problem. From this, we can solve the force vibration problem using Newmark-beta method.

Newmark-beta technique. The problem is solved here by using constant acceleration method as it is unconditionally stable. The stability criteria for constant acceleration method is given by

$$\frac{\Delta t}{T_n} \leq \frac{1}{\pi\sqrt{2}} \frac{1}{\pi\sqrt{\gamma - 2\beta}}$$

For $\gamma = \frac{1}{2}$ and $\beta = \frac{1}{4}$ this condition becomes $\frac{\Delta t}{T_n} < \infty$

3 Results and Discussion

The properties of FGM plates are graded through the thickness direction according to a volume fraction power law distribution (Fig. 3).

3.1 Free Vibration Analysis

The heterosis element is used in the code for free vibration analysis. For validation of the present code, the data available for the functionally graded plate aluminium

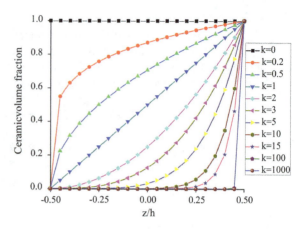

Fig. 3 Variation of volume fraction with the non-dimensional thickness

Table 1 Material properties of Aluminium Oxide–Titanium alloy FGM plate

Material	E(N/m2)	ρ(kg/m3)	υ
Ti-6A1-4V (ceramic)	122.56×10^9	4429	0.2884
Aluminium oxide	349.55×10^9	3750	0.26

Table 2 Variation of fundamental frequency with k values –simply supported plate-comparison

Power-law index	Fundamental frequency (Hz)		
	9-NHE	Simulation	He (2001)
k = 0 (Titanium)	141.34	146.69	144.25
k = 1	196.95	201.64	198.92
k = 15	245.41	250.56	247.3
k = 1000 (Alumina)	259.26	264.5	261.73

oxide –titanium alloy of size 0.4 m × 0.4 m × 0.005 m available in the literature of He et al. [12] is used. In numerical simulation by Abaqus, S8R5 element has been used. Table 1 shows the material properties. Table 2 validated the code with literature and simulation.

The present code is validated with results of He et al. (2001). The simulation results are also in good agreement with results obtained from FEM coding. This ensures the correctness of the formulation of the stiffness and mass matrix.

3.2 Free Vibration Analysis of C-Si C Plate

The analysis is done for C-Si C plate (0.5 × 0.5 × 0.001 m). Material properties are given in Table 3. Convergence results are shown in Fig. 4. First four mode of vibration shown in Fig. 5. Frequency of Vibration is minimum for carbon plate as shown in Table 4.

Table 3 Material properties C-Si C FGM plate

Material	E(G Pa)	ν	ρ (Kg/m^3)
Si-C (Ceramic)	320	0.3	3220
C(Metal)	28	0.3	1780

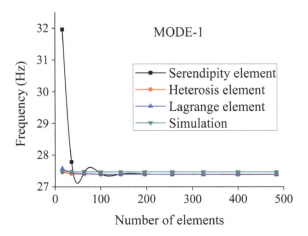

Fig. 4 Convergence of fundamental frequency of simply supported C-Si C FGM plate (k = 2)

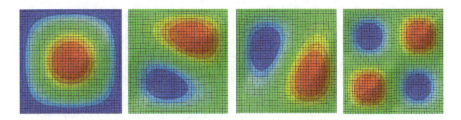

Fig. 5 First 4 mode shapes of simply supported C-Si C FGM plate (k = 2)-Simulation

3.3 Forced Vibration Analysis

Forced vibration analysis was carried out at the centre of the plate using harmonic sine loading and different impulse loadings.

Harmonic Sine Wave Loading. A harmonic force P(t) = P_0 sin(ωt) load is applied at the centre of the plate, where P_0 is the amplitude or peak value of the force and ω is the forcing frequency. $T = 2\pi/\omega$ is the forcing period of the FGM plate $P_0 = 1$ N and $\omega = 2\pi f$ where ω is circular frequency and f is natural frequency of the Plate. Figure 6 shows that maximum displacement at the centre of the plate increases with the material index (k value). Table 5 compares the un-damped and damped cases.

Table 4 Variation of the natural frequencies (Hz) of FGM simply supported Square Plate for different k values. 22 × 22 mesh-Heterosis element(FEM)

Mode no	k = 0 Si C	k = 2	k = 15	k = 1000 Carbon
1	37.91	27.39	22.2114	15.08
2	94.77	60.84	51.9549	37.7
3	94.77	60.84	51.9549	37.7
4	151.63	98.77	83.7656	60.33
5	189.55	121.26	103.718	75.41
6	189.55	123.04	104.522	75.41
7	246.41	158.98	135.449	98.03
8	246.41	158.98	135.449	98.03
9	322.27	202.97	174.955	128.21
10	322.27	202.97	174.955	128.21

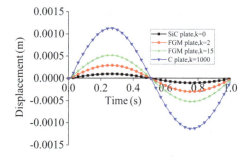

Fig. 6 Response of simply supported C-Si C FGM plate subjected to harmonic loading

Table 5 Displacement of simply supported C-Si C plate for different k values

Power-law index	Maximum displacement at centre (m)			
	CCCC		SSSS	
	Undamped	Damped	Undamped	Damped
k = 0 (Si C)	4.759×10^{-5}	4.73×10^{-5}	1.165×10^{-4}	1.006×10^{-4}
k = 2	1.722×10^{-4}	1.71×10^{-4}	3.142×10^{-4}	3.035×10^{-4}
k = 15	2.796×10^{-4}	2.68×10^{-4}	5.222×10^{-4}	5.201×10^{-4}
k = 1000(C)	5.442×10^{-4}	5.41×10^{-4}	1.229×10^{-3}	1.199×10^{-3}

Impulse loading. A very large force that acts for a very short time but with a time integral that is finite is called an impulse force. Impulse forces of rectangular, half-cycle sine, triangular shapes each with the same value of maximum force 1 N is applied at the centre of the plate The response behavior of FGM plate is studied for material index(k) value = 2. t_d is pulse duration. T_n is the natural time period of

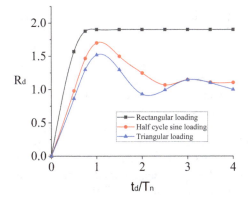

Fig. 7 Shock spectra of simply supported C-Si C FGM Plate (material index k = 2)

Table 6 Variation of deformation response factor (R_d) with t_d/T_n values (k = 2)

t_d/T_n	$R_d = u_0/u_{st0}$		
	Rectangular loading	Half sine loading	Triangular loading
0	0	0	0
0.5	1.569	0.982	0.863
0.75	1.876	1.469	1.3
1	1.901	1.701	1.52
1.5	1.901	1.5	1.298
2	1.901	1.25	0.93
2.5	1.901	1.071	0.996
3	1.901	1.15	1.148
3.5	1.901	1.111	1.103
4	1.901	1.108	1.003

vibration of the plate and u_{st0} is the static deflection of the plate. Static deflection is 1.438e-4 m and Natural time period is 0.0365 s. The shock spectra of the FGM plate with material index k = 2 is shown in Fig. 7. Variation of response factor with t_d/T_n, for k = 2 is given in Table 6. The present results are in good agreement with the available literature [13].

4 Conclusions

In the present investigation, a finite element formulation is developed for vibration analysis of FGM plates. The analysis is carried out by developing a computer program in MATLAB. A 9- noded heterosis element is used to model the FGM plate. The heterosis element exhibits improved characteristics as compared to the 8- noded serendipity and 9-noded Lagrange elements. It offers a high level of accuracy for extremely thin plate configurations. Convergence study has been carried out for ensuring the convergence of the numerical results. The results are also extracted from Abaqus CAE by using S8R5 shell elements and are in very good agreement with the developed elements. Free vibration analysis is done to study the different modes as well as frequencies. It is observed that free vibration response is minimum for carbon and maximum for Silicon carbide plate. The central deflection of the plate increases with increase in volume fraction index for all types of boundary conditions. From the shock spectra, it is clearly understood that if the pulse duration (t_d) is longer than $T_n/2$, the overall maximum deformation occurs during the pulse. Then the pulse shape is of great significance. For the larger value of t_d/T_n, the overall maximum deformation is influenced by the rapidity of the loading. The rectangular pulse in which the force increases suddenly from zero to maximum show the large deformation. The triangular pulse in which the increase in the force is initially slowest among the three pulses produces the smallest deformation. The half-cycle sine pulse in which the force initially increases at an intermediate rate causes deformation that for many values of t_d/T_n is larger than the response of the triangular pulse.

References

1. Jha DK, Kant T, Singh RK (2013) A critical review of recent research on function ally graded plates. Compos Struct 96:833–849
2. Wang CM, Reddy JN, Lee KH (2000) Shear deformable beams and plates relationships with classical solutions. Elsevier
3. Reddy BS, Kumar JS, Reddy EC, Reddy KVK (2014) Free vibration behaviour of functionally graded plates using higher-order shear deformation theory. J Appl Sci Eng 17(3):231–241
4. Vimal J, Srivastava RK, Bhatt AD, Sharma AK (2014) Free vibration analysis of moderately thick functionally graded skew plates. Eng Solid Mech 2:229–238
5. Gulshan MNA, Chakrabarti A, Prakash V (2014) Vibration characteristics of functionally graded material skew plate in thermal environment, world academy of science, engineering and technology. Int J Mech Mechatron Eng 8:142–153
6. Reddy JN (2000) Int J Numer Methods Eng 47:663–684
7. Hughes TRJ, Cohen M (1978) The heterosis finite element for plate bending. Comput Struct 9:445–450
8. Reddy N (1996) Mechanics of laminated composite plates. CRC Press
9. Lal R, Saini R (2013) Buckling and vibration of non-homogeneous rectangular plates subjected to linearly varying in-plane force. Shock Vib 20(5):879–894
10. Bathe KJ (1996) Finite element procedures. Prentice-Hall, Engle-wood Cliffs
11. Cook RD (2007) Concepts and applications of finite element analysis. Wiley
12. He XQ, Ng TY, Sivashanker S, Liew KM (2001) Active control of FGM plates with integrated piezoelectric sensors and actuators. Int J Solids Struct 38:1641–1655
13. Chopra AK (2007) Dynamics of structures, theory and applications to earthquake engineering. Pearson Education South Asia, New Delhi

ANN-Based Random First-Ply Failure Analyses of Laminated Composite Plates

Subrata Kushari, A. Chakraborty, T. Mukhopadhyay, S. R. Maity, and S. Dey

Abstract This paper presents the random first-ply failure analyses of laminated composite plates by using an artificial neural network (ANN)-based surrogate model. In general, materials and geometric uncertainties are unavoidable in such structures due to their inherent anisotropy and randomness in system configuration. To map such variabilities, stochastic analysis corroborates the fact of inevitable edge towards the quantification of uncertainties. In the present study, the finite element formulation is derived based on the consideration of eight-noded elements wherein each node consists of five degrees of freedom (DOF). The five failure criteria namely, maximum stress theory, maximum strain theory, Tsai-Hill (energy-based criterion) theory, Tsai-Wu (interaction tensor polynomial) theory and Tsai-Hill's Hoffman failure criteria are considered in the present study. The input parameters include the ply orientation angle, assembly of ply, number of layers, ply thickness and degree of orthotropy, while the first-ply failure loads for five criteria representing output quantity of interest. The deterministic results are validated with past experimental results. The results obtained from the ANN-based surrogate model are observed to attain fitment with the results obtained by Monte Carlo Simulation (MCS). The statistical results are presented for both deterministic, as well as stochastic domain.

Keywords First-ply failure · Laminated composite: Monte Carlo Simulation (MCS) · Artificial Neural Network (ANN) · Uncertainty quantification

S. Kushari (✉) · S. R. Maity · S. Dey
National Institute of Technology, Silchar, India
e-mail: subrata734@gmail.com

A. Chakraborty
Indian Institute of Technology, Guwahati, India

T. Mukhopadhyay
University of Oxford, Oxford, UK

1 Introduction

The probability of failure in a laminated composite can be quantified by the amount of load it can withstand until its fracture point. Due to the inheritance of randomness and uncertainty in material and geometrical properties, stochastic regime seems to be an efficient way of modelling the failure analysis as compared to the deterministic way. An alternative approach for the analyses necessitating monotonous model evaluation is the utilisation of approximation models, also referred to as metamodels [1]. A result found by Onkar et al. [2], where first-ply failure load is analysed to evaluate the mean and variance of the failure statistics showed that the stochastic finite element has high accuracy. An experimental investigation designed by Reddy and Pandey [3], who are considered as the pioneer of failure analysis of composites conducted computational, as well as numerical investigation on laminated composite, with different ply orientation, ply angle with in-plane and out-of-plane failure load.

Laminated composite plates failure can be seen by debonding or delamination, fibre pull-out, fibre breakage and matrix cracking. These modes of failure are major limitations of laminated composite plate. Delamination is the most common mode of failure among all other kinds of failure. These modes of failure are major limitations of laminated composite. Many researchers investigated on deterministic failure analysis of laminated composite plate. ANN incorporated in analysing the probability of delamination of laminated composite by Chakraborty [4], highlighted that the trained network can predict the delamination for different shape, size and location with less computational time. ANN is applied to predict the kinetic parameters like high-velocity impact on a carbon reinforced fibre composites (CRFC) [5]. In a particular case where split growth in the notched composite is analysed under constant amplitude fatigue by using ANN and power law, ANN proved to be a better predictive tool than power law [6]. End milling process of a Glass fibre reinforced plastic (GFRP) composites ANN provided significant performance increase in analysing the damage factor by developing five learning algorithms and training them which contributed in the reduction of cost as well as time in conducting experiment [7]. Prediction of impact location using ANN during damage based on the kinetic energy of an impact by incorporating the limited strain signatures as inputs provided a warning system in damage initiation [8]. In another instance, prediction for impact resistance of aluminium–epoxy-laminated composites were analysed and it was found that ANN can be used as a substitutive approach to evaluate the effect of bonding strength of laminated composites [9]. ANN also proved to a better option in analysing the failure of a cross-ply composite tube under torsion, as well as axial tension/compression compared to Tsai-Wu theory and tensor polynomial theory [10]. Some of the important works done on the first-ply failure of laminated composites and ANN are hereby mentioned [11–16]. The present study deals with uncertainty quantification for first-ply failure of laminated composite plates by incorporating ANN as surrogate model to reduce computational time.

2 Mathematical Formulation

In order to determine the first-ply failure of the laminated composite five failure criteria are taken into consideration as mentioned earlier. The finite element (FE) model is designed based on the failure criteria followed by this the surrogate model is implemented using ANN.

2.1 Failure Criteria for Laminated Composite

In the present work, a three-layered laminated composite as shown in Fig. 1, is considered to study the failure analysis. The orientation of the laminate is [45°, − 45°, 45°]. The five failure criteria are employed to analyse the first-ply failure load of the laminate and design the finite element model for the same.

Maximum Stress Theory. This theory involves two forms of stress (normal stress and shear stress) theories. It specifies that, when a material has exceeded its maximum stress enduring capacity in any of its axes, it fails. The mathematical formulation [17], can be expressed as

$$(\sigma_1^C)_u < (\sigma_1) < (\sigma_1^T)_u \tag{1}$$

$$(\sigma_2^C)_u < (\sigma_2) < (\sigma_2^T)_u \tag{2}$$

$$(\tau_{12})_u < (\tau_{12}) < (\tau_{12})_u \tag{3}$$

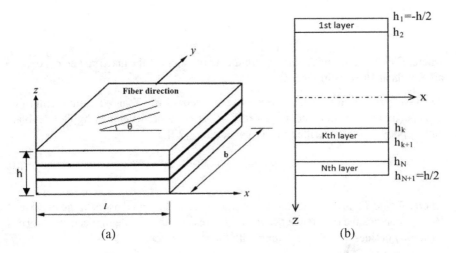

Fig. 1 a Isometric view of the laminate. b Layer thickness representation of the laminated plate

where σ_1, σ_2 represent the normal stresses in x-axis and y-axis, respectively. While τ_{12} represents the shear stress. σ^c and σ^T represent the compressive stress and tensile stress, respectively, through the laminate. Here the suffix 'u' is used to signify the ultimate stress point.

Maximum Strain Theory. This theory is based on the maximum normal strain theory of St. Venant and Tresca's strain equivalent of maximum stress theory for isotropic materials. According to this theory, when the shear and principal strain exceeds the ultimate strain the material tends to rupture or fail. The mathematical deduction [18], for the same can be expressed as

$$(\varepsilon_1^c)_u < (\varepsilon_1) < (\varepsilon_1^T)_u \qquad (4)$$

$$(\varepsilon_2^c)_u < (\varepsilon_2) < (\varepsilon_2^T)_u \qquad (5)$$

$$\gamma_{12} < \Gamma_{12} \qquad (6)$$

where, ε_1 and ε_2 represent the normal strains in x-axis and y-axis represents while γ_{12} represents the shear strain. ε^c and ε^T represent the compressive strain and tensive strain, respectively through the laminate and Γ_{12} represents the ultimate shear strain.

Tsai-Hill (Energy-Based Criterion) Theory. Tsai-Hill theory for the failure of laminate is a combination of two energy principle, the first one is the distortion energy (which is responsible for change of the shape) and the second one is dilation energy (which causes volumetric changes in the material). The failure of the material takes place when the following equation [19], holds true.

$$f(\sigma_{ij}) = F(\sigma_2 - \sigma_3)^2 + G(\sigma_3 - \sigma_1)^2 + H(\sigma_1 - \sigma_2)^2 + 2L\sigma_4^2 + 2M\sigma_5^2 + 2N\sigma_6^2 = 1 \qquad (7)$$

where, F, G, H, L, M and N signify strength parameters of the material and σ_4, σ_5, σ_6 are the shear stress components.

Tsai-Wu (Interaction Tensor Polynomial) Theory. The Tsai-Wu failure criterion is a special case of the general quadratic failure criteria developed by Gol'denblat and Kopnov. It can be written in a scalar form as [20]

$$F_i\sigma_i + F_{ij}\sigma_i\sigma_j \geq 1 \qquad (8)$$

where, F_i and F_{ij} are the first order and fourth order strength tensors of the material. Here σ_i denotes the difference between compressive and tensile induced stress. The term $\sigma_i\sigma_j$ defines an ellipsoid along with the stress space.

Tsai-Hill's Hoffman Failure Criteria. The Tsai-Hill's Hoffman criterion is a special condition of Tsai-Hill failure criteria. In Hoffman's failure criteria the difference between the strength of tension and compression is considered which is ignored in the case of Tsai-Hill failure criteria which is significant if brittle materials are considered. The modified criteria are established by adding the odd functions of the principal stress components (σ_1, σ_2 and σ_3) in the actual expression of Tsai-Hill criteria [21]. Thus

$$C_1(\sigma_2 - \sigma_3)^2 + C_2(\sigma_3 - \sigma_1)^2 + C_3(\sigma_1 - \sigma_2)^2 + C_4\sigma_1 \\ + C_5\sigma_2 + C_6\sigma_3 + C_7\sigma_4^2 + C_8\sigma_5^2 + C_9\sigma_6^2 = 1 \qquad (9)$$

Here C_1–C_9 denote the material parameters.

2.2 ANN-Based Surrogate Model

Artificial neural network (ANN) is analogous to the working phenomenon of a human brain based on which the present computational model is being prepared. The algorithm in ANN acquires its working procedure based on the input and internal hidden neuron configuration also known as weights after which the output of the data is compared with known correct values. The modelling of intricate association between input and output data involves non-linear statistical data modelling tool, which is further incorporated and executed through ANN. The input and output data are channelised through a training process which continues until a significant reduction in error is achieved. The input data moves forward on a layer basis, training the data in hidden networks, and simultaneously it is supervised to reduce the error through the back-propagation algorithm. The hidden layers can be more than one according to the design required. The principal objective of using ANN is its ability to compensate the computational time by developing an efficient model similar to the finite element model. The structure of input and output in an ANN is shown in Fig. 2.

ANN procures their result by following the patterns and relationships in data and learn through experience, instead of following a fixed set of programmable arrangement. The working process of an ANN is followed by a series of steps [22], as specified here. At first, the input data (x_n) is fed to the neurons in the input unit followed by calculating the output (y_n) from the hidden layer of neurons utilising the transfer function

$$y_n = \sum W_{nm}x_n + \psi \qquad (10)$$

$$H_m = \frac{1}{1 = \exp(-\alpha y_n)} \qquad (11)$$

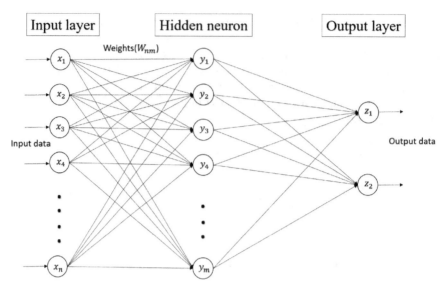

Fig. 2 Structure of an ANN

where W_{nm} denotes the connection weight for the neurons n and m, Ψ is the bias or threshold value for neuron m that can be observed as the non-zero offset in the data, H_m is the output of neuron m, and α signifies the non-linear parameter for the neuron's operation. Following this stage, compute the parameter to be studied (P_k) for the output neuron in a similar manner as mentioned in Eqs. (10) and (11). Since the algorithm will face an error at this stage, the error for each weight in the output value P_k and target output t_k is analysed by Eq. (13). This process is known as backpropagation.

$$\delta_k = (t_k - P_k)P_k(1 - P_k) \qquad (12)$$

where k is the output neuron. According to the error correction factor, the change in weights and bias are updated in this stage

$$W_{mk}^{new} = W_{mk}^{old} + \Delta W_{mk}(p) \qquad (13)$$

$$\Delta W_{mk}(p) = \eta \delta_k H_m + \mu \Delta W_{mk}(p-1) \qquad (14)$$

where W_{mk} signifies the adjusted value of the weight between output neuron k and hidden layer neuron m, p and $p-1$ refer to the present and previous cycles of correction, respectively. Also, η denotes the learning rate and μ *signifies momentum.*

Now the error for the hidden layer (δ_m) is calculated as Eq. (15) and an updated weight (W_{nm}) is formulated due to the hidden unit based on Eqs. (16) and (17).

Fig. 3 Flowchart of the first-ply failure analysis incorporating ANN as the surrogate model

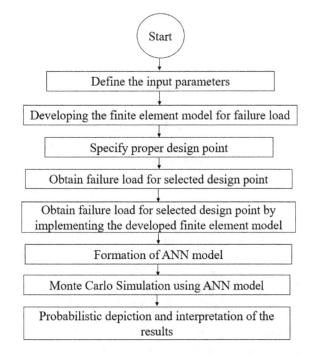

$$\delta_m = H_m \left(1 - H_m \sum \delta_k W_{mk}\right) \qquad (15)$$

$$W_{mk}^{new} = W_{mk}^{old} + \Delta W_{nm}(p) \qquad (16)$$

$$\Delta W_{mk}(p) = \eta \delta_m H_n + \mu \Delta W_{nm}(p-1) \qquad (17)$$

At the end terminating condition is checked after every new sample is calculated and finally a significant reduction in error is achieved.

The working module for the ANN-based random first-ply failure analysis can be depicted by the flowchart in Fig. 3.

3 Result and Discussion

In this section, the result obtained for the different failure criteria as mentioned above are discussed in brief. Firstly, a deterministic study is carried out for the analysis of first-ply failure loads with respect to the mentioned failure criteria and then the stochastic analysis is designed based on ANN as the surrogate model.

Table 1 Material properties of T300/5208 graphite-epoxy laminate [23] with ply orientation of [45°, −45°, 45°]

E_1 (GPa)	E_2 (GPa)	G_{12} (GPa)	G_{13} (GPa)	G_{23} (GPa)	μ	ρ (kg/m^3)
132.37	10.7	5.65	5.65	3.37	0.3	3202

The laminated composite considered is a three-layered T300/5208 graphite-epoxy laminate with ply orientation of [45°, −45°, 45°]. The mean value of the material properties for the specified material is specified in Table 1.

The plate has a dimension of 0.22 m length, 0.127 m breadth and 3 × 10^{-4} m as thickness. The plate is subjected to uniformly distributed load on the top surface in the z-direction. Three mesh size is considered for the verification of the deterministic model of sizes (2 × 2), (4 × 4) and (8 × 8) and they are simultaneously validated with the finite element model of Reddy and Pandey [3]. The mesh plane area is considered of (8 × 8) configuration comprising of 64 elements and 225 total number of nodes. The deterministic validation of the five different failure criteria from Reddy and Pandey [3], is shown in Table 2. The present deterministic validation of the failure modes is also validated by Karsh et al. [16], for the spatial vulnerability study for the first-ply failure for laminated composite.

Subsequently, the deterministic model is validated in Table 2, the ANN model is designed based on the parent MCS model. For the ANN model, the main MCS model which is of 10,000 samples sized is compared with the ANN-based MCS for three different sample size which is 64, 128 and 256 as depicted in Fig. 3.

As it can be perceived from Fig. 4, that out of the three sample size considered, 256 samples sized data converges with the parent MCS to a great extent for the five different failure criteria considered. For the lesser sample size data, the PDF tends to deviate from the parent MCS model. Thus, it can be concluded that as the sample size increases the ANN model tends to improve its accuracy comparatively.

Table 2 Validation of the present finite element model with experimental results [3] for in-plane loading of different failure criteria for the laminate with [45°, −45°, 45°] ply orientation

Failure theory	Failure load					
	(2 × 2)		(4 × 4)		(8 × 8)	
	Reddy et al. [3]	Present FE model	Reddy et al. [3]	Present FE model	Reddy et al. [3]	Present FE model
Max. stress	2854.40	3408.70	2164.32	2486.50	1908.16	1962.50
Max. strain	2947.68	3273.20	2268.60	2421.70	1940.48	1994.75
Tsai-Hill	2788.80	3091.40	1803.84	1897.91	1530.40	1563.70
Tsai-Wu	2886.72	3337.70	2218.88	2432.73	1917.76	1957.32
Hoffman	2850.24	3224.50	2156.80	2269.53	1905.76	1962.10

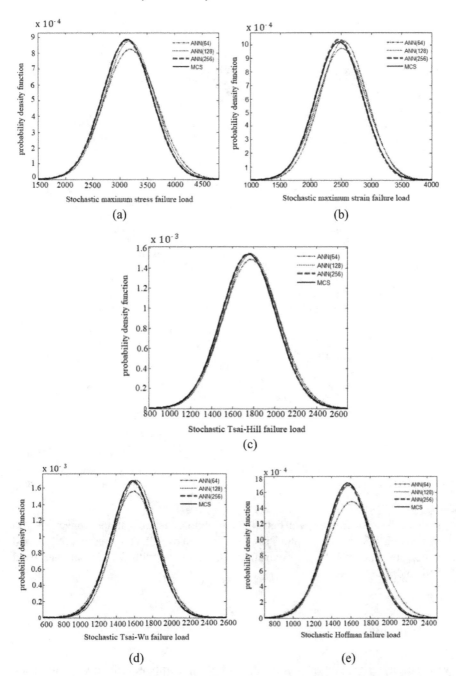

Fig. 4 Probability density functions (PDF) of parent MCS and ANN model considering first-ply failure for **a** Maximum stress. **b** Maximum strain. **c** Tsail-Hill. **d** Tsai-Wu and **e** Hoffman failure criteria

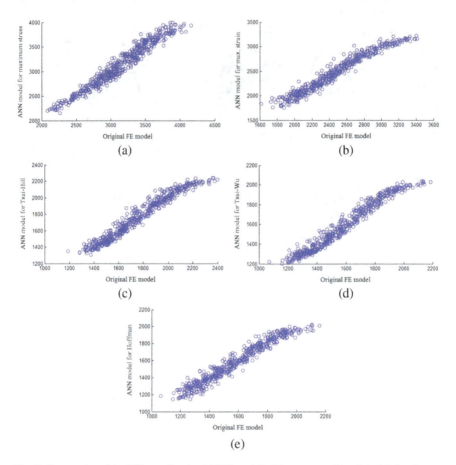

Fig. 5 Scatter plot of the 256 samples sized ANN model with respect to the original finite element (FE) model for **a** max. stress **b** max. strain. **c** Tsai-Hill. **d** Tsai-Wu and **e** Hoffman failure criteria

Since the 256 samples sized data of the ANN-based model almost converges with the parent MCS, the scatter plot shown in Fig. 5, shows that the present ANN-based model can substitute the time-consuming MCS model with the 256 samples sized ANN model.

4 Conclusion

The novelty of the present work is that ANN is incorporated along with stochastic finite element modelling for first-ply failure analysis of a three-layered laminated composite. It is concluded that as the sample size increases for the ANN-based model, the accuracy level of the model with respect to the parent MCS also increases. The scatter plot for the efficiently matched ANN model is depicted to conclude that the present parent MCS can be replaced with the efficient ANN-based model.

Acknowledgments The authors would like to acknowledge the Aeronautics Research and Development Board (AR&DB), Government of India (Project Sanction no.: ARDB/01/105885/M/I), for the financial support for the present research work.

References

1. Dey S, Mukhopadhyay T, Adhikari S (2017) Metamodel based high-fidelity stochastic analysis of composite laminates: a concise review with critical comparative assessment. Compos Struct 171:227–250
2. Onkar AK, Upadhyay CS, Yadav D (2007) Probabilistic failure of laminated composite plates using the stochastic finite element method. Compos Struct 77(1):79–91
3. Reddy JN, Pandey AK (1987) A first-ply failure analysis of composite laminates. Comput Struct 25(3):371–393
4. Chakraborty D (2005) Artificial neural network-based delamination prediction in laminated composites. Mater Des 26(1):1–7
5. Fernández-Fdz D, López-Puente J, Zaera R (2008) Prediction of the behaviour of CFRPs against high-velocity impact of solids employing an artificial neural network methodology. Compos A Appl Sci Manuf 39(6):989–996
6. Choi SW, Song EJ, Hahn HT (2003) Prediction of fatigue damage growth in notched composite laminates using an artificial neural network. Compos Sci Technol 63(5):661–675
7. Erkan Ö, Işık B, Çiçek A, Kara F (2013) Prediction of damage factor in end milling of glass fibre reinforced plastic composites using artificial neural network. Appl Compos Mater 20(4):517–536
8. Watkins SE, Akhavan F, Dua R, Chandrashekhara K, Wunsch DC (2007) Impact-induced damage characterization of composite plates using neural networks. Smart Mater Struct 16(2):515
9. Nazari A, Sedghi A, Didehvar N (2012) Modeling impact resistance of aluminum–epoxy-laminated composites by artificial neural networks. J Compos Mater 46(13):1593–1605
10. Lee CS, Hwang W, Park HC, Han KS (1999) Failure of carbon/epoxy composite tubes under combined axial and torsional loading 1. Experimental results and prediction of biaxial strength by the use of neural networks. Compos Sci Technol 59(12):1779–1788
11. Chen G, Wang H, Bezold A, Broeckmann C, Weichert D, Zhang L (2019) Strengths prediction of particulate reinforced metal matrix composites (PRMMCs) using direct method and artificial neural network. Compos Struct 110951
12. Solati A, Hamedi M, Safarabadi M (2019) Combined GA-ANN approach for prediction of HAZ and bearing strength in laser drilling of GFRP composite. Opt Laser Technol 113:104–115
13. Balcıoğlu HE, Seçkin AÇ, Aktaş M (2016) Failure load prediction of adhesively bonded pultruded composites using artificial neural network. J Compos Mater 50(23):3267–3281
14. Reddy AR, Reddy BS, Reddy KVK (2011) Application of design of experiments and artificial neural networks for stacking sequence optimizations of laminated composite plates. Int J Eng Sci Technol 3(6):295–310
15. Kam TY, Lai FM (1999) Experimental and theoretical predictions of first-ply failure strength of laminated composite plates. Int J Solids Struct 36(16):2379–2395
16. Karsh PK, Mukhopadhyay T, Dey S (2018) Spatial vulnerability analysis for the first ply failure strength of composite laminates including effect of delamination. Compos Struct 184:554–567
17. Tresca HE (1864) Sur l'ecoulement des corps solides soumis a de fortes pressions. Imprimerie de Gauthier-Villars, successeur de Mallet-Bachelier, rue de Seine-Saint-Germain, 10, près l'Institut

18. Sokolnikoff IS (1956) Mathematical theory of elasticity. McGraw-Hill Book Company
19. Mises RV (1913) Mechanik der festen Körper im plastisch-deformablen Zustand. Nachrichten von der Gesellschaft der Wissenschaften zu Göttingen, Mathematisch-Physikalische Klasse, pp 582–592
20. Tsai SW, Wu EM (1971) A general theory of strength for anisotropic materials. J Compos Mater 5(1):58–80
21. Hoffman O (1967) The brittle strength of orthotropic materials. Compos Mater 1(2):200–206
22. Guo D, Wang Y, Xia J, Nan C, Li L (2002) Investigation of $BaTiO_3$ formulation: an artificial neural network (ANN) method. J Eur Ceram Soc 22(11):1867–1872
23. Köhler L, Spatz HC (2002) Micromechanics of plant tissues beyond the linear-elastic range. Planta 215(1):466

Interaction of Higher Buckling Modes in Uniformly Compressed Simply Supported Unstiffened Plates

K. C. Kalam Aswathy and M. V. Anil Kumar

Abstract The effective width method (EWM) for calculating the ultimate strength of cold-formed steel (CFS) members assumes simply supported boundary conditions along the plate edges. Unstiffened plates, simply supported along three edges and unsupported along one longitudinal edge buckle in a single half-wave equal to plate length when subjected to uniform compression. It has been observed from the finite element analysis results that the simply supported unstiffened plates with large length to width ratios ($L/b \geq 10$) the deformation pattern changes from a single half-wave to a combination of three half-waves corresponding to the ultimate load. One of the possible reasons for this is the nearly same elastic buckling stresses in the first few modes leading to the interaction of higher buckling modes. At larger L/b ratios the interaction of the third buckling mode with fundamental mode was observed causing the reduction in the ultimate strength. A similar reduction in the ultimate strength due to buckling mode interactions are also found in CFS plain equal angle sections, as such sections composed of only unstiffened plates which buckle simultaneously. In sections like plain channels, where additional rotational restraint along the supported longitudinal edge is present, such mode interactions are found to not influence the strength of unstiffened elements in sections.

Keywords Local buckling · Unstiffened elements · Interaction

K. C. Kalam Aswathy (✉) · M. V. Anil Kumar
Indian Institute of Technology, Palakkad 678623, India
e-mail: 101703001@smail.iitpkd.ac.in

1 Introduction

Thin-walled CFS members are composed of plate elements which may buckle locally under compression before yielding. These plate elements may be classified as unstiffened or stiffened plate elements depending on whether one or both the longitudinal edges of the plate element are supported by the adjoining elements, respectively (Fig. 1). The simply supported unstiffened plates under uniform compression buckles in half-wavelength, L_{cr} equal to the plate length, L [1, 2]. The unstiffened element in cross-section buckles in multiple half-waves due to the additional rotational restraint from adjoining elements. The stress redistribution across the plate after local buckling enables it to carry additional loads and is termed as post-buckling capacity. Due to large transverse deformations in unstiffened plates, the stress redistribution occurs to a much larger extent resulting in higher post-buckling capacity than compared to stiffened plates [3, 4].

The effective width method (EWM) is one of the design methods for cold-formed steel members, where the cross-section is assumed as an assembly of plate elements. The non-uniform stress distribution across plate width (b) after local buckling is idealised as a uniform stress distribution over a reduced width of plate element termed as effective width (b_{eff}). The effective width of each plate element is calculated as a function of the elastic local buckling stress, f_{crl} and yield stress of the material, f_y as per Eq. 1 [5].

$$\frac{P_{ul}}{P_y} = \frac{b_{eff}}{b} = \begin{cases} 1 & \lambda_l < 0.673 \\ \left(1 - \frac{0.22}{\lambda_l}\right)\frac{1}{\lambda_l} & \lambda_l \geq 0.673 \end{cases} \quad (1)$$

where, P_{ul}—ultimate local buckling strength; P_y—yield strength of the element; b—plate width; b_{eff}—effective width; and $\lambda_l = \sqrt{f_y/f_{crl}}$—non-dimensional slenderness ratio.

The elastic local buckling stress, f_{crl} of a plate is computed using Eq. 2. The plate buckling coefficient, k accounts for the boundary conditions and stress distribution across the width. EWM assumes simply supported edge conditions and conserva-

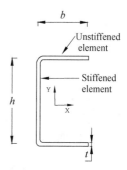

Fig. 1 Stiffened and unstiffened elements in a plain channel section

tively ignores the rotational restraint from adjoining elements by adopting k of 0.425 in Eq. 2. At larger aspect ratios (length to width ratio, L/b), it was observed that the deflected shape corresponding to ultimate load changes from a single half-wave [1] due to interaction of higher buckling modes.

$$f_{crl} = \frac{k\pi^2 E}{12(1-\nu^2)\left(\frac{b}{t}\right)^2} \quad (2)$$

where, E—Young's modulus; ν—Poisson's ratio and k—plate buckling coefficient.

This study evaluates the influence of higher mode interaction on the ultimate compressive strength of unstiffened elements and CFS sections containing such elements. The ultimate strength of the unstiffened plate was found to be reduced due to an increase in L/b due to the interaction of higher buckling modes. In addition to isolated unstiffened elements, the plain equal angle compression members are also considered in this study. The unstiffened elements in plain equal angle compression members are equivalent to simply supported unstiffened plates as both the elements buckles simultaneously. The unstiffened elements of plain channels, which locally buckles in multiple half-waves are also studied to understand these mode interactions at higher L/b ratios.

2 Finite Element Modelling

The unstiffened plate is numerically modelled in the finite element software ABAQUS [6] through the centre line using four-noded shell elements with linear shape function and reduced integration (S4R). The size of the finite element used was 6 mm × 6 mm based on mesh convergence studies. The eigenvalue buckling analysis on perfect geometry gives the critical buckling loads and corresponding buckling modes. The non-linear analysis is done on the imperfect plate by taking into account both material and geometric non-linearity. The buckling mode shapes from the linear buckling analysis are scaled and used as geometric imperfections for non-linear analysis using the *IMPERFECTION command in the input file [6]. The displacement controlled analysis using Newton–Raphson iteration scheme is used for non-linear analysis.

3 Unstiffened Plates

3.1 Validation of FE Model

The finite element (FE) model was validated by comparing with test results of simply supported unstiffened plates [2] under uniform compression. The dimensions, yield stress and average compressive test from test and FEA (f_{u-test} and f_{u-FEA},

Table 1 Comparison of FEA and test results for unstiffened plates [2]

b (mm)	t (mm)	L (mm)	f_y (MPa)	λ	f_{u-test} (MPa)	f_{u-FEA} (MPa)	% difference
175.2[a]	4.77	876	272.2	2.10	147.5	149.9	−1.58
125.6	4.77	628	272.2	1.51	158.2	163.7	−3.42
80.0	4.75	400	317.2	1.05	250.3	233.5	6.69
61.2	4.77	306	272.2	0.75	269.9	269.0	0.34
175.0	4.77	875	272.2	2.12	152.1	150.0	1.40
125.8	4.77	629	272.2	1.53	157.2	163.2	−3.82
100.3	4.77	501.5	272.2	1.22	197.1	201.7	−2.33
79.9	4.75	399.5	317.2	1.06	271.4	254.5	6.22
60.6[a]	4.77	303	317.2	0.81	285.1	274.2	3.83
Mean, μ							2.49
Standard deviation, σ							2.14

[a]Figure 2 shows comparison of compressive stress versus axial displacement plots

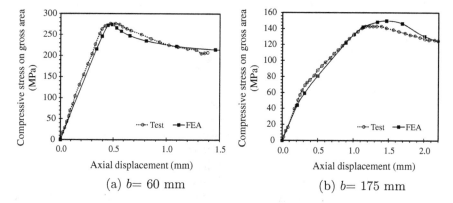

Fig. 2 Validation of FE model using test results [2]

respectively) are presented in Table 1. Figure 2 shows the plot of compressive stress of gross area versus axial displacement of two unstiffened plates of width 60 and 175 mm. The mean and standard deviation of 2.49 and 2.14 of the percentage difference between f_{u-test} and f_{u-FEA} and Fig. 2 confirms the accuracy of the FE model. The validated FE model was used to generate data of unstiffened plates required.

3.2 Elastic Buckling

The elastic buckling stress (f_{crl}) of uniformly compressed unstiffened plates with simply supported boundary conditions are calculated using the local buckling stress equation (Eq. 2), using a plate buckling coefficient, k of 0.425. At lower aspect ratio

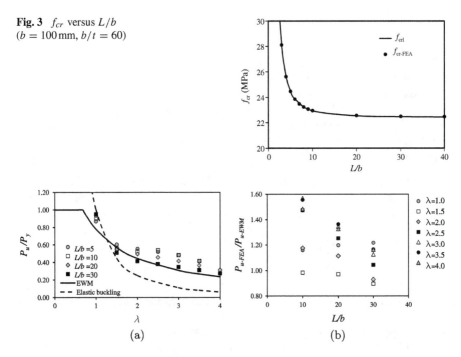

Fig. 3 f_{cr} versus L/b ($b = 100$ mm, $b/t = 60$)

Fig. 4 FEA results for unstiffened plates of various L/b ratios

(length to width ratio, L/b), k of $0.425 + (b/L)^2$ improves the accuracy of elastic buckling stress values [7]. Figure 3 plots the elastic buckling stress obtained from finite element analysis (FEA) (f_{cr-FEA}) and local buckling stress (f_{crl}) values against the L/b ratios for unstiffened plates with $b = 100$ and $b/t = 60$. Hence, in the range of aspect ratios used in practice ($L/b > 10$), the plate buckling coefficient, $k = 0.425$ used in EWM is accurate.

3.3 Ultimate Strength

The ultimate strength of plates is normally higher than their local buckling stresses due to post-buckling capacity. The effect of higher mode interaction in unstiffened plates with larger L/b ratio on the ultimate strength is studied using non-linear finite element analysis [6]. The first buckling mode is introduced as geometric imperfection using a scaling factor of $0.1t$, where t is the thickness of the plate. Figure 4 plots the ultimate strength normalised with yield strength (P_u/P_y) obtained from FEA against non-dimensional slenderness, λ for specimens of different L/b values. Equation 1 (denoted as EWM) and elastic buckling stress are also plotted in Fig. 4 for comparison.

The observations based on the study are as follows:

- The effect of aspect ratio (L/b) on elastic buckling stress is accurately accounted for by the plate buckling coefficient, k of $0.425 + (b/L)^2$ (Fig. 3).
- The normalised ultimate strength, P_u/P_y of unstiffened plates reduces with increase in aspect ratio as observed in Fig. 4.
- The effect of L/b ratio on P_u/P_y is prominent for $2.00 \leq \lambda \leq 3.50$ with maximum reduction in strength for λ value of 2.50.
- EWM safely predicts the ultimate strength of these elements. For lower values of aspect ratios, EWM is found to be conservative.

The presence of interaction of higher buckling modes in unstiffened plates with larger aspect ratios, L/b is demonstrated in Fig. 5, by plotting normalised out-of-plane deformation plot along the free longitudinal edge for specimens corresponding to ultimate load step. The first three buckling modes from eigenvalue buckling analysis are also plotted for comparison (Fig. 5a).

- The deformation corresponding to first buckling mode only is present in the unstiffened plates of λ of 1.50 (Fig. 5c), at ultimate load for aspect ratio of 10. But as the aspect ratio increases to 30, the presence of the third mode also is visible leading to the reduction in the strength.
- As the specimens become more slender ($\lambda = 3.50$) these mode interactions occur even for L/b of 10 (Fig. 5d).
- To understand the buckling mode interactions, fast Fourier transform (FFT) of the buckled mode shape at the ultimate step of the specimen with $L/b = 10$ and $\lambda = 3.50$ is performed (Fig. 5b). The maximum amplitudes occurring at frequencies of 7 and 18 correspond to the first and third modes, respectively. The higher magnitude corresponding to the frequency of 7 is due to the larger contribution of the first mode.

4 Cross-Sections Containing Unstiffened Elements

4.1 Plain Equal Angles

Stub plain equal angles are a peculiar class of cold-formed steel sections composed of only unstiffened plates. Due to simultaneous buckling of both flanges, stub plain equal angles and simply supported unstiffened plates show similar behaviour [8]. Angle sections with pinned end boundary condition, which are subjected to additional bending, are not considered in this study. The validation of the FE model of plain equal angle based on experimental results [9, 10] is provided in Table 2.

Due to very low torsional stiffness, these specimens are susceptible to torsional and flexural-torsional buckling. P_u/P_y for the equal angle specimens with aspect ratio, L/b ratios of 10, 20 and 30 are plotted in Fig. 6. The local (Eq. 1) and global (Eq. 3),

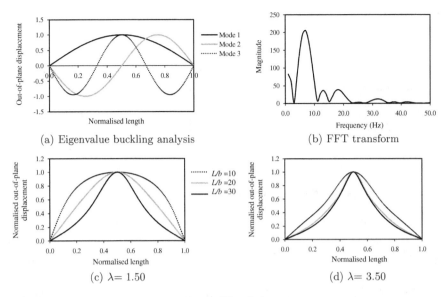

(a) Eigenvalue buckling analysis (b) FFT transform

(c) $\lambda = 1.50$ (d) $\lambda = 3.50$

Fig. 5 Out-of-plane displacement plots for unstiffened plates

Table 2 Comparison of FEA and test results for plain equal angles (a-[10], b-[9]) and plain channels (c-[11], d-[12])

Reference specimen	b (mm)	h (mm)	t (mm)	L (mm)	f_y (MPa)	λ	P_{u-test} (kN)	P_{u-FEA} (kN)	% difference
P1.9L250[a]	72.1	–	1.877	251	500	2.29	57.70	60.69	−5.18
P1.5L250[a]	71.7	–	1.495	249	530	2.91	39.60	42.75	−7.95
P1.2L250[a]	71.8	–	1.157	249.6	550	3.76	23.80	27.24	−14.45
LO2419[b]	47.27	–	2.34	2598	396	1.90	22.30	22.59	−1.30
SC 60 × 30[c]	41.48	78.38	1.219	253.49	226.08	1.30	32.93	31.19	5.28
SC 180 × 60[c]	76.30	221.28	1.219	634.49	226.08	2.99	37.83	36.77	2.78
P36F1000[d]	35.53	95.43	1.48	1000.2	550	1.73	59.00	64.72	−9.69
P36F2500[d]	35.64	95.83	1.48	2499.4	550	1.86	32.8	34.39	−4.85

buckling strengths are plotted in Fig. 6, for comparison. Similar to unstiffened plates, the buckling mode pattern involving three half-waves is also found in the flanges of equal angles. These mode interactions lead to strength reduction in these specimens similar to unstiffened plates.

$$\frac{P_{ue}}{P_y} = \begin{cases} 0.658^{\lambda_c^2} & \lambda_c < 1.5 \\ \dfrac{0.877}{\lambda_c^2} & \lambda_c \geq 1.5 \end{cases} \quad (3)$$

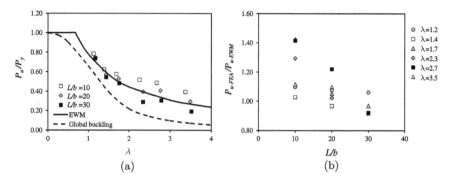

Fig. 6 FE results for plain equal angles of various L/b ratios

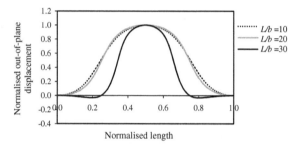

Fig. 7 Deformation plot along free longitudinal edge for equal angle specimens of $\lambda = 1.85$

The presence of higher mode interaction in plain equal-leg angle is confirmed by plotting the normalised out-of-plane displacement along the free edge at the ultimate step for the angle specimens of $\lambda = 1.85$ is given in Fig. 7. The buckling pattern involving three waves is evident for the specimen of L/b of 30 in Fig. 7.

4.2 Plain Channels

In plain channel section, adjoining element (wcb) offers additional rotational restraint to the unstiffened flanges. Hence, these elements buckle in multiple half-waves of half-wavelength $1.64b < L_{cr} < L$ [8]. The influence of the interaction of higher buckling modes due to larger L/b ratio on the ultimate strength of unstiffened plates with rotational restraint is studied using FEA of the plain channel (Fig. 1) specimens with $h = b$ and L/b of 10, 20 and 30, and presented in Fig. 8. The validation of the FE model for plain channel sections with fixed boundary conditions is presented in Table 2 [11, 12]. It is observed that the higher buckling mode interactions have very little influence on the normalised ultimate strength (P_u/P_y) of these specimens compared to the simply supported unstiffened plate. The out-of-plane deformation plot of the unstiffened flange at ultimate load also shows that higher buckling modes do

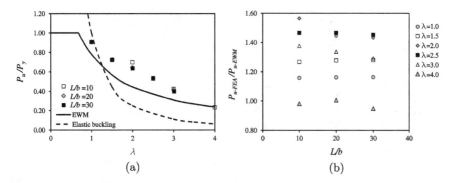

Fig. 8 FE results for plain channels of various L/b ratios

not influence the strength of these elements, unlike the simply supported unstiffened plates due to interaction with adjoining elements.

5 Conclusions

The unstiffened plates under uniform compression buckles in a single half-wave equal to plate length. At higher length to width ratios, L/b, the deformation pattern changes from a single half-wave to a combination of three waves which reduces the ultimate strength. A similar trend is also observed for slender specimens at smaller L/b ratios. A similar reduction in the ultimate strength due to buckling mode interactions are also found in CFS plain equal angle sections, as such sections composed of only unstiffened plates which buckle simultaneously. In sections like plain channels, where additional rotational restraint along the supported longitudinal edge is present, such mode interactions are found to not influence the strength of unstiffened elements in sections.

References

1. Bambach M, Rasmussen K (2004) Effects of anchoring tensile stresses in axially loaded plates and sections. Thin-Walled Struct 42(10):1465–1479
2. Bambach M, Rasmussen K (2004) Tests of unstiffened plate elements under combined compression and bending. J Struct Eng 130(10):1602–1610
3. Kalyanaraman V (1976) Performance of unstiffened compression elements. J Cent Cold-Form Steel Struct Libr 184
4. Winter G (1947) Strength of thin steel compression flanges. Trans ASCE 112:527
5. American Iron and Steel Institute (AISI) (2016) North American specification for the design of cold-formed steel structural members, S136-16, Washington, DC
6. ABAQUS. Version 6.8 [Computer Software]. Dassault Systemes Simulia, Providence, RI
7. Bulson PS (1970) The stability of flat plates. Chatto and Windus, London

8. Timoshenko SP, Gere JM (1961) Theory of elastic stability. Mc-Graw Hill, New York
9. Popovic D, Hancock GJ, Rasmussen KJ (1999) Axial compression tests of cold-formed angles. J Struct Eng 125(5):515–523
10. Young B (2004) Tests and design of fixed-ended cold-formed steel plain angle columns. J Struct Eng 130(12):1931–1940
11. Mulligan GP, Pekoz T (1984) Local buckling interaction in cold formed columns. J Struct Eng 110(11):2635–2654
12. Young B, Rasmussen KJ (1998) Tests of fixed-ended plain channel columns. J Struct Eng 124(2):131–139

Natural Frequency of Higher-Order Shear Deformable FGM Plates with Initial Geometric Imperfection Resting on Elastic Foundation

Mohammed Shakir and Mohammad Talha

Abstract In this paper, the natural frequency of higher order shear deformable FGM plates with initial geometric imperfections is investigated. The plates are resting on a Winkler–Pasternak elastic foundation. The effective material properties of the FGM plates are varied along the thickness direction using a simple power-law distribution. The formulations are based on Reddy's higher order shear deformation theory. A special function (i.e. the product of trigonometric and hyperbolic functions) is added to the transverse displacement to incorporate the initial geometrical imperfection. The fundamental equations are derived using a variational approach by adopting traction-free boundary conditions on the top and bottom faces of the plates. The results are obtained using the finite element method associated with C^0 continuous Lagrange's isoparametric elements. Comparison and convergence studies have been performed to confirm the efficacy of the present model. Numerical results cover the effects of different parameters like power-law index, plate aspect ratio, side-to-thickness ratio, imperfection size, and foundation parameters on the natural frequency of FGM plates.

Keywords Functionally graded plates · Geometric imperfection · Elastic foundation · Vibration analysis · Finite element method

1 Introduction

The extraordinary attributes of the functionally graded material (FGM) plates have been achieved in various applications like aircraft, spacecraft, marine, etc. because of the uniform gradation of material properties in the preferred direction [1]. However, structural deviations from the actual shape (termed as initial geometric imperfections) are inherited while heating at higher and cooling at low temperatures. Such

---

M. Shakir (✉) · M. Talha
Indian Institute of Technology, Mandi 175 005, India
e-mail: d18050@students.iitmandi.ac.in

© The Editor(s) (if applicable) and The Author(s), under exclusive license to Springer Nature Singapore Pte Ltd. 2021
S. K. Saha and M. Mukherjee (eds.), *Recent Advances in Computational Mechanics and Simulations*, Lecture Notes in Civil Engineering 103,
https://doi.org/10.1007/978-981-15-8138-0_13

imperfections show unavoidable local and global distributions over the structures and prominently affect the structural response [2].

On the other hand, the plates supported by elastic foundations also attracted many researchers to describe the interactive behavior of the plates in severe loading conditions [3], although limited studies are available comprising vibration characteristics of the FGM plates with initial geometric imperfections [4–10]. Kitipornchai et al. [4] examined the linear and nonlinear vibration responses of initially imperfect FGM laminated plates. Yang and Huang [5] described the nonlinear transient response of simply supported imperfect FGM plates in the thermal environment.

Gupta and Talha [6] analyzed the vibration response of geometrically imperfect FGM plates using a new non-polynomial-based higher order shear and normal deformation theory. Gupta and Talha [7] investigated the sensitivity of nonlinear flexural and vibration response of initially imperfect FGM plates.

Gupta and Talha [8] predicted the effect of geometric imperfection on the stability of FGM plates with porosity. Gupta and Talha [9] evaluated post-buckling and large amplitude vibration responses of imperfect FGM plates while considering microstructural defects. Tomar and Talha [10] determined the flexural and free vibration responses of geometrically imperfect functionally graded skew sandwich plates. Shakir and Talha [2] presented the transient response of geometrically imperfect FGM plates with material uncertainty. Based on the literature survey, it can be stated that the free vibration analysis of higher order shear deformable FGM plates with initial geometric imperfection, while the plates are resting on an elastic foundation, has not been reported yet to the best of the author's knowledge.

Therefore, the free vibration analysis of higher order shear deformable imperfect FGM plates resting on an elastic foundation has been explored in the present study. The FGM plate is assumed to be made of a mixture of the constituents (i.e. ceramic and metal), and the effective material properties are achieved as per power-law distribution associated with the Voigt model. The governing differential equation based on Reddy's HSDT is derived using the variational approach and the natural frequencies are assessed by employing the finite element method. The natural frequencies are evaluated to affirm the efficacy and performance of the proposed model are examined in terms of solution accuracy and rate of convergence. In addition, the influence of various parameters is also explored.

2 Mathematical Formulations

An FGM plate resting on a Winkler–Pasternak elastic foundation is considered with length 'a', width 'b', and height 'h' as shown in Fig. 1.

The plate is graded by varying volume fraction of one of the constituents (say ceramic) using power-law distribution function along the thickness direction 'z' as

$$V_c(z) = \left(\frac{1}{2} + \frac{z}{h}\right)^n ; (0 \leq n \leq \infty) \tag{1}$$

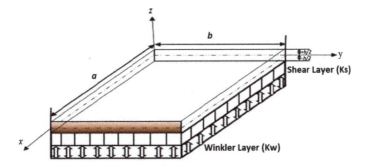

Fig. 1 An FGM plate resting on the elastic foundation

where n is the volume fraction index. For the $n = 0$ or $n \to \infty$, Eq. (1) represents the volume fraction of a fully ceramic or metal constituent, respectively.

For the appropriate plate kinematics, the displacement field for higher order shear deformation theory given by Reddy has been employed as [2]

$$u(x, y, x, t) = \bar{u}(x, y, t) + (z - c_1 z^3)\bar{\psi}_x(x, y, t) - c_1 z^3 \bar{\theta}_x(x, y, t)$$
$$v(x, y, x, t) = \bar{v}(x, y, t) + (z - c_1 z^3)\bar{\psi}_y(x, y, t) - c_1 z^3 \bar{\theta}_y(x, y, t)$$
$$w(x, y, x, t) = \bar{w}(x, y, t) + \hat{w} \qquad (2)$$

where $c_1 = 4/3h^2$; $\bar{\theta}_x = \partial \bar{w}/\partial x$; $\bar{\theta}_y = \partial \bar{w}/\partial y$, and $\hat{w}$ presents the initial imperfection induced in the transverse direction and is given as [4]

$$\hat{w} = \eta h \sec h[\delta_1(\alpha - \varphi_1)] \cos[\mu_1 \pi (\alpha - \varphi_1)] \sec h[\delta_2(\beta - \varphi_2)] \cos[\mu_2 \pi (\beta - \varphi_2)] \qquad (3)$$

where $\alpha = x/a$, $\beta = y/b$, and η is the imperfection amplitude. δ_1 and δ_2 indicate the imperfection degree of localization. μ_1 and μ_2 are the half-wave numbers of the imperfection along the x- and y-axes, respectively.

The present study employs the finite element method (FEM) to obtain natural frequencies of the imperfect FGM plates resting on elastic foundations. In order to implement the FEM, a C^0 isoparametric Lagrange's element is used to discretize the plate domain. The displacement field is written as [2]

$$\{q\} = \sum_{i=1}^{ne} N_i \{q\}_i; \; x = \sum_{i=1}^{ne} N_i x_i; \; y = \sum_{i=1}^{ne} N_i y_i \qquad (4)$$

where $\{q\} = \{\bar{u}, \bar{v}, \bar{w}, \bar{\psi}_x, \bar{\psi}_y, \bar{\theta}_x, \bar{\theta}_y\}^T$.

The strain energy for any particular element can be defined by [11]

$$U^{(e)} = \frac{1}{2}\int_v \{\varepsilon_i\}^T\{\sigma_i\}dv = \frac{1}{2}\int_v \{\varepsilon_i\}^T[Q_{ij}]\{\varepsilon_i\}dv = \frac{1}{2}\{q_i\}^T[K_i]\{q_i\} \quad (5)$$

The strain energy of the plate is given by

$$U = \sum_{v=1}^{ne} U^{(e)} = \frac{1}{2}\sum_{v=1}^{ne}\{q^e\}^T[K]^e\{q^e\} \quad (6)$$

The strain energy because of the foundation can also be expressed as [12]

$$U_f = \frac{1}{2}\int_A \left[k_w w^2 + k_s\left\{(w,_x)^2 + (w,_y)^2\right\}\right]dA$$

$$k_w^{(e)} = k_w \int_A [N]^T[N]dA, \quad (7)$$

$$k_s^{(e)} = k_s \int_A [(\partial N/\partial x)^T(\partial N/\partial x) + (\partial N/\partial y)^T(\partial N/\partial y)]dA$$

In a similar way, the kinetic energy is also derived as [11]

$$T = \frac{1}{2}\sum_{v=1}^{ne}\{q^e\}^T[m]^e\{q^e\} \quad (8)$$

The governing differential equation is derived using a variational approach and expressed for free vibration analysis as

$$[M]\{\ddot{q}\} + [K]\{q\} = 0 \quad (9)$$

where $[K]$ presents overall stiffness matrix which contains elastic stiffness matrix, matrices because of Winkler and Pasternak (shear layer) foundations, and is given by [12]

$$[K] = [\overline{K}] + [\overline{K}_w] + [\overline{K}_s] \quad (10)$$

3 Results and Discussion

The results comprise convergence, validation, and parametric studies of imperfect FGM plates resting on the elastic foundation for different boundary conditions, i.e. simply supported (SSSS) and clamped (CCCC) boundary conditions. A MATLAB code is developed to obtain the various results for the present mathematical model. The convergence and validation have been followed as stated in Example 1.

Example 1 In this example, the convergence and validation studies of simply supported ZrO_2/Al FGM plate ($E_c = 151 \times 10^9$ N/m^2, $\rho_c = 3000$ kg/m^3 for ceramic, and $E_m = 70 \times 10^9$ N/m^2, $\rho_m = 2707$ kg/m^3 for metal, [13]) resting on a Winkler–Pasternak elastic foundation ($K_s = K_w = 100$) is performed as presented in Table 1. The thickness and aspect ratios of the plate are taken as 20 and 1. The Poisson's ratio is assumed to be constant as 0.3 throughout. The frequency parameter ($\bar{\omega} = \omega_{min} h \sqrt{\rho_m/E_m}$) is obtained for different volume fraction indices (i.e. $n = 0$, 0.2, 1, 5) and compared. The response given in Table 1 gives a very good agreement with the exiting results [14] at 6×6 mesh size. Thus, the present mathematical model reveals good performance in terms of solution accuracy and convergence.

Parametric Study

Table 2 explores the vibration response (frequency parameter, $\bar{\omega} = \omega_{min} h \sqrt{\rho_m/E_m}$) of imperfect FGM plates with all edges simply supported (SSSS) and clamped (CCCC). The plate is graded by employing zirconia, ZrO_2 (ceramic) and aluminum, Al (metal). The responses are obtained for various imperfection modes with varying imperfection amplitudes, $\eta = 0$–1. The plate is considered with $a/h = 20$, $a/b = 1$, $n = 2$, $K_s = 100$, and $K_w = 100$. The sine-type and global-type imperfection modes show a significant effect on the structural responses and frequency parameter increases with an increase in the imperfection amplitude because imperfection dominates in the elastic stiffness. Moreover, CCCC boundary conditions present a higher structural response as compared to SSSS as expected.

The vibration response of FGM (ZrO_2/Al; $E_c = 151 \times 10^9$ N/m^2, $\rho_c = 3000$ kg/m^3, $E_m = 70 \times 10^9$ N/m^2, $\rho_m = 2707$ kg/m^3, [13]) plate for various imperfection amplitudes ($\eta = 0$–1) with varying foundation parameters ($K_s = k_s b^2/D$ and $K_w = k_w b^4/D$, where $D = E_m h^3/12(1 - \nu^2)$, [12]) is revealed as listed in Table 3. The study is performed for the plate with $a/h = 20$, $a/b = 1$, $n = 2$, and boundary conditions CCCC and SSSS. As in Table 3, the response increases with an increase

Table 1 Convergence and validation studies

Mesh size	Frequency parameter ($\bar{\omega}$)				
	$n = 0$	$n = 0.2$	$n = 1$	$n = 2$	$n = 5$
2×2	0.0426	0.0404	0.0394	0.0389	0.0390
3×3	0.0417	0.0404	0.0387	0.0382	0.0383
4×4	0.0415	0.0397	0.0385	0.0380	0.0382
5×5	0.0414	0.0395	0.0385	0.0379	0.0381
6×6	0.0413	0.0394	0.0384	0.0379	0.0381
Ref. [14]	0.0411	0.0395	0.0388	0.0386	0.0388
% diff.	−0.49	0.25	1.03	1.81	1.80

Table 2. Free vibration response for various imperfection modes

	η	Frequency parameter ($\bar{\omega}$)					
		Sine-type	G1	G2	G3	L1	L2
SSSS	0.0	0.0379	0.0379	0.0379	0.0379	0.0379	0.0379
	0.2	0.0444	0.0380	0.0379	0.0379	0.0384	0.0384
	0.5	0.0688	0.0385	0.0380	0.0379	0.0410	0.0410
	0.8	0.0982	0.0393	0.0381	0.0380	0.0454	0.0453
	1.0	0.1176	0.0400	0.0382	0.0380	0.0491	0.0489
CCCC	0.0	0.0508	0.0508	0.0508	0.0508	0.0508	0.0508
	0.2	0.0565	0.0509	0.0508	0.0508	0.0512	0.0512
	0.5	0.0792	0.0513	0.0509	0.0508	0.0535	0.0534
	0.8	0.1081	0.052	0.051	0.0508	0.0574	0.0573
	1.0	0.1278	0.0526	0.051	0.0509	0.0607	0.0606

Table 3 Free vibration response for various imperfection amplitudes along with different foundation parameters

	K_s, K_w	Frequency parameter ($\bar{\omega}$)				
		$\eta = 0.0$	$\eta = 0.2$	$\eta = 0.5$	$\eta = 0.8$	$\eta = 1.0$
SSSS	0, 0	0.0202	0.0305	0.0605	0.0924	0.1128
	0, 100	0.0226	0.0321	0.0613	0.0929	0.1132
	100, 0	0.0366	0.0432	0.0680	0.0977	0.1173
	100, 100	0.0379	0.0444	0.0688	0.0982	0.1176
CCCC	0, 0	0.0371	0.0446	0.0712	0.1022	0.1227
	0, 100	0.0385	0.0457	0.0719	0.1026	0.1231
	100, 0	0.0498	0.0556	0.0786	0.1076	0.1275
	100, 100	0.0508	0.0565	0.0792	0.1081	0.1278

in the imperfection amplitude and foundation parameters because of the increase in the elastic stiffness of the plate. The CCCC and SSSS boundary conditions express similar behavior, however, the CCCC boundary condition quantitatively indicates higher responses.

The behavior of ZrO_2/Al FGM plate for different side to thickness ratios is described as shown in Fig. 2. All edges clamped (CCCC) square plate ($a/b = 1$) is considered with volume fraction index (n), and foundation parameters (K_s, K_w) as 2, and (100,100), respectively. The sine-type imperfection mode is employed here. As shown in Fig. 2, the response increases with an increase in the imperfection amplitude, while it decreases with an increase in the thickness ratio because of the provided foundation.

Fig. 2 Vibration response for different side to thickness ratios

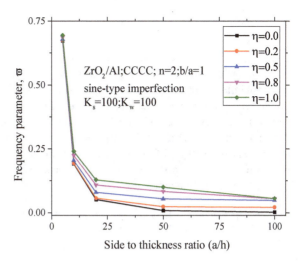

Figure 3 explores the vibration characteristics in terms of frequency parameter ($\bar{\omega} = \omega_{min} h \sqrt{\rho_m/E_m}$) of the ZrO$_2$/Al FGM plate for different aspect ratios. All edges of the imperfect plate ($\eta = 0.5$) are assumed to be clamped with thickness ratio (a/h), and foundation parameters (K_s, K_w) as 20, and (100, 100), respectively. The global type-G1 imperfection mode is taken into consideration. The aspect ratio (a/b) is varied from 1 to 3. As in Fig. 2, the response decreases with an increased volume fraction index and aspect ratio. Moreover, the highest response has been noted for the ceramic plate ($n = 0$) while the lowest for metal ($n \to \infty$).

Fig. 3 Vibration response for different aspect ratios

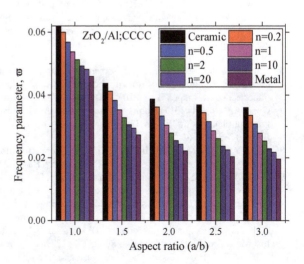

4 Conclusion

In the present study, the natural frequency of imperfect FGM plates resting on elastic foundation has been investigated. The conclusions have been made concerning different influencing parameters like volume fraction indices, foundation parameters, imperfection amplitudes, etc. The present mathematical model is well in terms of solution accuracy and convergence at a lesser mesh size. The ceramic plates show higher vibrations as compared to metallic plates. Imperfection and foundation dominate the elastic stiffness of the plate, and consequently, the vibration response significantly increases. Overall, the vibration response explores the significant changes for different imperfection amplitudes and foundation parameters.

Acknowledgements This work is supported by AR&DB, DRDO, Govt. of India through a project (**MSRB/TM/ARDB/GIA/16-17/0051; Dated Aug 24, 2016**), and the authors are very thankful for the same.

References

1. Gupta A, Talha M (2015) Recent development in modeling and analysis of functionally graded materials and structures. Progress in Aerosp Struct 79:1–14
2. Shakir M, Talha M (2019) On the dynamic response of imperfection sensitive higher order functionally graded plates with random system parameters. Int J Appl Mech 11(3):1950025
3. Lü CF, Lim CW, Chen WQ (2009) Exact solution for free vibrations of functionally graded thick plates on elastic foundations. Mech Adv Mater Struct 16(8):576–584
4. Kitipornchai S, Yang J, Liew KM (2004) Semi-analytical solution for nonlinear vibration of laminated FGM plates with geometric imperfections. Int J Solids Struct 41(9–10):2235–2257
5. Yang J, Huang XL (2007) Nonlinear transient response of functionally graded plates with general imperfections in thermal environments. Comput Methods Appl Mech Eng 196(25–28):2619–2630
6. Gupta A, Talha M (2016) An assessment of a non-polynomial based higher order shear and normal deformation theory for vibration response of gradient plates with initial geometric imperfections. Compos B Eng 107:141–161
7. Gupta A, Talha M (2017) Nonlinear flexural and vibration response of geometrically imperfect gradient plates using hyperbolic higher-order shear and normal deformation theory. Compos B Eng 123:241–261
8. Gupta A, Talha M (2018) Influence of initial geometric imperfections and porosity on the stability of functionally graded material plates. Mech Based Design Struct Mach 46(6):693–711
9. Gupta A, Talha M (2018) Imperfection sensitivity of the post-buckling characteristics of functionally gradient plates using higher-order shear and normal deformation theory. In: IOP conference series: materials science and engineering, vol 330, no 1, p 012091. IOP Publishing
10. Tomar SS, Talha M (2019) On the flexural and vibration behavior of imperfection sensitive higher order functionally graded material skew sandwich plates in thermal environment. Proc Inst Mech Eng Part C J Mech Eng Sci 233(4):1271–1288
11. Talha M, Singh BN (2010) Static response and free vibration analysis of FGM plates using higher order shear deformation theory. Appl Math Model 34(12):3991–4011
12. Matsunaga H (2000) Vibration and stability of thick plates on elastic foundations. J Eng Mech 126(1):27–34

13. Reddy JN, Chin CD (1998) Thermomechanical analysis of functionally graded cylinders and plates. J Therm Stresses 21(6):593–626
14. Baferani AH, Saidi AR, Ehteshami H (2011) Accurate solution for free vibration analysis of functionally graded thick rectangular plates resting on elastic foundation. Compos Struct 93(7):1842–1853

A Geometrically Inspired Model for Kirchhoff Shells with Cartan's Moving Frames

Bensingh Dhas and Debasish Roy

Abstract This work attempts at an entirely novel reformulation of Kirchhoff shells using the theory of moving frames which is a powerful alternative method to describe the geometry of smooth manifolds. In this setup, the defining geometrical features of the non-Euclidean manifold co-evolve with deformation, which means that the model is inherently equipped to accurately describe (possibly irreversible) changes in the referential geometry. Insights from the geometrically transparent kinematics as well as an analytical solution reported for thin strips undergoing very large cylindrical bending explicate on the far-reaching implications of the proposed shell model and its possible spin-offs.

Keywords Differentiable manifolds · Moving frames · Surfaces · Kirchhoff shells

1 Introduction

Thin structures are ubiquitous in nature and engineering; at small scales, they may manifest as thin-walled nanotubes and membrane boundaries of living cells. At larger length scales, they may appear as the fuselage of an aircraft and other engineered structures where weight is a critical factor. These structures have very high in-plane stiffness and very low out-of-plane bending and shear stiffness. The mechanical response of these structures are quite interesting even in the small strain regime (e.g. within elastic limits), due to the presence of multiple equilibrium solutions leading to abrupt changes in the load-deformation behavioural pattern. This behaviour is attributed to an interplay between the geometry of deformation and forces which

---

B. Dhas · D. Roy (✉)
Center of Excellence in Advanced Mechanics of Materials, Indian Institute of Science, Bangalore 560 012, India
e-mail: royd@iisc.ac.in

© The Editor(s) (if applicable) and The Author(s), under exclusive license to Springer Nature Singapore Pte Ltd. 2021
S. K. Saha and M. Mukherjee (eds.), *Recent Advances in Computational Mechanics and Simulations*, Lecture Notes in Civil Engineering 103,
https://doi.org/10.1007/978-981-15-8138-0_14

sustain them. A commonly employed mathematical model to capture such response is the shell theory developed by Koiter, which is a Kirchhoff type shell. This shell theory relies on the theory of surfaces developed by Gauss. The energy functional associated with this shell theory is quadratic in the mid-surface curvature.

A major difficulty in arriving at a well-rounded shell theory is caused by the distinction between the integrability (anti-derivative problem) and integration (quadrature problem) in dimensions higher than one [1]. A common manifestation of this fact is that, while a finite deformation (Kirchhoff) beam theory has no integrability constraints, one has to enforce the Gauss and Codazzi–Mainardi equations to ensure the compatibility between the metric and curvature tensors in shells [2]. A way around these compatibility equations is to formulate the equilibrium problem purely in terms of deformation, which would result in a system of fourth-order partial differential equations. The computer graphics community has developed a somewhat simplistic discrete approach to shells [3, 4], on a given triangulation and by assuming incompressibility, they restrict their degrees of freedom to be the dihedral angles between two triangles, which in turn are written in terms of the nodal coordinates of the triangle. Such an approach overlooks the forces that can be sustained by edges of a triangular shell element, leading to artificial stiffening effects in the prediction of load-deformation behaviour.

In this work, we discuss a reformulation of finite deformation Kirchhoff shell theory using Cartan's moving frame. This reformulation aims at clarifying the relationship between the equilibrium equations of a shell and the geometry of surfaces. An important outcome is that one can clearly distinguish the evolution of geometry from that of deformation, which is supposed to be a major challenge in developing solution procedures for the equation of motion for shells. In contrary, three dimensional elasticity assumes the geometry of the deformed configuration to be known and evolves only the deformation. Interestingly, the present reformulation should have a very important application in defect mechanics where the evolution of defects is associated with the geometry of the configuration.

2 Kinematics of Kirchhoff Shells

Kirchhoff shells are special models developed to approximate the 3D continuum mechanical response of solid bodies with one dimension much smaller than the other two. In other words, these are models for thin deformable bodies. Conventionally, two different approaches have been used to develop these models; the first is based on dimensional reduction (restricting the three-dimensional kinematics to two), while the second postulates an extended 2D continua without any reference to the 3D body. Both approaches have their own advantages and limitations; for details, see [5]. We follow the latter approach by postulating the mid-surface of the shell to be a surface embedded in 3D Euclidean space with the unit normal pointing in the thickness direction. Surfaces are non trivial examples of Riemann manifolds with non-zero curvature. The key invariants of a Riemann manifold are the metric and curvature tensors. For a surface, these tensors are of rank two, which are sometimes referred to as the first and second fundamental forms, respectively. The conventional approach to the theory of surfaces is to use Christoffel symbols to encode geometric information. In this work, we develop an alternate picture which is based on the theory of moving frames. We first discuss differential forms which are central to the theory of moving frames.

2.1 Differential Forms

Differential forms appear as a natural descriptor of many phenomena of interest in science and engineering. The theory of differential forms was developed initially by Hermann Grassmann and then by Ellie Cartan. Within differential geometry, differential forms have been a useful tool to study the geometry of differentiable manifolds. One of the classical examples of a physical theory that was effectively reformulated using differential forms is the theory of electromagnetism. This reformulation explicated the key geometric structure of the theory. An important consequence of this reformulation is the development of efficient numerical solution procedures that avoided spurious oscillations present in their conventional counterparts [6, 7].

We now present a brief introduction to differential forms; the material is standard and can be found in any introductory text in differential geometry, e.g. [8, 9]. Let $\mathcal{M}$ be a smooth manifold; at any point of $\mathcal{M}$, one may define the tangent space, which is a vector space. We denote it by $T_x\mathcal{M}$. The dual to this vector space is denote by $T_x^*\mathcal{M}$. If $\alpha \in T_x^*\mathcal{M}$ and $v \in T_x\mathcal{M}$, then the action $\alpha(v)$ produces a real number. If one picks at each point of $\mathcal{M}$ an element in $T_x^*\mathcal{M}$, we say that a one-form is defined on $\mathcal{M}$. Force is an important example whose differential geometric representation is a one-form; force acts on a velocity vector to produce power. Electric field is another example which can be represented as a one-form. An alternative way to understand an n-form is that they can be integrated over an n sub-manifold to produce a real number. In other words, an n-form is a map from n sub-manifolds to real numbers. The degree of a differential form is an important property; it is the dimension of the sub-manifold on which the form has to be integrated to produce a real number; it can vary from zero to the dimension of $\mathcal{M}$.

On the space of differential forms, one may define two important operations, viz., wedge product and exterior derivative. The wedge product denoted by $\wedge$ is a bilinear, antisymmetric operation. It takes two differential forms of degrees m and n to produce a differential form of degree $m + n$. The wedge product is a purely algebraic operation. The exterior derivative along with Stokes theorem defines a calculus (the exterior calculus) on the space of differential forms. Denoted by d, it takes a differential form of degree m and returns a differential form of degree $m + 1$.

Fig. 1 The line integral of the first structural equation along the curve Γ gives the difference between the position vectors between the start and end points of the curve

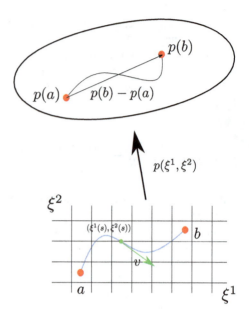

2.2 Theory of Moving Frames

Consider a thin elastic shell; we denote the mid-surface of the shell by $\mathcal{S} \subset \mathbb{R}^3$. At each point of the surface, one can define two types of vectors: one tangent and the other normal to the surface. A moving frame is an ordered triplet containing two orthonormal tangent vectors and the unit normal. We denote this triplet $\{e_1, e_2, e_3\}$ by $\mathcal{F}$. We choose the vector fields e_1 and e_2 such that they are the principal directions of the surface metric tensor. If $p \in \mathcal{S}$, the differential dp is given by

$$dp = \theta^i e_i; \quad i = 1, 2. \tag{1}$$

In the above equation, θ^i are scalar valued one-forms, which are also referred to as co-frames. To understand (1), we consider a coordinate system for the surface $\mathcal{S}$, see Fig. 1; the position vector p is now a smooth function of these coordinates (ξ^1, ξ^2). Consider a curve Γ parameterized by its arc-length s, $\Gamma : [0, L] \to \mathcal{S}$ and let v be a vector tangent to Γ; it belongs to the tangent space of $\mathcal{S}$. The integral of (1) along a curve leads to the displacement vector between the start and end points of the curve, which is given by

$$p(b) - p(a) = \int_a^b dp(v(s))ds$$
$$= \int_a^b (e_1 \theta^1(v) + e_2 \theta^2(v))ds. \tag{2}$$

Equation (1) is also called Cartan's first structural equation [9]. The vector fields e_i satisfy the relationship $\langle e_i, e_j \rangle = \delta_{ij}$; here, $\langle .,. \rangle$ is the usual inner product in $\mathbb{R}^3$. The orientation of the frame fields along the surface $\mathcal{S}$ is determined by the second structural equations given by

$$de_i = \omega_i^j e_j; \quad i = 1, \ldots, 3, \tag{3}$$

where ω_i^j is a matrix with one-forms as its entries. It is sometimes referred to as the connection matrix and its entries as connection one-forms. In the above equation, de_i is a vector-valued one-form; it acts on a tangent vector to give a vector. Since we have chosen an orthonormal frame, we also have $\omega_i^j = -\omega_j^i$. The connection one-forms can be used to compute the rate of change of vector fields on a manifold. If $w = \sum_{i=1}^{2} w^i e_i$ is an arbitrary vector field defined on the surface $\mathcal{S}$, then the covariant derivative of w in direction e_i is given as

$$\nabla_{e_i} w = dw^j(e_i)e_j + w^j \omega_j^k(e_i)e_k; \quad j, k = \{1, 2\}. \tag{4}$$

We refer to the covariant derivative of a vector field in the ith direction also by the symbol $w_{;i}$. Similar to the first structural equation, (3) can also be integrated along a curve; this integral gives the rotation of the frame $\mathcal{F}$ between the points $p(a)$ and $p(b)$. The relationship between the frame fields at a and b can thus be written as

$$e_i = \int_a^b e_j \omega_i^j(v) ds. \tag{5}$$

From (1) and (3), we see that the differentials of the position vector and the moving frame are given by five one-forms. An important point to note is that these equations now strongly depend on the curve Γ. However, the position vector and frame are all point-wise quantities, hence the differential one-forms which describe them must satisfy several exactness (Poincaré) relations to be compatible. These equations can be written as

$$ddp = 0; \quad dde_i = 0. \tag{6}$$

The first part of the compatibility equation (6) leads to the following system of exterior differential equations:

$$d\theta^i = \theta^j \wedge \omega_j^i; \quad i = 1, .., 3. \tag{7}$$

The second part of the compatibility equation leads to

$$d\omega_j^i = -\omega_k^i \wedge \omega_j^k. \tag{8}$$

This compatibility equation ensures that the surface is embedded in a space with zero curvature. We introduce the following definitions to enable a simple matrix representation for the structure equations and compatibility conditions: $\lambda^1 := \omega_1^3$, $\lambda^2 := \omega_2^3$ and $\lambda^3 := \omega_1^2$. In matrix form, the first structure equations can be written as

$$\begin{bmatrix} de_1 \\ de_2 \\ de_3 \end{bmatrix} = \begin{bmatrix} 0 & \lambda^3 & -\lambda^1 \\ -\lambda^3 & 0 & -\lambda^2 \\ \lambda^1 & \lambda^2 & 0 \end{bmatrix} \begin{bmatrix} e_1 \\ e_2 \\ e_3 \end{bmatrix}. \qquad (9)$$

Using the same definitions, the first compatibility equations can be written as

$$\begin{bmatrix} d\theta^1 \\ d\theta^2 \\ 0 \end{bmatrix} = \begin{bmatrix} 0 & \lambda^3 & -\lambda^1 \\ -\lambda^3 & 0 & -\lambda^2 \\ \lambda^1 & \lambda^2 & 0 \end{bmatrix} \wedge \begin{bmatrix} \theta^1 \\ \theta^2 \\ 0 \end{bmatrix}. \qquad (10)$$

It should be emphasized that the combining rule for matrix multiplication is the wedge product, not ordinary multiplication. The third relation in (10) leads to the condition $\theta^1 \wedge \lambda^1 + \theta^2 \wedge \lambda^2 = 0$. Using Cartan's theorem [9] on the just noted relationship, we conclude the existence of a unique symmetric matrix with real coefficients which relate λ^1, λ^2 and θ^1, θ^2,

$$\begin{bmatrix} \lambda^1 \\ \lambda^2 \end{bmatrix} = \begin{bmatrix} a_{11} & a_{12} \\ a_{12} & a_{22} \end{bmatrix} \begin{bmatrix} \theta^1 \\ \theta^2 \end{bmatrix}. \qquad (11)$$

In (8), the anti-symmetry of ω leads to only three independent equations. In matrix form, these equations can be written as

$$\begin{aligned} d\lambda^1 &= -\lambda^2 \wedge \lambda^3 \\ d\lambda^2 &= -\lambda^3 \wedge \lambda^1 \\ d\lambda^3 &= -\lambda^1 \wedge \lambda^2. \end{aligned} \qquad (12)$$

In terms of the vector-valued one-form dp, the first fundamental form can be given as

$$I = dp.dp. \qquad (13)$$

The dot product used to define the first fundamental form is understood as the dot product defined on the range of dp. The orthogonality between e_1 and e_2 leads to the following expression for the first fundamental form:

$$I = \theta^1 \otimes \theta^1 + \theta^2 \otimes \theta^2. \qquad (14)$$

We also introduce the following definition: $j := \theta^1 \wedge \theta^2$; j measures an infinitesimal area associated with the surface $\mathcal{S}$. The second fundamental form can now be written as

$$II = -de_3.dp$$
$$= \lambda^1 \otimes \theta^1 + \lambda^2 \otimes \theta^2. \tag{15}$$

Using the relationship given in (11), the above equation can be rewritten as

$$II = \begin{bmatrix} \theta^1 \\ \theta^2 \end{bmatrix}^t \begin{bmatrix} a_{11} & a_{12} \\ a_{12} & a_{22} \end{bmatrix} \begin{bmatrix} \theta^1 \\ \theta^2 \end{bmatrix}. \tag{16}$$

Note that the symmetry of the second fundamental form essentially follows from (16). The mean and Gaussian curvature are defined to be the trace and determinant of the coefficient matrix given in Eq. (11). We thus have

$$H = \frac{1}{2}(a_{11} + a_{22}) \tag{17}$$

$$K = a_{11}a_{22} - a_{12}^2. \tag{18}$$

We intend to write H and K in terms of the co-frame fields and connection one-forms. By taking a wedge product with θ^2 in the first relation of (11) and a wedge product with θ^1 in the second relation, we arrive at the following:

$$\lambda^1 \wedge \theta^2 = a_{11}\theta^1 \wedge \theta^2$$
$$\lambda^2 \wedge \theta^1 = a_{22}\theta^2 \wedge \theta^1. \tag{19}$$

Subtracting the second relation from the first, we have the following:

$$\lambda^1 \wedge \theta^2 - \lambda^2 \wedge \theta^1 = (a_{11} + a_{22})j. \tag{20}$$

Using the definition of H in the above equation, we arrive at the following expression for mean curvature:

$$H = \frac{1}{2j}(\lambda^1 \wedge \theta^2 - \lambda^2 \wedge \theta^1). \tag{21}$$

Similarly, taking a wedge product with ω_2^3 in the first relation in (11) leads to

$$\lambda^1 \wedge \lambda^2 = (a_{11}a_{22} - a_{12}^2)j. \tag{22}$$

Using the definition of Gaussian curvature in the above equation, we arrive at the following expression:

$$K = \frac{1}{j}(\lambda^1 \wedge \lambda^2). \tag{23}$$

2.3 Integration of Curves on the Surface

Once the compatibility between the deformation one-forms and connection one-forms is established, the position vector of each material point in the mid-surface of the shell can be computed using a simple integration. For any curve Γ in the reference configuration (chart), the deformation of Γ under the deformation map can be computed using the following formula:

$$p(\tau) = p_0 + \int_\Gamma dp(\tau) d\tau$$
$$= p_0 + \int_\Gamma \left(e_1 \theta^1(v) + e_2 \theta^2(v) \right) d\tau. \tag{24}$$

In the above equation, v is the vector tangent to the path Γ. Note that we have used only the definition of deformation differential to arrive at this formula. Moreover, to use this formula one should also know p_0, which can be the value of deformation at a Dirichlet boundary.

3 Energy Function of the Shell

Koiter proposed a stored energy density to capture the bending response of thin elastic structures; the energy density is given by

$$W = \frac{Eh^3}{24(1-\nu^2)}[(1-\nu)\text{Tr}(II^2) + \nu\text{Tr}(II)^2]. \tag{25}$$

Here, h is the thickness of the shell, E and ν are the Young's modulus and Poisson's ratio. Koiter's shell is a finite deformation theory applicable for large deformations, and small strain problems. Using the Cayley-Hamilton theorem, the above energy function can be written as

$$W = C_1 H^2 + C_2 K. \tag{26}$$

where C_1 and C_2 depend on material parameters like E, ν and h. The total energy of the shell due to bending may now be written as

$$I = \int_B C_1 H^2 j. \tag{27}$$

Using the definition of H given in (21), the total energy function can be written as

$$I = \int_B \frac{C_1}{4j} (\lambda^1 \wedge \theta^2 - \lambda^2 \wedge \theta^1)^2. \tag{28}$$

The square in the above equation makes sense since two forms are the highest degree forms on an embedded surface in 3D, hence represented as a real-valued function. The equilibrium configuration of a Kirchhoff shell is now given by the minimizer of (28), subject to the compatibility constrains (1), (8) and constraints introduced by mechanical equilibrium (linear and angular momentum balances). Note that the fields to be solved are not displacements but one-forms related to the deformed surface $\mathcal{S}$; this makes the formulation inherently mixed.

We introduce two second-order tensors $N^{\alpha\beta}$ and $M^{\alpha\beta}$ to characterize the internal forces and moments developed in the shell. The tensor $N^{\alpha\beta}$ is conjugate to the metric tensor while $M^{\alpha\beta}$ is conjugate to the curvature tensor. The force and moments developed in the shell may now be calculated as

$$n^\alpha = N^{\alpha\beta} e_\beta; \quad m^\alpha = M^{\alpha\beta} e_\beta \quad \alpha, \beta \in \{1, 2\}. \tag{29}$$

The energy function given in (27) is independent of the metric tensor and hence (27) does not produce internal forces. The bending moment tensor generated by (27) is given by

$$M = \frac{\partial W}{\partial II}$$
$$= 2C_1.H.I. \tag{30}$$

It should be mentioned that even though the n^α has no contribution from the stored energy, it may have contributions from forces arising from compatibility constraints.

4 Kinetics

Having discussed the kinematics, we now proceed to describe the equations of mechanical equilibrium. Assuming only elastic response, we keep at bay additional complexities due to thermodynamics. The equilibrium equations presented in this section can be found from multiple sources; e.g. [10–12]. Unlike 3D continuum mechanics, balances of linear and angular momenta lead to an independent system of partial differential equations. The balance of linear momentum can be written as

$$\frac{1}{j}(jn^\alpha)_{;\alpha} + f = 0. \tag{31}$$

In the above equation, f is the force applied to the surface of the shell and n^α are the internal forces generated by the shell to resist deformation. The balance of angular momentum can be written as

$$\frac{1}{j}(jm^\alpha)_{;\alpha} + (p_{,\alpha} \times n^\alpha) + l = 0. \tag{32}$$

Here, m^α and l are the internal and external couples. The derivatives in the first term of (31) and (32) are understood as covariant derivatives, while the derivative in the second term of (32) is the usual directional derivative. This distinction is important, since m^α is a vector-valued one-form (tangent object) and p is just a map between manifolds.

5 Pure Bending of a Thin Elastic Strip

Consider a thin rectangular elastic strip with one of its edges fixed. The coordinate system and the boundary conditions are shown in Fig. 2. A bending moment is applied at the edge opposite to the fixed edge. This bending moment causes the thin strip to bend in the form of a cylinder about a line parallel to the fixed edge. Even though the deformation is relatively large, the state of strain is homogeneous with unit stretch, the curvature is constant along x direction and zero curvature along the y direction. We now specialize the previously discussed theory for this problem, since the strip is known to bend only in one direction and does not stretch; the only unknown in the problem is $\lambda^1 = \lambda$; we assume the other connection one-forms to be zero. The deformation one-forms remain the same as in the reference and deformed configurations. We assume a Cartesian coordinate system for the reference configuration of the thin strip. In such a coordinate system, the deformation one-forms can be written as

$$\theta^1 = dx; \quad \theta^2 = dy. \tag{33}$$

The frame fields of the deformed configuration are denoted by e_1, e_2 and e_3; our assumption about deformation restricts the connection matrix to be

$$\omega = \begin{bmatrix} 0 & 0 & -\lambda \\ 0 & 0 & 0 \\ \lambda & 0 & 0 \end{bmatrix}. \tag{34}$$

The second fundamental form can now be computed as

$$II = -\lambda \otimes \theta^1. \tag{35}$$

Fig. 2 A thin cantilever strip subjected to end moment

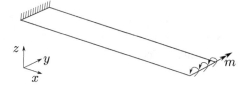

Using the first compatibility condition in the form of Cartan's lemma, we have

$$\lambda = a^{11}\theta^1. \tag{36}$$

In the above equation, a^{11} is a real-valued function dependent on both the directions x and y; using this relationship in II, we arrive at

$$II = a^{11}\theta^1 \otimes \theta^1. \tag{37}$$

Now the problem is to determine a^{11} from equations of compatibility, mechanical equilibrium and boundary conditions. Using the kinematic assumptions on the second compatibility condition leads to

$$d\lambda = 0$$
$$da^{11} \wedge \theta^1 + a^{11}d\theta^1 = 0.$$

From (33), we see that $dd\theta^1 = 0$; using this result in the above equation leads to

$$da^{11} \wedge \theta^1 = 0 \tag{38}$$

Now using the definition of differential for a real-valued function, we have $da^{11} = \frac{\partial a^{11}}{\partial x}dx + \frac{\partial a^{11}}{\partial y}dy$ using it in the above equation gives

$$\left(\frac{\partial a^{11}}{\partial x}dx + \frac{\partial a^{11}}{\partial y}dy\right) \wedge dx = 0. \tag{39}$$

And finally using the property of wedge product,

$$\frac{\partial a^{11}}{\partial y}dy \wedge dx = 0. \tag{40}$$

Since $dx \wedge dy$ is the volume-form associated with the deformation and can never be zero, the conclusion using this is

$$\frac{\partial a^{11}}{\partial y} = 0. \tag{41}$$

The above conclusion implies that a^{11} cannot vary along the y direction. We now proceed to use the mechanical equilibrium equation to arrive at the functional form of a^{11}, which now depends only on x. The mean curvature associated with the deformation can then be computed as

$$H = \frac{a^{11}}{j}\theta^1 \wedge \theta^2. \tag{42}$$

The bending moment tensor for the II given in (35) leads to

$$M = C_1 a^{11} \theta^1 \wedge \theta^2 I. \tag{43}$$

which leads to $m^1 = -C_1 a^{11} \theta^2$ and $m^2 = C_1 a^{11} \theta^1$. The moment equilibrium equations and the definition of covariant derivative then lead to

$$- da^{11}(e_1)\theta^2 + da^{11}(e_2)\theta^1 = 0. \tag{44}$$

Using (41) in the above equation leads to

$$\frac{\partial a^{11}}{\partial x} \theta^2 = 0. \tag{45}$$

In the above equation, the left-hand side can be equal to the zero function only if $\frac{\partial a^{11}}{\partial x} = 0$. Combining (41) and (45) leads to the conclusion that a^{11} has to be constant within the domain,

$$\lambda = c\theta^1, \tag{46}$$

where c is a constant. By applying the boundary condition that $m^1(e_1) = m$ on the free end, we conclude that

$$c = -\frac{m}{C_{1_A}} \tag{47}$$

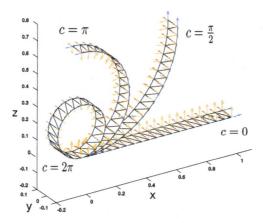

Fig. 3 The figure shows the deformed configuration of the thin elastic strip for different values of $c = 0, \frac{\pi}{2}, \pi$ and 2π. The frame fields are superimposed on the deformation, the tangent vectors are represented by red and blue arrows, while the yellow vectors represent the unit normal. The triangulation on the deformed configuration is only a visual aid; it has nothing to do with the solution procedure

Having computed the one-form λ, we now proceed to compute the frame field and position vector of the deformed configuration. These quantities are computed by integrating Cartan's first and second structure equations (1) and (3) along the curves in the reference configuration. The structure equations reduce to a system of linear ordinary differential equations when restricted along the curves. For simplicity, we consider straight lines originating from the fixed end of the elastic strip along the x direction. This choice is to take advantage of the displacement and rotation boundary conditions known at the fixed edge. These boundary conditions now become initial conditions for the restricted structure equations. A plot of the integrated position vector and frame field associated with the deformed configuration is shown is Fig. 3.

6 Conclusion

This chapter discussed a reformulation of Kirchhoff shell theory using the theory of moving frames. The main feature of this reformulation is that it exposes the compatibility conditions between the connection one-forms and deformation one-forms suitable for computation. In other words, they clearly separate geometry and deformation. The Kirchhoff shell theory is a special case in which deformation determines the geometry. We also demonstrated the theory for the case of pure bending of a thin elastic strip. Complicated loading and geometry await a numerical implementation using finite element exterior calculus or discrete exterior calculus.

Things become more involved and interesting when new degrees of freedom like director are bought in. Most often, these degrees of freedom have coupling between geometry and deformation. One such situation is the Reissner's shell theory. It will be an interesting exercise to develop the geometry of Reissner's shell using moving frames; such an exercise is kept as a scope for future work. Another interesting extension of the present work is to apply the present kinematic techniques to problems in defect mechanics where geometry of a given configuration is determined by the presence of defects like dislocation or disinclination. In such problems, deformation and geometry are two distinct entities with an energetic coupling between them.

References

1. Tao T, Differential forms and integration. https://www.math.ucla.edu/~tao/preprints/forms.pdf
2. Ciarlet PG (2005) An introduction to differential geometry with applications to elasticity. J Elast 78(1–3):1–215
3. Gingold Y, Secord A, Han JY, Grinspun E, Zorin D (2004) A discrete model for inelastic deformation of thin shells. In: ACM SIGGRAPH/Eurographics symposium on computer animation
4. Grinspun E, Hirani AN, Desbrun M, Schröder P (2003) Discrete shells. In: Proceedings of the 2003 ACM SIGGRAPH/Eurographics symposium on computer animation. Eurographics Association, pp 62–67
5. Naghdi PM (1973) The theory of shells and plates. Linear theories of elasticity and thermoelasticity. Springer, Berin, pp 425–640

6. Bossavit A (1988) Whitney forms: a class of finite elements for three-dimensional computations in electromagnetism. IEE Proc (Phys Sci, Meas Instrum, Manag Educ, Rev) 135(8):493–500
7. Arnold D, Falk R, Winther R (2010) Finite element exterior calculus: from hodge theory to numerical stability. Bull Am Math Soc 47(2):281–354
8. Tu L (2010) An introduction to manifolds. Universitext, Springer, New York
9. Guggenheimer H (1963) Differential geometry. McGraw-Hill series in higher mathematics. McGraw-Hill, New York
10. Simo JC, Fox DD (1989) On a stress resultant geometrically exact shell model. part i: formulation and optimal parametrization. Comput Methods Appl Mech Eng 72(3):267–304
11. Green AE, Zerna W (1992) Theoretical elasticity. Courier Corporation
12. Antman S (2006) Nonlinear problems of elasticity. Applied mathematical sciences. Springer, New York

Geomechanics and Geotechnics

3-Dimensional Analysis of Fixed-Headed Single Pile and 2 × 2 Pile Group in Multilayered Soil

Arindam Dey and Somenath Mukherjee

Abstract Apart from vertical loads from the supporting superstructure, pile foundations are often subjected to lateral loads and moments owing to the actions of wind, traffic, earth pressure, water wave, seismic forces or their combined action. The analysis and design of piles, subjected to such lateral forces, are very important for ensuring the serviceability of the supported structure. This article reports the lateral load behavior of a fixed-headed floating tip single pile and 2 × 2 pile group embedded in layered soil. For this purpose, nonlinear 3-dimensional Finite Element analysis is conducted using PLAXIS 3D. The multilayered soils are represented by Mohr–Coulomb models, while the pile is represented by the elastic embedded beam model. The 3D analysis provides detailed insight into the characteristic development of 3D stress-strain features surrounding the pile that governs its flexural response. The outcomes of the present study are compared to the results from simplified numerical approaches available in the literature, thereby highlighting the efficacy of the developed finite element model. The study revealed that the soil stratigraphy largely influences the flexural response of the pile. The influence of the pile group on the stress-strain behavior of the soil, flexural response of the pile, and the pile–soil interaction is also revealed.

Keywords Laterally loaded pile · Pile group · Nonlinear 3D finite element analysis · p-y curves · Stress-strain response

A. Dey (✉)
Indian Institute of Technology Guwahati, Guwahati 781039, India
e-mail: arindamdeyiitg16@gmail.com

S. Mukherjee
C. E. Testing Co. Pvt. Ltd., Kolkata, India

1 Introduction

Pile foundations are used in important structures like transmission line towers, bridges, chimneys, high-rise building, overhead water tanks, offshore structures like oil platforms and jetties, which are subjected to recognizable lateral forces and moments in addition to the vertical load. Lateral force arises due to wind, earthquake forces, wave action, traffic, lateral earth pressure, and their combinations. To ensure the stability and serviceability of various structures, it is essential to determine the allowable lateral load corresponding to an acceptable lateral deflection at the pile head or cutoff level of the pile. Based on the acceptable limits of the load, the flexural response of the pile foundation is estimated, which is further used to choose the design sections and reinforcements. The analysis of a laterally loaded pile is a complex problem since its flexural behavior is a function of pile–soil interaction.

Several investigators have conducted analytical and/or numerical studies to determine the flexural behavior of laterally loaded piles. Matlock and Reese [1] considered the nonlinear force-deformation characteristics of the soil to reach rational solutions for laterally loaded piles. The elastic theory was repeatedly applied such that the soil modulus constants are adjusted at each successive trial until a compatibility in the structure-soil-pile is achieved. The problem is highlighted on the beams on elastic foundation (BEF) approach and is solved with the aid of Finite Difference equations in their nondimensional form. Different variations of the soil modulus were considered with depth, and the flexural responses of the single rigid pile of varying lengths were reported. The concept was extended to develop a nondimensional analytical solution for exploring the effect of a two-layer soil system on the behavior of a laterally loaded pile [2]. The modulus of subgrade reaction was used to define the stiffness of the surface layer, while the stiffness of the surface layer was defined by the stiffness of the underlying layer. The complete range of the stiffness of two layers was investigated, which revealed that the surface layer poses a significant influence on the behavior of the pile-soil system. For single pile and pile groups embedded in cohesionless and cohesive soil, Broms [3, 4] presented methods to calculate the deflections at working loads, the ultimate lateral resistance, and moment distributions along the piles. Both unrestrained and restrained piles were considered. The lateral deflections were calculated using the concept of a coefficient of subgrade reaction and the ultimate lateral resistance was evaluated. Comparisons of the findings from the analytical approach to that of the experimental investigations revealed satisfactory agreement of deflections and maximum bending moments at working loads. Most of the piles encountered in practice are 'flexible' in nature. In such piles, the induced deformations and bending moments are confined to the upper part of the pile, and the overall length of the pile does not contribute when subjected to lateral loading. In such case, Randolph [5] presented simple algebraic expressions to provide the response of such flexible piles using the finite element method. The soil was represented by linearly varying subgrade modulus. Responses of both single and pile groups embedded in homogeneous soil layers were presented. Indian Standard Code IS 2911 [6] provides charts and tables which can be suitably used to ascertain the subgrade modulus and lateral pile capacity. However, it is applicable only for a single-layered soil, i.e. either sand or clay with constant subgrade reaction. For multilayered soil, no such provision is available in the code and rigorous methods are referred for the same.

Piles supporting bridge foundations often pass through multilayered soil and may or may not be subjected to scour. Depending on the characteristics of the soil strata, the loading conditions, and geometrical properties of the pile, the flexural response of the pile needs to be explored. Often such foundations are subjected to lateral loads and the failure of the same is governed by the serviceability criteria. In general, such foundations are considered to be within the serviceability limit state if the maximum lateral deflection of the pile head (or, at the cutoff level) is less than 1% of the pile diameter. Based on this concept, Yadav and Dey [7] explored the lateral load capacity of a single free-headed pile embedded in a stratified soil deposit that is subjected to scour. OASYS ALP v19.1, employing nonlinear Winkler spring and BEF approach, was used to investigate the influence of scour level on the degradation of the lateral load capacity of the pile and the variation of its flexural responses. For the situation where the pile foundation supporting the bridge pier is not subjected to scour, Mukherjee and Dey [8, 9] have used linear BEF and nonlinear p-y approach to analyze the flexural behavior and lateral load capacity of a fixed-headed floating tip single pile. The influence of pile diameter of the lateral load capacity and flexural response was successfully elucidated. However, it is to be noted that the above-adopted methods consider a lumped parameter modeling approach, wherein the stiffness of the surrounding soil and interaction with the embedded pile are considered through mechanical spring elements. Hence, in the above methods, there are no means to identify the evolution of the stress-strain behavior of the surrounding soil, and the response of only the pile is obtained as a function of the lumped parameters. Further, the above approaches follow a 2-dimensional load-deformation interaction in which the pile is considered as a plane strain element, which is an imprecise reality. It is understandable that a pile subjected to lateral load would exhibit 3-dimensional load dispersion and, accordingly, the induced stress-strain response in the soil would be three-dimensional, which would have different soil–structure interaction mechanisms leading to a different flexural response of the pile as observed in contrary to its 2D behavior. Hence, this article reports the application of 3D Finite Element analysis of a single flexible pile and pile groups subjected to lateral loading. The piles are considered fixed at the top with the aid of a pile cap, while the tip of the pile remains floating in the embedding medium. The influence of four different soil stratification systems on the flexural response and lateral load capacity of the piles is reported in the article. The article exhibits the evolution of the load-deformation and stress-strain mechanisms developing along various sections of the pile, which aids in the development of p-y curves for design purposes.

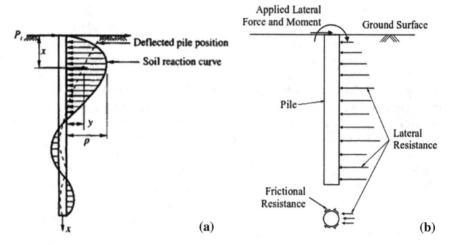

Fig. 1 a Typical representation of the deflection of a laterally loaded pile and the associated soil reaction curve **b** Generation of lateral and shear stresses around a laterally loaded pile

2 Load Transfer Mechanism of Laterally Loaded Pile

When external horizontal load acts on the pile head, the pile behaves as a transversely loaded beam, i.e. a part of the pile tries to shift horizontally in the direction of the applied load causing bending, rotation, and translation of pile. The pile presses upon the soil lying in the direction of applied load, thereby generating compressive and shear stresses in the soil. To satisfy equilibrium, the soil must provide a resistance along the entire pile shaft to balance the external horizontal force and moment. At every point along the length of the pile, the soil resistance (p) and the deflection of the pile (y) are related linearly or nonlinearly, and are generally expressed as $p = ky^n$, k is called the horizontal subgrade modulus and n is the coefficient of nonlinearity. Figure 1a exhibits a typical deflected shape of the pile and the associated soil reaction curve, while Fig. 1b highlights a schematic diagram of the lateral and frictional stresses generated along the pile shaft as the pile is pushed into the adjacent soil.

3 Problem Statement

Figure 2 depicts a single fixed-headed pile with floating tip embedded in stratified soil. Four variants of soil stratification are considered for the present study, which were encountered in in situ projects of bridge foundations. In all the cases, the piles of 1 m diameter is used, which is connected to a rigid pile cap at its head. The bottom of the pile cap rests at a depth of 2.5 m below the existing ground level (EGL), which marks the cutoff level of the pile. The pile cap is flushed at the EGL. The pile cap and the pile are made of M25 grade concrete. As per the standard practice, the serviceability

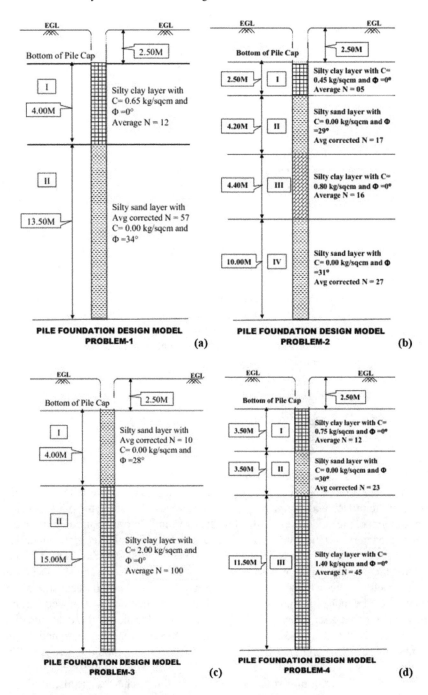

Fig. 2 Soil-pile configurations adopted in the present study

criterion is considered as 1% of pile diameter as the lateral deflection of the pile at its cutoff level. Further, for the same soil stratification and pile configurations, a 2 × 2 pile group is considered and the flexural behavior of the same is ascertained based on the earlier stated serviceability criterion.

4 Finite Element Modeling

A numerical model is a mathematical simulation to a real physical process. A numerical model is different from a physical model prototype or a full-scaled field modeling. In recent decades, the Finite Element method has been used increasingly for the analysis of stress, deformation, structural forces, bearing capacity, stability, and ground water flow in geotechnical engineering applications. The development in mathematics, along with the increase in the calculation power of computers, has led to the development of various robust and user-friendly software tools. In the present study, the flexural response of a laterally loaded single pile and pile groups is modeled using commercially available Finite Element program PLAXIS 3D vAE.01. PLAXIS 3D is used for three-dimensional analysis of ground water flow, stability, and deformation in various geotechnical engineering problems. PLAXIS 3D comprises advanced constitutive models for analysis of nonlinear, anisotropic, and time-dependent behavior of soil or rock.

4.1 Model Description: Domain Size, Meshing, and Boundary Fixity

For the present study, a model domain of 30 m (length) × 30 m (width) × 25 m (height) is considered to define the stratified soil layers. The pile is located at the center of the model, and comprises a pile cap to generate fixity at the pile head. The lateral extent of the model is so chosen that the stress and deformation contours generated by the laterally deforming pile will not be intersected by the vertical boundaries. Since the bottom of the pile is not expected to produce noticeable vertical deformation for the present problem, the total height of model is maintained constant at 25 m, which is more than the maximum length of the pile (23.6 m) as considered for Problem 2. The model domain is discretized into finite number of 10-noded tetrahedral elements that offer many nodes and Gauss points for the precise determination of displacements and stresses, respectively. PLAXIS 3D generates automatic finite element meshes with the aid of the 'robust triangulation technique', supported by 'enhanced mesh refinement' for refining meshes near structural elements, corners, and interfaces. PLAXIS 3D provides five basic meshing schemes specifically, 'very coarse', 'coarse', 'medium', 'fine', and 'very fine'. Each meshing scheme is defined by their relative element size, thereby controlling the basic size of the mesh element

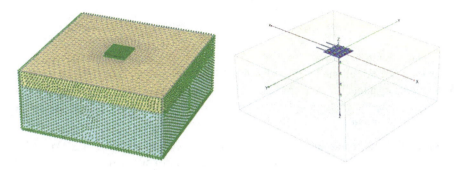

Fig. 3 Typical representation model domain, meshing, structural elements, soil stratification, and boundary fixities as used in the present study for Problem 3

in terms of the maximum and minimum dimensions of the model [10]. Each meshing scheme can be progressively refined with a user-defined mesh coarseness factor. The adopted mesh should be adequately and optimally fine to attain precise numerical outcomes. Hence, it is important to determine the optimal mesh size that satisfies the precision of the findings and remains computationally efficient at the same time. In such a scenario, a mesh convergence study is usually conducted to ascertain the optimal mesh discretization. Furthermore, standard fixity is used at the model boundaries, which allow the vertical lateral boundaries to move vertically without undergoing any lateral deformation, while the basal boundary is chosen to be completely non-yielding. Figure 3 exhibits a typical model domain, highlighting a typical soil stratification, structural components, mesh elements, and boundary fixities. For each of the problem highlighted in Fig. 2, similar PLAXIS 3D models were created.

4.2 Material Models and Model Parameters

In the present study, elastic-perfectly plastic Mohr–Coulomb (M-C) model has been considered for the soil model, while the pile is represented by an embedded beam model (EBM) and the pile cap is represented in terms of a rigid structural plate.

Material Parameters for Soil Layers: M-C model requires five inputs, specifically the strength parameters (cohesion c, angle of internal friction φ, and ψ as dilatancy angle) and the deformation parameters (elastic modulus E and Poisson's ratio v). Figure 2 highlights that for all the four problems, the soil is either silty clay or silty sand, provided along with their shear strength parameters and the average corrected N-value obtained from the Standard Penetration Test (SPT). Based on the available data, the modulus of elasticity of various soil layers are calculated as per Eqs. 1 and 2 [].

$$E_{s(\text{silty clay})} = 350c \quad (1)$$

$$E_{s(\text{silty sand})} = 400 + (10.5 \times N) \qquad (2)$$

The Poisson's ratio for silty sand and silty clay layers are considered 0.3 and 0.35, respectively. For all the layers, the soil is considered non-dilatant, and accordingly, the dilatancy angle is considered zero. Accordingly, for various problems, the corresponding parameters for soil layer are evaluated and used as input parameter in the PLAXIS 3D models.

Material Parameters for Pile and Pile Cap: The pile is represented by EBM, which is defined by its modulus of elasticity, unit weight, type of pile, and its diameter. In the present study, the pile is chosen as a massive circular pile of diameter 1 m, and made of M25 grade concrete. Accordingly, the unit weight of pile material is considered 26 kN/m^3 (as for a reinforced concrete member), and the modulus of elasticity of concrete is estimated as per Eq. 3, where f_{ck} is the characteristic compressive strength or grade of concrete in kN/mm^2 [6].

$$E_c(\text{kN/mm}^2) = 5000\sqrt{f_{ck}} \qquad (3)$$

The pile cap is represented by a structural plate of thickness 2.5 m, and is considered to be made of M36 grade concrete. Accordingly, the modulus of elasticity is calculated from Eq. 3, and the material property is considered isotropic.

4.3 Stages of Analysis

For all the 3D FE models developed for the above-stated problems, the analysis is carried out in three stages (or, phases as in PLAXIS 3D). The 1st phase comprises the in situ stress generation analysis, where the stresses are generated based on the at-rest (or, K0) condition. The displacements generated from this stage are not considered for the next stage of analysis. The 2nd stage comprises the activation of the pile cap and the pile (i.e. the structural elements) in the soil medium, and a plastic loading stage is carried out. Further, the displacements generated from this stage are not considered in the subsequent stage of analysis, while the generated stresses are incorporated. The 3rd stage comprises the activation of surface loading on the pile cap, and the analysis is continued until the stress generated within the soil exceeds its strength, or the maximum number of iterations is reached. Based on the outcome from this phase of analyses, the results corresponding to the achievement of the serviceability criterion at the cutoff level are extracted, and the inferences are presented in the next sections.

Fig. 4 Typical mesh convergence study conducted for Problem 1

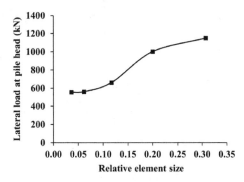

5 Results and Discussions

5.1 Convergence Study

As mentioned earlier, mesh convergence study is important to determine the optimal mesh size for a given problem for a precise as well as a computationally efficient solution. For the present study, various meshing schemes were explored, and the load level satisfying the serviceability criterion at pile cutoff level was measured. Figure 4 depicts the result of the mesh convergence study conducted for Problem 1, which reveals that adopting a 'fine' mesh is sufficient for obtaining results with sufficient precision for the present study. Similar mesh convergence studies were carried out for other problems as well, however, for the sake of brevity, the same are not presented here. Based on the convergence results, 'fine' mesh is considered for the analyses reported herein.

5.2 Flexural Response of Single Pile at Serviceability Limit of Cutoff Level

For each of the problems described in Fig. 2, the lateral load-deformation simulations for the single pile were carried out, and the load level pertaining to the serviceability criteria of 10 mm (1% of the pile diameter) at the cutoff level of the pile was ascertained. Henceforth, this particular load level is coined as the 'lateral load capacity (P_u) of the pile. For a single pile embedded in various soil configurations, Table 1 lists the major entities of the flexural response, namely the ultimate lateral load on the pile cap (P_u), maximum shear force (Q_{max}), maximum bending moment (M_{max}), maximum contact stress (p_{max}), and the length of fixity of pile from the existing ground level (L_{fx}). Following the earlier submission by Randolph [5], it can be observed that variations in soil stratification play a recognizable role in determining the flexural response of the pile. Figure 5 exhibits the flexural response of the single pile for various problems. It can be clearly noticed that even though the bending moment

Table 1 Flexural response parameters of pile upon attaining 1% serviceability criterion at the cutoff level and embedded in varying soil stratigraphies

Soil configuration	P_u (kN)	Q_{max} (kN)	M_{max} (kN-m)	p_{max} (kN/m^2)	L_{fx} (m)
P1	1782	1100	2582	57.5	6.1
P2	1620	1125	2827	35.1	5.1
P3	1782	1390	2816	72.1	5.0
P4	1794	903	2112	30.7	5.9

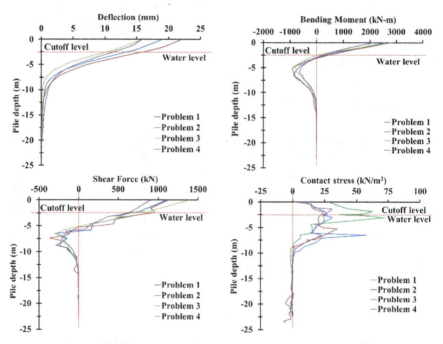

Fig. 5 Comparison of the flexural response at serviceability condition of a fixed-headed single pile embedded in varying soil stratigraphies

and shear force profiles are nearly similar for all the cases, the contact pressure distribution diagrams are different from each other. This is attributed to the variation in the soil stratification and their associated stiffness and strength parameters, thereby resulting in a variation in the deflection characteristics of the pile. A small change in the deflection magnitude can result in noticeable change in the soil-pile contact stress, which is largely evident from the contact stress at the tip of the pile. Although the tip deflection is barely recognizable, the negative contact stress generated at the tip indicated that the pile tip moved against the direction of applied load.

5.3 Lateral Stress-Deformation Behavior and Evolution of Mechanism of a Single Fixed-Headed Pile with Floating Tip

Figure 6 shows the stress-deformation behavior of the single pile for various soil stratigraphies considered in this study. Depending upon the variation in the properties of soil strata, the development of lateral deformation varies with the progressive increase in the lateral stress applied on the pile cap. As has been mentioned earlier, in each case, the pile capacity is ascertained as per the load level corresponding to the attainment of the serviceability criterion. However, for the sake of understanding the evolution of the mechanisms due to the progressive lateral displacement of the pile, the lateral load-deformation simulations were continued up to the limit of deformation of the pile head equal to its diameter. In order to portray the evolution, three different lateral deformation states were considered, namely lateral deformation equal to 1, 10, and 100% of the pile diameter, as marked in Fig. 6.

Figures 7 and 8 depict the typical evolution of the lateral deformation isosurfaces as the pile head is subjected to progressive lateral loading. It can be observed that at the 1% serviceability criterion, the deformation isosurfaces are nearly uniform around the pile and marginally more toward the direction of the applied load. As the lateral deformation progresses, the pile pushes more into the resistive passive zone of the soil, thereby releasing the stress behind the pile. Such a scenario is similar to the development of the passive pressure when a retaining wall is pushed into the backfill. Analogically, a passive mechanism develops around the pile toward the direction of the load, while an active mechanism develops behind it.

Figure 9 depicts a typical representation of the directional displacement of the soil around the pile at its cutoff level at various serviceability states. It is clearly noticeable that as the pile pushes into the soil due to the progressive application of the lateral load, the moving pile displaces the surrounding soil in all the orthogonal directions. This phenomenon indicates a progressively increasing mass of soil participating in resisting the pile movement (U_x), thereby increasing the passive zone. On the other

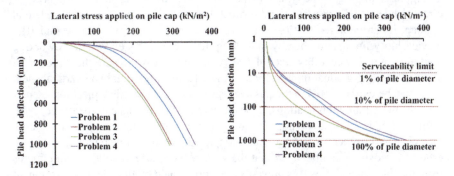

Fig. 6 Lateral stress-deformation behavior of the fixed-headed single pile in various soil stratigraphies and the demarcation of different serviceability states

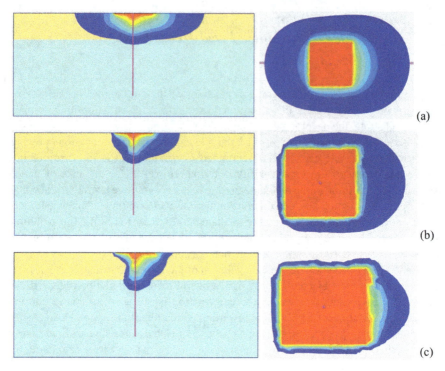

Fig. 7 Typical evolution of deformation isosurfaces with the progressive serviceability limit of a fixed-headed single pile in soil stratigraphy P1 **a** 1%, **b** 10%, **c** 100% of pile diameter

hand, the soil behind the pile acts as stress release under active condition, which gradually decreases with the pile movement. With the increase in lateral deformation of the pile, a significant orthogonal lateral displacement (U_y) and heaving (U_z) can be noticed in the passive region. Additionally, the settlement behind the pile can also be successfully noticed.

Figure 10 highlights the typical evolution of incremental deviatoric strain ($\Delta\gamma_s$), effective horizontal stress (σ_{xx}), and vertical shear stress (σ_{zx}) with progressive lateral loading. It can be noted that recognizable deviatoric strains generate beyond 10% serviceability limit state, thereby indicating that the pile is well within its working load limit at a serviceability state of 1%. As the pile moves toward the direction of applied load, the gradual mobilization of the effective horizontal and vertical shear stress can be clearly noted in the passive region. Further, for this typical representation, the maximum mobilization can be observed at the interface of the upper layers of the soil, indicating the contrast in strength properties of these layers. The finding may change for other soil stratigraphies, depending on the corresponding soil properties. Moreover, it can be noted that the vertical shear stress mobilization initiates from the pile tip, and traverses upwards to be dominated by the pile head movement at higher serviceability limits.

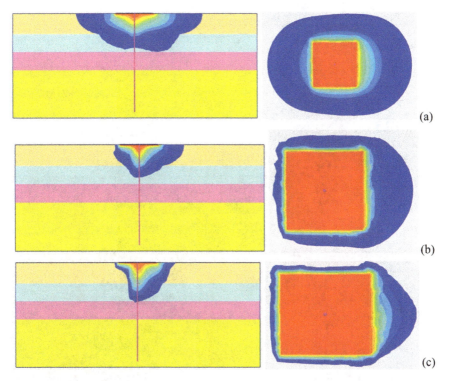

Fig. 8 Typical evolution of deformation isosurfaces with the progressive serviceability limit of a fixed-headed single pile in soil stratigraphy P2 **a** 1%, **b** 10%, **c** 100% of pile diameter

Figure 11 highlights the typical evolution of active and passive horizontal stress and its distribution at the cutoff level based on the in situ condition as well as for various serviceability states pertaining to different extents of pile movement. The horizontal stress at the in situ condition is uniform throughout the cross-section. However, with minimal pile movement and attainment of 1% serviceability state, the initiation of the passive stress can be clearly observed. Further lateral deformation of the pile indicates distinct transition of the active stress from behind the pile to the passive stress in the direction of pile movement. It can be observed that the within a certain zone behind the pile, the active stress nearly vanishes owing to the considerable release of stress in the lateral direction toward the application of load. At zones far away from the pile, as obvious, the in situ horizontal stress prevails. This observation provides the idea about the zone of soil that is influenced by the lateral deformation of the pile at its various serviceability states. The influence zone of the horizontal and shear stresses, as well as the deviatoric strain, is explicitly illustrated further through the cross-section profiles passing through the cutoff level of the pile, as depicted in Fig. 12.

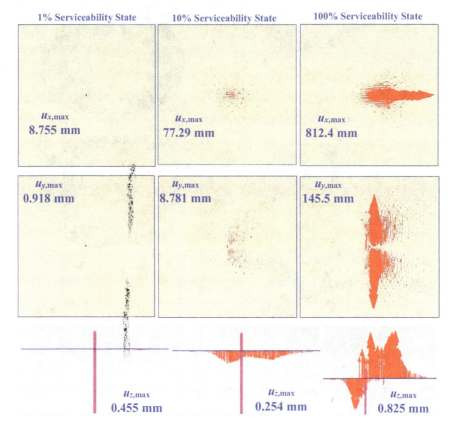

Fig. 9 Typical evolution of directional soil displacement with the progressive serviceability limit of a fixed-headed single pile in soil stratigraphy P2 leading to increasing passive zone

5.4 Flexural Characteristics, Lateral Stress-Deformation Behavior, and Evolution of Mechanism of a Fixed-Headed 2 × 2 Pile Group

The understanding of the lateral load-deformation behavior of a single fixed-headed pile is further extended to a fixed-headed 2 × 2 pile group, typically embedded in P2 soil stratigraphy. Figure 13 depicts a typical schematic of the problem statement considered herein; all the modeling aspects remain the same as described in Sect. 4. Figure 14 exhibits that for various serviceability levels, the lateral load capacity of the fixed-headed 2 × 2 pile group is recognizably higher. For attaining the 1% serviceability criterion at the pile head, the pile group requires a lateral load of 54 kN in comparison to 34 kN lateral load required by a single pile embedded in the same soil stratigraphy. For higher serviceability criteria of 10 and 100% pile diameters, the pile group requires 228 and 650 kN, in comparison to 119 and 295 kN required

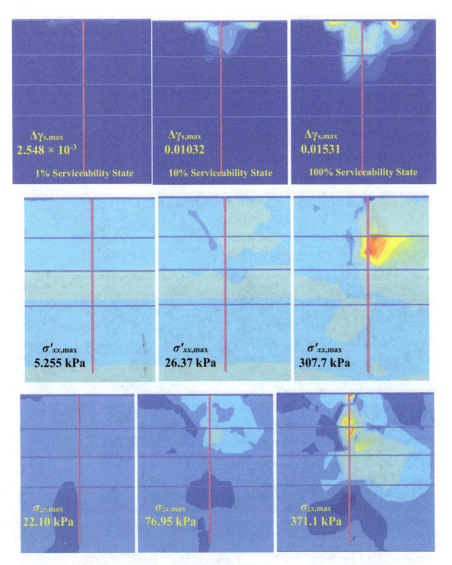

Fig. 10 Typical evolution of directional soil displacement with the progressive serviceability limit of a fixed-headed single pile in soil stratigraphy P2 owing to increasing passive zone

by the single pile for the same conditions. The lateral load capacity of the pile group, for this case, is nearly 1.6–2.2 times higher than that required by the single pile; this range is subjected to change depending upon soil stratigraphy and soil properties.

Figure 15 exhibits the isosurfaces of the incremental deviatoric strain developed with the progressive lateral loading and deformation of the pile group. It can be clearly observed that with an increase in the lateral loading and deformation of the pile group, the deviatoric strain develops all around the pile group, almost like a

Fig. 11 Typical evolution of active and passive stress at cutoff levels along with the different serviceability states of a fixed-headed single pile

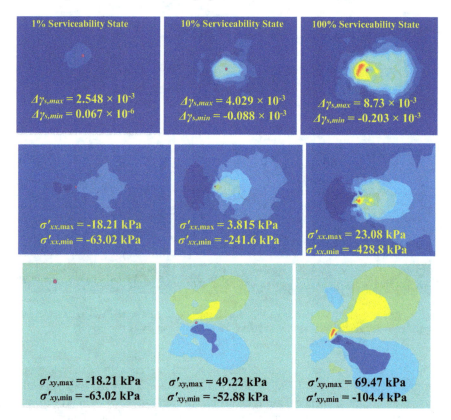

Fig. 12 Typical evolution of deviatoric strain, effective horizontal stress, and horizontal shear stress at the cutoff level for progressive lateral loading of the 2×2 pile group

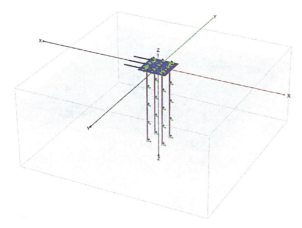

Fig. 13 Schematic representation of a fixed-headed 2 × 2 pile group, with floating tips, embedded in P2 soil stratigraphy

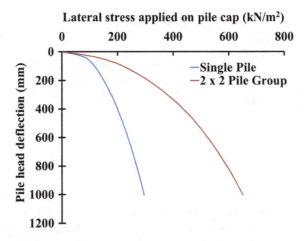

Fig. 14 Schematic representation of a fixed-headed 2 × 2 pile group, with floating tips, embedded in P2 soil stratigraphy

ring around the periphery of the group. Further, it can be noted that within the pile group, the soil does not develop recognizable deviatoric strain, thereby indicating that the volume of soil encased within the piles acts like a block material and moves in unison with the pile during the deformation of the group. This observation confirms the 'block failure' action of the pile group, which is a significantly contributory phenomenon to increase the lateral load capacity of the pile group.

Figure 16 exhibits the development of the effective horizontal stress on the cross-section plane passing through the cutoff level. It can be observed that with the increase in the lateral load, the stress distribution around individual piles gradually increases in size and overlaps at higher load levels. The overlap of these stress isobars around the

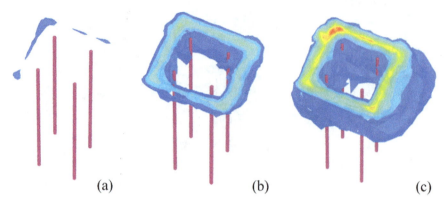

Fig. 15 Evolution of isosurfaces of incremental deviatoric strain with progressing lateral loading of a fixed-headed 2 × 2 pile group for varying serviceability criterion **a** 1%, **b** 10%, **c** 100%

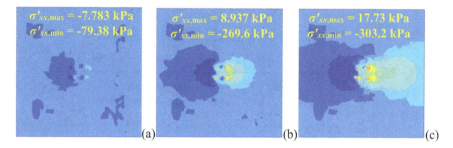

Fig. 16 Evolution of effective horizontal stress at cutoff level with progressing lateral loading of a fixed-headed 2 × 2 pile group for varying serviceability criterion **a** 1%, **b** 10%, **c** 100%

pile results in higher resisting capacity of the soil in the passive zone located toward the direction of the applied load. Hence, the lateral load capacity of the pile group increases noticeably than that of a single pile. Based on the above two observations, it is quite evident that soil–pile interaction plays a significant role in enhancing the performance and lateral load capacity of group piles embedded in stratified layers. Further discussion on the same is beyond the scope of the article.

6 Conclusions

This paper discusses the 3D finite element modeling for the determination of the flexural characteristics of a fixed-headed single pile and a 2 × 2 pile group subjected to lateral load and embedded in stratified soil deposits. The lateral load capacity of the piles is determined from various serviceability limit criteria based on the deformation of the pile recorded at the pile head or the pile cutoff level. The evolution of the stress-strain mechanisms in the pile and the surrounding soil with the gradual attainment of the higher level serviceability criteria is elucidated in this paper. Based on the outcomes from the present study, the following conclusions are drawn:

- 3D Finite Element studies are essential to decipher the lateral load bearing capacity of pile foundations. Conventional estimates of lateral capacity using a plane strain condition fail to portray the three-dimensional stress-strain dissipation as the pile is pushed in the embedding soil medium.
- Among the various flexural characteristics, the contact stress developed at the pile–soil interface is most sensitive to the changes in the pile deformation. A minor change in the deformation can lead to a noticeable change in the contact stress and its profile.
- The lateral loading on a pile generates passive stress zone in the direction of load, and active stress zone behind the pile. A distinct transition from the active pressure from behind the pile to the passive pressure in front of the pile is observed.
- At higher serviceability criterioa, the active stress behind the pile might be much diminished, indicating a detachment of the pile from the adjacent soil, thereby releasing all the tensile stress developed in the pile–soil interface in the active zone.
- For higher serviceability state, the passive resisting zone in front of the pile increases due to the significant displacement of the soil in all orthogonal directions, leading to substantial heaving in the direction of the load, and subsidence away from it.
- The increase in the vertical shear stress with the increase in the lateral load leads to higher soil–pile contact interaction, thereby leading to higher mobilization of friction along the pile surface.
- The lateral deformation of the pile group generates incremental deviatoric strain isosurfaces around the periphery of the piles, there clearly highlighting the development of block mechanism within the soil surrounded by the piles. The block moves along with the pile, and leads to higher bearing capacity owing to its contribution in the passive resistance.
- The overlap of the effective horizontal stress isobars leads to the significant increase in the lateral resistance of the pile against deformation, as the pile has to displace a higher volume of soil to attain a particular serviceability criterion.

References

1. Matlock H, Reese LC (1960) Generalized solutions for laterally loaded piles. J Soil Mech Found Div Proc ASCE 86(SM5):63–91
2. Davisson MT, Gill HL (1963) Laterally loaded piles in layered soil system. J Soil Mech Found Div Proc ASCE 89(SM3):63–94
3. Broms B (1964) Lateral resistance of piles in cohesionless soils. J Soil Mech Found Div Proc ASCE 90(SM3):123–156

4. Broms B (1964) Lateral resistance of piles in cohesive soils. J Soil Mech Found Div Proc ASCE 90(SM2):27–63
5. Randolph M (1981) The response of flexible piles to lateral loading. Geotechnique 31(2):247–259
6. IS 2911 Part I/Sec 2 (2010) Design and construction of pile foundations part 1 concrete piles, Section 2 bored cast in-situ concrete piles. BIS, India
7. Yadav AK, Dey A (2013) Lateral load capacity of piles in stratified deposits subjected to scour. In: Proceedings of The Indian Geotechnical Conference: IGC Ganga 2013, Roorkee. IGS Roorkee, Roorkee, pp 1–10
8. Mukherjee S, Dey A (2016) Analysis of laterally loaded fixed headed single floating pile in multilayered soil using BEF approach. In: Proceedings of the National Seminar on Geotechnics for Infrastructure Development, Kolkata. IGS Kolkata, Kolkata, pp 1–12
9. Mukherjee S, Dey A (2018) Analysis of laterally loaded fixed headed single pile in multilayered soil using P-Y approach. In: Anirudhan IV, Banerjee S (eds) Geotechnical applications, Lecture Notes in Civil Engineering, vol 13, Springer, Singapore, pp 41–50
10. PLAXIS (2017) Reference Manual. 3D Version AE.01. PLAXIS, Delft, The Netherlands

Significance of Interface Modeling in the Analysis of Laterally Loaded Deep Foundations

Ramyasri Rachamadugu and Gyan Vikash

Abstract Interface modeling is one of the important components in the numerical modeling of soil–structure interaction problem. It is equally important as material and geometrical modeling. Inaccurate modeling of the interface between the soil and the structure may lead to unreliable results. However, in many cases, interface modeling is overlooked and the interface between the soil and the structure is generally modeled as a rigid connection. In the present paper, we have shown the influence of interface modeling in the analysis of laterally loaded deep foundations. A finite element model of laterally loaded foundation–soil system is developed wherein the foundation is modeled as a linear elastic system and the soil as nonlinear which is defined by the multi-yield surface plasticity model. The interface between the soil and the foundation is modeled using zero thickness contact element, which is defined, by constitutive relationships capable to define sliding and separation mechanisms at the interface. The results obtained from the proposed finite element model are then compared with the conventional approach of interface modeling. The present study indicates that the conventional approach overestimates the lateral load capacity of deep foundation as well as it is unable to explain the mechanisms of the deformation of laterally loaded foundation–soil system.

Keywords Interface modeling · Lateral loading condition · Soil–well foundation system · Nonlinear material modeling

1 Introduction

Laterally loaded foundation compresses the surrounding soil present at the front side and creates expansion in soil located on the rare side. This interaction between soil and the foundation generates traction along the contact plane. Resolving components of traction will result in the normal and tangential stresses at the interface. Generated

---

R. Rachamadugu · G. Vikash (✉)
Shiv Nadar University, Greater Noida 201314, India
e-mail: gyan.vikash@snu.edu.in

normal stresses are responsible for pressure buildup and tangential stress induces shear between the soil and foundation particles [7]. Tensile normal stress at the rare side of the foundation results in separation. Compressive normal stresses at the front side of the foundation lead to the penetration of soil particles in the foundation or vice versa, whereas both tensile and compressive tangential stress results in sliding at the interface [9, 10]. The major factors that influence soil–foundation interaction are soil nonlinearity and soil type. This interaction between soil and the foundation at the interface is inevitable and is responsible for permanent deformations in a laterally loaded soil–foundation system due to separation and sliding. But the existing design methods preferred for determining lateral response either neglect the interface interaction in design procedure [3] or state that the influence of the interaction is less significant on the response of foundation–soil system [9, 10] or assume no loss in contact between soil and the foundation at the interface. Some of the analytical methods consider interface modeling as linear and nonlinear elastic soil [1, 2]. This elastic approach cannot capture the permanent deformations which occur in foundation–soil system due to separation and sliding at the interface. There exists a need for the development of the finite element method in which accurate modeling of interface elements is included with soil nonlinearity in determining lateral response of the foundation. This helps in understanding the contribution of separation and sliding at the interface in governing the deformation mechanism of the foundation–soil system. Hence, the objective of this study is to observe the influence of interface modeling with the inclusion of soil nonlinearity and soil type in the prediction of a lateral response of the foundation–soil system. This is done by developing a Two-dimensional finite element model using OpenSees [8], a finite element solver and GiD version 14, a pre- and post-processor software with contact elements to capture interface mechanism for deep foundation embedded in nonlinear soils. This study considers Well foundations as it is mostly preferred in India [5, 9–11]. The influence of the interface in lateral response prediction for deep foundation can be observed by comparing the results estimated with and without contact elements.

2 Problem Statement

The present study considers two cases in which a Well foundation of double-D cross-section with slenderness ratio (L/D) as 2 embedded in soft clay and loose sand as shown in Fig. 1. Embedded depth of the well is 24 m and the foundation's width is 12 m. The dredge hole is filled with the sand with a modulus of elasticity of 50 MPa. Bottom plug and the top plug having thickness 5 m and 1 m, respectively, are made of M25 grade concrete. The lower portion of the bottom plug is approximated with a straight curve as shown in Fig. 1. Well cap and well steining of the foundation having thickness 2 m and 1.5 m, respectively, are made of M35 grade concrete. In the present study, the maximum scour level is considered at 12 m depth (L_s) from the riverbed. Lateral load (P) is applied at the top of the well cap as shown in Fig. 1.

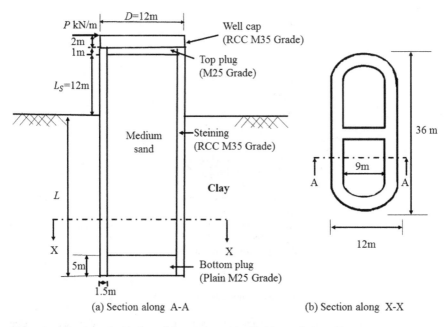

Fig. 1 Details of the double-D well foundation embedded in nonlinear soils

3 Numerical Modeling

The well–soil system considered herein is a two-dimensional (2-D) plain strain problem; finite domain of the soil is determined from the sensitivity analysis. Sensitivity analysis is performed with widths 11D, 21D, 31D and 41D and with depths of 3D, 4D, 5D and 6D, by considering fixed boundary conditions at the bottom of the soil domain and roller supports at the sides. Sensitivity analysis is also performed determining the optimum mesh size with the consideration of gradually varying mesh size of width from either sides of the foundation of 0.08 m to 5 m at the ends with varying height of mesh size from 0.25 m, 0.5 m, 1 m, 1.5 m and 2 m. The result of the sensitivity analysis indicated convergence for the domain width of 31D and depth 6D and optimum gradually varying mesh size of width from either sides of the foundation as 0.08 m to 5 m at the ends with a height of the mesh size of 1 m width. It considers two unit foundation with sand as one unit and soil domain as another unit. These units are meshed using four nodded quadrilateral elements and connected with interface elements. Nonlinear material modeling is done considering pressure-independent multi-yield surface modeling for clay, and pressure-dependent multi-yield surface modeling for sand. The properties of the soil are given in Table 1. In the present study, well foundation and the soil filled in the dredge hole of the foundation are modeled as a linear elastic material, which is defined by three parameters which are, modulus of elasticity (E), Poisson ratio (γ) and density (ρ) as stated in Table 2.

Table 1 Constitutive model parameters of soil

Loose sand	Soft clay
$\rho(t/m^3)$–1.7	$\rho(t/m^3)$–1.3
γ-0.34	γ-0.4
G_r(kPa)–5.5 × 10^4	G_r(kPa)–6 × 10^4
$\emptyset$-30^0	c- 18 kPa
d-0.5	
$\emptyset_{pT}$-29^0	
c_o –0.21	

ρ : Bulk mass density of soil
γ: Poisson's ratio
G_r: Reference low strain shear modulus specified at reference mean effective confining pressure 80 kPa.
γ_{max}: Octahedral shear strain at which the maximum shear strength is obtained is considered as 0.1 in this study
d: A positive constant defining variations of G as function of instantaneous effective confinement p' as $G = G_r \left(\frac{P'}{Pr'}\right)^d$
$\emptyset_{pT}$: Phase transformation angle
c: cohesion
c_o: A nonnegative constant defining the rate of shear-induced volume decrease (contraction) or pore pressure buildup

Table 2 Mechanical properties of foundation elements

Materials	M35	M25	Medium sand
E (GPa)	29.5	25	0.05
γ	0.15	0.15	0.3
ρ(kg/m^3)	2500	2500	1900

*Determination of E for the given concrete grade is calculated as stated in [4]

A larger value corresponds to faster contraction rate.

The interface between the foundation and the surrounding soil is modeled using a zero thickness contact elements. The master node of zero thickness element is attached to the foundation and slave node to the soil. Stiffness of the zero thickness element in numerical analysis is assigned by performing the Penalty method. In the penalty method, iterations are performed to determine the optimum value of stiffness for the contact element. Ideally at the interface, there should exist no penetration of soil into the foundation or vice versa and rigid sliding. But in reality, Impermeability and zero resistance for sliding is not possible. There always exists a normal traction which leads to some amount of penetration (Kolay et al. 2013), and resistance for sliding is offered by friction or cohesion between soil and the foundation. Hence, the objective in the modeling interface is to determine maximum penetration and

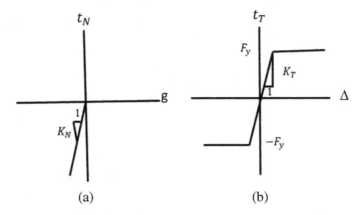

Fig. 2 Constitutive relationship considered for contact element **a** Separation **b** Sliding

minimum resistance offered by soil and the foundation for rigid sliding at the interface. Consider amount of penetration as g, then value of g should have be than or equal to zero. Positive values of g indicate more penetration, zero value of g indicates contact and negative values of g indicate no contact. Penetration is mainly due to pressure; constraint on g values is active in pressure-dependent material and inactive in pressure-independent material. If the limiting value is considered as the initiation point for rigid sliding, then the limiting value for cohesive soil is suggested as $\sqrt{3}$ times undrained strength by von Mises yield criteria. For cohesionless soil, limiting value is the product of friction coefficient and normal traction. Therefore, interface modeling should be able to capture zero stiffness for tensile normal stress for separation and should detect the penetration for compressive normal stress for positive values of g. It should also predict initiation of sliding when tangential traction reaches the limiting value. Including all the constraints on separation and sliding, interface modeling is done using zero thickness contact elements.

To define sliding and separation mechanisms at the interface, the following constitutive relationships, as shown in Fig. 2, are used in the present study. Figure 2a shows an elastic perfectly plastic constitutive relationship with tangential stiffness K_T and failure stress F_y. This constitutive relationship is considered to model the sliding mechanism. Figure 2b shows the constitutive relationship considered for the separation mechanism. This figure shows that the contact element resists the normal compressive stress with stiffness K_N. However, it is having zero stiffness for tensile condition. As a result, separation begins at the interface when deformation along the normal direction becomes tensile. K_N in the contact element is assigned perpendicular and K_T in the parallel direction to the interaction plane. Three parameters, K_N, K_T and F_y are required to define the contact element. Among these three, value of F_y for clay is calculated following von Mises failure criteria which define F_y as $F_y = \sqrt{3}S_u$, where S_u is undrained strength. In undrained condition, value of S_u of saturated clay is equal to the value of undrained cohesion (c). In case of sand, F_y is calculated following Mohr Coulomb's criteria which states F_y as $F_y = \mu t_N$, where

Table 3 Iterations performed in Penalty method

Iterations	Normal penalty (K_N)	Tangential penalty (K_T)
1	E_s	0.01XE_s
2	10X E_s	0.01X (10X E_s)
3	100X E_s	0.01X (100X E_s)
4	200X E_s	0.01X (200X E_s)
5	100X E_s	0.1X (100X E_s)
6	100X E_s	1X (100X E_s)

friction coefficient (μ) is determined as tan ($\frac{2}{3}\emptyset$), $\emptyset$ is friction angle, t_N is the normal traction.

Other two parameters, K_N and K_T, will be determined using the Penalty method. In case of cohesive soils, due to dominance of shear stress, separation depends on sliding, hence K_T values are fixed and K_N values are estimated through convergence. Converge is considered in cohesionless soils, separation initiates sliding because of variation in normal stress due to pressure buildup during lateral loading condition [9, 10]. Therefore, K_N values are fixed and K_T values are determined from iteration. To fix K_N, it depends on the amount of penetration and t_N where t_T is the tangential traction. Fixing of K_N value can be done only by resisting the amount of penetration. Hence, it is very important to consider the allowable amount of penetration as it determines the stiffness of contact element for separation, which consequently leads to sliding. The literature considers 0.2 mm as allowable penetration. This determines t_N as a 0.0002 times K_N. This ensures all the input parameters for the modeling contact element are in terms of K_N and K_T, and the values of K_N and K_T are obtain by performing iteration. Iteration is started with the assumption that contact elements have same stiffness of that surrounding soil. Hence, the initial iteration has K_N and K_T values equal to Young's modulus of soil (E_s). Trails of iteration with different K_N and K_T are mentioned in Table 3. Results in Fig. 3 indicate convergence in well cap displacement values for trials 4, 5 and 6 in all the cases. Hence, values of

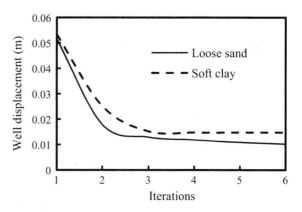

Fig. 3 Variation in well cap displacement obtained for different number of iterations

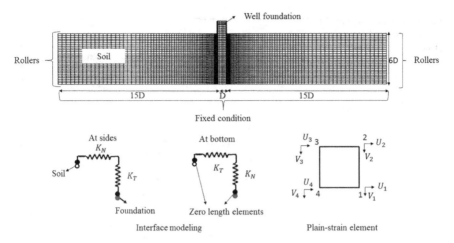

Fig. 4 Two-dimensional finite element model of well–soil system

K_N and K_T in trial 6 are considered as optimum stiffness value of contact elements for both sandy and clay soil. In this study, the interface between filled sand and well foundation is assumed as a perfect bonding condition. The perfect bonding condition indicates no separation and sliding at te hsoil–foundation interface and to model this, equal constraints in both X and Y directions are assigned to both soil and foundation nodes at the interface. This ensures displacements of soil and foundations are equal in X and Y directions. Based on the results obtained, a 2-D finite element model for well–soil system is developed which is shown in Fig. 4.

4 Result and Discussion

Figure 5a, b shows the curve between lateral load and ground level deflection of the well foundation in loose sand and soft clay with interface modeling and perfect bonding condition. The lateral load (P) is increased incrementally with an increment of 250 kN till the ground level deflection of the well foundation (Y) reaches 0.15D, which is equal to 1.8 m [6].

Figure 5 indicates that the trend of the P–Y curve of the well foundation embedded in soft clay is similar to that embedded in loose sand. It can be observed from the plot in Fig. 5, the estimation of magnitude of the ultimate load with the perfect bonding condition is high compared to interface modeling. However, it can be observed from Fig. 5; the P–Y curve is nonlinear in case of interface modeling and linear in perfect bonding condition. Nonlinear P–Y curve consists of three phases which are marked as zone 1(OA)—initial linear portion; zone 2 (AB)—intermediate nonlinear portion; and zone 3 (BC)—later linear portion of the curve as shown in Fig. 6. It is important to note that the right side of the base of the foundation gets penetrated into the soil

Fig. 5 a, b Lateral load (*P*) and ground level deflection (*Y*) of the well foundation embedded in loose sand and soft clay considering both perfect bonding condition and interface elements

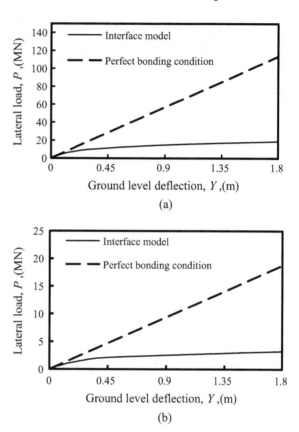

below the foundation with an equal amount of separation on the left side. Figure 6b, c depicts the evolution of Δ_h and Δ_v along the *P–Y* curve, respectively, where Δ_h is the magnitude of separation between the foundation and the soil measured on the left side of the foundation, and Δ_v is the magnitude of the separation measured at the base of the foundation.

It can be observed from Fig. 6 that both Δ_h and Δ_v are almost negligible in the initial linear portion of the *P–Y* curve. Δ_h increases with increase in the load in the nonlinear potion of the curve (AB). In this zone, the separation at the base (Δ_v) is not significant. However, the amount of both Δ_h and Δ_v increase on further increase in the load. The amount of Δ_h and Δ_v further increase and become significant as the load–deformation curve approaches the later linear portion of the curve (BC). From these observations, it can be inferred that the deformation of laterally loaded well–soil system embedded in nonlinear soil is governed by relative sliding and separation at the interface of the soil–foundation system which eventually lead to its ultimate state. Figure 7 indicates the displacement contour plots obtained with interface modeling and perfect bonding condition for foundation embedded in soft clay for lateral load of 5 MN.

Fig. 6 Variation of Y, Δ_h and Δ_v with P for well foundation in loose sandy soil

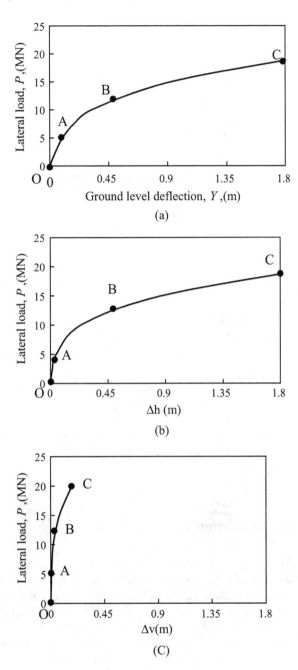

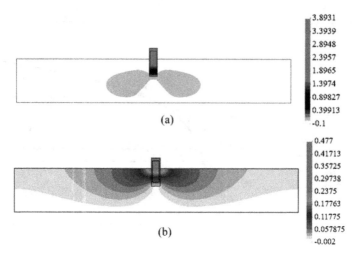

Fig. 7 Displacement contour plots of foundation embedded in soft clay for a applied load of 5 MN at well cap with (**a**) Interface bonding (**b**) Perfect bonding condition

This Fig. 7a indicates displacement contour plots of localization of displacement contour lines near the foundation for numerical analysis performed with interface modeling. This indicates separation on the left side as well as at the base. Figure 7b depicts that displacement contour plots cover the whole domain for perfect bonding condition and states no separation between the foundation and soil either at the sides or at the base of the well foundation. This is due to the consideration of soil resistances at the interface to ensure no loss of contact in the perfect bonding condition. Hence, it is clear from the observations that the perfect bonding condition overestimates the lateral deflection of the foundation and is unable to explain the deformation mechanism in the laterally loaded well–soil system.

5 Conclusion

The analysis done in the present study indicates the importance of interface modeling in predicting the lateral load–deformation curve of a well foundation embedded in nonlinear soils. In case of the perfect bonding condition, the load–deformation curve is linear, whereas in case of interface modeling, the load–deformation curve smoothly varies from initial linear part to later linear part of the curve. This study highlights the role of separation and sliding at the interface in reducing the lateral capacity of the system to reach the ultimate state. The results obtained in the present study indicate a need of considering interface modeling while predicting the lateral response of a soil–foundation system accurately for both sand and clay soil types. Load–deformation curve obtained from numerical analysis considering interface modeling gives the designer a clear idea on choosing the design load in which the soil well foundation system can withstand without permanent deformation. Hence, this study attempts to answer the question on how significant the role of interface modeling is in understanding the lateral deformation mechanism of a soil–well foundation system and gives a clear idea on interface element modeling in two-dimensional finite element numerical analysis. However, the study can be further extended with the consideration of the influence of slenderness depth and multilayer arrangement with interface modeling in three-dimensional numerical analysis for realistic prediction.

References

1. Gerolymos N, Gazetas G (2006) Winkler model for lateral response of rigid caisson foundations in linear soil. Soil Dyn Earthq Eng 26(5):347–361
2. Gerolymos N, Gazetas G (2006) Development of Winkler model for static and dynamic response of caisson foundations with soil and interface nonlinearities. Soil Dyn Earthq Eng 26(5):363–373
3. IRC:78–1983 Standard specifications and code of practice for road bridges, Section: VII, foundations and substructure. IRC, New Delhi
4. IS 456 (2000) Plain and reinforced concrete. Code of practice (4th rev). IS, New Delhi
5. Jain SK (2016) Earthquake safety in India: achievements, challenges and opportunities. Bull Earthq Eng 14(5):1337–1436. https://doi.org/10.1007/s10518-016-9870-2
6. Kumar ND, Rao SN (2012) Lateral load: deflection response of an embedded caisson in marine clay. Marine Georesour Geotechnol 30(1):1–31
7. Kolay C, Prashant A, Jain SK (2013) Nonlinear dynamic analysis and seismic coefficient for abutments and retaining walls. Earthq Spectra 29(2):427–451
8. Mazzoni S, McKenna F, Scott MH, Fenves GL (2006) OpenSees command language manual. Pacific Earthquake Engineering Research (PEER) Center, p 264
9. Mondal G, Prashant A, Jain SK (2012) Significance of Interface nonlinearity on seismic response of well-pier system in cohesionless soil. Earthq Spectra 28(3):1117–1145. https://doi.org/10.1193/1.4000074
10. Mondal G, Prashant A, Jain SK (2012b) Simplified seismic analysis of soilwellpier system for bridges. Soil Dyn Earthq Eng 32(1):42–55. http://dx.doi.org/10.1016/j.soildyn.2011.08.002
11. Ponnuswamy S (2008) Bridge engineering (2nd ed). McGraw Hill Professionals

Micromechanical Modeling of Material Cross-anisotropy

Geethesh Naiyyalga and Mousumi Mukherjee

Abstract Micromechanical behavior of granular materials is an important aspect for various engineering fields like soil mechanics, powder technology and mineral processing. The anisotropic material structure of the granular deposits is generally represented by a fabric tensor which takes into account the distribution of normal vectors at the interparticle contacts. In this regard, a truncated spherical harmonic function, restricted up to second-order terms, is often used to define the fabric tensor; however, this constrains the generic description of material anisotropy. While getting deposited under gravity, granular solids often exhibit a cross-anisotropic fabric structure with a preferential particle contact orientation favored by the direction of deposition. In the present study, the stiffness tensor of a cross-anisotropic fabric structure has been explored taking into account both the second and fourth order truncation of the spherical harmonic function. It has been noticed that the fourth order truncation gives a better control in capturing the cross-anisotropic nature of the granular fabric. Further, the bounds on fabric parameters are examined along with the influence of their respective magnitudes on the elastic stiffness moduli of the granular packing.

Keywords Granular material · Cross-anisotropy · Fabric tensor · Spherical harmonic expansion · Hertz–Mindlin contact model · Micromechanics

1 Introduction

The mechanical behavior of granular materials is of significant concern for various engineering fields like soil mechanics, powder technology and mineral processing. The macroscopic stress-strain response of such granular assembly is greatly influenced by the microscale properties like force-deformation characteristics of interparticle contact, particle contact orientation, number of interparticle contact and particle size distribution. In this regard, various micromechanical models are available in

G. Naiyyalga · M. Mukherjee (✉)
School of Engineering, IIT Mandi, Mandi 175 005, India
e-mail: mousumi@iitmandi.ac.in

© The Editor(s) (if applicable) and The Author(s), under exclusive license to Springer Nature Singapore Pte Ltd. 2021
S. K. Saha and M. Mukherjee (eds.), *Recent Advances in Computational Mechanics and Simulations*, Lecture Notes in Civil Engineering 103,
https://doi.org/10.1007/978-981-15-8138-0_17

the literature [3, 5–7, 9, 10, 12], which explicitly takes into account interparticle properties and particle packing structure. The anisotropic material structure of the granular deposits in these works has been represented by a fabric tensor which takes into account the distribution of normal vectors at the interparticle contacts [2, 4]. Spherical harmonic function is generally used to define such fabric tensor [8]. A truncated form of the harmonic function, restricted up to second-order terms, is usually employed for this purpose; however, this constrains the generic description of material anisotropy [11].

In the present study, the elastic stiffness tensor of a cross-anisotropic fabric structure has been explored within the micromechanical framework taking into account both the second and fourth order truncation of the spherical harmonic function. While getting deposited under gravity, the granular solids often exhibit such cross-anisotropic fabric structure with a preferential particle contact orientation favored by the direction of deposition. The granular assembly has been represented here by a network of rigid spherical particles which conforms Hertz–Mindlin type contact model. The constitutive relation for the representative volume element (RVE) has been derived from the force–displacement relationship of interparticle contacts assuming a kinematic hypothesis [5], which enforces the local strain at the interparticle contact to be equal to the overall strain of the RVE. The expressions for elastic stiffness moduli for the RVE have been established for the granular packing exhibiting a cross-anisotropic fabric structure. Further, the influence of fabric parameters on these cross-anisotropic elastic moduli has been explored and bounds for the fabric parameters are also examined.

2 Description of Granular Assembly

2.1 Strain and Stress Within a Volume Element

The granular system can be idealized as a network of rigid convex particles, which support the loads applied at the boundary, through resistance at interparticle contacts. Each particle along with its surrounding particles forms a local particle group and several of such local particle groups further manifest a representative volume element (RVE), which generally characterizes the global macroscopic response of the assembly. Following the kinematic hypothesis of Chang et al. [2], a uniform strain field can be assumed which results in the local strain to be equal to the global engineering strain (ε_{ij}) of the RVE and further be related to the incremental relative contact displacement (u_j^c) by following

$$\Delta u_j^c = \Delta \varepsilon_{ij} l_i^c \tag{1}$$

where l_i^c is the branch vector joining the centroids of two adjacent particles. The energy conservation principle for the RVE along with the imposed kinematic

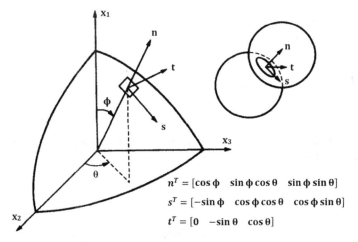

Fig. 1 Local coordinate system at an interparticle contact

constraint furnishes following expressions for macroscopic global stress (σ_{ij}) and stiffness tensor (C_{ijkl})

$$\Delta \sigma_{ij} = \frac{1}{V} \sum_{c=1}^{N} \Delta f_j^c l_i^c \tag{2}$$

$$\Delta \sigma_{ij} = C_{ijkl} \Delta \varepsilon_{kl} = \left[\frac{1}{V} \sum_{c=1}^{N} K_{jl}^c l_k^c l_i^c \right] \Delta \varepsilon_{kl} \tag{3}$$

where f_j^c, N and V are the interparticle contact force, total number of contacts in RVE and volume of RVE, respectively. In a local coordinate system, defined by three orthogonal unit vectors n, s and t at the contact plane (Fig. 1), the contact stiffness tensor K_{jl}^c can be expressed in terms of contact normal (k_n^c) and shear (k_s^c) stiffness as

$$K_{ij}^c = k_n^c n_i^c n_j^c + k_s^c \left(s_i^c s_j^c + t_i^c t_j^c \right) \tag{4}$$

2.2 Representation of Packing Structure

The packing structure in this micromechanical approach is statistically represented by the distribution of the branch vector, which coincides with the direction of contact normal for a packing with equal-sized spherical particles ($l_k^c = 2r n_k^c$). In case of RVE with large number of contacts, the summation of a function, e.g. stiffness tensor

C_{ijkl} over all the contacts can be written as (in spherical co-ordinate system which is defined by ϕ, θ as depicted in Fig. 1)

$$C_{ijkl} = \frac{1}{V} \sum_{c=1}^{N} K_{jl}^{c} l_{k}^{c} l_{i}^{c} = \frac{r^2 N}{\pi V} \int_{0}^{2\pi} \int_{0}^{\pi} n_i(\phi, \theta) K_{jl}(\phi, \theta) n_k(\phi, \theta) \xi(\phi, \theta) \sin\phi \, d\phi \, d\theta \tag{5}$$

where the influence of directional distribution is considered through the density distribution factor $\xi(\phi, \theta)$

$$1 = \int_{\Omega} \xi(\phi, \theta) d\Omega = \int_{0}^{2\pi} \int_{0}^{\pi} \xi(\phi, \theta) \sin\phi \, d\phi \, d\theta \tag{6}$$

A Spherical Harmonics Expansion (SHE) is often employed to represent the orientational distribution of interparticle contacts in three dimensions [8]

$$\xi(\phi, \theta) = \frac{1}{4\pi} \left\{ 1 + \sum_{k=2}^{\infty} \left[a_{k0} P_k(\cos\phi) + \sum_{m=1}^{k} P_k^m(\cos\phi)(a_{km} \cos m\theta + b_{km} \sin m\theta) \right] \right\} \tag{7}$$

where $P_k(\cos\phi) = k$ is the Legendre polynomial, $P_k^m(\cos\phi) = $, associated Legendre function, a_{k0}, a_{km} and b_{km} are fabric parameters. For ensuring the centro-symmetric condition [$\xi(\phi, \theta) = \xi(\pi - \phi, \theta + \pi)$], only even harmonics of the density function $\xi(\phi, \theta)$ are admissible. For simplicity, a truncated form of the expression consisting of second order terms ($k = 2$ and $m = 2$) has been applied by the past researchers [4],

$$\xi(\phi, \theta) = \frac{1}{4\pi} \left\{ 1 + \frac{1}{4} a_{20}(3\cos 2\phi + 1) + 3\sin^2\phi(a_{22}\cos 2\theta + b_{22}\sin 2\theta) \right\} \tag{8}$$

However, such truncation imposes constraint on capturing material symmetry condition, and a higher order truncation can become more effective in mimicking the material symmetry or particle contact distribution in general.

3 Distribution Density Function Considering Higher Order Truncation

3.1 Generic Fabric Representation

Truncating the function in Eq. 7 up to the fourth order results in

$$\xi = \frac{1}{4\pi} \left\{ \begin{array}{l} 1 + a_{20}P_2(\cos\phi) + P_2^2(\cos\phi)(a_{22}\cos2\theta + b_{22}\sin2\theta) \\ + a_{40}P_4(\cos\phi) + P_4^2(\cos\phi)(a_{42}\cos2\theta + b_{42}\sin2\theta) \\ + P_4^4(\cos\phi)(a_{44}\cos4\theta + b_{44}\sin4\theta) \end{array} \right\} \quad (9)$$

The Legendre polynomial and its associated functions are obtained from Rodrigues formula [1].

$$P_2(\cos\phi) = \frac{1}{4}(3\cos2\phi + 1),$$

$$P_2^2(\cos\phi) = 3\sin^2\phi,$$

$$P_4(\cos\phi) = \frac{1}{8}\left(\frac{9}{8} - 15\cos2\phi + \frac{35}{8}\cos4\phi + \frac{35}{2}\cos2\phi\right),$$

$$P_4^2(\cos\phi) = \frac{15}{2}(7\cos^2\phi - 1)\sin^4\phi,$$

$$P_4^4(\cos\phi) = 105\sin^4\phi \quad (10)$$

Substituting Eq. 10 into Eq. 9, the interparticle contact density distribution function takes the following form:

$$\xi = \frac{1}{4\pi} \left\{ \begin{array}{l} 1 + a_{20}\frac{1}{4}(3\cos2\phi + 1) + 3\sin^2\phi(a_{22}\cos2\theta + b_{22}\sin2\theta) \\ + a_{40}\frac{1}{8}\left(\frac{9}{8} - 15\cos2\phi + \frac{35}{8}\cos4\phi + \frac{35}{2}\cos2\phi\right) \\ + \frac{15}{2}(7\cos^2\phi - 1)\sin^4\phi(a_{42}\cos2\theta + b_{42}\sin2\theta) \\ + 105\sin^4\phi(a_{44}\cos4\theta + b_{44}\sin4\theta) \end{array} \right\} \quad (11)$$

Substituting Eq. 11 into Eq. 5 and subsequently integrating the terms, the constitutive tensor can be obtained. The elements of which in Voigt notation are listed below

$$C_{11} = \frac{4r^2N}{315V}[(63 + 36a_{20} + 8a_{40})k_n + 2(21 + 3a_{20} - 4a_{40})k_s]$$

$$C_{12} = \frac{4r^2N}{3465V}[(231 + 33a_{20} + 198a_{22} - 44a_{40} + 300a_{42})(k_n - k_s)]$$

$$C_{13} = \frac{4r^2N}{3465V}[(231 + 33a_{20} - 198a_{22} - 44a_{40} - 300a_{42})(k_n - k_s)]$$

$$C_{15} = C_{16} = 0$$

$$C_{14} = \frac{8r^2N}{1155V}[(33b_{22} + 50b_{42})(k_n - k_s)]$$

$$C_{22} = \frac{4r^2N}{1155V}\left[\begin{array}{l}(231 - 66a_{20} + 396a_{22} + 11a_{40} - 320a_{42} + 3080a_{44})k_n \\ + (154 - 11a_{20} + 66a_{22} - 11a_{40} + 100a_{42} - 3080a_{44})k_s\end{array}\right]$$

$$C_{23} = \frac{4r^2N}{315V}[(21 - 6a_{20} + a_{40} - 840a_{44})(k_n - k_s)]$$

$$C_{25} = C_{26} = 0$$

$$C_{24} = \frac{4r^2N}{1155V}[3080b_{44}(k_n - k_s) + 33b_{22}(6k_n + k_s) + 10b_{42}(-16k_n + 5k_s)]$$

$$C_{33} = \frac{4r^2N}{1155V}\left[\begin{array}{l}(231 - 66a_{20} - 396a_{22} + 11a_{40} + 320a_{42} + 3080a_{44})k_n \\ + (154 - 11a_{20} - 66a_{22} - 11a_{40} - 100a_{42} - 3080a_{44})k_s\end{array}\right]$$

$$C_{35} = C_{36} = 0$$

$$C_{34} = \frac{4r^2N}{1155V}[3080b_{44}(-k_n + k_s) + 33b_{22}(6k_n + k_s) + 10b_{42}(-16k_n + 5k_s)]$$

$$C_{44} = \frac{2r^2N}{315V}[2(21 - 6a_{20} + a_{40} - 840a_{44})k_n + (63 - 9a_{20} - 2a_{40} + 1680a_{44})k_s]$$

$$C_{45} = C_{46} = 0$$

$$C_{55} = \frac{4r^2N}{3465V}\left[\begin{array}{l}(231 + 33a_{20} - 198a_{22} - 44a_{40} - 300a_{42})k_n \\ + 99a_{20}k_s + 2(693 - 297a_{22} + 88a_{40} + 930a_{42})k_s\end{array}\right]$$

$$C_{56} = \frac{2r^2N}{1155V}[132b_{22}k_n + 200b_{42}k_n + 99b_{22}k_s - 310b_{42}k_s]$$

$$C_{66} = \frac{4r^2N}{3465V}\left[\begin{array}{l}(231 + 33a_{20} + 198a_{22} - 44a_{40} + 300a_{42})k_n \\ + 99a_{20}k_s + 2(693 + 297a_{22} + 88a_{40} - 930a_{42})k_s\end{array}\right] \quad (12)$$

Based on the concept of symmetry, different types of material anisotropy can be inferred for the particles in contact. The type of material symmetry can be mathematically dealt with the understanding of the fourth-order stiffness tensor obtained from the constitutive relation between stress and engineering strain. Observing a reflection symmetry in x_2–x_3 plane, granular material conforms to monoclinic material symmetry. Representation of the symmetry in Voigt notation is shown as

$$C_{\alpha\beta} = \begin{bmatrix} C_{11} & C_{12} & C_{13} & C_{14} & 0 & 0 \\ C_{12} & C_{22} & C_{23} & C_{24} & 0 & 0 \\ C_{13} & C_{23} & C_{33} & C_{34} & 0 & 0 \\ C_{14} & C_{24} & C_{34} & C_{44} & 0 & 0 \\ 0 & 0 & 0 & 0 & C_{55} & C_{56} \\ 0 & 0 & 0 & 0 & C_{56} & C_{66} \end{bmatrix} \quad (13)$$

With the help of Voigt notation, orthotropic material symmetry can be represented as in Eq. 14. However, in order to represent this kind of material symmetry, a few of the fabric parameters should be omitted, i.e. $b_{22} = b_{42} = b_{44} = 0$.

$$C_{\alpha\beta} = \begin{bmatrix} C_{11} & C_{12} & C_{13} & 0 & 0 & 0 \\ C_{12} & C_{22} & C_{23} & 0 & 0 & 0 \\ C_{13} & C_{23} & C_{33} & 0 & 0 & 0 \\ 0 & 0 & 0 & C_{44} & 0 & 0 \\ 0 & 0 & 0 & 0 & C_{55} & 0 \\ 0 & 0 & 0 & 0 & 0 & C_{66} \end{bmatrix} \quad (14)$$

It is important to mention that Eq. 13 boils down to the same expressions of the stiffness tensor for the monoclinic case as derived by Chang et al. [2] with second-order truncated form of distribution density function when $a_{40} = a_{42} = a_{44} = b_{42} = b_{44} = 0$.

3.2 Cross-anisotropic Fabric

Fabric Parameters and Contact Orientation

Particles in contact often exhibit cross-anisotropic material symmetry, and this behavior can be captured through the expression given in Eq. 14 when the fabric parameters $a_{22} = a_{42} = a_{44} = 0$. For such material symmetry, the elements of stiffness tensor reduce to $C_{22} = C_{33}$, $C_{12} = C_{13}$, $C_{55} = C_{66}$ and $C_{44} = \frac{1}{2}(C_{22} - C_{23})$. In addition to the above requirements, positive definiteness of the strain energy can be ensured by following constraints on the fabric parameters ($\alpha = \frac{k_s}{k_n}$)

$$a_{20} < \frac{105 + 210\alpha}{6(14 - 5\alpha)}; a_{40} < \frac{63(12 + 13\alpha)}{4(-1 + \alpha)(-14 + 5\alpha)} \quad (15)$$

The bounds on these two fabric parameters are further depicted in Fig. 2 in terms of the shear to normal stiffness ratio. It is important to note that $\alpha = 1$ creates

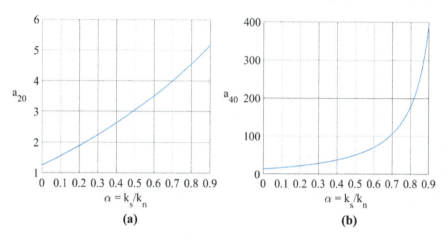

Fig. 2 Upper bound on **a** a_{20} and **b** a_{40} with respect to α

singularity in the bounds of fabric parameter a_{40} and hence, in Fig. 2 the value of α has been restricted to 0.9.

An illustration of the interparticle contact distribution density for a cross-anisotropic material fabric is shown in Fig. 3, where $a_{20} = -0.11$ and $a_{40} = 2.027$. The contact distribution density function has been plotted for both second and fourth order truncation for the comparison purpose. The x_1 direction being the axis for material anisotropy, a uniform contact distribution with circular functional representation can be observed in the x_2–x_3 plane as shown in Fig. 3a. This will lead to a constant elastic property in the x_2–x_3 plane. On the contrary, a non-uniform contact distribution can be noticed in Fig. 3b for the x_1–x_3 and x_1–x_2 planes, which indicate

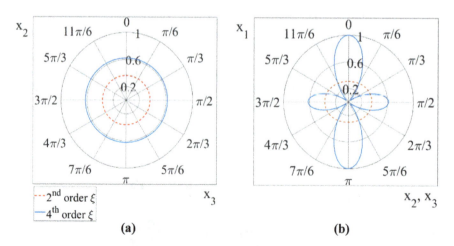

Fig. 3 Distribution density of interparticle contact in **a** x_2–x_3 plane, **b** x_1–x_3 plane and x_1–x_2 plane ($a_{20} = -0.11$ and $a_{40} = 2.027$)

that the elastic property will differ based on the orientation of the particle contact in this plane. It can also be noticed that the fourth order truncation will give a better control in capturing the experimentally obtained particle contact distribution and hence, will mimic the cross-anisotropic material response more accurately.

Highlighting the role of the fabric parameters in the evolution of particle contact, Fig. 4 depicts the change in shape of the distribution density of interparticle contact in (a) x_2–x_3 plane, (b) x_1–x_3 plane and x_1–x_2 plane with different fabric parameter values. Chang et al. [3] showed how the distribution density function represents cross-anisotropy with $a_{20} = 0.8$. The same material symmetry is depicted in Fig. 4a, b with $a_{40} = 0$ for the fourth order truncated SHE. Figure 4c, d shows the change in contact orientation distribution when $a_{40} = 2.027$ and the coefficient $a_{20} = 0.8$. In this case, one can observe that the contacts are more aligned toward the x_1 direction. The influence of the fabric parameter a_{40} can further be observed in Fig. 4e, f when $a_{20} = 0.8$ and $a_{40} = -0.4$. In comparison to the second order truncated SHE, the contact direction for the fourth order truncation now shows a reduced preference over the x_1 direction. Figure 4g, h relates the intensity of the contact distribution oriented along the x_2–x_3 plane when $a_{40} = 0$ and with $a_{20} = -0.11$. It clearly shows that, similar to a_{40}, a negative magnitude of a_{20} reflects more bias in particle contact distribution toward the x_2 and x_3 direction.

Linear Elastic Material Parameters

In case of linear elasticity, the generalized constitutive relation for an orthotropic material can be expressed as

$$\begin{Bmatrix} \varepsilon_1 \\ \varepsilon_2 \\ \varepsilon_3 \\ \gamma_{23} \\ \gamma_{13} \\ \gamma_{12} \end{Bmatrix} = \begin{bmatrix} \frac{1}{E_1} & \frac{-v_{21}}{E_2} & \frac{-v_{31}}{E_3} & 0 & 0 & 0 \\ \frac{-v_{12}}{E_1} & \frac{1}{E_2} & \frac{-v_{32}}{E_3} & 0 & 0 & 0 \\ \frac{-v_{13}}{E_1} & \frac{-v_{23}}{E_2} & \frac{1}{E_3} & 0 & 0 & 0 \\ 0 & 0 & 0 & \frac{1}{G_{23}} & 0 & 0 \\ 0 & 0 & 0 & 0 & \frac{1}{G_{13}} & 0 \\ 0 & 0 & 0 & 0 & 0 & \frac{1}{G_{12}} \end{bmatrix} \begin{Bmatrix} \sigma_1 \\ \sigma_2 \\ \sigma_3 \\ \tau_{23} \\ \tau_{13} \\ \tau_{12} \end{Bmatrix} \quad (16)$$

where E, v and G are the Young's modulus, Poisson's ratio and shear modulus in the respective direction. Additional constraints are applicable in case of cross-anisotropy $v_{12} = v_{13}$; $v_{21} = v_{31}$; $v_{23} = v_{32}$; $G_{12} = G_{13}$; $E_2 = E_3$. The expressions for these elastic moduli for a granular material can further be deduced based on the relations derived in Eq. 12. The Young's modulus in the x_1 direction is obtained as

$$E_1 = \frac{4k_n A r^2 N}{5\{(4 - 21 + 6a_{20} - a_{40})k_n - 3(7 + a_{20})k_s + 4a_{40}k_s\}V} \quad (17)$$

where $A = 2(a_{20}(-10 + 7a_{20}) - 5(7 + 2a_{40}))k_n - 5(21 + 3a_{20} - 4a_{40})k_s$.

Similarly, Young's modulus in the x_2 and x_3 directions is given by the following:

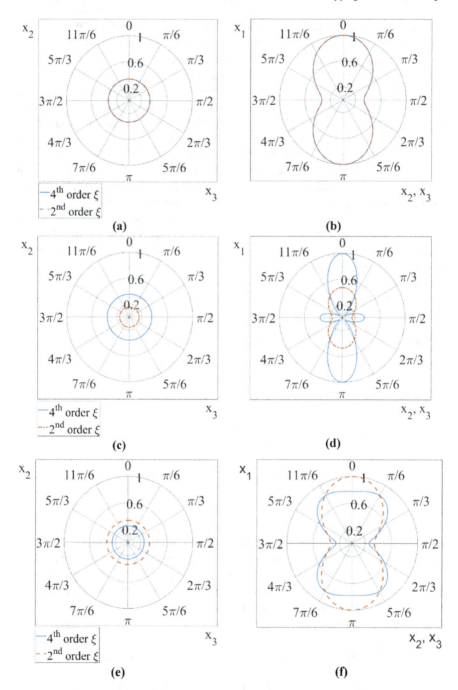

Fig. 4 Distribution density of interparticle contact considering second and fourth order truncation of SHE in **a, c, e, g** x_2–x_3 plane; **b, d, f, h** x_1–x_3 plane and x_1–x_2 plane for different sets of fabric parameter values

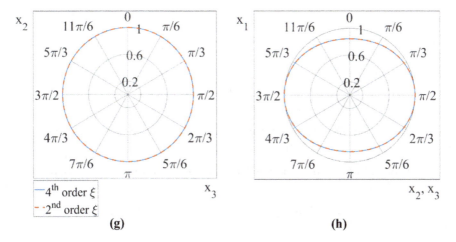

Fig. 4 (continued)

$$E_2 = \frac{-4k_n\{2(-21 + 6a_{20} - a_{40})k_n + (-63 + 9a_{20} + 2a_{40})k_s\}Ar^2N}{5(B + C + D)V} \quad (18)$$

where

$$B = \{-3528 + 657a_{20}^2 + 12a_{20}(-84 + a_{40}) - a_{40}(861 + 8a_{40})\}k_n^2$$

$$C = \{-6174 + 9a_{20}(-133 + 22a_{20}) + 9(63 + 2a_{20})a_{40} + 16a_{40}^2\}k_n k_s$$

$$D = (21 + 3a_{20} - 4a_{40})(-63 + 9a_{20} + 2a_{40})k_s^2$$

Shear modulus and Poisson's ratio for the cross-anisotropic material is given as

$$G_{23} = \frac{2\{2(21 - 6a_{20} + a_{40})k_n + (63 - 9a_{20} - 2a_{40})k_s\}r^2N}{315V} \quad (19)$$

$$G_{13} = \frac{4\{(21 + 3a_{20} - 4a_{40})k_n + (126 + 9a_{20} + 16a_{40})k_s\}r^2N}{315V} \quad (20)$$

$$\nu_{12} = \frac{-(21 + 3a_{20} - 4a_{40})(k_n - k_s)}{4(-21 + 6a_{20} - a_{40})k_n - 3(7 + a_{20})k_s + 4a_{40}k_s} \quad (21)$$

$$\nu_{31} = \frac{(21 + 3a_{20} - 4a_{40})(k_n - k_s)(X + Y)}{P + Q + R} \quad (22)$$

$$\nu_{23} = \frac{(k_n - k_s)(J + K)}{P + Q + R} \quad (23)$$

where

$$X = 2(-21 + 6a_{20} - a_{40})k_n$$

$$Y = (-63 + 9a_{20} + 2a_{40})k_s$$

$$P = \left(-3528 + 657a_{20}^2 + 12a_{20}(-84 + a_{40}) - a_{40}(861 + 8a_{40})\right)k_n^2$$

$$Q = (-6174 + 9a_{20}(-133 + 22a_{20}) + 9(63 + 2a_{20})a_{40} + 16a_{40}^2)k_n k_s$$

$$R = (21 + 3a_{20} - 4a_{40})(-63 + 9a_{20} + 2a_{40})k_s^2$$

$$J = \left(-882 + 225a_{20}^2 - 12a_{20}(21 + a_{40}) + a_{40}(-399 + 8a_{40})\right)k_n$$

$$K = (21 + 3a_{20} - 4a_{40})(-63 + 9a_{20} + 2a_{40})k_s$$

In connection with Eq. 16, Young's modulus and Poisson's ratio for isotropic material symmetry are obtained as

$$E_{iso} = E_1 = E_2 = \frac{4k_n(2k_n + 3k_s)Nr^2}{3(4k_n + k_s)V}; \quad \nu = \nu_{12} = \nu_{31} = \nu_{23} = \frac{k_n - k_s}{4k_n + k_s} \quad (24)$$

The cross-anisotropic elastic moduli are normalized with respect to their isotropic counterparts and are plotted in Fig. 5 for different particle contact stiffness. In addition, the influence of different contact orientation on such modulus has also been depicted in the figure by considering two different sets of fabric parameter magnitude. Figure 5a, which has been plotted with $a_{20} = 0.8$ and $a_{40} = 0$, shows a decrease in the normalized Young's modulus (E_1/E_{iso}) in the x_1 direction with an increase in the contact stiffness ratio (α), whereas a reverse trend is observed for normalized Young's modulus (E_2/E_{iso}) in the x_2 direction. In case of the second parameter set, $a_{20} = 0.8$ and $a_{40} = -0.4$, the variation of E_1/E_{iso} over α is noticed to reduce considerably and a higher change in the magnitude of E_2/E_{iso} can also be noted from Fig. 5c. Such variation is expected due to the less preferred orientation of particle contact distribution in the x_1 direction for the second case (Fig. 4e, f). In comparison to Young's modulus, the change in fabric parameter magnitude can be observed to have a greater impact on the Poisson's effect in different planes (Fig. 5b, d).

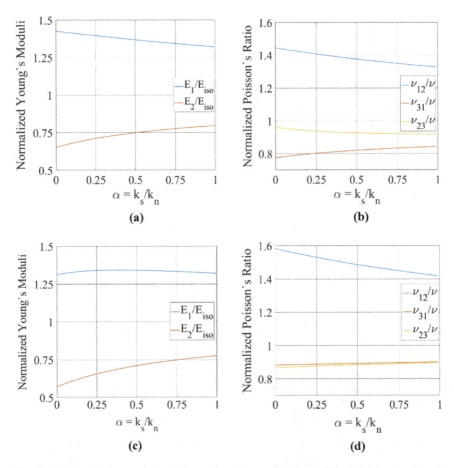

Fig. 5 Variation in **a, c** normalized Young's modulus and **b, d** normalized Poisson's ratio with particle contact stiffness, respectively, for two sets of fabric parameter values

4 Conclusion

A micromechanical approach has been taken in the present study to derive the elastic stiffness tensor of a granular system in terms of contact stiffness and particle contact orientation. The anisotropic material structure of the granular assembly has been denoted by a fabric tensor taking into account the distribution of normal vectors at the interparticle contacts. Further, a truncated spherical harmonic function has been considered to represent such fabric tensor. In this regard, a more generic form of an elastic stiffness tensor has been derived for a granular network with monoclinic material symmetry by extending the truncation of the spherical harmonic function up to the fourth order. The stiffness tensor for a reduced material symmetry, i.e. cross-anisotropy, has been explored taking into account both the second and fourth order truncation. It has been noticed that the fourth order truncation gives a better control in capturing the cross-anisotropic nature of the granular fabric. Further, the bounds on the fabric parameters are examined, and the expressions for elastic stiffness moduli has been established for a granular packing exhibiting cross-anisotropic fabric structure. Such framework of material anisotropy representation, taking into account the mapping between micro and macro properties, can easily be extended for elasto-plastic constitutive relations and further can be implemented within any numerical framework for solving complex boundary value problems with a greater flexibility of material modeling.

References

1. Abramowitz M, Stegun IA (1965) Handbook of mathematical functions. Dover Publications, New York
2. Chang CS, Chao SC, Chang Y (1995) Estimates of mechanical properties of granulates with anisotropic random packing structure. Int J Solids Struct 32(14):1989–2008
3. Chang CS, Hicher PY (2005) An elastic-plastic model for granular materials with microstructural consideration. Int J Solids Struct 42:4258–4277
4. Chang CS, Misra A (1990) Application of uniform strain theory to heterogeneous granular solids. J Eng Mech ASCE 116(10):2310–2328
5. Chang CS, Misra A (1990) Packing structure and mechanical properties of granulates. J Eng Mech ASCE 116(5):1077–1093
6. Chang CS, Yin ZY (2010) Micromechanical modeling for inherent anisotropy in granular materials. J Eng Mech 136(7):830–839
7. Jenkins JT, Strack ODL (1993) Mean-field inelastic behavior of random arrays of identical spheres. Mech Mater 16:25–33
8. Kanatani K (1984) Distribution of directional data and fabric tensors. Int J Eng Sci 22(2):149–164
9. Kruyt NP, Rothenburg L (2002) Micromechanical bounds for the effective elastic moduli of granular materials. Int J Solids Struct 39(2):311–324
10. Liao CL, Chang TP, Young D, Chang CS (1997) Stress-strain relationship for granular materials based on hypothesis of best fit. Int J Solids Struct 34(31–32):4087–4100
11. Rahmoun J, Kondo D, Millet O (2009) A 3D fourth order fabric tensor approach of anisotropy in granular media. Comput Mater Sci 46:869–880
12. Rothenburg L, Selvadurai APS (1981) Micromechanical definitions of the Cauchy stress tensor for particular media. In: Selvadurai APS (ed) Mechanics of structured media. Elsevier, Amsterdam, pp 469–486

Stability of an Unsupported Elliptical Tunnel Subjected to Surcharge Loading in Cohesive-Frictional Soil

Puja Dutta and Paramita Bhattacharya

Abstract This study investigates the stability of an unsupported elliptical tunnel subjected to surcharge loading in cohesive-frictional soil. Non-circular tunnels, like an elliptical tunnel, provide more space and structural stability due to their smooth tunnel profile. The analysis has been performed assuming a plane strain condition by using lower bound limit analysis in conjunction with finite elements and second-order conic programming (SOCP). Limit analysis determines the collapse load at which the soil mass undergoes an unrestricted plastic deformation. The soil medium around the tunnel has been modelled as a homogenous and isotropic Mohr–Coulomb material obeying an associative flow rule. SOCP enables the usage of the non-linear Mohr–Coulomb yield criteria, which is expressed as a set of second-order cones and overcomes the difficulty of singular apex point of the Mohr–Coulomb yield function and is robust and efficient in solving a huge number of variables. In this study, a uniform surcharge loading is applied to the ground surface, and the stability is determined for smooth interface condition between the loading surface and the underlying soil. The results are presented in the form of dimensionless stability charts. The stability number in this analysis is indicated as σ_s/c obtained for different combinations of soil cover ratio H/B, aspect ratio of tunnel D/B, $\gamma B/c$ and ϕ where H is soil cover depth of tunnel, B and D are width and height of tunnel, respectively, and c, γ and ϕ are cohesion, unit weight and peak friction angle of soil. It has been noted that the stability number increases with increase in the peak friction angle of soil and soil cover depth ratio and decreases with an increase in D/B and $\gamma B/c$. The size of the plastic zone increases as H/B keeps increasing and extends downwards enclosing the entire tunnel.

Keywords Tunnel · SOCP · Limit analysis · Stability number · Soil plasticity

---

P. Dutta (✉) · P. Bhattacharya
Indian Institute of Technology Kharagpur, Kharagpur 721 302, India
e-mail: pujadutta@iitkgp.ac.in

© The Editor(s) (if applicable) and The Author(s), under exclusive license to Springer Nature Singapore Pte Ltd. 2021
S. K. Saha and M. Mukherjee (eds.), *Recent Advances in Computational Mechanics and Simulations*, Lecture Notes in Civil Engineering 103,
https://doi.org/10.1007/978-981-15-8138-0_18

1 Introduction

With the increase in urbanization around the world, the need for land for building infrastructure has also been growing. Inadequate land within the built-up areas has become a major concern. Tunnelling provides an opportunity to use the underground spaces preventing the acquisition of land and stripping the population away from the urban locations. Examining stability of tunnels thus becomes an important parameter. Several investigations regarding the stability of tunnels have been carried out earlier. Atkinson and Potts [2] examined the stability of circular tunnels in cohesionless soil using large-scale tests and limit analysis. Davis et al. [5] used limit bound theorems to study the stability of a circular tunnel in soft clay. Leca and Dormieux [6] assessed the stability of shallow circular tunnels in cohesionless soil using the limit analysis approach. Using the limit analysis approach provided by Sloan [15], Sloan and Assadi [16] and Wilson et al. [17] determined the stability of square and circular tunnels in soft clay in terms of support pressure, Yamamoto et al. [18, 19] determined the stability of single circular and square tunnels under surcharge loading, Sahoo and Kumar [12, 13] investigated the stability of unsupported circular tunnel, Chakraborty and Kumar [4] and Sahoo and Kumar [14] examined the stability of circular tunnels under seismic forces, Yamamoto et al. [20, 21] computed the maximum surcharge load which can be supported by dual circular and square tunnels with zero lining pressure along the tunnel periphery and Abbo et al. [1] determined the stability charts for wide rectangular tunnel in soft clay. Non-circular tunnels like elliptical-shaped tunnels have also gained popularity due to their smooth tunnel profile. Yang et al. [22] studied the stability of an unsupported elliptical tunnel in cohesive-frictional soil by using upper bound limit analysis; Zhang et al. [23] studied the effect of tunnel spacing on the stability of dual unlined elliptical tunnel in cohesive-frictional soil; and Zhang et al. [24] studied the stability of an elliptical tunnel in cohesionless soils in terms of support pressure. Other than these, a few researchers have also carried out work on the stability of tunnels by using centrifuge and numerical modelling as Lee et al. [7] conducted centrifuge tests and numerical simulations to determine the surface settlements, pore water generation and arching effects during tunnelling, Chehade and Shahrour [3] studied the effect of the interaction between twin tunnels on soil settlement and internal forces, and Osman [11] studied the ground movement due to the presence of circular tunnel in clay based on kinematic plastic solution.

The present work deals with the stability of elliptical tunnel subjected to surcharge loading embedded in cohesive-frictional soil with smooth interface condition. The study has been carried out by using lower bound limit analysis with finite elements and second-order conic programming. Lower bound limit analysis provides the safe estimate of the collapse load. In this method, soil is assumed to be perfectly plastic and hence, there is no requirement of solving the elastic part of the load-settlement curve unlike in elasto-plastic analysis, thereby reducing the computational time.

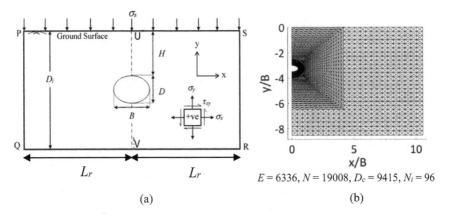

Fig. 1 a Schematic diagram of the problem with stress boundary condition, **b** typical finite element mesh for $D/B = 0.5$, $H/B = 3.0$

2 Problem Definition

An elliptical tunnel of width B and height D is placed at a cover depth H below the ground surface in soil as shown in Fig. 1a. The soil mass is assumed to obey Mohr–Coulomb yield criterion and an associated flow rule. To perform a plane strain analysis, the length of tunnel is assumed to be very long in comparison to its width. A continuous surcharge loading σ_s has been applied at the ground surface. The interaction between the loading surface and the soil has been modelled as a smooth interface condition. c and ϕ are cohesion and internal friction angle of soil, respectively, and γ is the unit weight of soil. The stability number is presented as σ_s/c for different combinations of H/B, D/B, $\gamma B/c$ and ϕ.

3 Domain, Boundary Conditions and Finite Element Mesh

The domain and the stress boundary conditions are presented in Fig. 1a. The domain has been discretized into three-noded triangular elements. The entire soil domain is symmetric about an axis passing through the centre of the domain coinciding with the y-axis. Therefore, only one half of the total domain, i.e. SUVR has been chosen for the present analysis. Along the tunnel periphery, the normal and shear stresses are zero and the shear stress is zero along the ground surface and at the plane of symmetry. The bottom horizontal boundary VR and vertical boundary RS are kept far away from the tunnel such that yielded elements should not touch the boundaries and a further increase in the domain does not cause any change in the value of σ_s.

The horizontal extent (L_r) and the vertical dimensions of the domain (D_i) are kept equal to (i) $7B$ and $5B$, respectively, for $H/B = 1$ and (ii) $14B$ and $11B$, respectively, for $H/B = 5$. Typical finite element mesh for $D/B = 0.5$ and $H/B = 3$ is presented in Fig. 1b where parameters E, N, D_c and N_i define the total number of elements, nodes, discontinuities and nodes along the ground surface, respectively.

4 Analysis

The magnitude of maximum surcharge loading has been obtained by using lower bound limit analysis in conjunction with finite elements as formulated by Sloan [15] and second-order conic programming (SOCP) as proposed by Makrodimopoulos and Martin [8]. The stresses are assumed to vary linearly throughout each element. The nodal stresses σ_x, σ_y and τ_{xy} are the basic unknown variables. To carry out the lower bound limit analysis, the stress field should satisfy the element equilibrium conditions, discontinuity equilibrium, stress boundary conditions and the stress field nowhere violates the yield condition. Statically admissible stress discontinuities are permitted along the common edges shared by two adjacent triangular elements, hence the normal and shear stresses are made continuous along the common edges shared by two adjacent elements.

Under the plane strain condition, the Mohr–Coulomb yield criterion with tensile normal stress taken to be positive can be presented as

$$\sqrt{(\sigma_x - \sigma_y)^2 + (2\tau_{xy})^2} \leq [2c\cos\phi - (\sigma_x + \sigma_y)\sin\phi] \qquad (1)$$

Taking new variables s_{xx}, s_{xy} and s_{aux}, the non-linear Mohr–Coulomb yield criterion given in Eq. 1 is converted into second-order conic constraint,

$$\sqrt{s_{xx}^2 + s_{xy}^2} \leq s_{aux} \qquad (2)$$

The relationship between basic stress variables (σ) and conic variables (x_{SOCP}) for ith node can be written as

$$[A_{SOCP}^i]\{\sigma^i\} + \{x_{SOCP}^i\} = \{b_{SOCP}^i\} \qquad (3)$$

where the symbol $\sigma^i = \{\sigma_x^i, \sigma_y^i, \tau_{xy}^i\}^T$ and x_{SOCP}^i consist of a set of 3-dimensional second-order cones as $\{x_{SOCP}^i\} = \{s_{xx}, s_{xy}, s_{aux}\}^T$ and

$$[A_{SOCP}^i] = \begin{bmatrix} -0.5 & 0.5 & 0 \\ 0 & 0 & -1 \\ 0.5\sin\phi & 0.5\sin\phi & 0 \end{bmatrix} \quad \{b_{SOCP}^i\} = \{0 \; 0 \; c\cos\phi\}^T \qquad (4)$$

SOCP enables the usage of the non-linear Mohr–Coulomb yield criteria, which is expressed as a set of second-order cones and overcomes the difficulty of singular apex point of the Mohr–Coulomb yield function. It is efficient in solving a huge number of variables and takes less time in computation. A code in MATLAB (2015) has been developed for the present analysis. Using optimization toolbox MOSEK (8.1.0.54), conic optimization has been carried out in the present analysis.

The lower bound of σ_s/c is obtained by maximizing the normal compressive load along the ground surface subjected to a set of equality constraints formed by element equilibrium, discontinuity equilibrium, stress boundary condition and Eq. 4 and second-order conic constraint as given in Eq. 2.

5 Results and Comparison

5.1 Comparison of Present Result with Available Literature

For the verification of the present formulation, the present mesh has been used to evaluate values of σ_s/c for circular tunnel with $D/B = 1$. Variation of σ_s/c with H/B for $D/B = 1.0$, $\gamma B/c = 0$–3 and $\phi = 0°$ and $\phi = 10°$ has been computed and compared in Fig. 2 with the lower bound solutions given by Yamamoto et al. [18]. The trend of the present results matches well with the solutions reported by Yamamoto et al. [18].

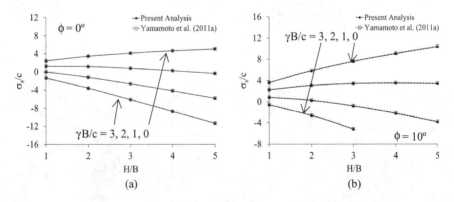

Fig. 2 Comparison of present results of σ_s/c with H/B for $D/B = 1.0$ with Yamamoto et al. [18] for **a** $\phi = 0°$ and **b** $\phi = 10°$

5.2 Variation of σ_s/c with H/B

The lower bound solution of σ_s/c has been obtained for different combinations of (i) D/B equal to 0.5, 0.75, 1.5 and 2, (ii) H/B equal to 1 to 5, (iii) $\gamma B/c$ varying from 0 to 3 with an interval of 1, and (iv) ϕ between 0° and 30° with an interval of 10°. The stability charts are presented in Figs. 3, 4, 5 and 6.

The positive value of stability number (σ_s/c) indicates that the ground surface can support up to a maximum value of normal compressive stress σ_s such that the tunnel remains stable, while the negative value implies that the ground surface can only support tensile load and hence no surcharge load can be applied to the soil surface for the given condition. The stability number increases with an increase in ϕ when there is no change in D/B, H/B and $\gamma B/c$. The increase in the aspect ratio of tunnel D/B also decreases the value of σ_s/c. For all those cases, where the ground surface can support a surcharge load, the stability number has been found to increase with an increase in soil cover depth ratio (H/B). The increase in the $\gamma B/c$ parameter decreases the amount of maximum surcharge load that can be applied at the soil surface.

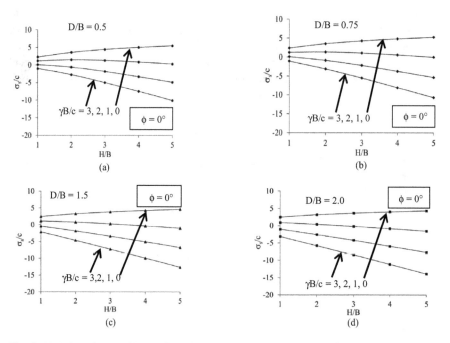

Fig. 3 Variation of σ_s/c with H/B for elliptical tunnel with **a** $D/B = 0.5$, **b** $D/B = 0.75$, **c** $D/B = 1.5$ and **d** $D/B = 2.0$ embedded in cohesive soil with $\phi = 0°$

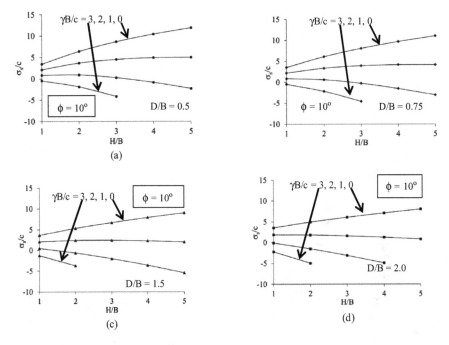

Fig. 4 Variation of σ_s/c with H/B for elliptical tunnel with **a** $D/B = 0.5$, **b** $D/B = 0.75$, **c** $D/B = 1.5$ and **d** $D/B = 2.0$ embedded in cohesive-frictional soil with $\phi = 10°$

5.3 Proximity of the Stress State to Yield

The proximity of the stress state at a point with respect to the shear strength is determined in terms of a ratio a/d where $a = (\sigma_x - \sigma_y)^2 + (2\tau_{xy})^2$ and $d = [2c\cos\phi - (\sigma_x + \sigma_y)\sin\phi]^2$. The value of a/d becomes equal to unity at a point where plastic failure occurs and remains smaller than unity for all non-yielding points. The stress contour plots are provided for $\gamma B/c = 2$ and $\phi = 20°$. Figure 7a, b shows the contour plot for $D/B = 0.5$ and 1.5, respectively, for $H/B = 2$, and Fig. 7c for $D/B = 0.5$ and $H/B = 7$. The size of the plastic zone increases with increase in the aspect ratio of the tunnel. With increase in H/B, the size of the plastic zone increases and extends downwards enclosing the complete tunnel.

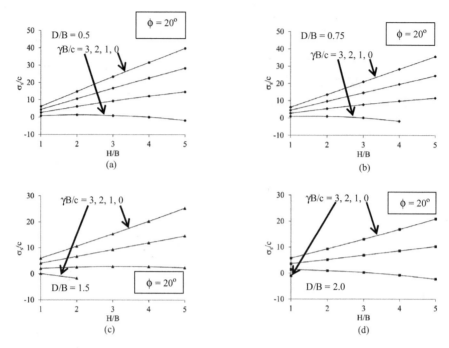

Fig. 5 Variation of σ_s/c with H/B for elliptical tunnel with **a** $D/B = 0.5$, **b** $D/B = 0.75$, **c** $D/B = 1.5$ and **d** $D/B = 2.0$ embedded in cohesive-frictional soil with $\phi = 20°$

6 Conclusion

Using lower bound limit analysis with finite elements and second-order conic programming, the stability of single elliptical tunnel in cohesive-frictional soil subjected to surcharge loading with smooth interface condition has been investigated. The stability has been observed to depend upon various factors such as (a) soil cover depth ratio H/B, (b) aspect ratio of tunnel D/B, (c) ϕ and (d) $\gamma B/c$. Stability charts are provided for a tunnel embedded in soil with $\phi \leq 30°$. The aspect ratio of the tunnel has an impact on the stability of the tunnel. The stability charts show that the increase in D/B decreases the amount of maximum surcharge load when there is no change in soil properties and soil cover depth. The stability number has been observed to increase with an increase in the soil friction angle and decrease in $\gamma B/c$. The maximum amount of surcharge load has also been noted to increase with an increase in the depth of the tunnel from ground surface. The solutions obtained in the present work are expected to be useful to the practising engineers to determine

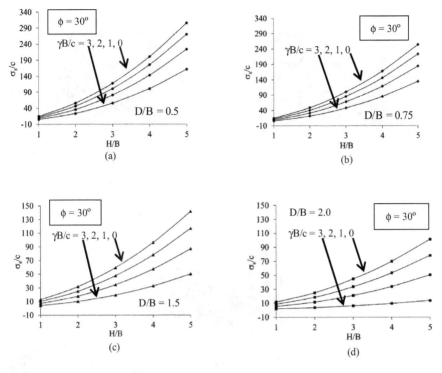

Fig. 6 Variation of σ_s/c with H/B for elliptical tunnel with **a** $D/B = 0.5$, **b** $D/B = 0.75$, **c** $D/B = 1.5$ and **d** $D/B = 2.0$ embedded in cohesive-frictional soil with $\phi = 30°$

the magnitude of maximum surcharge load and can be applied to the soil surface ensuring the stability of the tunnel.

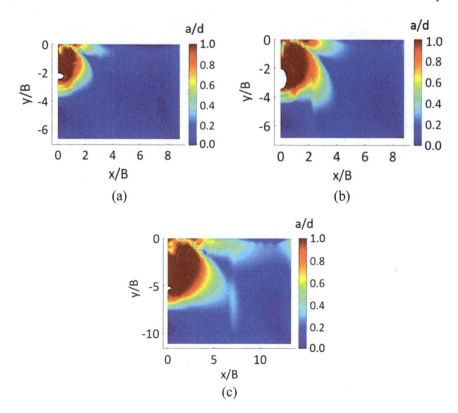

Fig. 7 Proximity of the stress state to failure of elliptical tunnel in cohesive-frictional soil with $\gamma B/c = 2$, $\phi = 20°$ for **a** $D/B = 0.5$ and $H/B = 2$; **b** $D/B = 1.5$ and $H/B = 2$ and **c** $D/B = 0.5$ and $H/B = 5$

References

1. Abbo AJ, Wilson DW, Sloan SW, Lyamin AV (2013) Undrained stability of wide rectangular tunnels. Comput Geotech 53:46–59
2. Atkinson JH, Potts DM (1977) Stability of a shallow circular tunnel in cohesionless soil. Geotechnique 27(2):203–215
3. Chehade FH, Shahrour I (2008) Numerical analysis of the interaction between twin-tunnels: influence of the relative position and construction procedure. Tunn Undergr Space Technol 23:210–214
4. Chakraborty D, Kumar J (2013) Stability of a long unsupported circular tunnel in soils with seismic forces. Nat Hazards 68:419–431
5. Davis EH, Gunn MJ, Mair RJ, Seneviratne HN (1980) The stability of shallow tunnels and underground openings in cohesive material. Geotechnique 30(4):397–416
6. Leca E, Dormieux L (1990) Upper and lower bound solutions for the face stability of shallow circular tunnels in frictional material. Geotechnique 40(4):581–606
7. Lee CJ, Wu BR, Chen HT, Chiang KH (2006) Tunnel stability and arching effects due to tunnelling in soft clayey soil. Tunn Undergr Space Technol 21:119–132
8. Makrodimopoulos A, Martin CM (2006) Lower bound limit analysis of cohesive-frictional materials using second-order cone programming. Int J Numer Meth Eng 66:604–634

9. MATLAB (2015) Computer software. MathWorks, Natick, MA
10. MOSEK Aps version 8.1.0.54 [Computer Software]. MOSEK, Copenhagen, Denmark
11. Osman AS (2010) Stability of unlined twin tunnels in undrained clay. Tunn Undergr Space Technol 25(3):290–296
12. Sahoo JP, Kumar J (2012) Seismic stability of a long unsupported circular tunnel. Comput Geotech 44:109–115
13. Sahoo JP, Kumar J (2013) Stability of long unsupported twin circular tunnels in soils. Tunn Undergr Space Technol 38:326–335
14. Sahoo JP, Kumar J (2014) Stability of a circular tunnel in presence of pseudostatic seismic body forces. Tunn Undergr Space Technol 42:264–276
15. Sloan SW (1988) Lower bound limit analysis using finite-elements and linear programming. Int J Numer Anal Methods Geomech 12:61–77
16. Sloan SW, Assadi A (1991) Undrained stability of a square tunnel in a soil whose strength increases linearly with depth. Comput Geotech 12:321–346
17. Wilson DW, Abbo AJ, Sloan SW, Lyamin AV (2011) Undrained stability of a circular tunnel where the shear strength increases linearly with depth. Can Geotech J 48:1328–1342
18. Yamamoto K, Lyamin AV, Wilson DW, Sloan SW, Abbo AJ (2011) Stability of a circular tunnel in cohesive–frictional soil subjected to surcharge loading. Comput Geotech 38:504–514
19. Yamamoto K, Lyamin AV, Wilson DW, Sloan SW, Abbo AJ (2011) Stability of a single tunnel in cohesive–frictional soil subjected to surcharge loading. Can Geotech J 48:1841–1854
20. Yamamoto K, Lyamin AV, Wilson DW, Sloan SW, Abbo AJ (2013) Stability of dual circular tunnels in cohesive-frictional soil subjected to surcharge loading. Comput Geotech 50:41–54
21. Yamamoto K, Lyamin AV, Wilson DW, Sloan SW, Abbo AJ (2014) Stability of dual square tunnels in cohesive-frictional soil subjected to surcharge loading. Can Geotech J 50:829–843
22. Yang F, Zhang J, Yang J, Zhao L, Zheng X (2015) Stability analysis of unlined elliptical tunnel using finite element upper-bound method with rigid translatory moving elements. Tunn Undergr Space Technol 50:13–22
23. Zhang J, Yang F, Yang J, Zheng X, Zeng F (2016) Upper-bound stability analysis of dual unlined elliptical tunnels in cohesive-frictional soils. Comput Geotech 80:283–289
24. Zhang J, Yang J, Yang F, Zhang X, Zheng X (2017) Upper-bound solution for stability number of elliptical tunnel in cohesionless soils. Int J Geomech 17(1):06016011

PFC3D Modeling of Rock Fragmentation by Pressure Pulse

Jason Furtney and Jay Aglawe

Abstract Unconventional hydrocarbon resources like tight oil make up a large fraction of the global hydrocarbon reserve. Tight oil exists in low permeability formations that require permeability enhancement for economic production. Chemically induced pressure pulses are being investigated as an alternative to hydraulic fracturing to enhance the permeability of these reservoirs. A numerical investigation of rock damage and fragmentation due to a chemically induced pressure pulse around a well is carried out using Itasca's discrete element code *PFC3D*. Itasca's material-modeling support package is used to create synthetic rock specimens that are subject to explosive loading. Python scripting, which is embedded in *PFC* Furtney et al.(Proceedings of the fourth Itasca symposium on applied numerical modeling (Lima, Peru). Minneapolis, Itasca International, Inc, [1]), is used to apply the gas pressure and generate the visualization. Qualitative features of the laboratory experiments matched with *PFC3D* results. Fragmentation of a block specimen expected at a range of peak pressure and pressure rise time values was analyzed. As expected, and as observed in the experiments, the number of radial fractures increases as the rise time decreases. The intensity of the fracturing and near hole crushing increase as the peak pressure increases. Permeability change is estimated from the predicted fracturing. Insights gained from numerical modeling have helped the development of the chemical stimulation methodology.

Keywords Discrete element · Fracturing · Rock fragmentation

J. Furtney (✉)
Itasca Consulting Group, Minneapolis, MN, USA
e-mail: jason.furtney@itascainternational.com

J. Aglawe
Itasca India Consulting Private Limited, Nagpur 440 010, India

1 Introduction

Chemically induced pressure pulses are being investigated as an alternative to hydraulic fracturing to stimulate unconventional hydrocarbon reservoirs. The concept involves pumping a mixture of chemicals into a well bore; a rapid exothermic reaction is triggered by the formation temperature or by the introduction of a catalyst. The reaction creates a pulse of pressure that induces tensile hoop stresses in the formation leading to radial fracturing around the well. It is envisioned that this will be a lower-cost option for stimulation than traditional hydraulic fracturing. A series of laboratory-scale experiments were conducted by the client to assess the potential of these chemical reactions to fragment rock. A series of tests were conducted on cubes of rock 25 cm in scale. The chemical reaction can be configured to have different rise times and peak pressures. This work seeks to answer the question: what peak pressure and rise time are optimal for improving permeability without excessively damaging near wellbore rock?

2 Numerical Modeling

Itasca's discrete element software *PFC3D* is used to represent the rock and model the fracturing process. The Itasca material-modeling support package [2] is used to create specimens of spherical-grain, parallel-bonded synthetic material, which are subject to explosive loading. Three specimens are created: a coarse-grained 25 by 25 by 10 cm box, a fine-grained 25 by 25 by 10 cm box, and a coarse-grained 25 cm cube. The coarse models have an average particle diameter of 8.5 mm, and the fine model has an average particle diameter of 4.25 mm. The coarse model particle diameters were chosen to result in a specimen with approximately 10,000 particles.

2.1 Micro-properties

Each specimen consists of a packing of spherical particles connected by parallel bonds. Table 1 gives the material micro-properties. The essential micro-properties are density (1960 kg/m^3), modulus (1.5e9 Pa), normal to shear stiffness ratio (1.5), cohesion (20e6 Pa), and tensile strength (1e6 Pa). A full description of all the micro-properties in Table 1 is given in [2].

These micro-properties give rise to the following macroscopic properties: a Young's modulus of 1.82e9 Pa, a (static) direct tensile strength of 0.74e6 Pa, and an unconfined compressive strength of 3.4e6 Pa. The material specimens are created without a hole. The material specimens are initially under zero stress.

Table 1 Micro-properties of parallel-bonded material

Property	Value
Common group:	
N_m	SS_ParallelBonded
T_m, α, C_ρ, $\rho_v \left[\frac{kg}{m^3}\right]$	2, 0.7, 1, 1960
S_g, T_{SD}, $\{D_{\{l,u\}}\}$ [mm], $\phi\}$, D_{mult}	0, 0, {8.0,9.0,1.0}, 1.0
Packing group:	
S_{RN}, P_m [kPa], ε_P, $\varepsilon_{\lim}$, $n_{\lim}$ C_p, n_c	10000, 100, 1×10^{-2}, 8×10^{-3}, $2 \times 10^6$1, 0.35
Parallel-bonded material group:	
Linear group:	
E^* [GPa], κ^*, μ	1.5, 1.5, 0.4
Parallel-bond group:	
g_i [mm], $\bar{\lambda}$, $\bar{E}^*$ [GPa], $\bar{\kappa}^*$, $\bar{\beta}$	0.5, 1.0, 1.5, 1.5, 1.0
$(\bar{\sigma}_c)_{\{m,\,sd\}}$ [MPa], $(\bar{c})_{\{m,\,sd\}}$ [MPa], $\bar{\phi}$ [degrees]	{1.0, 0}, {20.0, 0}, 0
Linear material group:	
E_n^* [GPa], κ_n^*, μ_n	1.5, 1.5, 0.4

2.2 Pressure Pulse

For this type of loading, the tensile strength is expected to be the most important parameter. Additional work can be performed for more complete material property calibration using flat jointed material, which would allow matching both the UCS and tensile strength.

Physically, the pressure in the hole is controlled by several interactive processes, including the chemical reaction, the product equation of the state, the deformation of the rock, and the gas flow into the newly created fractures. The goal of this modeling effort is to examine the mechanical response of rock due to loading from a pressure pulse. As a simplification, the pressure in the hole is assumed to follow a predefined path. As a modeling simplification, the pressure is assumed to be the same everywhere in the hole at a given time. Pressure is defined as a piece-wise function of time. An initial region of linear increase to a given peak pressure occurs over a given rise time followed by a region of exponential decay.

$$p(t) = \begin{cases} \frac{p_p t}{t_r} & \text{when } t < t_r \\ p_p e^{\lambda(t-t_r)} & \text{else} \end{cases}$$

$$\lambda = \ln\left(\frac{1}{2}\right)/t_d \tag{1}$$

where p_p is peak pressure, t_r is rise time, and t_d is the time for the pressure to drop to half of the peak value (t_d is taken as five times the characteristic time, discussed below). Figure 1 shows the applied pressure p as a function of time t.

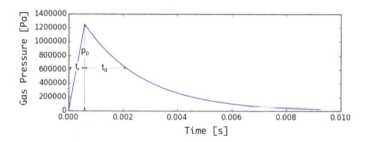

Fig. 1 Predefined pressure pulse used in the study

3 Results and Discussions

3.1 Parameter Study on Coarse 25 by 25 by 10 Cm Model

A parameter study is conducted to show the effect peak pressure and rise time have on the block fragmentation. A 3.8 cm diameter hole is created, parallel to the z-axis, along the entire length of the specimen by deleting the *PFC3D* particles in this region. Particles bordering the inside of this hole are tagged for pressure application. A force equal to pressure divided by particle diameter squared is applied to these tagged particles in the radial direction away from the hole centerline. The tagged particles are used as the pressure application site throughout the model run. As the position of the tagged particles changes, the local radial direction used for pressure application is recalculated.

The peak pressure and the rise time are varied in a 5 by 5 matrix of model runs. Dimensionless quantities for peak pressure and rise time are introduced. The peak pressure is scaled by a characteristic pressure, p_c, taken as the micro tensile strength (1e6 Pa). Rise time is scaled by a characteristic time, t_c, taken as the time required for a p-wave to travel the specimen length. A one-dimensional estimate of the sound speed c is used, $c = \sqrt{\text{modulus/density}} = 875$ m/s. The length of the specimen is 25 cm, which gives a characteristic time, t_c, of 2.9e-4 s.

A matrix of 25 models is run with peak pressures of 0.75, 1.0, 1.25, 1.5, and 2.0 times p_c and rise times of 0.1, 0.5, 1, 2, and 10 times t_c.

Figure 2 shows cross-sections of the middle 2 cm of the 25 *PFC3D* models; the view is down the z-axis along the hole. Parallel bond breaks for each of the 25 cases in the parameter study are shown. Black dots represent bond breaks that separate two fragments. Red dots represent bond breaks that are internal to a fragment. Fragments are defined as groups of *PFC* particles that are connected via intact parallel bonds. The fragment plots in Fig. 3 are composed of concave triangular surface meshes of the individual fragments. A three-step process generates the surface meshes: (1) a Delaunay tetrahedralization of the centroids of all the particles in a fragment is found; (2) triangular faces are removed if they have an edge length greater than four times the average particle radii; and (3) triangular faces on the interior of the model are removed. The resulting surface mesh connects the centroids of particles on the exterior of a fragment.

The results of this work show the variation in fracturing with changes in dimensionless rise time and dimensionless peak pressure. Additional model runs with a range of dimensional parameters could determine the range of values over which this dimensionless scaling would be applicable.

Figure 3 shows the fragments created in each of the 25 cases. The following conclusions are drawn from this study:

- Peak pressures below p_c result in a few bond breaks.
- Peak pressures above 1.5 p_c result in a region of intense bond breaking adjacent to the hole.

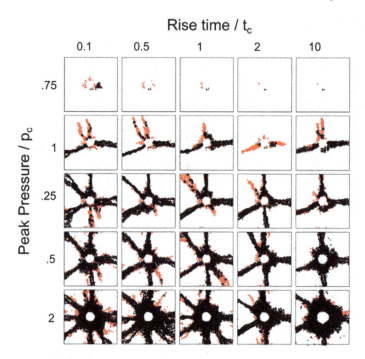

Fig. 2 Bond breaks predicted for the 25 cases in the parameter study

- Faster pressure rise times give a larger number of discrete radial fractures.

The case with rise time of 2 t_c and peak pressure of 1.25 p_c is considered the base case. This base case is used in the next two sections to investigate the role of model resolution and three-dimensional geometries. Figure 4 shows the pressure and the total number of bond breaks as a function of time for this base case. The number of bond breaks increases quickly for a period of time near the peak pressure. Bond breaking continues at a lower rate after the peak pressure. 90% of the bond breaks occur by 5 ms after the peak pressure.

The modeling in this work focuses on the damage processes occurring during and shortly after this sharp pressure increase. The duration of the model runs is 10 ms during which most of the fragmentation occurs. If the pressure increase occurs over a period of seconds or minutes, the physical processes occurring would be closer to hydraulic fracturing by its nature. Modeling of hydraulic fracturing requires a different methodology that incorporates fracture flow and leak-off. Additional information about the peak pressure and rise times coming from an investigation of the chemical kinetics could help in clarifying the nature of the processes.

In this modeling, the synthetic rock specimen is under zero stress at the beginning of the model runs. If the material is confined, a greater gas pressure would be needed to cause fracturing and damage.

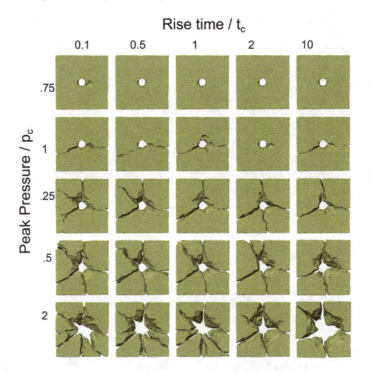

Fig. 3 Block fragments created in the 25 cases in the parameter study

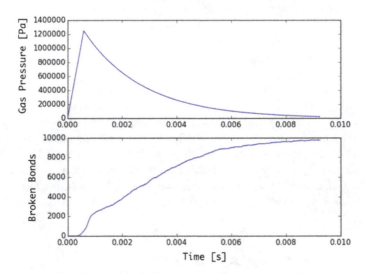

Fig. 4 Pressure and bond breaks shown as a function of time for the base case

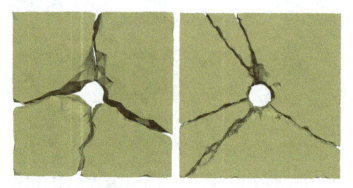

Fig. 5 Comparison between fragmentation in the coarse model (left) and fine model (right)

3.2 Model Resolution Effects

The base case from the previous section (rise time of 2 t_c and peak pressure of 1.25 p_c) is used to investigate the effect of particle size on the model result. A specimen with the same 25 cm by 25 cm by 10 cm shape is created with half the particle diameter. The finer specimen has around 100,000 particles. The material properties, hole diameter, and loading procedure are the same as the parameter study described above. Figure 5 provides a comparison between the fragments created in the coarser and fine base cases. The finer case, shown on the right, has five radial fractures with two groups of two closely spaced radial fractures. The coarse base case shown on the left has four approximately equally spaced radial fractures. Although the fine and coarse models have some qualitative similarities, more investigation is required to understand the effect of particle resolution on fracturing.

It was expected by Itasca that the finer model would show a proportionally larger number of bond breaks because the effective mode 1 fracture toughness of the finer model is lower. The fracture toughness of bonded particle models is discussed by Potyondy and Cundall [3]. This paper shows that mode 1 fracture toughness is proportional to the square-root of particle radius, and so it is expected that the toughness of the fine material is less than that of the coarse material by a factor of $1/\sqrt{2}$. For the case of halving the particle diameter, this scaling does not have a strong effect. For larger changes in particle size, this scaling may need to be applied to have material specimens with a constant fracture toughness.

3.3 Three-Dimensional Effects

The two previous sections consider a three-dimensional model subject to simplified loading conditions. In these models, the hole extends the length of the specimen and the applied forces are limited to acting in the *x-y* plane. This section considers a 25 cm

Fig. 6 Fragments created in the 25 cm cube model with the hole in the middle of the model only

cube specimen with a 7.8 cm long, 3.8 cm diameter hole. The hole is located in the center of the specimen and is aligned along the z-axis. The three-dimensional model has around 31,000 particles. The rise time and material properties are the same as in the base case described above. The pressure loading is the same and, in addition, the gas applies a force upward and downward at the top and bottom of the hole.

Figure 6 shows the fragments created in this case. Three images are shown, with the leftmost showing all the fragments, the middle showing the fragment nearest the observer removed, and the rightmost showing only the lower fragment. Radial fractures occur around the hole circumference, and cone-shaped cracks occur above and below the hole. Similar patterns of fracturing are observed in rock blasting.

A peak pressure of 2.5 p_c is used in this case relative to the 1.25 p_c in the base case. In this case, the hole is limited to the central 7.8 cm of the specimen. A greater peak pressure is needed to create qualitatively similar fractures to the base case for two reasons. First, both radial and cone fractures are created resulting in greater fracture surface area. Second, these fractures are created by proportionally less reaction products. The surface area over which the pressure pulse is applied is smaller, which results in less mechanical work done by the gas on the rock.

4 Conclusions and Future Work

Itasca's discrete element code *PFC3D* was able to reproduce the qualitative features of the laboratory experiments for the fragmentation of a block specimen expected at a range of peak pressure and pressure rise time values. *PFC3D* has tremendous potential to capture the complex damage mechanisms and fragmentation occurring in the rock mass due to high values of pressure pulse applied to the walls of the borewell.

The present modeling considers relatively small specimens. Work is underway to develop a version of *PFC3D* with message passing interface (MPI) that will lead to an order of magnitude improvement in model size and calculation speed. With larger models, the interaction between newly created fractures and pre-existing discontinuities will be studied. The peak pressures in the present work are intentionally kept at or below the compressive strength of the rock or avoid compressive shear failure. In the related problem of rock blasting with commercial explosives, the peak pressure is greater than an order of magnitude or more and leads to local compressive shear failure. The discrete element method with the linear contact model does not represent compressive shear failure well. Work is underway to develop a new contact model that can accurately describe this type of material deformation. Such a model will allow a broader class of dynamic fracturing problems to be studied.

References

1. Furtney JK, Purvance M, Emam S (2016) Using the Python Programming Language with Itasca Software. In: Proceedings of the fourth Itasca symposium on applied numerical modeling (Lima, Peru). Minneapolis, Itasca International, Inc
2. Potyondy D (2016) Material-modeling support in PFC [fistPkg21]. Itasca Consulting Group, Inc., Technical Memorandum ICG7766-L, Minneapolis, Minnesota
3. Potyondy DO, Cundall PA (2004) A bonded-particle model for rock. Int J Rock Mech Min Sci 41(8):1329–1364

Modeling of Seismic Actions on Earth Retaining Structures

Rohit Tiwari, Nelson Lam, and Elisa Lumantarna

Abstract Present work deals with numerical modeling of seismic actions on earth retaining structures. Finite element (FE) analyses have been carried out on scaled-down retaining wall models. The capability of finite element models has been evaluated for the replication of shaking table experiment results. It was observed that FE models are highly sensitive to assigned nonlinear material models, especially for hardening and softening behavior of backfill soil. A detailed and simplified FE modeling procedure is explained for the simulation of seismic actions on earth retaining structures.

Keywords Retaining wall · Finite element modeling · Shaking table experiment · Similitude

1 Introduction

Earth retaining structures such as retaining walls, bridge abutment, and basement walls are key elements of a modern infrastructure system. Satisfactory performance of earth retaining structures during an earthquake could be ensured by an accurate and realistic assessment of earthquake actions on them. Earthquake response of earth retaining structures could be simulated using finite element (FE) modeling techniques; however, outcomes of FE analyses should be validated with experimental results, in order to ensure accuracy of FE analysis results. The present study deals with FE modeling of seismic actions on retaining wall. Shaking table experiments have been performed on scaled-down retaining wall model at The University of Melbourne. Consolidated drained (CD) triaxial test and one-dimensional (1D) compression tests have been performed on backfill soil material, used in shaking table experiment. Series of pluck tests have also been performed during shaking table

R. Tiwari (✉) · N. Lam · E. Lumantarna
The Department of Infrastructure Engineering, The University of Melbourne, Parkville, VIC 3010, Australia
e-mail: tiwarir@student.unimelb.edu.au

experiments for understanding free vibration response of scaled-down retaining wall model. Observations from shaking table experiment and results from geotechnical testing of backfill soil have been used to develop a FE model of retaining wall and analyzed against shaking table base excitations, good agreement has been observed between shaking table experiment and FE model results.

2 Literature Review

Various studies have been performed for understanding the seismic behavior of fixed base earth retaining structures. Mononobe and Matsuo [10]-modified Coulomb earth pressure theory (MO method) and suggested a linearly increasing seismic earth pressure along retaining wall depth. Matuo and Ohara [8] performed shaking table experiment on scaled-down model of quay wall to study earthquake-induced lateral pressure on quay wall, they observed higher seismic pressure near retaining wall top. Whitman and Liao [14] studied seismic displacement of gravity retaining wall. Wood [16], Veletsos et al. [13] analyzed retaining walls to understand dynamic soil pressure on retaining walls. Sherif and Fang [12], Choudhury and Chatterjee [3], Lew et al. [7], and Wilson and Elgamal [15] studied seismic soil pressure on retaining walls and observed nonlinear dynamic pressure distribution along with retaining wall height. Mikola and Sitar [9] used a centrifuge to perform experiments on scaled-down fixed base retaining wall models; to understand the amount of soil pressure on retaining walls, their outcomes matched well with results of MO method. Cakir [2] studied seismic behavior of fixed base cantilever retaining wall using the FE model and calibrated a 2 degrees-of-freedom retaining wall backfill soil model for finding seismic displacement of retaining wall. However, the role of backfill soil on the seismic displacement of earth retaining structure and dynamic nature of seismic earth pressure is not addressed in the literature, which is essential for developing a simplified solution for analyzing the seismic response of earth retaining structures. Therefore, a detailed experimental and FE investigation is required for the understanding role of backfill soil and time-dependent nature of seismic pressure on earthquake response of earth retaining structures.

3 Shaking Table Experiment on Fixed Base Retaining Wall Model

3.1 Details of Scaled-Down Model and Instrumental Setup

Shaking table experiment has been performed at The University of Melbourne, on a 2m x 2m shaking table. scaled-down model of fixed base retaining wall is considered with a scale factor of 10, Aluminum has been chosen as retaining wall material in

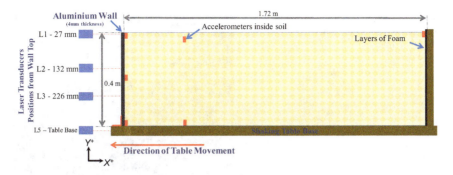

Fig. 1 Geometrical details and instrumentation of the scaled down retaining wall model for the shaking table experiment

order to get a better-deflected shape of the deformed retaining wall. The height, width, and thickness of scaled-down retaining wall is kept as 0.4 m, 0.4 m, and 0.004 m, respectively. Figure 1 shows details of scaled-down retaining wall model, along with sign convention adopted in the present study. The base of retaining wall is fully constrained, length of backfill soil behind the retaining wall is 1.72 m, a smooth surface wooden box is fabricated for retention of backfill soil along its length. Multiple layers of high-density foam have been applied on the back of the wooden frame to minimize boundary effects, sandpaper is applied at the model base and back face of retaining wall to initiate friction.

Crushed rock has been used as backfill soil material for all shaking table tests. Maximum dry density of backfill soil is estimated as 1790 kg/m^3, total height of backfill soil is 0.4 m, backfill soil has been placed in four layers with 0.1 m maximum layer thickness, density of backfill soil is controlled by compacting it in each layer, using shaking table vibrations, which minimize over compaction in bottom layers.

Laser sensors with high sample frequency have been used for capturing displacement of retaining wall. Acceleration of retaining wall and backfill soil is captured using bidirectional and triaxial accelerometers. Figure 1 also shows details of the instrumental setup. Figure 2 shows the photograph of fabricated scaled-down retaining wall model before test.

3.2 Pulses Used for Shaking Table Experiment

Figure 3 shows different pulses considered in the present investigation. First two pulses are half-cycle sine pulse (1a, 1b), second two pulses are one-cycle sine pulse (2a, 2b), and the third set contains multiple pulses (3a, 3b). Maximum amplitude and duration for every pulse are shown in Fig. 3. Shaking table experiments have been performed for all pulses. Few quarter cycle pulses have also been used for pluck test, to understand free vibration response of retaining wall–backfill system.

Fig. 2 The scaled down model setup for the shaking table experiment

3.3 Results of Shaking Table Experiment With Different Pulses

Figure 4 shows relative displacement at retaining wall top for all six pulses considered in shaking table test. During the initial phase of loading (base moving x^{-iv} direction) movement of retaining wall towards backfill soil have been observed in all case, which is due to the inertia of retaining wall and backfill soil. However, soon after inertia phase ends, active state displacement of retaining wall has been observed in all cases which persist for entire test duration; this confirms the generation of active failure wedge in backfill soil. Separation of backfill soil and retaining wall has also been observed during the initial phase of loading for all cases.

Figure 5 shows response spectrum acceleration for backfill soil base, middle, and top for multiple pulses only. Amplification of lateral acceleration toward backfill soil top has been observed for both multiple pulses, which is responsible for higher retaining wall deformation at top.

3.4 Geotechnical Investigations of Backfill Soil

In order to understand constitutive behavior of backfill soil (crushed rock), some geotechnical investigations have been performed, sieve size analysis has been performed for finding particle size gradation curves, based on which the mean particle size of crushed rock is observed as 7 mm. It should be noted here that backfill soil behind retaining wall is a confined condition; therefore, 1D compression test has been

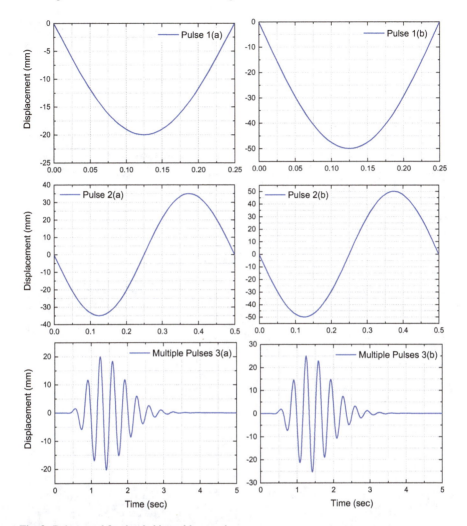

Fig. 3 Pulses used for the shaking table experiment

performed for finding constrained modulus of crushed rock for different confinements [6], for scaled-down model the total backfill height is 0.4 which generates around 7.0-kPa compression at backfill base. Based on results of (1D) compression test, constrained modulus of backfill soil for 7.0-kPa confinement was estimated as 2.5 MPa.

Consolidated drained (CD) test has been performed for crushed rock at 34, 68, and 136 kPa confinements, respectively (ASTM, D7181-11 [4]). Based on CD test results, angle of internal friction and dilation angle of crushed rock has been estimated as 44° and 19°, respectively.

252   R. Tiwari et al.

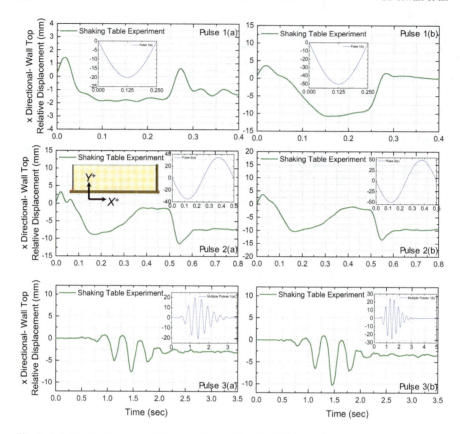

Fig. 4 Relative displacement at the retaining wall top for different pulses

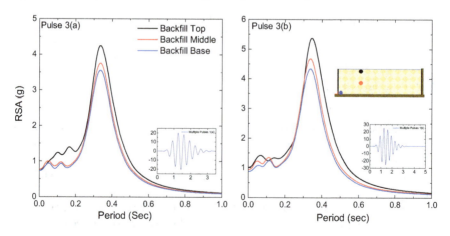

Fig. 5 Response spectrum acceleration for the backfill soil base, middle and top

4 FE Modeling of Shaking Table Experiment

4.1 Geometrical Modeling, Surface Interaction, and Boundary Conditions

Two-dimensional (2D) geometry of scaled-down retaining wall model is developed in FE simulation software Abaqus. Geometrical details of 2D FE model is shown in Fig. 6. Four-node continuum plain strain element (CPE4R) with reduced integration, hourglass, and distortion control scheme has been chosen for FE meshing. Nonlinear finite element analyses have been carried out using Abaqus explicit scheme, which is suitable for large deformation problems (Abaqus/Explicit User's Manual, version 6.13. [1]). The denser mesh is adopted for retaining wall–backfill soil interface and backfill soil–model base interface; for higher accuracy of results, no mesh convergence study has been performed for validation investigations. The interaction has been modeled as frictional contact in the tangential direction and hard contact in the normal direction. The coefficient of friction between retaining wall–backfill soil and foam–backfill soil is 0.64 and 0.1, respectively. It should be noted here that during high-amplitude input pulses, friction between backfill soil and model base highly influence overall displacement response of retaining wall [5]. Therefore, FE models have been calibrated for the coefficient of friction between backfill soil and model base.

Two types of boundary conditions have been adopted in all FE models, the first boundary condition was applied during the initial phase (no external loading) to satisfy equilibrium conditions, in which vertical boundaries were restrained in lateral "x" direction, the model base was restrained in vertical "y" direction. Geostatic stresses and void ratio have also been specified in backfill soil during the initial phase. During the loading phase (dynamic explicit step), the base of the scaled-down model was allowed to move in "x" direction, however, constrained in "y" direction. All pulses and accelerograms have been applied at the model base, during the dynamic explicit step. Stresses are positive for increasing "x" and "y" directions.

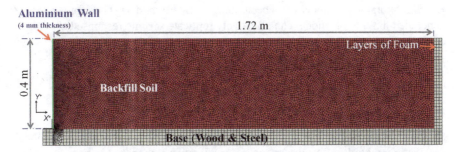

Fig. 6 FE model of the shaking table experiment on the scaled down retaining wall

Table 1 Material properties considered in FE simulations

Sr. no.	Material	Density (Kg/m^3)	Elastic modulus (GPa)	Poisson's ratio
1.	Retaining Wall	2700	69	0.33
2.	Backfill Soil	1790	0.00290	0.45
3.	Angle, Base	7800	200	0.3
4.	Base Wood	1000	100	0.3
5.	Foam	2000	0.1	0.4

4.2 Constitutive Modeling of Materials

Table 1 shows material properties considered in FE modeling, retaining wall has been modeled with elastic aluminium material. Mohr–Coulomb model was used to model the yielding of backfill soil. Hardening of cohesionless backfill soil with Mohr–Coulomb model could be simulated with the help of cohesive yield stress of soil against calculated plastic strains, CD triaxial test data has been used to simulate hardening of backfill soil material, as explained by Potts and Zdravkovic [11].

4.3 Validation of the Shaking Table Experiment Results With FE Simulations

As explained earlier shaking table experiment has been carried out for six pulses. Displacement of the shaking table base has also been captured using laser sensors, which was applied to the FE model as input motion. Figure 7 shows a comparison of relative displacement (at retaining wall top, relative to base) observed from the shaking table experiment and FE simulations, a good agreement has been observed between experiment and FE simulation results. Frequency domain analyses have also been carried out for acceleration data obtained from pluck test results, based on which the natural frequency (first mode) of retaining wall is estimated as 19.53 Hz. Similar natural frequency has been obtained by FE models. Therefore, it can be concluded that the FE model can effectively replicate seismic response behavior of earth retaining structures.

5 Conclusion

Shaking table experiment has been performed on scaled-down retaining wall model for six different pulses. Displacement and acceleration response of retaining wall has been studied and frequency domain analyses have also been carried out in order to understand displacement-based earthquake response of retaining walls. Separation

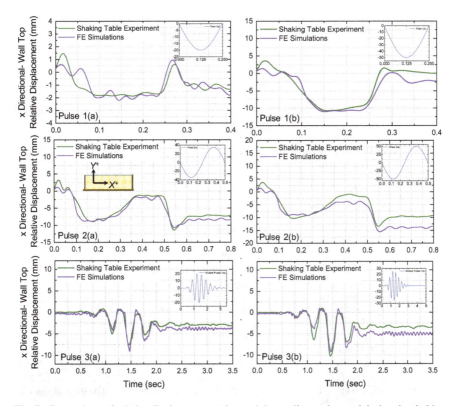

Fig. 7 Comparison of relative displacement at the retaining wall top, observed during the shaking table experiment and the FE investigations

of retaining wall and backfill soil, amplification of acceleration has been observed during the shaking table experiment. The capability of FE software Abaqus has been evaluated for replication of shaking table experiment results. Mohr–Coulomb material model has been calibrated with triaxial test data. Good agreement has been observed between shaking table experiment and FE simulation results. Further, investigations are in progress for achieving a simplified model for seismic analysis of earth retaining structures.

Acknowledgments Bushfire and Natural Hazards Cooperative Research Centre, Melbourne, Australia.

References

1. Abaqus/Explicit User's Manual, version 6.13 (2013) Dassault Systèmes Simulia Corporation, Providence, Rhode Island, USA
2. Cakir T (2013) Evaluation of the effect of earthquake frequency content on seismic behavior of cantilever retaining wall including soil–structure interaction. Soil Dyn Earthq Eng 45:96–111
3. Choudhury D, Chatterjee S (2006) Displacement-based seismic active earth pressure on rigid retaining walls. Electr J Geotech Eng 2(0660)
4. D7181-11 (2011) Method for consolidated drained triaxial compression test for soils. ASTM International, West Conshohocken, USA
5. Hashemnia K, Pourandi S (2018) Study the effect of vibration frequency and amplitude on the quality of fluidization of a vibrated granular flow using discrete element method. Power Technol 327:335–345
6. Kim H-K, Santamarina JC (2008) Sand-rubber mixtures (large rubber chips). Can Geotech J 45:1457–1466
7. Lew M, Sitar N, Atik LA (2010) Seismic earth pressures: fact or fiction. In: 2010 Earth retention conference (ER2010) proceedings, pp 656-673. ASCE, Bellevue, Washington, USA
8. Matuo H, Ohara S (1960) Lateral earth pressure and stability of quay walls during earthquakes. In: 2nd World Conference on earthquake engineering proceedings, Tokyo, Japan
9. Mikola M, Sitar N (2013) Seismic earth pressures on retaining structures in cohesionless soils. California Department of Transportation (Caltrans), USA
10. Mononobe N, Matsuo H (1929) On the determination of earth pressures during earthquakes. In: World engineering conference proceedings, vol 9, paper No 388
11. Potts DM, Zdravkovic L (1999) Finite element analysis in geotechnical engineering, 1st edn. Thomas Telford Publishing, London
12. Sherif MA, Fang YS (1984) Dynamic earth pressures on walls rotating about the top. Soils and Found 24(4):109–117
13. Veletsos AS, Adel HY (1994) Dynamic modeling and response of soil-wall systems. J Geotech Eng 120(12):2155-2179
14. Whitman RV, Liao S (1985) Seismic design of gravity retaining walls. US Army Corps of Engineers, Washington, USA
15. Wilson P, Elgamal A (2015) Shake table lateral earth pressure testing with dense c–φ backfill. Soil Dyn Earthq Eng 71:13–26
16. Wood JH (1973) Earthquake-induced soil pressures on structures. Ph. D. thesis, California institute of technology, Pasadena, California, USA

Computational Structural Dynamics

Detection of Damages in Structures Using Changes in Stiffness and Damping

Uday Sinha and Sushanta Chakraborty

Abstract Structural damage identification has been gaining a lot of importance for almost two decades. In this paper, a fibre-reinforced plastic (FRP) plate has been considered under the free–free support condition. Damage is assumed to be confined within a small area of the plate locally. Experimental data has been numerically simulated for both damaged and undamaged plate. Both stiffness and damping changes are considered in the present study. Rayleigh's proportional damping has been used for modelling the damping in the plate. For the model updating process, Inverse Eigen sensitivity Method (IEM) has been implemented. The objective function is based on the variation between frequency response function (FRF) values. The damaged area is modelled with degraded values of elastic moduli. Further, the global modal damping factors of the damaged model have been increased to account for the increase in damping due to damage. A two-stage damage identification approach has been adopted. Damage site is located first using a damage index based on mode shape curvature difference. Subsequently, in the second stage, the damage is quantified employing model updating technique. The estimated reduced stiffness and increased damping parameters are found to be converging to actual values in these numerically simulated examples.

Keywords Damage identification · Vibration-based methods · Finite element model updating · FRP · Inverse eigen sensitivity method · Rayleigh's proportional damping · FRF

U. Sinha (✉) · S. Chakraborty
IIT Kharagpur, Kharagpur 721302, India
e-mail: udaysinha58@gmail.com

1 Introduction

Damage identification in structures has achieved a lot of importance in the research community, as damage in a structure can lead to its unpredictable failure and may result in human fatalities as well as economic losses. Regular and systematic inspection of structures is necessary to prevent the occurrence of these catastrophic events. Damage is generally referred to some undesirable changes (like changes in material properties, loosening of connections or changes in the boundary conditions) that can alter the performance or serviceability of the structure. The dissipation characteristics also undergo substantial changes. Vibration-based methods are popular global damage detection techniques that use traditional vibration testing equipment and modal analysis techniques to extract damage-sensitive structural dynamic behaviour. Damage to any structure will cause some changes in physical parameters (mass, stiffness and damping) which will alter the dynamic responses (natural frequencies, mode shapes and its curvature, FRFs).

In the early works of damage identification in structures, natural frequencies were primarily used. Salawu [20] have provided a useful review of works that adopted shifts in natural frequency for detecting damage. Mode shape curvature (MSC) has been extensively used to detect local damages. Pandey et al. [16] for the first time successfully applied MSC difference technique for damage detection in a beam analytically. The damage was realised as a reduction in material elastic modulus locally at a cross section. Abdel Wahab and De Roeck [1] introduced a damage indicator function calculated by mode shape curvature difference between an undamaged and the damaged structure averaged over several modes and successfully detected damages in bridge structure. Rucevskis et al. [19] applied the MSC-based damage detection using the average summation of the curvature difference of several modes. Roy [18] applied MSC for locating damage in a 16-storey shear building.

Damage in a structure has also been seen to cause alterations in the dissipative properties. Advances in the vibration measurement technology have facilitated a much more accurate assessment of damping and studies that correlate damping with damage have emerged as complementary to methods using solely frequencies and mode shapes. Williams and Salawu [21] advocated that damping changes can be a possible indicator for damage in situations where frequency and mode shapes deemed insensitive to damage. Kyriazoglou et al. [8] studied the variations in specific damping capacity for damage identification in composite beams. The damage was provided using quasi-static loading and fatigue. Kıral et al. [7] studied the variations in natural frequency due to damage was not consistent, whereas the damping ratios always increased with damage severity and was found to be location-sensitive. Montalvão et al. [11] proposed a damping damage indicator (DaDI) for damage localisation, basically delamination detection in carbon fibre-reinforced plastic plate. Montalvão et al. [12, 13] extended the study and proposed a multi-parameter damage indicator (MuDI). The formulation of this indicator was a combination of DaDI and a frequency damage indicator.

Through finite element model updating technique, the discrepancies between the responses of the finite element model and responses directly measured from a real/experimental structure is being minimised by adjusting the model parameters. Basically, finite element model updating is applied as an inverse problem to correct the uncertain initially assumed model parameters. This technique can be put to use for damage detection in structures if appropriate damage-sensitive stiffness or damping parameters can be defined within the finite element modelling framework. The process of model updating is, therefore, better posed as an optimisation process. The residuals between the responses of the numerical model and experimental observation are minimised by adopting proper objective functions. The success of such optimisation will depend upon the appropriate selection of damage-sensitive model parameters and an efficient gradient-based or evolutionary minimisation algorithm, which are eventually iterative algorithms.

The sensitivity-based methods are the most popular methods among various model updating techniques based on measured dynamic data. Chen and Garba [3] provided a mathematical formulation to calculate eigen sensitivities using matrix perturbation method and subsequent model updating. Mottershead and Friswell [14] provided a review of early works on finite element model updating. Friswell [6] gave an outline of the practice of inverse methods in damage detection through vibration data. A basic introduction to model updating through sensitivity methods is compiled by Mottershead et al. [15].

Sensitivity-based model updating technique has also been applied to update damping parameters. Lin and Zhu [9] through a sensitivity-based model updating method, updated the generalised proportional and non-proportional damping matrix for a truss structure. Arora et al. [2] used complex FRFs using complex updated parameters of a fixed–fixed beam assuming structural as well as viscous damping. Pradhan and Modak [17] used normalised FRFs derived from the complex FRFs to identify the damping matrix. Mondal and Chakraborty [10] estimated the proportional viscous damping parameters through sensitivity-based model updating approach for a fibre-reinforced plastic (FRP) plate. Several studies exist in current literature to update stiffness as well as damping parameters; however, their extension for effective damage detection still remains unexplored, particularly for FRP type of layered structures.

In the present study, damage detection is carried out for a numerically modelled FRP rectangular plate. Both stiffness and damping changes have been taken into consideration. Damping modelling has been carried out using Rayleigh's proportional damping considering viscous damping. A numerically simulated damage plate has been considered, where damage has been imparted locally into some elements with degraded elastic moduli. For the damaged plate, global increase in the modal damping factors has been simulated. A two-stage damage identification process has been adopted. First, the damage localisation is carried out through a damage index that takes into account the mode shape curvature difference between the undamaged and damaged plates. In the second stage, quantification of damage is done using finite element model updating through Inverse Eigen sensitivity method (IEM). The objective function is based on the difference between the FRF magnitude of damaged and undamaged plates. In the present investigation, IEM is applied for the updating process to quantify the changes in elastic and damping parameters simultaneously.

2 Mathematical Formulations

2.1 The Inverse Eigen Sensitivity Method

First-order approximation of a Taylor series expansion of a function f of n variables $(r1, r2,\ldots, rn)$ can be written as follows [3]:

$$f(r_1, r_2, \ldots, r_n) \cong f(\bar{r}_1, \bar{r}_2, \ldots, \bar{r}_n) + \sum_{i=1}^{n} \left(\frac{\partial f}{\partial r_i}\right)_{r_i - \bar{r}_i} (r_i - \bar{r}) \quad (1)$$

where the Taylor series is expanded in the neighbourhood of $r = r_0$

If more functions are involved, Eq. 1 can be expressed in a matrix form as follows:

$$\{f(r)\} = \{f(\bar{r})\} + \left[\frac{\partial f(r)}{\partial r}\right]_{r=\bar{r}} (r - \bar{r}) \quad (2)$$

$\left[\frac{\partial f}{\partial r}\right]_{r=\bar{r}}$ is the first-order sensitivity matrix, r_i are the parameter that needs to be identified, $\bar{r}_i$ is the initial assumed value.

Also, $f_i(r) =$ Measured responses of the structure, and
$f_i(\bar{r}) =$ Responses of the finite element model of the structure.
The above equation can be expressed as follows:

$$\{\Delta f\} = [S]\{\Delta r\} \quad (3)$$

A residual error vector can be written as follows:

$$\{\Delta\} = \{\Delta f\} - [S]\{\Delta r\} \quad (4)$$

An initial solution is performed, the sensitivity values are calculated and a new $\{r\}$ is generated using an iterative approach as follows:

$$\{r\}_{i+1} = \{r\}_i + \{\Delta r\}_i \quad (5)$$

The iteration is carried out by updating the selected model parameters until the square of error minimises to a selected margin ε.

$$\{\Delta\}^T\{\Delta\} \leq \varepsilon \quad (6)$$

2.2 Damage Index Based on Mode Shape Curvature (MSC) Difference

The mode shape curvatures in both x and y directions of the plate are calculated by applying central difference on the displacement mode shapes. The formulae are given as follows:

$$\varphi^{''}(x) = \frac{\varphi(x + h_x) - 2\varphi(x) + \varphi(x - h_x)}{l_x^2} \quad (7)$$

$$\varphi^{''}(y) = \frac{\varphi(y + h_y) - 2\varphi(y) + \varphi(y - h_y)}{l_y^2} \quad (8)$$

where l_x and l_y are the grid spacing in the x and y-direction, respectively.

The mode shape curvature difference is calculated from the following formula considering both x and y-direction [22] and is given as:

$$MSC_d^n = \left| \varphi^{''}(x)^{ud} - \varphi^{''}(x)^d \right| + \left| \varphi^{''}(y)^{ud} - \varphi^{''}(y)^d \right| \quad (9)$$

where $\varphi^{''}(x)^{ud}$ and $\varphi^{''}(x)^d$ are the mode shape curvature in the x-direction for undamaged and damaged plates, respectively. And, $\varphi^{''}(y)^{ud}$ and $\varphi^{''}(y)^d$ are the mode shape curvature in y-direction for undamaged and damaged plates, respectively.

A modified damage index (DI) based on the work of Rucevskis et al. [19] has been adopted for identifying the damage location. The DI is defined as a squared-average summation of curvature difference considering all modes and normalised with respect to the largest value of each mode.

$$DI = \left(\frac{1}{N} \sum_{n=1}^{N} \frac{MSC_d^n}{\text{absmax}(MSC_d^n)} \right)^2 \quad (10)$$

where N is the total number of modes considered.

2.3 Frequency Response Functions (FRFs) Considering Viscous Damping

Synthesis of FRF data is carried out using the mode superposition method [5]. Assuming viscous damping with the modal damping factor being ξ_r, the acceleration FRF can be expressed as follows:

$$H_{ij}(\omega) = -\omega^2 \sum_{r=1}^{N} \frac{\varphi_{ir}\varphi_{jr}}{\omega_r^2 - \omega^2 + 2i\omega\omega_r \xi_r} \tag{11}$$

where H_{ij} is the acceleration response at the ith d.o.f. for a sinusoidal exciting force at the jth d.o.f.

Considering Rayleigh's damping, the modal damping factor is expressed as follows:

$$\xi_r = \frac{1}{2}\left(\frac{a_0}{\omega_r} + a_1 \omega_r\right) \tag{12}$$

Here, a_0 and a_1 are the mass and stiffness proportional Rayleigh's damping coefficients. Including the damping coefficients in the formulation of frequency response function [10]:

$$H_{ij}(\omega) = -\omega^2 \sum_{r=1}^{N} \frac{\varphi_{ir}\varphi_{jr}}{(\omega_r^2 - \omega^2 + i\omega(a_1\omega^2 + a_0))} \tag{13}$$

2.4 Correlations for Comparing FRFs

The shape of FRFs is correlated through a Cross Signature Assurance Criterion (CSAC) represented by the position and magnitude of the resonance peaks.

$$CSAC(\omega_n) = \frac{\left|H_{\exp_p}^T(\omega_n) H_{nu_p}(\omega_n)\right|^2}{\left(H_{\exp_p}^T(\omega_n) H_{\exp_p}(\omega_n)\right)\left(H_{nu_p}^T(\omega_n) H_{nu_p}(\omega_n)\right)}, n = 1, 2 \ldots, N \tag{14}$$

where $H_{\exp}$ is the experimental FRF data, H_{nu} is the numerical FRF dat, and N represents the frequency points/range considered.

The Cross Signature Scale Factor (CSF) is a correlation function that estimates the differences in the FRF amplitude and is given as follows:

$$CSF(\omega_n) = \frac{2\left|H_{\exp_p}^T(\omega_n) H_{nu_p}(\omega_n)\right|}{\left(H_{\exp_p}^T(\omega_n) H_{\exp_p}(\omega_n)\right) + \left(H_{nu_p}^T(\omega_n) H_{nu_p}(\omega_n)\right)}, n = 1, 2 \ldots, N \tag{15}$$

CSAC and CSF together gives a correlation function known as Cross Signature Correlation (CSC) function coined by Dascotte and Strobbe [4].

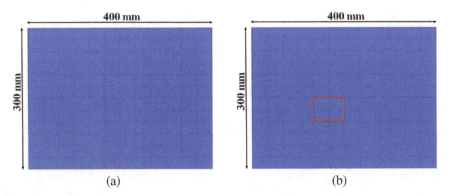

Fig. 1 Figure showing (a) undamaged plate, (b) damaged plate

3 Numerical Problem

A rectangular FRP composite plate having dimensions 400 mm × 300 mm and thickness of 10 mm is considered with free–free support condition (Fig. 1). The plate is modelled as an equivalent single-layer plate model in ABAQUS using eight-noded shell elements (S8R).

The aim of this problem is to identify the damage location and quantify the changes in the elastic parameters and global damping parameters due to the induced damage. The in-plane elastic parameters for the intact plate are considered as $E_x = 37$ GPa, $E_y = 35$ GPa, $G_{xy} = 6.5$ GPa; the Poisson's ratio is taken as 0.162. The out-of-plane elastic modulus is taken as $G_{xz} = 5$ GPa and $G_{yz} = 5$ GPa. The mass density is taken as 1960 Kgm^{-3}. Rayleigh's damping has been considered in this problem. For the intact plate, the values of damping coefficients considered are $a_0 = 3$ and $a_1 = 2\text{E}^{-6}$. Eigen solution is performed to obtain the natural frequencies, mode shapes.

Numerically simulated damage plate is created by local reduction of stiffness for some elements. The damaged elements were degraded values of elastic moduli. For the damaged portion, the in-plane elastic moduli have been reduced as $E_x = 10$ GPa, $E_y = 8$ GPa. For simulating enhanced damping for the damaged plate, the damping parameters have been increased as $a_0 = 6$ and $a_1 = 4\text{E}^{-6}$. The elements that have been chosen as damaged elements with degraded values of elastic moduli are shown in Fig. 1b. FRFs have been generated for both intact plate and damaged plate through mode superposition as described in Sect. 2.3.

4 Result and Discussions

Natural frequencies and modal damping factors for the undamaged and damaged plates have been provided in Table 1. It shows the reduction in natural frequencies due to the induced damage. Also, the enhanced damping due to damage is reflected by the increase in modal damping factors.

Table 1 Frequencies and modal damping factors of undamaged and damaged plates

Mode No.	Frequency (Hz)		Modal damping factor	
	Undamaged	Damaged	Undamaged	Damaged
1	159.84	157.83	0.025	0.045
2	282.83	274.25	0.026	0.049
3	428.87	427.15	0.033	0.063
4	490.97	469.76	0.036	0.068
5	579.41	576.95	0.041	0.079
6	782.18	769.93	0.052	0.102
7	858.39	834.09	0.057	0.11
8	907.45	904.56	0.06	0.118

The drive point FRF curve is generated at node 16 for both damaged and undamaged cases using Eq. 16. Figure 2 compares both FRFs. The SAC value is found to be 63.8% when the FRF of undamaged and damaged plots are correlated, and the FRF magnitudes are correlated using CSC values with respect to frequency is shown in Fig. 3.

For locating the damaged area, Damage Index at each grid point using Eq. 10 has been estimated. Element numbers 77-78-89-90 are identified as damaged elements, as shown in Fig. 4. After locating the damage elements, two subsets of parameters are taken for updating. One with the elements of identified damaged site and other with remaining elements, thereby limiting the number of parameters to be applied for the updating process. IEM, which is adopted for the updating process, does not yield good results while working with too many parameters due to the ill-conditioning.

The second stage involves quantifying the damage parameters through model updating. The objective function is formed taking the difference of FRFs between the damaged and undamaged plates which is expressed as follows:

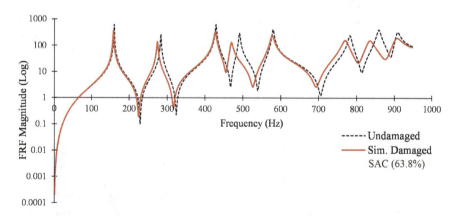

Fig. 2 Comparison of FRFs of undamaged and damaged plates

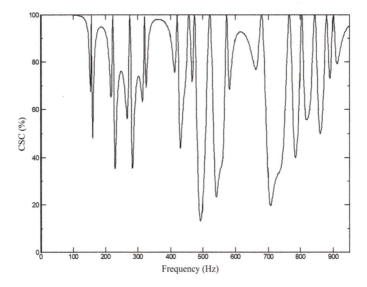

Fig. 3 Cross signature correlation (CSC) along frequency

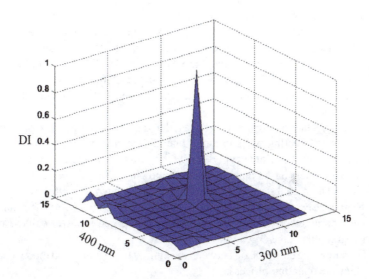

Fig. 4 Damage location based on DI using mode shape curvature difference

$$E = \|H_d - H_{un}\|^2 \tag{16}$$

where, H_d, H_{un} are the damaged and undamaged FRFs, respectively. The FRFs are generated using the formulation given in using Eq. 13.

The first-order sensitivities of the frequency response functions with respect to the stiffness and damping parameters are expressed as follows:

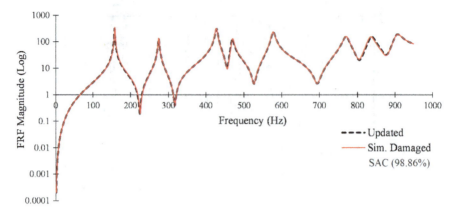

Fig. 5 Comparison of FRFs of updated and damaged plates

$$S = \left[\frac{\partial H}{\partial r_n} \right] \quad (17)$$

Here, r_n are the stiffness and damping parameters.

The iterative procedure for updating the parameters using IEM is given in Sect. 2.1. The parameters with their initial guess values are subjected to suitable changes within reasonable selected bounds. The similarity between the FRFs is quantitatively verified during each iteration step through the SAC values. The SAC value of the updated FRF is found to be 98.86%. The plot of the updated FRF is shown in Fig. 5. And, the CSC plot shows a good correlation between the updated FRF and the FRF of the damaged pate shown in Fig. 6.

The FRF correlations show a well-behaved updating procedure. The procedure of locating damage, creating two subsets of material parameters and taking the global damping parameters has yielded good convergence of the parameters. After updating, the elastic parameters have converged to the following values, $E_x = 10.8$ GPa and $E_y = 7.89$ GPa. The global damping parameters have updated to the following values, $a_0 = 5.83$, and $a_1 = 3.97$ E^{-6}. Figures 7 and 8 shows the convergence curves of the updated parameters. The elastic parameters converged to lower values, and the damping parameters converged to higher values.

Table 2 shows the comparison between the natural frequencies and modal damping factors after the updating process.

5 Conclusion

The purpose of the study is to establish a damage identification strategy and including finite element model updating into its framework. A rectangular FRP plate structure has been considered for the study. The damaged plate has been modelled with some

Detection of Damages in Structures Using Changes ...

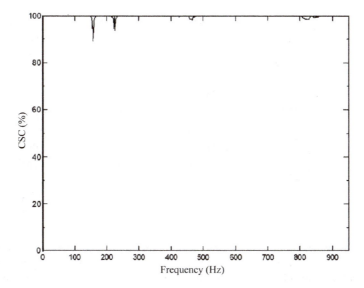

Fig. 6 Cross signature correlation (CSC) for the updated FRF along with frequency

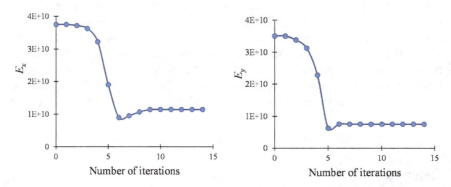

Fig. 7 Convergence curves of the updated elastic parameters

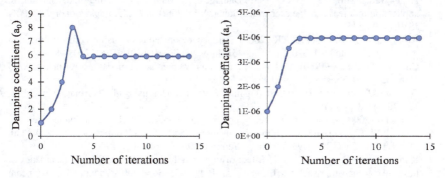

Fig. 8 Convergence curves of the updated damping parameters

Table 2 Natural frequencies and modal damping factors of the updated and damaged plates

Mode No.	Frequency (Hz)		Modal damping factor	
	Damaged	Updated	Damaged	Updated
1	157.83	157.13	0.045	0.049
2	274.25	274.17	0.049	0.052
3	427.15	427.69	0.063	0.064
4	469.76	468.82	0.068	0.069
5	576.95	577.45	0.079	0.08
6	769.93	769.88	0.102	0.102
7	834.09	835.62	0.11	0.11
8	904.56	904.49	0.118	0.117

elements having reduced stiffness and those elements were assigned with degraded elastic moduli. Also, enhanced damping for the damage case has been assumed. Now, involving model updating only by inverse estimation of parameters to detect damage leads to ill-conditioning and does not provide accurate results. Therefore, to avoid involving a large set of parameters, a two-stage damage identification strategy has been adopted. First damage localisation is done through a damage index estimated from the MSC difference between the damaged and undamaged plates. In the second stage, finite element model updating through IEM, now involving a reduced parameter set, has been applied for quantifying the damage. The process successfully located the damage site and through the sensitivity-based model updating process, the stiffness and damping parameters converged very close to the induced damage values.

References

1. Abdel Wahab MM, De Roeck G (1999) Damage detection in bridges using modal curvatures: application to a real damage scenario. J Sound Vib 226(2):217–235. https://doi.org/10.1006/JSVI.1999.2295
2. Arora V, Singh SP, Kundra TK (2009) Damped model updating using complex updating parameters. J Sound Vib 320(1–2):438–451. https://doi.org/10.1016/J.JSV.2008.08.014
3. Chen JC, Garba JA (1980) Analytical model improvement using modal test results. AIAA J 18(6):684–690. https://doi.org/10.2514/3.50805
4. Dascotte AE, Strobbe J (1999) Updating finite element models using FRF correlation functions. In: 17th IMAC, pp 1169–1174
5. Ewins DJ (2000) Modal testing : theory, practice, and application. Research Studies Press
6. Friswell MI (2007) Damage identification using inverse methods. Philos Trans R Soc A Math Phys Eng Sci 365(1851):393–410. https://doi.org/10.1098/rsta.2006.1930
7. Kıral Z, Murat B, Kıral G (2012) Effect of impact failure on the damping characteristics of beam-like composite structures. Compos Part B 43:3053–3060. https://doi.org/10.1016/j.compositesb.2012.05.005

8. Kyriazoglou C, Le Page BH, Guild FJ (2004) Vibration damping for crack detection in composite laminates. Compos A Appl Sci Manuf 35(7–8):945–953. https://doi.org/10.1016/j.compositesa.2004.01.003
9. Lin RM, Zhu J (2006) Model updating of damped structures using FRF data. Mech Syst Signal Process 20(8):2200–2218. https://doi.org/10.1016/J.YMSSP.2006.05.008
10. Mondal S, Chakraborty S (2018) Estimation of viscous damping parameters of fibre reinforced plastic plates using finite element model updating. Int J Acoust Vib 23(1). https://doi.org/10.20855/ijav.2018.23.11064
11. Montalvão D, Ribeiro AMR, Duarte-Silva J (2009) A method for the localization of damage in a CFRP plate using damping. Mech Syst Signal Process 23(6):1846–1854. https://doi.org/10.1016/j.ymssp.2008.08.011
12. Montalvão D, Ribeiro AMR, Duarte-Silva JAB (2011) Experimental assessment of a modal-based multi-parameter method for locating damage in composite laminates. Exp Mech 51(9):1473–1488. https://doi.org/10.1007/s11340-011-9472-5
13. Montalvão D, Karanatsis D, Ribeiro AM, Arina J, Baxter R (2015) An experimental study on the evolution of modal damping with damage in carbon fiber laminates. J Compos Mater 49(19):2403–2413. https://doi.org/10.1177/0021998314547526
14. Mottershead JE, Friswell MI (1993) Model updating in structural dynamics: a survey. J Sound Vib 167(2):347–375. https://doi.org/10.1006/jsvi.1993.1340
15. Mottershead JE, Link M, Friswell MI (2011) The sensitivity method in finite element model updating: a tutorial. Mech Syst Signal Process 25(7):2275–2296. https://doi.org/10.1016/J.YMSSP.2010.10.012
16. Pandey AK, Biswas M, Samman MM (1991) Damage detection from changes in curvature mode shapes. J Sound Vib 145(2):321–332. https://doi.org/10.1016/0022-460X(91)90595-B
17. Pradhan S, Modak SV (2012) A method for damping matrix identification using frequency response data. Mech Syst Signal Process 33:69–82. https://doi.org/10.1016/J.YMSSP.2012.07.002
18. Roy K (2017) Structural damage identification using mode shape slope and curvature. J Eng Mech 143(9):04017110. https://doi.org/10.1061/(ASCE)EM.1943-7889.0001305
19. Rucevskis S, Janeliukstis R, Akishin P, Chate A (2016) Mode shape-based damage detection in plate structure without baseline data. Struct Control Health Monit 23(9):1180–1193. https://doi.org/10.1002/stc.1838
20. Salawu OS (1997) Detection of structural damage through changes in frequency: a review. Eng Struct 19(9):718–723. https://doi.org/10.1016/S0141-0296(96)00149-6
21. Williams and Salawu OS (1997) Damping as a damage indication parameter. In: Proceedings of the 15th international modal analysis conference, pp 1531–1536
22. Wu D, Law SS (2004) Damage localization in plate structures from uniform load surface curvature. J Sound Vib 276(1–2):227–244. https://doi.org/10.1016/J.JSV.2003.07.040

A Data-Based Technique for Damage Detection Handling Environmental Variability During Online Structural Health Monitoring

K. Lakshmi and Junia Blessy

Abstract The success of data-based online damage detection techniques depends upon the ability to detect the deviation from the previous measurements of the healthy system, changes in the material and/or geometric properties, boundary conditions, and system connectivity. Most of the data-based techniques extract features like frequencies, mode shapes, etc., for further processing. However, the random excitation, the varied environmental conditions, and the undesired measurement noise, often bring in the stochasticity in the features extracted from the vibration data. Also, the environmental variability due to temperature alters the material property of the structure, and creates an effect which is similar to that of the real damage. This fact emphasizes the need for techniques to differentiate the effects of environmental variability from damage, during diagnosis, using the output-only vibration data. In this paper, a data-based technique, which can effectively handle the environmental variability, as well as capable of locating the region of damage in the structure, using the acceleration time-history data, is presented. In this paper, Mahalanobis Squared Distance (MSD), which is popularly used in novelty detection, is used to handle the effect of environmental or operational variability (EOV) and simultaneously perform the damage detection, by treating the acceleration time-history of sensor nodes as feature vectors. With the confirmation of the presence of damage, subsequently, the spatial domain of damage is identified, by performing a sequential elimination approach, while forming the feature vectors for MSD evaluation. The results of the numerical studies using a synthetic data and benchmark data show that the proposed data-based technique using MSD is efficient in eliminating the variability and precisely locating the damage region, spatially on the structure.

Keywords Structural health monitoring · Damage location · Mahalanobis squared distance · Environmental and operational variability · Acceleration responses · Measurement noise

---

K. Lakshmi (✉) · J. Blessy
CSIR-SERC, CSIR Complex, Taramani, Chennai 600113, India
e-mail: lakshmik@serc.res.in

© The Editor(s) (if applicable) and The Author(s), under exclusive license to Springer Nature Singapore Pte Ltd. 2021
S. K. Saha and M. Mukherjee (eds.), *Recent Advances in Computational Mechanics and Simulations*, Lecture Notes in Civil Engineering 103,
https://doi.org/10.1007/978-981-15-8138-0_22

1 Introduction

Most of the civil engineering structures are exposed to daily and seasonal variations of its ambiance such as temperature, the wind, and operational loadings leading to nonstationary dynamic and quasi-static responses. These ambient variations can potentially mask any changes in structural response due to damage. Also, the responses measured from the structure are sensitive to the changes in operational and environmental conditions irrespective of damage or structural degradation [1]. Hence, the influence of the variability due to operational and environmental changes is a key issue in SHM and is gaining attention among the research community [2]. In fact, the difficulty in differentiating the effect of operational and environmental variations from that of the actual damage in a structure poses itself as one of the biggest challenges for implementing the SHM technology from research to the real practice. The two difficult tasks are, to perfectly capture the time-varying nature of the data and to filter it by protecting the damage information at the same time. Generally, temperature is the most profound ambient force affecting the structural response, as it alters the stiffness of the structure and/or the boundary conditions of a structure itself [3–5].

Based on the fact that the temperature variation changes the frequency [6] and its effect on the measured dynamics response has been addressed in several studies. Yan et al. [7] established the variation of the Young's modulus of steel and concrete with the temperature. Cornwell et al. [8] found that the fluctuations in modal frequencies are as a result of the thermal gradient across the deck of the Alamosa Canyon bridge and observed a bilinear relationship with temperature. Moorty and Roeder [9] established that the vibration records obtained from the analytical model and the measured values of Sutton Creek Bridge in Montana, USA, indicate the expansion of the bridge deck with an increase in temperature. Doebling and Farrar [10], tested the Alamosa Canyon Bridge in New Mexico, USA, and proved that there was approximately 5% variation in the first mode frequency of the bridge in the 24-h cycle.

A considerable amount of literature relevant to the methods to handle the effect of environmental variations in structures is available. One of the famous methods is to model the parameters, which are monitored with respect to the environmental/operational factors considered to have an effect. If a model is employed to predict the values of those parameters, the error would become a damage-sensitive feature. The main disadvantage of this method is that if a reasonable confidence to be developed in the effectiveness of the model prediction, all the factors which affect the monitored parameters must also be monitored. The techniques, where the measurements of the variability are unavailable, have also been explored. One of the approaches was explored by Surace and Worden [11], where a long span of response data representing the varied environmental and operational conditions of the structure, is used as the baseline data. Any new recording can be compared with the baseline to find the abnormality. Obviously, this approach needs a large storage data leading to the reduced feature sensitivity to damage. Manson [12] used principal component analysis (PCA) to extract the first few principal component scores whose variances are high and related to the signatures of changing environmental conditions. Kullaa [13] proposed a method based on PCA and factor analysis to use the linear projection and thereby extracting the environmental variability. He concluded that this technique can be used on more complex datasets, as both PCA and FA admit nonlinear generalizations. The other promising method which is also related to PCA and Factor analysis is using Mahalanobis Squared Distance (MSD) [14], which is a popular novelty detection technique.

In this paper, MSD is used to handle the effect of environmental or operational variability (EOV) and simultaneously perform the damage detection, by treating the acceleration time-history of sensor nodes as feature vectors. The eigenvalue decomposition of the covariance matrix used in MSD, transforms the feature vectors to the space of independent variables. By this process, MSD becomes the sum of the independent terms. Using the property that the independent terms with higher eigenvalues contribute to low MSD and vice versa, the effect of environmental variability is filtered out, exposing the presence of damage. With the confirmation of the presence of damage, subsequently, the spatial domain of damage is identified, by performing a sequential elimination approach, while forming the feature vectors for MSD evaluation.

The effectiveness of the proposed technique is validated using a numerical example of a simply supported beam, with simulated environmental variability and measurement noise. Also, the proposed technique is validated using the vibration data of the simulation benchmark problem, a simply supported beam with a spring, with environmental and operational variability. The results of the studies show that the proposed data-based technique using MSD is efficient in eliminating the variability and precisely locating the damage region, spatially on the structure.

2 Handling EOV Using Mahalanobis Squared Distance

Mahalanobis squared distance is a classic discordancy measure used to identify the outliers. The MSD indicates how far away a specific measurement is from the mean of the training data samples, relative to the size of the samples [15]. The mean and covariance matrix could be inclusive or exclusive measures and so the statistics may or may not have been computed from data where outliers are already present.

The outlier analysis for each sample of the multivariate feature vector, $\{x_\zeta\}$, to compute the MSD is given by

$$M_\zeta^2 = (\{x_\zeta\} - \{\bar{x}\})^T [C]^{-1} (\{x_\zeta\} - \{\bar{x}\}) \tag{1}$$

The features in the vector may not be statically independent and the covariance matrix not being diagonal. Hence, it is possible to perform a transformation of the feature vector so that the covariance matrix is diagonalized, by computing the eigenvectors $\{V_i\}$ and eigenvalues σ_i^2 of $[C]$

$$[C]\{V_i\} = \sigma_i^2\{V_i\} \tag{2}$$

The orthogonality properties being given by

$$[V]^T[C][V] = [R] \tag{3}$$

The spectral decomposition of the covariance matrix is given by

$$[C] = [V][R][V]^T \tag{4}$$

Assuming the following transformation,

$$\{\tau_i\} = [V]^T\{x_i\} \tag{5}$$

The mean and the covariance matrix is calculated as below:

$$\{\bar{\tau}\} = \frac{1}{N}\sum_{i=1}^{N}\{\tau_i\} = [V]^T\{\bar{x}\} \tag{6}$$

$$[C]_\tau = \frac{1}{N-1}\sum_{i=1}^{N}(\{\tau_i\} - \{\bar{\tau}\})(\{\tau_i\} - \{\bar{\tau}\})^T \tag{7}$$

Using the inverse transformation,

$$\{x_i\} = [V]\{\tau_i\} \tag{8}$$

And therefore the MSD reduces to

$$M_\zeta^2 = \sum_{i=1}^{n}\frac{1}{\sigma^2}(\tau_{\zeta i} - \bar{\tau}_i)^2 \tag{9}$$

which shows that the MSD can be decomposed into a sum of independent contributions from each component of the transformed variables $\tau_{\zeta i} = \{V_i\}^T\{x_\zeta\}$ and the contributions being weighed by the inverse of the associated eigenvalues σ_i^2 which can be interpreted as the variances of the new, transformed variables.

The number of feature vector being large, the total variability in the feature vector can be explained by a small number of transformed features usually called the principal component scores. This can occur only when the eigenvalues are equal to zero, the associated eigenvectors which form the null space of the training data. An effective null space is defined by putting a threshold on the singular values, assuming that the singular values below this threshold are only non-zero due to the presence of noise in the training data.

A practical way to determine the number 'p' of vector in the principal subspace is to define the following indicatorand to determine 'p' as the lowest integer such that I > e (%), where e is a threshold value (99.9%). Assuming that the principal 'p' components have been identified the MSD can be decomposed into two parts,

$$I = \frac{\sum_{i=1}^{p} \sigma_i^2}{\sum_{i=1}^{n} \sigma_i^2} \qquad (10)$$

$M_{1\zeta}^2$—MSD of $\{x_\zeta\}$ projected on the principal component.
$M_{2\zeta}^2$—MSD of $\{x_\zeta\}$ projected on the null space of the principal component.

$$M_\zeta^2 = \sum_{i=1}^{p} \frac{1}{\sigma_i^2}(\tau_{\zeta i} - \bar{\tau}_i)^2 + \sum_{i=p+1}^{n} \frac{1}{\sigma_i^2}(\tau_{\zeta i} - \bar{\tau}_i)^2 = M_{1\zeta}^2 + M_{2\zeta}^2 \qquad (11)$$

2.1 Detection of Presence of Damage

Assuming that very large variability will exist in the feature vector extracted from the healthy condition due to environmental effects and this variability being important than any other sources such as noise will belong to a set of principal 'p' components. Mahalanobis distance can scale each independent component with respect to the inverse of its variance, the distance will have low sensitivity to the environmental changes. By including the feature vector measured in all possible environmental condition in the computation of the covariance matrix, the MSD is made insensitive to the environmental conditions. Therefore, MSD, evaluated for a new sample from a healthy structure, shows a low value, when compared to a sample with the influence of damage, for which it is not trained. Based on this measure, the presence of damage is clearly indicated by MSD. In this study, the acceleration time-history data, collected from the sensors on the structure is adopted as the feature vector to evaluate the MSD. Therefore, the size of the vector $\{x_\zeta\}$ becomes ($n \times 1$), where n is the number of sensors used to obtain the acceleration responses.

2.2 Detection of Location of Damage

In addition to the presence of EOV, a damage, also affects dynamic response, especially in the sensor data, collected near the location of damage. In view of this, the location of damage is identified by sequentially removing the acceleration data of small group of sensors while forming the feature vectors to evaluate MSD.

If the feature vectors (i.e., acceleration time-history responses in this work), related to the group of sensors, placed in the region of damage, are eliminated, during the evaluation of MSD, the magnitude of MSD is supposed to fall lower and comparable to that of a healthy training condition. This will give a confirmation that the eliminated feature vectors belong to the region of damage. Thus, MSD is used to identify both the presence and the spatial region of damage.

3 Numerical Study

Numerical studies are conducted using a numerical model of a simply supported beam and a simulation benchmark problem of a simply supported beam with a spring to validate the proposed technique to handle environmental variability and to detect the presence and location of damage.

3.1 Simply Supported Beam

The first numerical example considered is a simply supported beam girder shown in Fig. 1 with a span of 5.5 m and discretized into 20 elements. The material and geometrical properties are also shown in the Fig. 1. The beam is excited using a random dynamic loading which is stochastic in nature. The acceleration time-history response is computed on every node using finite element analysis with Newmark's time marching scheme with random loads and environmental conditions. The first seven frequencies of the beam are 38 Hz, 151 Hz, 339 Hz, 457 Hz, 603 Hz, 917 Hz, and 942 Hz, respectively. The sampling rate is considered as 1000 samples per second. This simulates the data collected on a bridge girder on several time instant with varied traffic levels.

One of the main issues related to structural damage diagnostic techniques, when applied to real situations, is their sensitivity to noise. In view of this, it was decided to add white Gaussian noise to the acceleration time-history response generated by the finite element code. The white Gaussian noise is added in the form of SNR (signal-to-noise ratio) to the acceleration time-history before they are processed. SNR defines the amplitude of the noise with respect to that of the clean signal. When the noise level is given by a particular value of SNR, it means that a noisy signal with such an

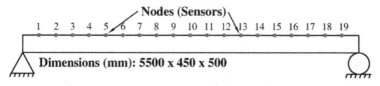

Fig. 1 Simply supported beam

SNR has been added to the time series of each node. Moreover, the noisy sequences affecting different nodes are uncorrelated, in this way severe experimental conditions were simulated.

The structure is assumed to be healthy and the acceleration time series are generated from the undamaged structure at varied levels of random loads (various operational conditions) and with varied temperatures ranging from -15° to 50° centigrade.

Now, current datasets consisting of acceleration data from the current condition of the structure, which is considered to be healthy for a specified time instance and beyond which damage has originated are generated. The treatment of current datasets, as the combination of acceleration time series data corresponding to healthy and damaged condition of the structure, paves way for the clear identification of the time instance of damage. Current data subsets (i.e., 5-s data) are generated with varied load levels, varied temperatures and noise levels. The current data consists of both data of a healthy structure as well as data of a structure with simulated damages in some of the elements. Hypothetical damage scenarios are assumed by means of reduction of stiffness of some elements. Damage is simulated into simply supported beam girder after obtaining 4000 samples of current data (i.e., after 4 s) by reducing the stiffness of element 6 by 20%, varying both the temperature and SNR.

Initially, the feature vector is made of the baseline acceleration data of all 19 sensors. This baseline feature vectors are used to obtain the covariance matrix, eigenvalues, and eigenvectors, which transform, a new sample to the component space to evaluate the MSD. With the arrival of a new current data, they are standardized with the mean and standard deviation of the baseline data.

The MSD is evaluated for each sample of a current data. The evolution of MSD for the 6000 samples of acceleration responses from a healthy state of the structure is shown in Fig. 2a. Similarly, the evolution of MSD over 6000 samples of current data, with simulated damage at 4000th sample, is shown in Fig. 2b. From Fig. 2b, it can be seen clearly that the presence of damage is revealed at the 4000th sample and the time instant of damage can be calculated from the sampling rate of the data. In this example, it is 4 s.

Once the presence and the time instant of damage are revealed, the location of damage is to be identified. A sequential process of eliminating the data related to a group of sensors, while forming the feature vectors is adopted. In view of this, three sensors are eliminated in every iteration. For example, sensor nodes 1–3, shown in Fig. 1, are eliminated in the first iteration and thus forming a feature vector of size 16 × 1. With the reduced feature vector, MSD is calculated and the iterations are incremented. When the MSD is evaluated by ignoring the acceleration data from sensor nodes 4–6, for the formation of feature vector, its evolution is found to be similar to a baseline data, indicating that the location of damage is in the region between the fourth and the sixth sensor nodes. The MSD values of the reduced sensors is shown in the Fig. 2c. From Fig. 2, it can be concluded that MSD is not only capable of distinguishing the effect of EOV and damage, it can also locate the damage region, spatially on the structure, using the acceleration data as the feature vector.

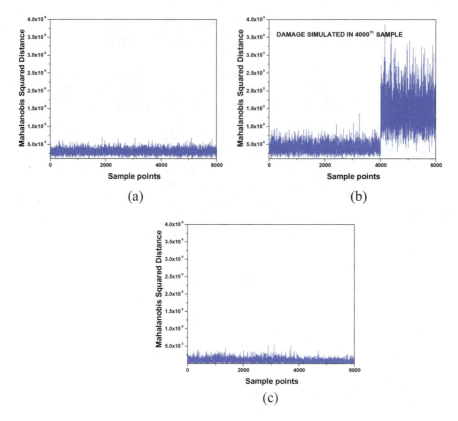

Fig. 2 MSD of simulated acceleration response of a simply supported numerical beam: **a** Baseline data **b** Current data with simulated damage at 4000th sample. (c) Current data with reduced feature vectors, when the acceleration data of group of sensors, near damage, is eliminated

3.2 Benchmark Structure: Beam with Variable Spring

The simulation benchmark problem [16] of a simply supported beam with a variable spring is considered as the second example to validate the proposed acceleration data-based technique. The simply supported beam is of length 1.4 m, with a uniform rectangular cross section of 50 mm × 5 mm as shown in Fig. 3.

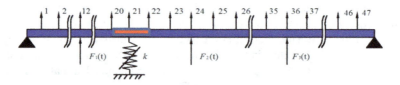

Fig. 3 A beam with a spring

The beam is supported with a spring 612.5 mm from the support, with the spring constant k depending nonlinearly on temperature [16]:

$$k = k_0 + aT^3 \tag{12}$$

where $k_0 = 100$ kN/m, $a = -0.8$ (with compatible units) and T is a temperature with a uniform random distribution between -20 and $+40$ °C. The beam is divided into three sections of equal length. The Young's modulus E_i in the ith section has linear relationship with a corresponding independent and dimensionless environmental variable.

$$E_i = E_0 + V_i Z_i; \quad i = 1, 2, 3 \tag{13}$$

where $E_0 = 207$ GPa, Z_i are standardized Gaussian variables: $Z_i \in N(0, 1)$, and the standard deviations V_i of different sections are: $V_1 = 5$ GPa, $V_2 = 3$ GPa, and $V_3 = 7$ GPa. The structure is modeled with 144 simple beam elements and a single spring element. Independent random excitations with different amplitudes in each measurement excite the structure at three points.

Thus, the environmental or operational variability is introduced by three factors [16]:

(1) a variable spring which has a nonlinear relationship between temperature and the spring constant, shown in Eq. 12.
(2) three regions with independently varying Young's moduli as shown in Eq. 13, and
(3) the distribution of random loads $F_1(t)$, $F_2(t)$ and $F_3(t)$, at three points as shown in Fig. 3.

Damage is a decrease in the beam depth at the spring support in two elements along a total length of 19.4 mm. This type of damage could represent local corrosion around the spring joint. Sensor 21 is located in the middle of the damaged region as shown in Fig. 3. A damage scenario, where the height of the damaged beam is 4 mm is considered for this study. The MSD values of the samples of a baseline data and current data with simulated damage at 11500th sample are shown in Fig. 4a and 4b, respectively. From the figure, it can be verified that the MSD based technique is able to indicate the damage and its time instant using the acceleration data as feature vector.

To detect the region of damage, the sequential approach of forming the feature vectors by eliminating the data from a predefined group of sensors, described in Sect. 2.2, is carried out. Accordingly, six number of sensor data are ignored in every trial while forming the feature vectors for MSD evaluation. When the acceleration data of the sensor nodes 19-24, are not considered as feature vectors, the MSD values fell low and showed a pattern similar to a baseline data. This can be verified from the Fig. 4c by comparing with Fig. 4a. From the above observation, it can be seen that the region of damage is identified by sequential elimination of feature vectors

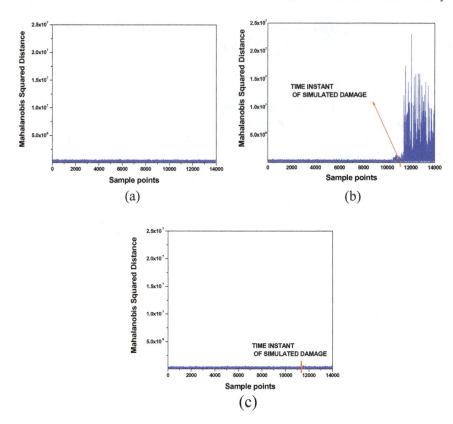

Fig. 4 Mahalanobis squared distance of acceleration response of simply supported beam with a spring benchmark problem: **a** Baseline data **b** Current data with simulated damage at 11,500th sample. **c** Current data with reduced feature vectors, when the acceleration data of group of sensors, near damage, is eliminated

used in the evaluation of MSD. Therefore, it can be concluded that MSD is able to expose the time instant and also the region of damage.

4 Conclusion

In this paper, a data-based technique to filter out the effect of environmental variability and to detect damage is presented. The technique uses directly the acceleration data from all the sensor nodes on a structure without extracting any feature out of it. Mahalanobis squared distance, a famous novelty detection technique is shown to be performing well using the acceleration data, in filtering out the EOV effects existing in the training data and exposing the time instant of damage. The location of damage is identified using a sequential elimination process, where the acceleration data of a group of sensors are eliminated at every iteration before MSD is evaluated. The numerical study and the validation of the simulation benchmark results show that the proposed data-based technique identifies the time instant and location of damage using the acceleration data from the structure, by filtering out the EOV and handling the measurement noise. The limitation of the proposed technique is that it indicates the region of damage rather than the exact elemental location of damage. Toward the above limitation, a more refined sequential approach is taken up as a future work by the authors. Also, the validation of the technique using simulation studies of minor damage like cracks and the experimental work are considered as future works.

References

1. Farrar CR, Sohn H, Worden K (2001) Data normalization: a key for structural health monitoring. Technical report, Los Alamos National Laboratory
2. Worden K, Farrar CR, Manson G, Park G (2007) The fundamental axioms of Structural Health Monitoring. Proc R Soc A 463(2082):1639
3. Sohn H, Dzwonczyk M, Straser EG, Kiremidjian AS, Law KH, Meng T (1999) An experimental study of temperature effect on modal parameters of the Alamosa Canyon Bridge. Earthq Eng Struct Dynam 28(8):879–897
4. Alampalli S (2000) Effects of testing, analysis, damage, and environment on modal parameters. Mech Syst Signal Process 14(1):63–74
5. Peeters B, De Roeck G (2001) One-year monitoring of the Z24-Bridge: environmental effects versus damage events. Earthq Eng Struct Dynam 30:149–171
6. Zhou Y, Sun L (2019) Effects of environmental and operational actions on the modalfrequency variations of a sea-crossing bridge: A periodicity perspective. Mech Syst Signal Process 131:505–523
7. Yan AM, Kerschen G, De Boe P, Golinval JC (2005) Structural damage diagnosis under varying environmental conditions—Part I: A linear analysis. Mech Syst Signal Process 19:847–864
8. Cornwell P, Farrar CR, Doebling SW, Sohn H (1999) Environmental variability of modal properties. Exp Tech 23(6):45–48
9. Moorty S, Roeder CW (1992) Temperature-dependent bridge movements. ASCE J Struct Eng 118:1090–1105
10. Doebling SW, Farrar CR (1997) Using statistical analysis to enhance modal-based damage identification. In: Proceedings of DAMAS 97: structural damage assessment using advanced signal processing procedures, University of Sheffield, UK, 199–210, 1997
11. Surace C, Worden K (1997) Some aspects of novelty detection methods. In: Proceedings of the third international conference on modern practice in stress and vibration analysis, Dublin (1997)
12. Manson G (2002) Identifying damage sensitive environment insensitive features for damage detection. In Proceedings of international conference on identification in engineering systems, Swansea, UK, pp 187–197
13. Kullaa J (2004) Structural health monitoring under variable environmental or operational conditions. In: Proceedings of second European workshop on structural health monitoring, Munich, Germany, pp 1262–1269
14. Worden K, Manson G, Fieller NRJ (2000) Damage detection using outlier analysis. J Sound Vib 229(3):647–667 (2000)
15. Dervilis N, Cross EJ, Barthorpe RJ, Worden K (2014) Robust methods of inclusive outlier analysis for structural health monitoring. J Sound Vib 333(20):5181–5195
16. Kullaa J (2014) Benchmark data for structural health monitoring. In: 7th European workshop on structural health monitoring, La Cité, Nantes, France

Damage Detection in Presence of Varying Temperature Using Mode Shape and a Two-Step Neural Network

Smriti Sharma and Subhamoy Sen

Abstract The dynamic characteristics of any structural system get affected not only due to damage but also from variations in ambient uncertainty. Thus, false positive or negative alarm may be signalled if temperature effects are not taken care off. The difficulty lies in correlating response measurements to corresponding damage patterns in the presence of varying temperature. This study employs machine learning algorithm to filter out the temperature effect from the measured mode shapes. A two-stage data-driven approach has been developed in which damage detection and localization are performed in consequence. For detection, a model to correlate mode shapes and temperature is formulated using an Auto-Associative Neural Network (AANN) and a temperature-invariant prediction error is defined as Novelty Index (NI). NIs are further classified to corresponding damage cases employing a fully connected layer network. With numerical experiments, the algorithm presented excellent efficiency and robustness against varying temperature in detecting damage.

Keywords Structural Health Monitoring (SHM) · Damage Detection · Neural Network · Auto-Associative Neural Network (AANN) · Fully connected layer network

1 Introduction

Vibration-based damage detection is one of the promising fields in Structural Health Monitoring (SHM) in order to obtain a reliable, prompt and efficient approach to ensure the safety of the structures. A momentous condition assessment relies majorly on the measured modal properties such as natural frequencies and mode shapes. Changes in the modal properties may possibly reflect the health of any structure.

S. Sharma (✉) · S. Sen
i4S Laboratory, Indian Institute of Technology Mandi,
Kamand, Himachal Pradesh 175005, India
e-mail: d17007@students.iitmandi.ac.in

© The Editor(s) (if applicable) and The Author(s), under exclusive license to Springer Nature Singapore Pte Ltd. 2021
S. K. Saha and M. Mukherjee (eds.), *Recent Advances in Computational Mechanics and Simulations*, Lecture Notes in Civil Engineering 103,
https://doi.org/10.1007/978-981-15-8138-0_23

Thus, damage in the structure can ideally be sensed through variations in modal properties. During the normal operation of a structure, damage sensitive features, such as modal information of dominant modes, are extracted first in order to characterize and benchmark the healthy state. This can later be compared with the same collected once when the structure is perceived to be damaged for further confirmation of the hypothesis and subsequent localization of damage.

Unfortunately, not only damage but also effects of varying environmental factor, particularly temperature, may potentially alter the vibrational properties of the structure [1, 4, 8, 13, 14]. False positive or false negative damage may be signalled when the environmental effect on the changes of dynamic properties of a structure is not accounted for. Isolation of damage-sensitive features from the overall variation of vibrational properties due to all environmental agents as well as damage poses a major challenge for successful condition assessment of any structure.

In this study, we will be, however, focusing on the damage identification in the presence of temperature effects only. In practice, temperature affects both materials as well as geometric stiffness properties of any structure. Change in temperature may cause changes in material properties at various levels which indirectly affects the stiffness and subsequently the modal properties. However, temperature effects on geometric stiffness depend on structural configuration and thereby the impacts are very much case-specific. It is determined by element geometry, stresses developed and is independent of elastic properties. This component of stiffness is mostly ignored in most of the modelling approaches and therefore, temperature effects are taken into condition only partially (through materials). In this study, we have taken this aspect into the modelling that enhances the reality replication prospective, discussed in later sections.

Further, in this article, an interrelationship between mode shapes and temperature and damage is formed using machine learning algorithms in order to predict damage under varying temperature conditions. In the context of modal information-based damage detection, majorly the emphasis has been on the modal frequencies for detection and/or localization of damage in structures. However, in some circumstances, the modal frequencies' sensitivity towards damage is not detectable. For these cases, the damage-induced variation in frequencies is often less than the precision at which the modal frequencies are estimated from noisy response data. In comparison, although it has been established that the mode shapes are most suitable to identify damage localization, they did not receive much consideration because of the difficulty in their identification with high precision. Previous studies also showed that thermal effect on mode shapes is mostly negligible [15] which is, however, true for the only case when the structure is under uniform temperature rise and the geometric stiffness is not taken under consideration. Also, determination of mode shapes requires simultaneous measurement of vibration at a higher number of points of structure that demands dense instrumentation which is, however, a costly affair. In this study, we, however, employed mode shapes as damage-sensitive features since the changes induced by damage for certain kind of structures are very small and as such difficult to be identified through modal frequencies.

Machine learning algorithms have been broadly used by researchers from the last few decades to develop a wide range of parametric and non-parametric vibration-based structural damage detection techniques. Among parametric methods, neural networks are very proficient for pattern recognition and classification. ANNs can model any complex relationship given that it has been trained with a rich dataset encompassing all aspects concerned. Over the years, ANN has been proved to be more efficient than traditional mathematical models [2] since the later sometimes gets crippled with non-realistic assumptions. In the context of health monitoring, existing studies concluded that ANNs are very much capable of providing correct damage identification when modal information from the structure is simulated numerically and are error-free. The performance of the same with real data is although subjective to the precision of the data being used for training the network.

This article presents a two-stage damage-based algorithm that is capable of filtering out the effect of temperature in vibration-based structural damage detection where vibrational properties and temperature are measured. The proposed approach employs two consequent neural networks for detection and localization of damage. Firstly, for detection, a correlation model between mode shapes and temperature is formulated using an Auto-Associative Neural Network (AANN). AANN considers identified mode shapes of the healthy system under different temperatures as input while the same has been attempted to be predicted as output. The unique feature of AANN is that input and output parameters are always same. AANN correlates the mode shapes with temperature and simulated mode shapes themselves under the assumption of healthy state. However, any violation will invalidate the prediction model and will cause a large prediction error. A predictor model is simulated for a batch of different damage conditions under different temperatures and mode shapes are put through the AANN model in order to obtain the residual error for each damage case. However, these prediction errors are temperature-invariant that can be used for identifying the location of damage through fully connected network.

Fully connected network is just like a traditional network meant for classification [6, 9]. They are used for various applications and the major advantage of this network is that they are "structure agnostic". The first version of a fully connected neural network was "Perception" created in the 1950s. It works in the same principle as the traditional multi-layer perceptron neural network. This network consists of input, output and hidden layers. These fully connected layers do the classification based on the features hidden in the input. Each individual layer within the fully connected layer network consists of several neurons. Typically, for such network, neurons between two adjacent layers are fully pairwise connected, but neurons within a single layer share no connection. These are also called "universal approximators" since they are capable of learning any function.

The following section demonstrates the effects of temperature in the context of a beam-like structure which is followed by the detailed methodology for the proposed two-stage damage detection technique. Next, numerical experimentation has been presented in order to validate the proposed technique. For this, a numerical beam is selected and different damage scenarios involving different temperatures and locations are considered. The predictive ability and possibility of false alarm associated with the proposed algorithm have been investigated in this endeavour.

2 Temperature Effect on Dynamic Properties of a Beam-Like Structure

The effect of temperature on structural stiffness is the most significant out of the three major parameters (mass, stiffness, damping) with mass being mostly unaffected. Stiffness is the property of structure that characterizes structural response under an externally applied load. In order to understand the change in the behaviour (static or dynamic) of the structure under varying temperature, one, therefore, should put emphasis on understanding the effect of temperature on stiffness.

There are numbers of factors that affect the stiffness: geometry, material and boundary conditions are a few of them. Obviously, change in material properties cause the change in stiffness, although not always linearly or monotonically. Elastic modulus, Poisson's ratio and density can be considered as the specifics of the material those define its stiffness. In this study, we have considered elastic modulus, thermal coefficient and Poisson's ratio to be temperature dependant. For the relevant interrelation we have taken the basis from the article of [11].

Besides affecting material properties, that directly affect the material stiffness of a structure, temperature affects the structure in a complicated manner which leads to changes in structural configuration, dimension and boundary condition. For an example, for beam-like structures that can be considered as an idealization of bridges, a significant temperature change can cause supports to move due to its axial elongation or sometimes changes in boundary condition if the elongation is restrained. This further adds on to the effect of temperature on the material stiffness.

If, because of certain restrained degree/s of freedom, the structure undergoes a change in shape, nonlinear behaviour arises which further brings in geometrical nonlinearity into the system. Due to this nonlinearity, another stiffness effect comes into play originating from the impact of prestress (thermal stress if the cause is temperature) on the structural geometry. For example, due to restraining an axial compression/elongation a tensile/compressive axial force may develop in member, which in turn increases/decreases its bending stiffness, respectively. The stiffness that reflects this kind of stress stiffening effect is called geometric stiffness. In this study, we have taken this aspect into the modelling that enhances the reality replication perspective. Two types of heating regimes are studied in reference to our context.

- Mean temperature rise (uniform over the length) $= \Delta T$
- Effective thermal gradient through the depth $T, y = (T2 - T1)/d$

where, $T2$ and $T1$ are the temperatures of a beam on its top and bottom surfaces, respectively, and T, y is the gradient of temperature over the beam depth d.

However, temperature effects on geometric stiffness depend on the structural configuration and thereby the impacts are very much case-specific. For example,

Fig. 1 Idealized support conditions

in a beam with both ends fixed, the compensatory action of mean temperature rise and axial restraints leads to zero deflection in the beam and in this process an axial prestress gets developed causing the bending stiffness to decrease. For the same beam with one end axial degree of freedom released, prestress will never develop causing no change in geometric stiffness. Also, if the temperature rise is not uniform over the cross section of the beam, the thermal gradient induces a curvature due to the differential in the induced prestress. However, for both end fixed condition, this curvature gets nullified by the action of end moments. This temperature-induced moment, in turn, affects the geometric stiffness. Again for the one end fixed and the other end laterally free beam, this moment will still be developed since both the ends are capable of generating the compensatory moment. The same is not true for a simply supported beam, in which, the beam will deflect causing no change in geometric stiffness. This case specificity condition of the temperature effects on geometric stiffness makes the problem challenging as well as interesting. In real structures such as bridges, the superstructure consists of girders/beams with deck slab provided with some support conditions. In the support system, typically one end is kept as fixed and another end is kept as transversely restrained by means of roller bearings. However, the rotational degrees of freedom, in the end, are mostly (can be considered as) restrained. From the practical point of view, the boundaries are never fully restrained but can only be finitely restrained, which can be modelled using a lateral spring of linear/nonlinear nature. In our study, to replicate the real scenario, we have considered a special support condition which is restrained axially by a translational spring in place of roller bearing which ensures finite deformation, as well as this beam end, provides the required rotational rigidity to ensure bowing effect to impact the beam stiffness. The relevant figure of the idealized support has been given in (cf. Fig. 1).

2.1 Thermal Expansion/Contraction-Induced Changes in Stiffness

If for a beam with a linear spring as lateral restraint, temperature is increased, the impending impact of thermal stress can be calculated by equating the thermal force to the spring force. In our study, we have considered a beam which is restrained axially by a translational spring of stiffness k_t. Thermal expansion develops axial

Fig. 2 Uniform heating of beam with finite axial restraint

compression as shown (cf. Fig. 2). The total displacement produced in this beam is due to thermal expansion of the beam end and contraction of the lateral spring which are compensating for each other. This force induced in the beam is obviously compressive in nature since it is compensating for a thermal elongation. The resulting thermal stress can then be calculated as follows:

$$\sigma_c = \frac{E\alpha \Delta T}{1 + EA/k_t L} \quad (1)$$

where E is material elasticity, α is thermal elongation coefficient, L is the length of the beam, A is the cross-sectional area and σ_c is the thermal stresses induced in the member.

2.2 Thermal Bowing Induced Changes in Stiffness

In case of thermal bowing effect (T, y), varying temperature over the section of the beam induces a curvature of amount $\phi = \alpha T, y$. This curvature reduces the horizontal distance between the ends of the beam, which results in contraction strain, $\epsilon_\phi = 1 - \frac{sin(l\phi/2)}{l\phi/2}$.

Considering the beam is rotationally restrained at the ends, a moment will generate at the fixed ends of the beam. However, this moment is difficult to incorporate in the equations of geometric stiffness, which can only cater for axial prestress. This can, however, be represented through a restoring tensile force that gets induced in order to restrict the beam from bending due to thermal curvature. The resulting tensile force can be quantified in a similar way as for the case of expansion strain (cf. Fig. 3).

Fig. 3 Bowing effect of beam with finite axial restraint

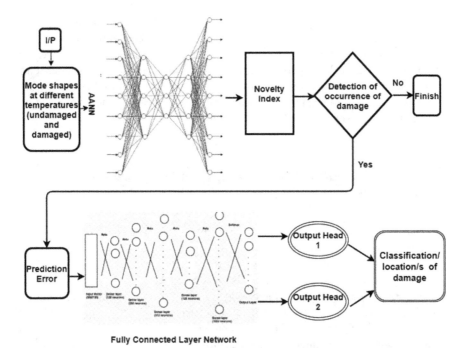

Fig. 4 Implementation of AANN and Fully Connected Layer Network

Thus it is evident, that the above-mentioned temperature parameters affect the structural stiffness significantly. The incorporation of these impacts is important to enhance the predictive ability of the numerical model that will later be employed for the proposed model-based damage detection technique. This numerical model ideally should be a replica of the physical model which in turn helps us to understand the intricate functioning and interrelationship between all the contributing aspects and their variations. Subsequently, this numerical model is used for detecting damage in the structure using machine learning algorithms.

3 Methodology

3.1 Damage Detection Using Machine Learning Algorithms

Several machine learning have been reported in the literature for detection and localization of damage [5, 7, 10, 12, 16]. Neural Network works as a functional abstraction of the biologic neural structures of human nervous systems. ANN generalizes the knowledge implicit in the training samples and becomes capable of providing efficient solutions to new situations. It is, therefore, provides a powerful tool for detection and localization of damage. Machine learning algorithms can be employed to understand and model a functional relationship between changing environmental conditions, especially temperature and interrelation between the modal properties [7]. In this study, machine learning algorithms are implemented in a sequence of steps: first, each algorithm is trained and its parameters are adjusted under varying environmental conditions (structure assumed to be in a healthy state), followed by validation of the developed model.

3.2 Auto-Associative Neural Network (AANN) and Fully Connected Layer Network

Recent trends in neural network research concern more about the detection of novelty [16]. Among the various options available within the neural network, AANN has been identified as one of the promising networks for novelty detection [2, 16]. AANN is a type of network whose target output is considered to be same as input [5]. Its network architecture consists of three hidden layers with an internal "bottleneck" layer. Within the three layers, the first layer is termed as de-mapping layer where nodes are usually higher in number than input layer and similarly, the third layer is called mapping layer having equal number of nodes as in de-mapping layer. Thus, AANN network firstly de-constructs the input into nodes larger in the dimension of input layers. Subsequently, the information in all these nodes is compressed in the bottleneck layer. Next, the same procedure is repeated but in reverse order to get back the inputs (see Fig. 5). Once this network is trained, this learns the interrelation between the input parameters themselves. This correlation is obviously valid as long as the inputs originate from the same source. Any alterations in the source will eventually cause an alteration in this correlation structure and subsequent error in the predicted output.

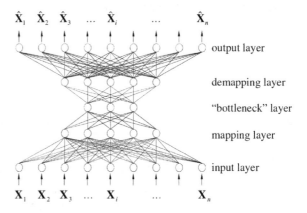

Fig. 5 Schematic diagram of the AANN model

In this study, AANN is applied as a structural damage alarm. A set of measured mode shapes obtained from a healthy state under different temperature conditions is used to train the network. A five-layer network has been employed to predict the same as output data. A multi-layer perceptron with Levenberg–Marquardt feedforward backpropagation is utilized to train the AANN model in this study. Tan-sigmoid transfer functions are employed for the hidden nodes as a nonlinear activation function for all layers. Once trained, this network can perfectly return the same output mode shapes as provided in the network from a healthy state. On the other hand, if mode shapes from a damaged system are fed into this network, the network should not return the same mode shapes as output. Damage obviously alters the correlation between different modes and it gets reflected from the AANN output.

To detect the occurrence of such mismatch that can be attributed to a possible damage, the AANN prediction error, becomes very handy and can be used as a novelty index as follows:

$$\varepsilon^k = \phi_i^k(T) - \phi_o^k(T) \qquad (2)$$

where ϵ^k denotes AANN prediction error for k^{th} measurement. For a measured mode shape $\phi_i(T)$ used as input, $\phi_o(T)$ gives the AANN prediction as follows: $\phi_o(T) = AANN(\phi_i(T))$. Clearly, as $\phi_i(T)$ and $\phi_o(T)$ are functions of ambient temperature T, the error term becomes inherently a function of T and thus not temperature invariant.

To take the ambient temperature into consideration, temperature has been additionally incorporated along with the mode shapes as input. This is turn includes the correlation of mode shapes and temperature into the AANN model. The model is trained with undamaged mode shapes for which prediction error is considerably low and for the same reason, the prediction error will be large for the damaged system. Moreover, this prediction error is not only high in magnitude but also demonstrates a special pattern that is unique to individual damage cases.

Following, a fully connected layer is employed for classifying this prediction error to corresponding damage cases. In this study, the errors from the AANN are trained with a basic fully connected layers network. ReLU function is used as an activation function within the hidden layers of the this network model. Traditionally, sigmoid and tanh have been the popular choices for activation function. However, recently it has been observed that ReLU function has a better response to model non-linearity through neural network model [3]. For output layers, the softmax function is employed for classification through computing the likelihood for an input to be classified under a specific class (cf. Fig. 6).

4 Numerical Experiment

The numerical analysis is performed on a steel beam (cf. Fig. 7) to validate the proposed algorithm. The geometric properties of steel beam are listed in Table 1. The beam is divided into ten segments. Single and multiple damages (maximum two) are

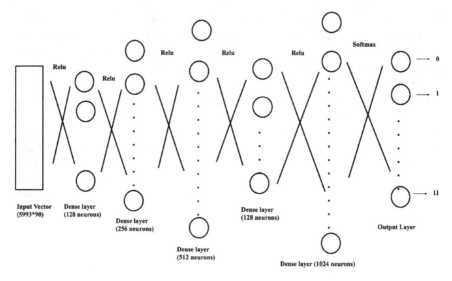

Fig. 6 Fully Connected Layer Network

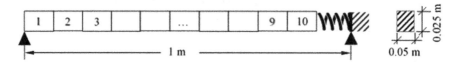

Fig. 7 Schematic diagram of the numerical beam

Table 1 Details of steel beam

Structure	Steel beam
Dimension (m)	$1 \times 0.05 \times 0.025$
Density (kg/m^3)	7850
T	$-30°C–70°C$
A (m^2)	0.125
E (GPa)	201.04e9
Poisson's ratio	0.3262
G (GPa)	7.62e10
Iz (m^4)	6.51e-08
Iy (m^4)	2.60e-07
J (m^3)	3.26e-07
Thermal coefficient (°C)	1.23e-05

introduced randomly in different segments in the form of stiffness reduction varying from 50–70% of the original stiffness values. Different damage cases involving different damage severity and location under different temperatures are studied in this attempt. Mode shapes are measured under different temperature conditions from damaged and undamaged states. With the geometric stiffness taken into consideration, the mode shapes are found to be sensitive to temperature changes (cf. Fig. 8). Also, with damage as well the mode shapes are found to vary with temperature. This makes the problem of detecting damage more complicated.

To tackle the above-mentioned problem, AANN model is used. Firstly, a set of undamaged test cases under different temperature conditions are simulated and corresponding mode shapes are recorded. These mode shapes are then put through the AANN model. The model is trained with undamaged mode shapes for which prediction error is found to be considerably low. Eventually, since the AANN model is trained with mode shapes from healthy state only, the prediction error is expected to be large for mode shapes collected from damaged systems.

It has been observed that AANN yields significantly large error in prediction from the damaged cases which in turn points at the possibility of the presence of damage. However, no information on their location can be obtained from this algorithm. Next, for the localization purpose, the fully connected layer network has been employed with the AANN prediction errors corresponding to all case studies (damaged and undamaged cases combined) as training input, while the corresponding damage locations considered as training outputs or output classes. For input, a total of 5993 samples are used, out of which 5393 are used for training and 600 are used for testing the efficacy of the proposed algorithm. Out of the training samples, 540 samples are used for validation.

Five dense layers are used, each having different numbers of neurons as 128, 256, 512, 128 and 1024, respectively (cf. Fig. 6). ReLU is used as the activation function between these dense layers and at the output layer, softmax function is used for classification purposes. Two output heads are used those refer two possible damage locations in the beam. The beam segments are numbered as 1–10 through which the damage can be located and 0 is denoted as no-damage. Two output heads successfully identified the damage location for ten segments properly with accuracy varying between 90 and 94% (cf. Figs. 9 and 10). Loss percentage and accuracy percentage for training and validation sample is found to be consistent for both the output heads which proves the robustness of this network in damage localization (cf. Figs. 9 and 10).

Normalized confusion matrices, cf. Figs. 11 and 12, show the maximum number of samples that are actually detected accurately from the first and second output heads, respectively. This confusion matrix is in a form of a table that is typically used to describe the performance of a classification model on a set of test data for which true values are known. It is basically a summary of predictive ability of the network for locating a damage in the beam. The confusion matrix shows the ways in which our classification model is confused about when it makes predictions. The ratio of correct predictions over total test cases is summarized by diagonal elements of the normalized confusion matrix.

Fig. 8 Mode shapes variation for undamaged and damaged system under different temperature

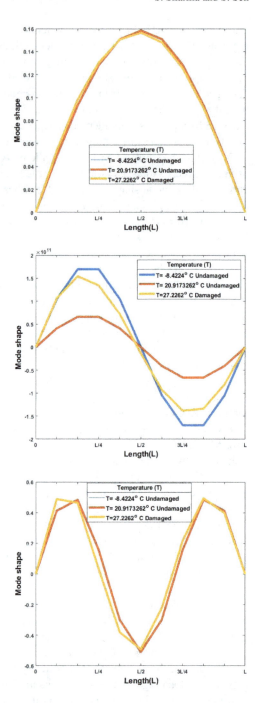

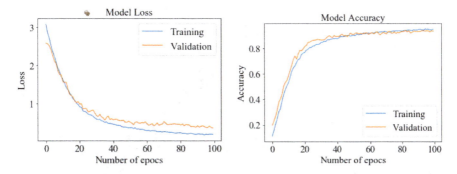

Fig. 9 Loss and accuracy for first output head

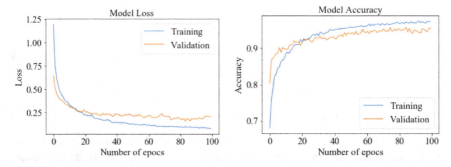

Fig. 10 Loss and accuracy for second output head

Fig. 11 Normalized confusion matrix for first output head

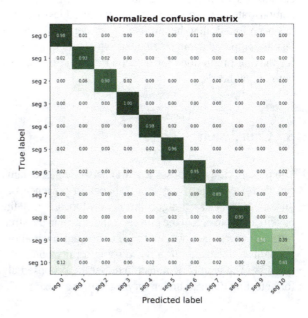

Fig. 12 Normalized confusion matrix for second output head

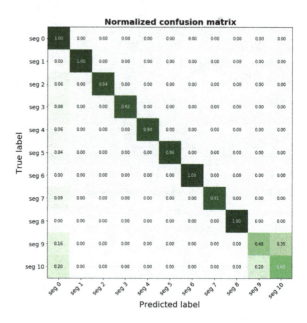

Figures 11 and 12 demonstrate that most of the damaged and undamaged cases are localized accurately by the proposed classification network. While high numbers in the diagonals ensure that damage/undamaged cases are detected with precision causing less numbers of negative false alarm, the low values in the off diagonals demonstrate that proposed algorithm caused less numbers of positive false alarm. It has been found, that the number of correct predictions for the segments 9 and 10 is less as compared to other segments which might be due to the model that is confused for making predictions for detecting damage in between the segments 9 and 10 in both output heads.

5 Conclusion

The present study proposes a neural network-based damage detection algorithm that can detect damage using mode shapes in presence of temperature variation. For this, an AANN based algorithm is developed that defines a prediction error-based novelty index to detect the presence of damage in the structure. The prediction error is further found to be invariant to the temperature variation. This aspect is further used for damage localization purpose using a fully connected layer-based classification network. The results demonstrated that the proposed algorithm proves to be robust for detecting damage under varying temperature.

Acknowledgements Financial support received from Indian Institute of Technology Mandi, HP, India under Seed Grant scheme through grant file no. IITM/SG/SUS/66 is gratefully acknowledged.

References

1. Alampalli S (1998) Influence of in-service environment on modal parameters. In: Proceedings-SPIE the international society for optical engineering, vol 1. Citeseer, pp 111–116
2. Bakhary N, Hao H, Deeks AJ (2007) Damage detection using artificial neural network with consideration of uncertainties. Eng Struct 29(11):2806–2815
3. Cha Y-J, Choi W, Büyüköztürk O (2017) Deep learning-based crack damage detection using convolutional neural networks. Comput-Aided Civ Infrastruct Eng 32(5):361–378
4. Farrar CR, Jauregui DA (1998) Comparative study of damage identification algorithms applied to a bridge: I. experiment. Smart Mater Struct 7(5):704
5. Figueiredo E, Park G, Farrar CR, Worden K, Figueiras J (2011) Machine learning algorithms for damage detection under operational and environmental variability. Struct Health Monit 10(6):559–572
6. Iliadis M, Spinoulas L, Katsaggelos AK (2018) Deep fully-connected networks for video compressive sensing. Digit Signal Process 72:9–18
7. Jin C, Jang S, Sun X, Li J, Christenson R (2016) Damage detection of a highway bridge under severe temperature changes using extended kalman filter trained neural network. J Civ Struct Health Monit 6(3):545–560
8. Kullaa J (2009) Eliminating environmental or operational influences in structural health monitoring using the missing data analysis. J Intell Mater Syst Struct 20(11):1381–1390
9. Li J, Li B, Jizheng X, Xiong R, Gao W (2018) Fully connected network-based intra prediction for image coding. IEEE Trans Image Process 27(7):3236–3247
10. Mehrjoo M, Khaji N, Moharrami H, Bahreininejad A (2008) Damage detection of truss bridge joints using artificial neural networks. Expert Syst Appl 35(3):1122–1131
11. Reddy JN, Chin CD (1998) Thermomechanical analysis of functionally graded cylinders and plates. J Therm Stresses 21(6):593–626
12. Santos A, Figueiredo E, Silva MFM, Sales CS, Costa JCWA (2016) Machine learning algorithms for damage detection: Kernel-based approaches. J Sound Vib 363:584–599
13. Wahab MA, De Roeck G (1997) Effect of temperature on dynamic system parameters of a highway bridge. Struct Eng Int 7(4):266–270
14. Xia Y, Hao H, Zanardo G, Deeks A (2006) Long term vibration monitoring of an rc slab: temperature and humidity effect. Eng Struct 28(3):441–452
15. Zhang D, Bao Y, Li H, Jinping O (2012) Investigation of temperature effects on modal parameters of the china national aquatics center. Adv Struct Eng 15(7):1139–1153
16. Zhou HF, Ni YQ, Ko JM (2011) Eliminating temperature effect in vibration-based structural damage detection. J Eng Mech 137(12):785–796

Breathing Crack Localization Using Nonlinear Intermodulation Based Exponential Weighting Function Augmented Spatial Curvature Approach

J. Prawin

Abstract Localization of breathing crack using vibration-based measurements is a highly challenging inverse problem. In this paper, we present the conventional spatial curvature-based damage detection approach augmented with specially formulated exponential weighting functions for breathing crack identification using the vibration responses obtained spatially across the structure under bitone harmonic excitation. The proposed exponential weighting function provides an enhanced damage diagnosis as it exploits nonlinear sensitive damage features such as superharmonics and intermodulations, induced by the opening and closing behaviour of the breathing crack. The robustness and effectiveness of the proposed damage diagnostic scheme is validated through both numerical and laboratory-level investigations.

Keywords Exponential weighting function · Sidebands · Bitone harmonic excitation · Spatial curvature

1 Introduction

Most of the damages in the structural elements reflect in the form of visible cracks and these cracks will generally be open and remain open during vibration. This is true only when the structure is in the linear state even after damage. On the other hand, subtle cracks such as breathing crack wherein the edges of the crack come into and out of contact during vibration exhibit nonlinear behaviour once damage sets in, even though the structure is linear before damage.

Localization of breathing crack is generally identified by the comparison of the Fourier power spectrum amplitudes of superharmonics of various sensors obtained spatially across the structure [1]. The energy of superharmonics is found to higher around the spatial location of the damage when compared to the other locations. It should be mentioned here that the input amplitude of excitation, excitation frequency

J. Prawin (✉)
CSIR-Structural Engineering Research Centre, CSIR Campus, Chennai, TN, India
e-mail: prawinpsg@gmail.com

© The Editor(s) (if applicable) and The Author(s), under exclusive license to Springer Nature Singapore Pte Ltd. 2021
S. K. Saha and M. Mukherjee (eds.), *Recent Advances in Computational Mechanics and Simulations*, Lecture Notes in Civil Engineering 103,
https://doi.org/10.1007/978-981-15-8138-0_24

has to be properly chosen to observe these superharmonics induced by the breathing crack. Apart from this, it is essential to ensure that there exist no pseudo-harmonics while exciting the structure with breathing crack with pure single-tone harmonic excitation. In view of this, instead of exciting the structure with single-tone harmonic excitation for breathing crack identification, various researchers have developed the damage diagnostic technique using the vibration time history measurements obtained under bitone harmonic excitation. The bitone harmonic excitation refers to the harmonic excitation of the structure with two input excitation frequencies simultaneously similar to single-tone harmonic excitation. The two input excitation frequencies are popularly referred in the literature as pumping and probing frequency. It should be mentioned here that probing frequency is always much larger than pumping frequency. The presence of the breathing crack in case of the bitone harmonic responses is identified by the presence of sidebands apart from the superharmonics of two input excitation frequencies. The sidebands are formed due to the coupling of pumping and probing frequencies and these sidebands are also called as intermodulations [2–4].

Most of the researchers employ only the first sideband on either side of probing frequency for breathing crack identification [2–4]. They conveniently ignore the higher order sidebands due to the following reasons: the amplitude of the linear or two input excitation harmonic components is often three to four orders higher in magnitude than that of higher order intermodulation components. The energy levels of these higher order sideband components and the noise components of the response are more or less with the same energy levels. Hence, it is highly challenging to isolate these higher order components as they often get buried in the noise components. However, ignoring these higher order sideband components in the damage diagnostic process may often fail to detect the presence of the breathing crack in the structure, especially when the crack is in the incipient stage [5–9]. Therefore, it is essential to devise appropriate techniques to extract these higher order harmonics from the measured time history signals in order to robustly detect the breathing cracks at their incipient stage.

In the present work, a traditional spatial curvature-based damage diagnostic technique augmented with an exponential weighting function is proposed in this paper for localization of breathing crack. The proposed exponential weighting function takes into account nonlinear sensitive damage features such as intermodulations/sidebands and provides an enhanced damage diagnosis. The proposed approach considers all the higher sidebands for accurate localization of breathing crack in contrast to the earlier works. The robustness and effectivenss of the proposed damage diagnostic scheme is verified through numerical and laboratory-level investigations.

2 Breathing Crack Identification with an Enhanced Exponential Weighting Function

All the existing conventional spatial curvature-based damage diagnostic techniques are applicable only for open crack models due to the fact that it ignores the nonlinear sensitive features such as superharmonics and sidebands. In order to use the curvature-based approach [5, 6] for breathing crack identification, we introduce appropriate exponential weighting functions which help in isolating the benign (or noise) components from the nonlinear intermodulation components and also enrich their resolution

The damage localization index is defined as the multiplication of the exponential weighting function ($J(\omega)$) and absolute difference of the second spatial derivative of the Fourier power spectrum response of the undamaged ($\alpha''_{\text{undameged},ij}(\omega_k)$) and damaged states ($\alpha''_{\text{damaged},ij}(\omega_k)$) of the structure [4]. It is given by

$$NDLI(i) = \frac{DLI(i)}{\max(DLI(i))}; \quad i = 1, 2, 3, \ldots N \qquad (1)$$

$$DLI(i) = \sum_{j=1}^{N} \sum_{k=1}^{W} J(\omega_k) \left| \alpha''_{\text{undamaged},ij}(\omega_k) - \alpha''_{\text{damaged},ij}(\omega_k) \right| \qquad (2)$$

where NDLI(i), DLI(i) indicate the normalized damage localization index and damage localization index of the ith degree of freedom. The indices W and N denote the number of frequencies and the total number of degrees of freedom.

The basic objective of formulating the weighting function is that it should enhance the resolution of nonlinear features (i.e. nonlinear intermodulation) buried in noise. Accordingly, the shape and trend of the weighting function are chosen. The zone factor is introduced into the weighting function to accommodate the shift in frequencies around nonlinear intermodulations associated with probing excitation frequency due to breathing crack. With this, the exponential weighting functions, are formulated as follows:

$$J(\omega_k) = \max(J_j(\omega_k)); \quad j = 1, 2, \ldots, 2 \times n; \qquad (3a)$$

$$J_j(\omega) = \exp\left(-\frac{z \times \left(\omega - \omega_{\text{mod}}^{j^2}\right)}{\left(\omega_{\text{mod}}^{j_{\text{prob}2}}\right); \omega_{\text{mod}}^{\text{prob}_{\text{pump}}}}\right) \qquad (3b)$$

The zone factor Z indicates the region of influence around every sideband. The proposed exponential weighting function exhibits maximum value of 1 at the sideband frequency and less than 1 in the zone of influence of each sideband. The exponential weighting function values are almost zero at other frequencies. The parameter

'n' represents the number of sidebands. The location at which the normalized damage localization index exhibits value of one is the spatial location of the damage.

3 Numerical Investigations

A simply supported steel beam shown in Fig. 1 is chosen as the numerical example, to test the proposed breathing crack localization technique. The span of the beam is 0.7 m and has an area of $4e^{-4} m^2$ and moment of inertia as $0.667e^{-8} m^4$. The first few natural frequencies of the underlying linear healthy beam are found to be 89.671, 354.689, 799.16, 1419.46, 1711.46 and 2218.17 Hz. The beam is idealized with 20 elements. Breathing crack is modelled through a unit step function. The detailed formulations related to breathing crack idealization can be found in Prawin et al. [4].

The simply supported beam is subjected to varied excitation amplitudes of 0.1 and 100 N with excitation frequencies of 90 and 1710 Hz near the centre. The beam is simulated with breathing crack near 1/5 span of the beam (i.e. at element no. 4). The simulated breathing crack depth is 0.07 h, where h is the depth of the beam. The spectral density plot of the cracked structure obtained under varied amplitudes of excitation is shown in Fig. 2a. The zoomed plot of Fig. 2a is presented in Fig. 2b. The spectral density plot corresponding to 0.1 N, presented in Fig. 2a shows two peaks at 90 and 1710 Hz. This confirms that the cracked structure behaves linearly under 0.1 N excitation as the structure vibrates only at its excitation frequencies. In contrast to 0.1 N excitation, the spectral density plot shown in Fig. 2a, corresponding to 100 N excitation vibrates not only at its excitation frequencies, i.e. 90 and 1710 Hz but also at its super harmonics of pumping frequency, i.e. 180, 270, 360 Hz ($n\omega_{pump}$), super harmonics of probing frequency, i.e. 3420, 5130 Hz ($n\omega_{prob}$) and intermodulation frequencies, i.e. 1800, 1890 Hz ($\omega_{prob} \pm n\omega_{pump}$) and so on. Further, it is evident from Fig. 2b that the peak at fundamental excitation harmonics exhibits very high magnitude when compared to the magnitude of the peaks at their sidebands. The presence of both actual harmonic excitation and its corresponding superharmonics and intermodulations/sidebands concludes the nonlinear behaviour of the beam induced by breathing cracks.

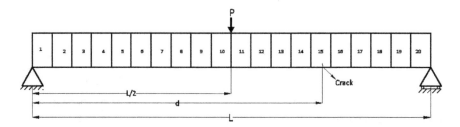

Fig. 1 Simply supported beam with a breathing crack

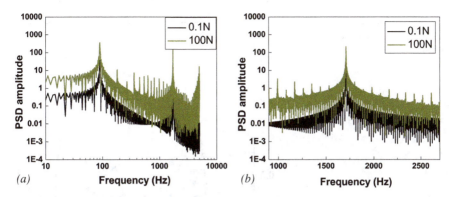

Fig. 2 Fourier power spectrum. **a** Full plot. **b** Zoomed plot

The exponential weighting function for varied values of zone factor is shown in Fig. 3. Lower the zone factor, higher the zone of influence and higher the magnitude of damage localization index. In order to overcome the influence of zone factor, the normalized damage localization index given in Eq. (1) is preferred in the present work. Low zone factor of <10 is considered.

Numerical simulation studies have been carried out by considering the breathing crack being present at three different spatial locations (i.e. element: 5, 9 and 15) spatially across the beam. The breathing crack depth is same for all the three test cases and chosen as 0.07 h. The results of the damage localization index estimated for these three test cases are given in Fig. 4. It can be concluded from Fig. 4 that the maximum value of the damage localization index in all the three test cases occurs at the two sensor nodes corresponding to the damaged element.

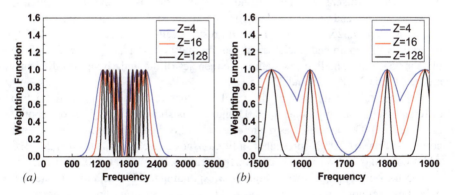

Fig. 3 Exponential weighting function. **a** Full plot. **b** Zoomed plot

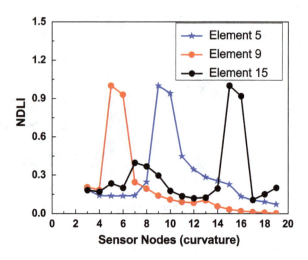

Fig. 4 Normalized damage localization index—varied breathing crack locations

4 Experimental Validation

A 1-m aluminium multi-crack cantilever beam, i.e. two cracks at two different locations is considered as the experimental example. The similar experimental specimen was earlier employed by Prime and Shevitz [5] and Douka et al. [6] for the identification of breathing crack. Four pieces of two aluminium alloy beams are bonded together as illustrated in Fig. 5 to create the test specimen. Breathing behaviour is induced in the test beam by the contact of the faces of the top plates. The breathing crack is induced at two different locations from the fixed end. The first breathing crack is around 0.2 m from the fixed end which is between sensors 2 and 3, and it is much closer to sensor 2. The second breathing crack is around 0.7 m which is between sensor 5 and sensor 6 and much closer to sensor 6. The top and bottom plate thicknesses are 9.525 mm and 3.175 mm, respectively. The crack depth is 0.25 h, where h is the combined thickness of the top and bottom plate. The width of the plate is 25.4 mm. The first few natural frequencies of the underlying healthy beam are 10.38, 65.03, 182.138 and 357.15 Hz.

The spectral density plot of the cracked structure obtained under these two different excitation amplitudes of 0.03 and 5 N is shown in Fig. 6. The response measured at 0.03 N is chosen as the reference healthy data. The spectral density plot corresponding to 0.03 N, presented in Fig. 6, shows two peaks at 10 and 200 Hz, which coincides with the input excitation frequencies. In contrast to 0.03 N excitation, the spectral density plot shown in Fig. 6, corresponding to 5 N excitation vibrates not only at its excitation frequencies but also at their corresponding superharmonics and intermodulation frequencies. Further, it can be observed from Fig. 6 that the peak at fundamental excitation harmonics exhibit a very high magnitude when compared to the magnitude of the peaks at their sidebands and superharmonics.

The damage localization index is estimated using Eqs. (1)–(3a, 3b) and the corresponding plot is shown in Fig. 7. The damage localization index plot given in Fig. 7

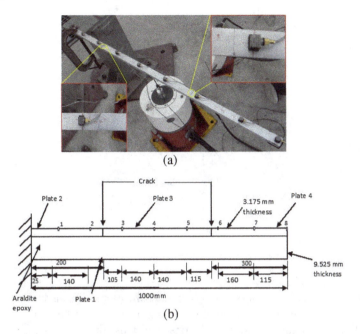

Fig. 5 Multiple breathing crack experimental specimen

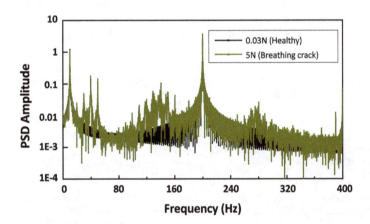

Fig. 6 Fourier power spectrum

shows two peaks. The first peak is around sensors 2 and 3. The second peak is at sensor 6. These damage localization index peaks clearly reflect the actual breathing crack locations of the considered beam. Therefore, we can clearly conclude through numerical and experimental simulation studies that the damage localization index based on exponential weighting function augmented spatial curvature approach for breathing crack localization can localize the multiple cracks present in the structure.

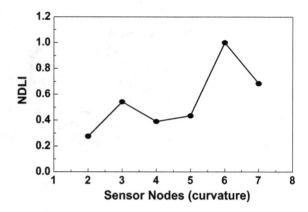

Fig. 7 Damage localization index—experimental multiple breathing crack specimen

5 Conclusion

An exponential weighting function augmented traditional spatial curvature approach is presented in this paper for breathing crack localization using the bitone harmonic responses. The proposed exponential weighting function-based approach takes into account all the nonlinear intermodulations/sidebands for breathing crack identification and provides an enhanced damage diagnostic scheme. Both numerical and experimental studies concluded that the proposed normalized damage localization index is capable of identifying the spatial location of breathing crack in the structure. The proposed approach has the ability to identify more than one breathing crack in the structure.

Acknowledgements The authors express acknowledgements to the technical staff of ASTAR and SHML Lab of CSIR-SERC for their help during laboratory testing.

References

1. Bovsunovsky A, Surace C (2015) Non-linearities in the vibrations of elastic structures with a closing crack: a state of the art review. Mech Syst Signal Process 62:129–148
2. Sampaio RPC, Maia NMM, Silva JMM (1996) Damage detection using the frequency-response-function curvature method. J Sound Vib 226:1029–1042
3. Kim D-G, Lee S-B (2010) Structural damage identification of a cantilever beam using excitation force level control. Mech Syst Signal Process 24:1814–1830
4. Prawin J, Rama Mohan Rao A (2019) A method for detecting damage-induced nonlinearity in structures using weighting function augmented curvature approach. Struct Health Monitor 18(4):1154–1167
5. Prime MB, Shevitz BW (1996) Linear and nonlinear methods for detecting cracks in beams. In: Proceedings of 14th international modal analysis conference, pp 1437–1443
6. Douka E, Hadjileontiadis LJ (2005) Time-frequency analysis of the free vibration response of a beam with a breathing crack. Nondestruct Test Eva 38:3–10

7. Klepka A, Staszewski WJ, Jenal RB et al (2012) Nonlinear acoustics for fatigue crack detection–experimental investigations of vibro-acoustic wave modulations. Struct Health Monitor 11(2):197–211
8. Lim HJ, Sohn H, DeSimio MP et al (2014) Reference-free fatigue crack detection using nonlinear ultrasonic modulation under various temperature and loading conditions. Mech Syst Signal Process 45(2):468–478
9. Kim GW, Johnson DR, Semperlotti F et al (2011) Localization of breathing cracks using combination tone nonlinear response. Smart Mater Struct 20(5):055014

Error in Constitutive Equation based Approach for Isotropic Material Parameter Estimation in Frequency-Domain Elastodynamics

Shyamal Guchhait and Biswanath Banerjee

Abstract A new version of error in constitutive equation (ECE)-based material parameter identification technique for linear elastic structure in frequency-domain elastodynamics has been proposed in this article. The inverse identification problem is solved by minimizing the ECE cost functional. The ECE functional measures the error in constitutive equation due to two incompatible stress and strain fields. This incompatibility is produced due to the generation of these two fields following dissimilar constraints. The stress field is dynamically admissible and the strain field is kinematically admissible with the measured displacement data. First, the strain field is generated by using a simple penalization technique with weak incorporation of full or partial noisy measured displacement data. This penalization technique also acts as a regularization to tackle the ill-posedness of the inverse problem. Then the stress field is generated by solving a linear system of equations. Thus, in the proposed methodology, the generation of incompatible stress and strain field is uncoupled in nature which reduces the numerical computational cost in contrast to standard modified error in constitutive equation (MECE)-based method. Afterward, explicit linear update formulas are formed for isotropic material model. In numerical examples, identification of heterogeneous isotropic material parameters is performed for 3D structures. The numerical experimentation shows that the proposed method can effectively identify the elastic material parameter distribution in a few number of iterations. The present method can be utilized in large-scale elastic parameter estimation problem because of its low computational cost.

Keywords Frequency-domain elastodynamics · Inverse problem · Identification of elastic material parameters · Error in constitutive equation functional

---

S. Guchhait (✉)
CSIR-Structural Engineering Research Centre, CSIR Campus, Chennai 600 113, India
e-mail: shyamal.guchhait@gmail.com

B. Banerjee
Indian Institute of Technology Kharagpur, Kharagpur 721 302, India

1 Background

1.1 Forward Problem

The methodology of inverse identification of elastic material parameters is very much dependent on the quality of measurement data. At the same time, in all the iterations of the proposed inverse update method, the two incompatible stress and strain fields are needed to be generated with which we proceed for the material update step. These two steps of the proposed technique are directly governed by the basic equilibrium equation of the system. In this context, we first describe here the basic equation of motion of linear elastodynamics problem in frequency domain as follows:

$$\nabla \cdot \boldsymbol{\sigma} + \boldsymbol{b} = -\rho \omega^2 \boldsymbol{u} \quad \text{in} \quad \Omega^0 \tag{1a}$$

$$\boldsymbol{u} = \bar{\boldsymbol{u}} \quad \text{on} \quad (\Gamma_u^0), \tag{1b}$$

$$\boldsymbol{\sigma} \cdot \boldsymbol{n} = \bar{\boldsymbol{t}} \quad \text{on} \quad (\Gamma_t^0), \tag{1c}$$

$$\boldsymbol{\sigma} = \mathbb{C} : \epsilon[\boldsymbol{u}], \tag{1d}$$

where $\boldsymbol{u}$ denotes the displacement field, $\boldsymbol{b}$ represents the body force, $\bar{\boldsymbol{t}}$ denotes the tractive force, ϵ represents strain tensor, $\boldsymbol{\sigma}$ is the stress tensor, $\boldsymbol{n}$ represents outward unit normal $\bar{\boldsymbol{u}}$ which is the essential boundary condition. (Γ_u^0) denotes the region for essential boundary condition and (Γ_t^0) represents region of traction boundary condition satisfying the $(\Gamma_u^0) \cup (\Gamma_t^0) = (\partial \Omega^0)$ and $(\Gamma_t^0) \cap (\Gamma_u^0) = \emptyset$ criteria. Here $\mathbb{C} := C_{pqrs}$ represents constitutive elasticity tensor.

1.2 Material Parameter Estimation Technique Based on MECE Functional

The ECE methodology is fully dependent on the formation of a cost functional which evaluates the discrepancy in constitutive equation between a dynamically admissible stress field and a kinematically admissible strain field. In the field of linear elastic material parameter identification, ECE cost functional or its different variants have been utilized by several researchers. It was first proposed by Ladevèze and Leguillon [2] in error estimation of finite element method. Lately, a variant of the conventional ECE cost functional, namely, modified error in constitutive equation (MECE) has been developed by Feissel and Allix [1] to include the effect due to addition of data corruption term in penalty format. MECE has been successfully used for large-scale elastic material parameter identification by Banerjee et al. [3]. Recently, ECE-based inverse estimation technique has been explored for linear anisotropic material parameters [4] and for damage detection of isotropic and composite plate [5]. The MECE cost functional, considering the error in measured and computed displacement as a penalty term, can be given as

$$\Lambda_{MECE}(\sigma, u, \mathbb{C}) = \frac{1}{2} \int_{\Omega^0} (\sigma - \mathbb{C} : \epsilon(u))^T (\mathbb{C})^{-1} (\sigma - \mathbb{C} : \epsilon(u)) \, d\Omega^0 + \frac{\alpha}{2} \|u(\mathbb{C}) - u^m\|^2_{L_2(\Omega^0_m)}. \tag{2}$$

Minimization of the above MECE cost functional is generally performed by following an alternate minimization strategy. At first, appropriate Lagrangian is constructed for the MECE cost functional maintaining the equilibrium equation. Then optimality conditions are used to find out the incompatible pair of stress and strain fields which generates the constitutive discrepancy. After that material parameters are updated by using the incompatible pair of stress and strain fields considering standard ECE functional or its different variants. The above two steps are performed in an iterative way until the stopping criteria is achieved.

Now, in MECE inverse identification technique, we have to solve a coupled problem with primary and Lagrangian variables for generation of two incompatible stress and strain fields in frequency-domain elastodynamics problem [3]. The discrete version of the coupled problem in MECE can be written as

$$\begin{bmatrix} \overline{K} - \omega^2 \overline{M} & \overline{K} \\ -\alpha B & \overline{K} - \omega^2 \overline{M} \end{bmatrix} \begin{Bmatrix} u \\ w \end{Bmatrix} = \begin{Bmatrix} P \\ -\alpha B u^m \end{Bmatrix}, \tag{3}$$

where $\overline{K}$ and $\overline{M}$ represent global stiffness and mass matrix. B is the diagonal Boolean matrix and w is the Lagrangian variable. The above equation maintains the unique solvability criteria when ω is not an eigenfrequency of the system [3]. However, solving of the above equation for large-scale system is computationally costly. If iterative solution strategy is adopted, then designing an efficient and effective pre-conditioner for the above coupled problem is not a trivial task. Exploiting the block structure of the above equation Banerjee et al. [3] have used successive over-relaxation (SOR) strategy to solve the above equation and thus reducing the computational cost. In this article, we propose a new variant of ECE, which uncouples the generation of incompatible stress and strain fields and thus reduces the computation cost greatly.

2 Material Parameter Estimation Technique Based on Error in Constitutive Equation

2.1 Discrete Version of Mechanical Fields

We adopt here the standard FEM discretization technique for finding out the incompatible mechanical fields. First, we have taken up the penalization problem and have found out the discrete form as follows:

$$\begin{aligned} (\overline{K} - \omega^2 \overline{M}) u_D + \alpha B (u_D - u^m) &= P \\ \Rightarrow (\overline{K} - \omega^2 \overline{M} + \alpha B) u_D &= \alpha B u^m + P, \end{aligned} \tag{4}$$

where B represents diagonal Boolean matrix which takes out the measured displacement data taken sparsely at the chosen node points. Next, the static problem of is discretized with the above penalized displacement field (u_D) to get the discrete field v

$$\overline{K}v = \omega^2 \overline{K} u_D + P. \tag{5}$$

2.2 Generation of Uncoupled Mechanical Fields from MECE Method

Now, we will describe how we can arrive at the uncoupled form of generating incompatible stress and strain fields in the proposed method from the coupled version of standard MECE. The discrete version of the coupled problem in MECE is given in Eq. (3). Now, let us assume $u = u_D$ and $v = u_D + w$, i.e., $w = v - u_D$. So, the coupled problem of MECE can be written in u_D and v as

$$\begin{bmatrix} \overline{K} - \omega^2 \overline{M} & \overline{K} \\ -\alpha B & \overline{K} - \omega^2 \overline{M} \end{bmatrix} \begin{Bmatrix} u_D \\ v - u_D \end{Bmatrix} = \begin{Bmatrix} P \\ -\alpha B u^m \end{Bmatrix}. \tag{6}$$

From the above equation, we get the following two equations:

$$\overline{K}v = \omega^2 \overline{M} u_D + P \tag{7}$$

$$\left(\overline{K} - \omega^2 \overline{M} + \alpha B\right) u_D = \alpha B u^m + \left(\overline{K} - \omega^2 \overline{M}\right) v. \tag{8}$$

Now, when $\mathbb{C} \to \mathbb{C}^*$, i.e., $u_D \to u^m$, $w \to 0$, i.e., $u_D \to v$. Applying this condition, we get the modified form of P from Eq. (7) as

$$P = \left(\overline{K} - \omega^2 \overline{M}\right) v. \tag{9}$$

Putting this form of P in Eq. (8), we get discrete penalized equation of u_D as follows:

$$\left(\overline{K} - \omega^2 \overline{M} + \alpha B\right) u_D = \alpha B u^m + P. \tag{10}$$

So, from Eqs. (10) and (7), we get two discrete uncoupled form of equations as

$$\left(\overline{K} - \omega^2 \overline{M} + \alpha B\right) u_D = \alpha B u^m + P \tag{11}$$

$$\overline{K}v = \omega^2 \overline{M} u_D + P. \tag{12}$$

In the present method, the incompatible pair of strain and stress fields is generated from the above two equations which is uncoupled in nature. Whereas, in the MECE technique, the generation of two incompatible mechanical fields is coupled

(Eq. 3). With this point, it is clear that the present method significantly reduces the numerical cost in generating the incompatible fields especially for large-scale parameter identification problem.

2.3 Material Parameter Update

Estimation of linear elastic material parameters is performed by minimizing the error in the constitutive relation. The cost functional for this updating step of material parameters is taken as

$$U(\mathbb{C}) = \frac{1}{2} \int_{\Omega^0} |\Sigma|_F^2 \, d\Omega^0 := \frac{1}{2} \int_{\Omega^0} |\sigma_S - \mathbb{C} : \epsilon(u_D)|_F^2 \, d\Omega^0. \quad (13)$$

Now, update equations of the material parameters are fount out by taking derivative of the above cost functional with respect to C_{ijkl} as

$$\frac{\partial U}{\partial C_{ijkl}} = \int_{\Omega^0} \frac{1}{2} \left(2\Sigma : \left[\frac{\partial \Sigma}{\partial C_{ijkl}} \right] \right) d\Omega^0 = \int_{\Omega^0} (\sigma_S - \mathbb{C} : \epsilon(u_D))^T : \frac{\partial \mathbb{C}}{\partial C_{ijkl}} : \epsilon(u_D) \, d\Omega^0. \quad (14)$$

It can be observed here that in contrast to the ECE cost functional, the above functional does not possess the normalization term $(\mathbb{C})^{-1}$. This facilitates to get the linear explicit update formula for anisotropic elastic material parameters.

A. Isotropic Material Model

In this section, we will develop the explicit update equations of isotropic material parameters of bulk modulus (K) and shear modulus (G). Now, the constitutive relationship for this isotropic material model can be written as

$$\sigma = (K - \frac{2G}{3})(tr\epsilon)I - 2G\epsilon. \quad (15)$$

Now, using Eq. (14) with respect to bulk and shear modulus, we get following two equations:

$$\int_{\Omega^0} \left(\Sigma : \left[\frac{d\Sigma}{dK} \right] \right) d\Omega^0 = \int_{\Omega^0} \left[(tr\epsilon) \left(\sigma - (K - \frac{2G}{3})(tr\epsilon)I - 2G\epsilon \right) : I \right] d\Omega^0 = 0$$

$$\int_{\Omega^0} \left(\Sigma : \left[\frac{d\Sigma}{dG} \right] \right) d\Omega^0 = \int_{\Omega^0} \left[2 \left(\sigma - (K - \frac{2G}{3})(tr\epsilon)I - 2G\epsilon \right) : (\frac{1}{3}tr\epsilon I - \epsilon) \right] d\Omega^0 = 0. \quad (16)$$

We get the following explicit update formulas for bulk and shear modulus by solving the above equations:

$$K = \int_{\Omega^0} \frac{tr(\sigma)\{3(tr\epsilon^2) - (tr\epsilon)^2\} + (tr\epsilon)(tr(\sigma\epsilon))\{(trI) - 3\}}{3(tr\epsilon)\{(tr\epsilon^2)(trI) - (tr\epsilon)^2\}} d\Omega^0 \quad (17a)$$

$$\mu = \int_{\Omega^0} \frac{1}{2} \frac{tr(\sigma\epsilon)(trI) - (tr\sigma)(tr\epsilon)}{(tr\epsilon^2)(trI) - (tr\epsilon)^2} d\Omega^0. \quad (17b)$$

In this case, only a single measured displacement field is needed for the update of bulk and shear modulus simultaneously.

3 Numerical Experiments

In this section, we will evaluate the numerical performance of the proposed inverse reconstruction method with measured displacement data in frequency domain. Here, we have synthetically generated the sparse noisy measured displacement data by first solving a forward problem in a finer mesh and then polluting it by artificial Gaussian random noise with zero mean and unit variance. In all the following numerical examples, inverse reconstruction of material parameters is executed element-wise. In each iteration of the proposed reconstruction technique, we update the parameters only when the positive-definiteness of the elasticity tensor keeps unchanged. By this way, we numerically maintain the positivity constraints of the material parameters.

3.1 Identification of Isotropic Parameters of 3D Cube

In this example, we have taken up the problem of identification of heterogeneous isotropic elastic material parameters of 3D solid cube. Each side of the cube is having a dimension of 1 m. The isotropic material properties of 3D solid cube are taken from the work by Fathi et al. [6]. One central spherical inclusion of 0.15 m radius is embedded in the solid cube. The elastic properties of the background and the central inclusion are given in Table 1. The mass density of the isotropic solid cube is taken as 2000 kg/m^3. The cube is fixed at the lower face (at y = 0 m). A pressure load of 1 kPa is applied at the top face (y = 1.0 m). Simultaneously, at the other remaining four sides of the cube, 0.5 KPa pressure is given to excite the bulk and shear response of the structure. One single 1 Hz (Table 2) is used here to generate the measured displacement data as well as for the inverse reconstruction scheme. We have considered a total of 132651 number of nodes and 125000 number of 3D brick elements for this cube problem. Thus, the total unknown material parameters become 250000. The reference and the reconstruction plots of spatial distributions of bulk modulus and shear modulus are shown in Figs. 1 and 2, respectively. It is obvious from these reconstruction figures that the proposed reconstruction technique is able to capture the inclusions qualitatively as well as quantitatively. The demarcation of the central inclusion can be clearly noticed in these figures. The number of iterations (n), the

Table 1 Material properties of 3D isotropic cube

Parameters	G (MPa)	K (MPa)
Background	80.0	133.0
Inclusion	160.0	267.0

Table 2 Numerical data for 3D isotropic cube

Loading frequency	δ	α_0	β_1	n
1.0 Hz	0%	1×10^7	0.25	17
	1%	1×10^5	0.025	30

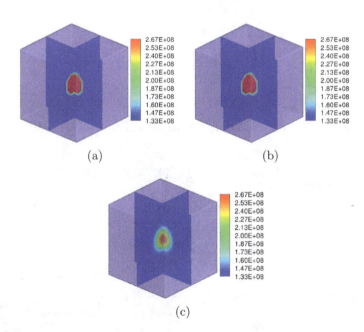

Fig. 1 Reconstruction of bulk modulus for 3D isotropic material model: **a** reference, **b** $\delta = 0\%$, and **c** $\delta = 1\%$

initial values (α_0), and multiplication factors (β_1) of the penalty parameter for noisy and noiseless cases are given in Table 2. Figure 3 shows the relative misfit plot for this problem (1% noisy case) which clearly depicts that the proposed inverse algorithm is successful in detection of inclusion of elastic isotropic material parameters within very less number of iterations.

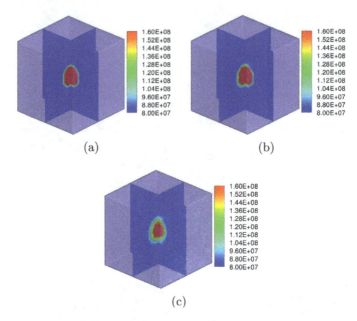

Fig. 2 Reconstruction of shear modulus for 3D isotropic material model: **a** reference, **b** $\delta = 0\%$, and **c** $\delta = 1\%$

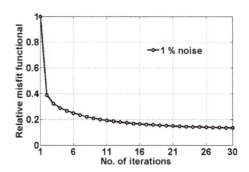

Fig. 3 Relative misfit plot of 3D isotropic cube problem

3.2 Parameter Identification of Isotropic Semi-circular Cylindrical Shell

In this problem, we have considered more complex geometry of semi-circular cylindrical shell for the identification of heterogeneous elastic material parameter. The geometry of semi-circular cylindrical shell (Fig. 4) is identical to the model used by Sze et al. [7] with the exception of the boundary conditions. In this damage, detection problem of shell structure one square damage (Fig. 5a) is introduced at the middle of the shell geometry. Damage is introduced by reducing the 25% of Young's modulus (E) keeping Poisson's ratio constant over the domain. The elastic properties of the

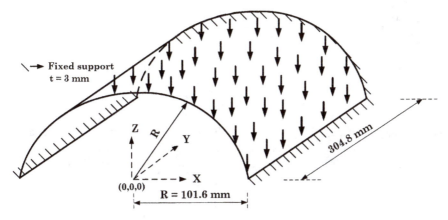

Fig. 4 Schematic of semi-circular cylindrical shell

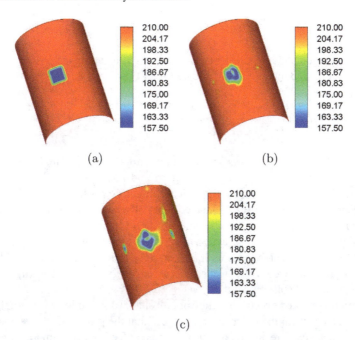

Fig. 5 Reconstruction of E for isotropic semi-circular cylindrical shell: **a** reference, **b** $\delta = 0\%$, and **c** $\delta = 1\%$

Table 3 Material properties of isotropic semi-circular cylindrical shell

Parameters	$E\ (GPa)$	ν
Background	210.0	0.3
Inclusion	157.50	0.3

background and damaged zone are given in Table 3. Now, the element-wise damaged modulus of elasticity can be represented as $E_d^e = d^e E^e$. The constitutive matrix of isoparametric degenerated shell structure in the damaged case can be presented as

$$C_{damaged} = d^e C^e = \frac{d^e E^e}{(1-\nu^2)} \begin{bmatrix} 1 & \nu & 0 & 0 & 0 \\ \nu & 1 & 0 & 0 & 0 \\ 0 & 0 & \frac{(1-\nu)}{2} & 0 & 0 \\ 0 & 0 & 0 & \frac{5(1-\nu)}{12} & 0 \\ 0 & 0 & 0 & 0 & \frac{5(1-\nu)}{12} \end{bmatrix}, \quad (18)$$

where d^e represents elemental damage index which is a damage indicator. We get the update of d^e and then E^e using update Eq. (14) as follows:

$$d^e = \frac{\{\sigma_i\}^T [C^e] \{\epsilon_i\}}{\{\epsilon_i\}^T [C^e]^T [C^e] \{\epsilon_i\}} \quad \text{and} \quad E^e_{updated} = d^e E^e. \quad (19)$$

As, only Young's modulus is updated here, only a single loading at a particular frequency is considered. The loading pattern and boundary conditions are given in Fig. 4. The system is uniformly loaded vertically at a frequency 20 Hz (Table 4) to generate the measured displacement data. The structure is modeled in FEM using four-noded degenerated isoparametric shell element having five degrees of freedom at each node. Reconstructions of Young's modulus is performed here for 60×60 FEM mesh which leads to 3600 material parameters as unknowns. For this numerical problem, we have considered only vertical component of displacement field as measured displacement data.

The reference and the reconstruction plots of spatial distributions of E are shown in Fig. 5. It is clear from the reconstruction figure that the proposed inverse technique is successful to capture the inclusion position perfectly. The demarcation of the inclusion also can be clearly observed in these figures.

The number of iterations (n), the initial values (α_0), and multiplication factors (β_1) of the penalty parameter for noisy and noiseless case are given in Table 4. Figure 6 shows the relative misfit plot for this problem. It can be observed from this figure that the proposed algorithm is successful in heterogeneous reconstruction of isotropic material parameters within a very few number of iterations.

Table 4 Numerical data for semi-circular cylindrical shell problem

Loading frequency	δ	α_0	β_1	n
20.0 Hz	0%	1×10^5	1.0	11
	1%	1×10^2	0.079	19

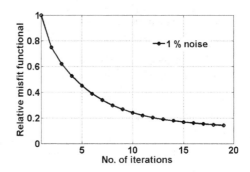

Fig. 6 Relative misfit plot of isotropic semi-circular cylindrical shell problem

4 Conclusion

We have developed here a variant of ECE-based material parameter estimation technique for linear elastic structure. In this proposed technique, material parameters are inversely identified by minimizing the cost functional which is the measure of error in constitutive relation due to two differently admissible strain and stress fields. Strain field is kinematically admissible with the measurement data and stress field is dynamically admissible. Strain field is generated with the weak imposition of the sparse noisy measured displacement data by the penalization technique. Penalization parameter plays an important role here to tackle the ill-posedness of the inverse problem. Stress field is calculated solving a static problem. So, in the proposed method, incompatible stress and strain fields are generated in uncoupled way which is in contrast to standard MECE technique. Thus, the numerical computational cost of the present method reduces compared to MECE method. Afterward, material parameters are updated with linear explicit update formulas. In numerical examples, we have done the experimentation on heterogeneous isotropic material parameter identification of 3D structures especially of circular cylindrical shell. The limited numerical experimentation shows that the proposed method can effectively identify the distribution of elastic material parameters in a few number of iterations. The present method can be utilized in large-scale elastic parameter estimation problem because of its low computational cost.

References

1. Feissel P, Allix O (2007) Modified constitutive relation error identification strategy for transient dynamics with corrupted data: the elastic case. Appl Mech Eng 196:1968–1983. https://doi.org/10.1016/j.cma.2006.10.005
2. Ladevèze P, Leguillon D (1983) Error estimate procedure in the finite element method and applications. SIAM J Numer Anal 20:485–509. https://doi.org/10.1137/0720033
3. Banerjee B, Walsh TF, Aquino W, Bonnet M (2013) Large scale parameter estimation problems in frequency-domain elastodynamics using an error in constitutive equation functional. Comput Meth applied mechanics and engineering. 253:60–72. https://doi.org/10.1016/j.cma.2012.08.023
4. Guchhait S, Banerjee B (2016) Anisotropic linear elastic parameter estimation using error in the constitutive equation functional. Proc R Soc A: Math Phys Eng Sci 472:20160213. https://doi.org/10.1098/rspa.2016.0213
5. Guchhait S, Banerjee B (2018) Constitutive error based parameter estimation technique for plate structures using free vibration signatures. J Sound Vib 419:302–317. https://doi.org/10.1016/j.jsv.2018.01.020
6. Fathi A, Loukas FK, Babak P (2015) Full-waveform inversion in three-dimensional PML-truncated elastic media. Comput Methods Appl Mech Eng 296:39–72. https://doi.org/10.1016/j.cma.2015.07.008
7. Sze KY, Liu XH, Lo SH (2004) Popular benchmark problems for geometric nonlinear analysis of shells. Finite Elem Anal Des 40:1551–1569. https://doi.org/10.1016/j.finel.2003.11.001

A Novel Sloshing Damper for Vibration Control of Short Period Structures

Anuja Roy, Atanu Sahu, and Debasish Bandyopadhyay

Abstract Tuned sloshing damper is a well-established and popular passive control device for vibration mitigation of structures subjected to excitations such as wind, earthquake, wave, etc. However, being a long period system, the applicability of this type of control device is restricted to flexible structures. To overcome this, a new kind of translational spring-connected sloshing damper is introduced in the present research paper and its applicability to short period structures is explored. First, a time-domain formulation of a linear single-degree-of-freedom (SDOF) structure with spring-connected sloshing damper system is developed. A nonlinear model based on the shallow water wave theory is utilized for modelling the liquid motion in the sloshing damper. Further, a numerical study on the performance of the damper system attached to a short period structure subjected to harmonic input is carried out. The performance of the damper system is examined on the basis of reduction in the peak value of the structural displacement. The performance of the proposed damper system indicates that the spring-connected sloshing damper has great potential as a vibration mitigation device for short period structures.

Keywords Spring-connected sloshing damper · Vibration control · Short period structures · Passive control device

1 Introduction

Tuned sloshing damper (TSD), being cost-effective and low-maintenance dynamic vibration absorber, is a well-established and popular passive control device for vibration mitigation of lightly damped structures subjected to excitations such as wind,

---

A. Roy (✉) · D. Bandyopadhyay
Jadavpur University, Salt Lake Campus, LB-8, Sector - III, Kolkata 700098, India
e-mail: anuja.civil@gmail.com

A. Sahu
National Institute of Technology, Silchar, Cachar, Assam 788010, India

© The Editor(s) (if applicable) and The Author(s), under exclusive license to Springer Nature Singapore Pte Ltd. 2021
S. K. Saha and M. Mukherjee (eds.), *Recent Advances in Computational Mechanics and Simulations*, Lecture Notes in Civil Engineering 103,
https://doi.org/10.1007/978-981-15-8138-0_26

earthquake, wave, pedestrian-induced vibration, etc. The TSD offers certain advantages over the more common tuned mass damper (TMD) such as easy installation especially in case of existing structures, easy adjustment of natural frequency, efficiency even in case of low-amplitude vibrations, elimination of low friction bearing surface, non-requirement of threshold level excitation, etc.

TSDs have already been successfully installed in tall airport towers, commercial towers, skyscrapers and bridges [1–4] due to its efficiency in controlling unwanted vibration in flexible structural systems. TSD itself is a long period system and consequently it is suitable for such flexible structures. However, the use of TSD as an effective passive control device is not viable for relatively short period structures. This is due to the constraint in obtaining a feasible configuration of TSD from the criteria of tuning the damper frequency to the structural frequency.

Different numerical models based on shallow water wave theory have been proposed by different researchers along with the methods for solving such equations following Lax Finite Difference Method, Random Choice method, etc. [5–7]. Experimental studies have also been conducted to validate the numerical models of TSD [4, 8–10]. However, the use of such TSDs is reliable to be used in relatively flexible systems having the natural time period more than 1 s. A very limited number of research works are available on the modelling of TSDs having nonconventional arrangements. A numerical model was proposed by Lu et al. [5] where the TSD is subjected to the rotational motion in addition to the horizontal excitation. Following this, Samanta and Banerji [10] introduced a rotational spring at the bottom of a liquid damper to increase its effectiveness employing the idea of matching the phase of lateral and rotational motions. However, fixing a rotational spring at the bottom of a TSD and maintaining its stability in operating condition is often a difficult task. From these perspectives, a translational spring-connected sloshing damper system is proposed in the present research paper for vibration control of short period structures which are also vulnerable to damage by external excitations. A time-domain formulation of a linear single-degree-of-freedom (SDOF) structure with the proposed damper system is developed and presented. To model the motion of the sloshing liquid in the damper, a detailed nonlinear modelling based on the shallow water wave theory proposed by Sun et al. [7] is used. Numerical simulations are carried out to show the effectiveness of the proposed damper system connected to a stiff (short period) structure subjected to harmonic base input excitation. The reductions in the peak and RMS values of the structural displacement are considered as the basis for investigating the performance of the damper system. Further, an attempt is made to study the sensitivity of the performance of the damper system due to variation in its tuning ratio.

2 Modelling of Structure with Damper and Equations of Motion

The model of a spring-connected sloshing damper-SDOF structural system subjected to a horizontal base acceleration $\ddot{z}(t)$ is shown in Fig. 1. The mass, stiffness and damping of the linear SDOF structural system are denoted by m_s, k_s and c_s, respectively. ζ_s and $\omega_s \left[= \sqrt{k_s/m_s} \right]$ represent the damping ratio and the natural frequency of the same, respectively. A detailed nonlinear model of liquid sloshing, based on the established shallow water wave theory, originally proposed by Sun et al. [7] is used to model the liquid motion in the damper. This model is briefly reproduced here. The sloshing damper is a rigid rectangular tank of length $2a$, width b and height H, partly filled with liquid of density ρ and kinematic viscosity ν having a free upper surface. The undisturbed liquid depth is h. The liquid is assumed as incompressible and irrotational. The liquid pressure remains constant at the liquid free surface. The liquid particle motion is presumed to develop only in the vertical plane along the direction of excitation. The influence of internal friction between the fluid particles is significant only in the boundary layer near the solid boundaries for low-viscosity liquids. As regard to sloshing of liquid, the effect of wave breaking is taken into consideration. The sloshing damper is connected to the structure by a translational spring element. The stiffness and damping of this element are denoted by k_d and c_d, respectively. The mass of the damper system, inclusive of the liquid mass and the container mass of the sloshing damper, is denoted by m_d. The frequency of the

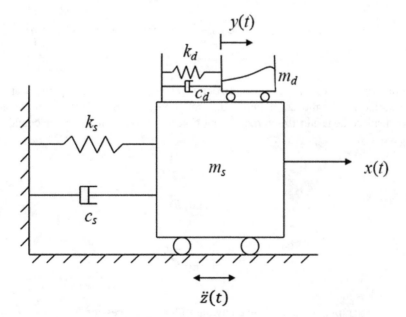

Fig. 1 Schematic diagram of the proposed damper system attached to the structure

combined damper system is estimated assuming the liquid in the container to be relatively still, i.e. when, in fact, the device represents a mass damper.

Considering the dynamic equilibrium of the spring-connected sloshing damper-SDOF structural system, subjected to horizontal base acceleration, $\ddot{z}(t)$, the following equation of motion normalized with respect to the structural mass is obtained:

$$\{\ddot{x}(t) + \ddot{z}(t)\} + 2\zeta_s \omega_s \dot{x}(t) + \omega_s^2 x(t) = 2\zeta_d \omega_d \mu_d \dot{y}(t) + \omega_d^2 \mu_d y(t) \quad (1)$$

where $c_d/m_d = 2\zeta_d \omega_d$, $k_d/m_d = \omega_d^2$ and μ_d is the ratio of the damper mass m_d to structural mass m_s.

The normalized equation of motion of the spring-connected sloshing damper system is

$$\ddot{y}(t) + \ddot{x}(t) + \ddot{z}(t) + 2\zeta_d \omega_d \dot{y}(t) + \omega_d^2 y(t) + \frac{F(t)}{m_d} = 0 \quad (2)$$

Here, $F(t)$ is the horizontal force arising out of the hydrostatic pressure on left and right walls of the damper container.

The governing equations describing the liquid motion, obtained from continuity equation and Navier–Stoke's equation, are as follows:

$$\frac{\partial \eta}{\partial t} + h\sigma \frac{\partial [\phi u(\eta)]}{\partial x} = 0 \quad (3)$$

$$\frac{\partial u(\eta)}{\partial t} + (1 - T_H^2) u(\eta) \frac{\partial u(\eta)}{\partial x} + C_{fr}^2 g \frac{\partial \eta}{\partial x} + gh\sigma\phi \frac{\partial^2 \eta}{\partial x^2} \frac{\partial \eta}{\partial x} = -C_{da} \lambda u(\eta) - \ddot{y} \quad (4)$$

where $u(\eta)[= u(x, \eta, t)]$ is the horizontal velocity of the liquid particle at the free surface, g is the acceleration due to gravity, y and $\ddot{y}$ are the horizontal base displacement and the horizontal base acceleration of the damper container, respectively.

$$T_H = \tanh[k(h + \eta)] \quad (5)$$

$$\sigma = \frac{\tanh(kh)}{kh} \quad (6)$$

$$\phi = \frac{\tanh[k(h + \eta)]}{\tanh(kh)} \quad (7)$$

Here, k is the wave number.

λ is the damping parameter given by the following expression:

$$\lambda = \frac{1}{(\eta + h)} \frac{1}{\sqrt{2}} \sqrt{(\omega_l \nu)} \left[1 + \left(\frac{2h}{b}\right) + S \right] \qquad (8)$$

S is the surface contamination factor. ω_l is the fundamental linear sloshing frequency of the water in the TLD tank given by Lamb [11]

$$\omega_l = \sqrt{\left\{ \frac{\pi g}{2a} \tanh(\pi \Delta) \right\}} \qquad (9)$$

where $\Delta [= h/2a]$ is the liquid depth ratio. The wave breaking coefficients C_{fr} and C_{da} are given by the following expressions:

$$C_{fr} = 1.05 \qquad (10)$$

$$C_{da} = 0.57 \sqrt{\left[\frac{2 \Delta h \omega_l}{\nu} (y)_{max} \right]} \qquad (11)$$

The horizontal force due to hydrostatic pressure coming from sloshing damper is the difference between the hydrostatic pressure on left and right walls of the container:

$$F(t) = \frac{\rho g b}{2} \left[(\eta_r(t) + h)^2 - (\eta_l(t) + h)^2 \right] \qquad (12)$$

where $\eta_r(t)$ and $\eta_l(t)$ are the free surface elevations at the right and left walls of the damper container, respectively.

3 Numerical Simulation Procedure

Equations (1)–(4) need to be solved simultaneously to obtain the controlled structural response. For this, Eqs. (3) and (4) are discretized with respect to x into difference equations. The Runge–Kutta–Gill method is used to solve the difference equations for obtaining the horizontal velocity of the liquid $u(t)$ and the liquid free surface elevation $\eta(t)$. The resulting horizontal force from the damper container due to the hydrostatic pressure is then computed using Eq. (12). The same numerical technique, i.e. Runge–Kutta–Gill method, is employed to calculate the displacement responses of the structure and the damper system in time domain.

4 Numerical Results

An example structure having short time period is considered in the present numerical investigation to examine the performance of the damper system for the passive control of short period structures. The natural time period of the example structure is 0.9 s. The value of the structural damping, ζ_s, is considered as 1%. The value of mass ratio, μ_d, is assumed to be 1%, i.e. 0.01. The tuning ratio of the damper system, γ, is taken as 0.99 following Den Hartog's solution for optimum tuning ratio of tuned mass damper [2]. The value of k_d is considered as 240 N/m and c_d is taken as zero for the present simulation. Assuming a liquid depth ratio of 0.05 and considering the above-mentioned value of mass ratio, the length of the damper container is obtained. The mass of the damper container is assumed to be same as the liquid mass. Water is used as the damper liquid. The input, given to the structure, is harmonic. The amplitude of the harmonic input is 0.03 m/s^2. The input frequency is equal to the structural frequency, i.e. 6.981 rad/s.

The performance criteria of the damper system are the reductions in the peak and RMS values of the structural displacement. The proposed damper system has been validated considering infinite rigidity of the translational spring, where it will act as the traditional TSD [9]. A numerical illustration of the response reductions achieved by the damper system is presented in Table 1. It is evident from Table 1 that substantial reduction has been achieved by the proposed damper system. The sample time histories of the uncontrolled and controlled structural displacements for 1% mass ratio are presented in Fig. 2.

An investigation on the sensitivity of the control performance of the damper system for the example short period structure to tuning ratio is also carried out. In order to do this, the frequency of the damper system is varied over a range and the RMS structural displacements are obtained. In this process, the stiffness of the connecting spring is kept constant while the mass of the damper system is varied. The liquid depth ratio is kept constant at a value of 0.05 and the variation in the mass of the damper system is obtained by changing the length of the damper container. The range of tuning ratio considered is from 0.8 to 1.25. This study is conducted for the same amplitude of harmonic input, as used before, i.e. 0.03 m/s^2. Results for the RMS structural displacement reductions achieved by the proposed damper system

Table 1 Percent response reduction by the damper system

Input	Mass ratio (%)	Tuning ratio	Percentage reduction in structural displacement	
			Peak	RMS
Harmonic	1	0.99	78.58	86.70
	2		84.25	90.40
	3		85.31	92.63

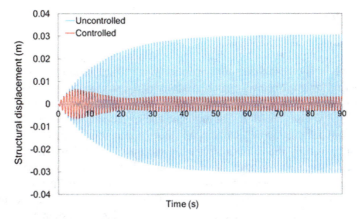

Fig. 2 Displacement time history of the structure to harmonic input of amplitude $= 0.03$ m/s^2 for $\gamma = 0.99$ and $\mu = 1\%$

for varying tuning ratio are presented in Fig. 3. It is observed that the performance of the damper system is extremely sensitive to tuning ratio. The damper system achieves very high response reductions near optimal tuning ratio which is obtained as 1.03, close to unity, in the present case.

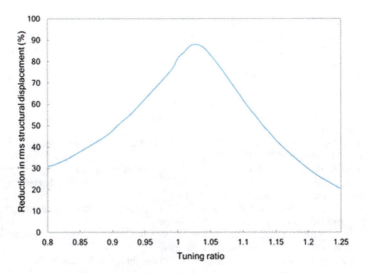

Fig. 3 Sensitivity of damper performance to tuning ratio for harmonic input of amplitude $= 0.03$ m/s^2

5 Conclusion

In this work, a novel translational spring-connected sloshing damper is introduced and the applicability of this damper system as an effective vibration control device for a short period structure is investigated. An attempt is also made to study the sensitivity of the performance of the damper system to tuning ratio. The performance of the damper system is evaluated in time domain. The results in time-domain study show that substantial reduction of 78% in peak value of structural displacement response may be achieved by the damper system in the tuned condition. The reduction in case of RMS structural displacement as obtained is 86% (For 1% mass ratio). The performance of the damper system is observed to be highly sensitive to tuning ratio and the optimal tuning ratio is obtained as 1.03 for the case under consideration. To conclude, the numerical study indicates that this new damper system can be used as a potential control device for the vibration suppression of stiff structures.

Acknowledgment The financial support provided by RUSA 2.0 Scheme at Jadavpur University during this research is gratefully acknowledged by the first author (A. Roy).

References

1. Tamura Y, Fujii K, Ohtsuki T, Wakahara T, Kohsaka R (1995) Effectiveness of tuned liquid dampers under wind excitation. Eng Struct 17(9):609–621
2. Soong TT, Dargush GF (1997) Passive energy dissipation systems in structural engineering. Wiley, New York
3. Nakamura SI, Fujino Y (2002) Lateral vibration on a pedestrian cable-stayed bridge. Struct Eng Int 4:295–300
4. Malekghasemi H, Ashasi-Sorkhabi A, Ghaemmaghami AR, Mercan O (2015) Experimental and numerical investigations of the dynamic interaction of tuned liquid damper–structure systems. J Vibr Control 21(15):2707–2720
5. Lu ML, Popplewell N, Shah AH, Chan JK (2004) Nutation damper undergoing a coupled motion. Modal Anal 10(9):1313–1334
6. Samanta A, Banerji P (2006) Efficient numerical schemes to analyse earthquake response of structures with tuned liquid dampers. In: 13th symposium on earthquake engineering. Indian Institute of Technology Roorkee, pp 1372–1381
7. Sun LM, Fujino Y, Pacheco BM, Chaiseri P (1992) Modelling of tuned liquid damper (TLD). J Wind Eng Indus Aerodyn 41–44, 1883–1894
8. Chavan SA (2002) Tuned liquid damper for structural control: experimental studies and numerical analysis. M. Tech Project, IIT, Bombay
9. Roy A, Zhang Z, Ghosh AD, Basu B (2019) On the nonlinear performance of a tuned sloshing damper under small amplitude excitation. J Vibr Control 1–11. https://doi.org/10.1177/1077546319867232
10. Samanta A, Banerji P (2008) Structural control using modified tuned liquid dampers. In: 14th world conference on earthquake engineering, pp. 1–8, Beijing, China
11. Lamb H (1932) Hydrodynamics, 6th edn. Cambridge University Press, London

Modeling and Simulation in Earthquake Engineering

Fracture Mechanics Based Unilateral and Bilateral Earthquake Simulations: Application to Cable-Stayed Bridge Response

K. S. K. Karthik Reddy and Surendra Nadh Somala

Abstract The phenomena of forward-directivity effects cause pulse-type earthquake ground motions that result in significant damage to structures. Forward directivity ground motions can be facilitated by typically simulating unilateral ruptures and occasionally by bilateral ruptures. Traditional analysis methods do not employ the dynamics of fault rupture hence are inadequate to capture the full effects of these pulse-type ground motions. Computational seismology overcomes this limitation and plays an important role to simulate dynamic earthquake ruptures. The objective of this paper is to use an open-source code SPECFEM3D to generate synthetic field vector data to improve the understanding of pulse-type ground motions generated using dynamic simulations. The software was used to generate synthetic earthquakes of moment magnitude, $M_w = 7$ with a strike-slip mechanism. Two cases were considered with nucleation at the end and in the center of the fault to generate unilateral and bilateral ruptures. The generated ground motions are then interpreted to comprehend the concept of directivity. Later, the seismic response of the bridge is evaluated for selected stations around the fault. The behavior of the bridge in terms of displacement field is evaluated which showed a similar response for stations located at a distance of 2 and 10 km in front of the fault. Further inference of bridge response is drawn by comparing the Fourier amplitude spectrum of velocities at these particular stations. The peak amplitude frequencies of the velocity fields at these stations lie in the regime of natural frequencies of the bridge which caused it to resonate in turn exhibiting high displacements at stations in front of the fault.

Keywords Directivity ground motions · Unilateral and bilateral ruptures · Effect on suspension bridge · Dynamic rupture · SPECFEM3D · Location of asperity

---

K. S. K. Karthik Reddy (✉) · S. N. Somala
Indian Institute of Technology, Hyderabad, India
e-mail: ce17resch01003@iith.ac.in

© The Editor(s) (if applicable) and The Author(s), under exclusive license to Springer Nature Singapore Pte Ltd. 2021
S. K. Saha and M. Mukherjee (eds.), *Recent Advances in Computational Mechanics and Simulations*, Lecture Notes in Civil Engineering 103,
https://doi.org/10.1007/978-981-15-8138-0_27

1 Introduction

Amid all-natural calamities, earthquakes are causing most of the socio-economic loss around the globe due to their unpredictable nature of occurrence and propagation. Though the earthquakes are inevitable, their effects can be minimized through a better understanding of their characteristics such as faulting style, magnitude, the zone of influence, the activity of plate movement, and soil characteristics. Most of the earthquakes are of tectonic origin and takes place in the faults.

The motion of the fault occurs when the shear stress on the fault plane overcomes the strength of the material or the friction which holds the ruptured surface together. The total energy is dissipated into the surrounding medium in the form of elastic waves, called seismic energy [12]. The response of civil infrastructure is regulated in terms of response spectrum where the ground motion is converted to response spectra and an equivalent force and is applied to the structure.

Major modification in response spectra is suggested while considering the importance of the building and the type of soil profile but minute care is taken to account for directivity ground motions. Directivity is a consequence of a fault rupturing wherein earthquake ground motion in the direction of rupture propagation is more severe than that in other directions from the earthquake source [13].

2 Characteristics of Near Field Ground Motions and Its Influence on Cable-Stayed Bridges

The evidence of directivity ground motions was observed for 1994 Northridge, California, 1999 Chi-Chi, Taiwan, and 1999 Kocaeli and Duzce, Turkey, etc. The near field stations which are located in the direction of fault rupture and slip are classified as forward directivity stations [2].

The peculiar characteristics of these ground motions can be identified by exploring the velocity-time histories which are associated with large amplitude and long period [3]. The peaks in response spectra are not concentrated at a lower time period for directivity ground but might shift to higher time periods >0.6 s depending on the magnitude of the earthquakes [8]. It is also recommended to understand the manifestation of backward directivity, which is inconsistent with the phenomena of forward directivity, wherein the velocity pulses are less significant though the backward directivity station is located at an identical distance like forward directivity station. Hence the response spectra need to polarize with both amplification and de-amplification factors for forward and backward stations correspondingly [7].

In the case of unilateral ruptures, the conflicting behaviour of backward and forward directivity ground can be sensed, whereas in the case of bilateral ruptures both stations represent similar field vectors. The reason for the conflicting behaviour of field vectors at stations in front and behind the fault for unilateral ruptures is a result of positing the asperity at one end of the fault and allowing the rupture propagation only in one direction and arresting the rupture propagation in the opposite direction. The velocity pulses resulting from rupture propagation superimpose and give rise to a single velocity pulse of high amplitude and time period. In case of a bilateral rupture, the position of asperity is located at the centre of the fault giving liberty for the rupture to propagate in both the direction affecting the stations in front and behind the faults in the same manner.

To evaluate the aggressive impact of directivity ground motions on structures with the high time period the sensitivity of cable-stayed bridges to large amplitude oscillations was studied by [14] through shake table tests on Puqian cable-stayed bridge, China. The replicated prototype model is reduced by a scale of 20 and is subjected to near field ground motions from 1999 Chi-Chi, Taiwan earthquake. the finding through shake table tests revealed the severe damage by the near field ground motions to the tower components. A similar study has been carried out [9] through modeling of Fatih Sultan Mehmet Bridge, Istanbul in MULSAP software, the behavior of the bridge was gauged through bending moment and displacements along the length of the bridge for various near and far-field ground motions depicting the vigorous reaction of the bridge to near field ground motions. Li et al. [5] proposed a new method, record decomposition incorporation (RBI) to mimic artificial near-fault pulse-type ground motions using simple equivalent pulses, the validation of the fitted pulses can be supported by the similar response of Sutong cable-stayed bridge, China modeled in ANSYS to observed time history.

3 Fracture Mechanics Approach to Earthquake Simulations

The ground motion of earthquakes is forecasted by physics-based ground motion simulations using numerical methods by assimilating the physics of the earthquake source and the out coming propagation of seismic waves. Stochastic simulation methods combine the physics and kinematics of the earthquake source employing indirect approaches. These approaches signify the physics of earthquakes but do not essentially solve the accepted mathematical concepts that describe the physics of source dynamics and wave propagation [11]. The finite-difference (FD) was first introduced by [1] in seismology. These methods were also used to describe the propagation of waves in stratified media and sedimentary basins to assess ground-motion amplification [7]. The initial 3D simulations were performed by [4] who simulated and valid M_w 4.4 aftershock of the 1989 Loma Prieta, California, earthquake using a point source model, with an increase in the computational facility, vigorous models were created and simulated in the spectrum of obligatory frequencies [6, 10].

4 Methodology

4.1 Elasto-Dynamic Equation

Earthquake ground-motion simulation involves obtaining the solution of the linear momentum equation, which can be written in Cartesian coordinates and indicial notation as

$$\frac{\partial \tau_{ij}}{\partial x_j} dv + F_i = \rho \frac{\partial^2 u_i}{\partial t^2} \tag{1}$$

τ_{ij} represents the Cauchy stress tensor, ρ is the mass density, and F_i and u_i are the body forces and displacements in the ith direction within a bounded domain. In earthquake simulations, our objective is to calculate ground displacements varying with time due to conditions on the fault plane which can be done numerically with the help of the representation theorem.

4.2 Dynamic Rupture Modeling Using the Spectral Element Method (SEM)

Dynamic rupture modeling is done using the SPECFEM3D package. It uses SEM, which is a higher-order Finite Element Method (FEM). SEM can handle complex geometries like intermittent faults, heterogeneous 3D velocity structure and it gives accurate results with good convergence rate in case of seismic wave propagations. It provides the flexibility to incorporate realistic boundary conditions (absorbing and free surface) in the simulation model.

In our model, we consider two layers for modeling the fault surface. Let the two layers be $\Sigma+$ and $\Sigma-$ as shown in the Fig. 1 are fault surfaces which move in opposite directions because of stress Ti acting on surfaces, having displacements U+ and U− respectively. Slip (s) or discontinuity at a location (say origin) on fault will be $s = $ U+U−

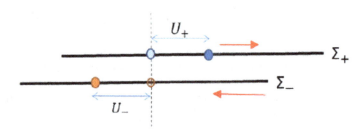

Fig. 1 Representing fault plane Σ as a double-layered surface

In Eq. (1) we can neglect body forces, then we can rewrite it in the form of Eq. (2), and then a weak form can be obtained. We solve the weak form of Eq. (2), with initial conditions on fault and Eqs. (3)–(5) as boundary conditions.

$$\rho \frac{\partial^2 u_i}{\partial t^2} - \frac{\partial \tau_{ij}}{\partial x_j} = 0 \quad (2)$$

Any point on fault plane stress cannot exceed the strength of fault, which is expressed numerically as Eq. (3), where T^T is tangential traction, T^N is normal traction, μ is the static frictional coefficient as defined in slip weakening friction law.

$$|T^T| - \mu |T^T| \leq 0 \quad (3)$$

The slip will be zero until stress in material reaches yield, and once slip starts left-hand side of Eq. (3) will be 0, and slip should always be in the direction of tangential traction, as it is the driving force for rupture, giving rise to two more conditions

$$|\dot{s}|(|T^T| - \mu |T^N|) = 0 \quad (4)$$

$$\dot{s}|T^T| - |\dot{s}|T^T = 0$$

The fault geometry and earth domain are discretized into small hexahedral elements using the Trelis package. The global matrix equation obtained after the discretization of weak form. Lagrange's functions are used as shape functions for u_i, Gauss-Lobatto-Lagrange (GLL) quadrature is used to integrate the weak-form over the elemental domain. By using GLL quadrature there will be only diagonal terms in the mass matrix, so calculating the inverse of the mass matrix becomes very simple efficient, reducing memory requirements for simulation significantly.

$$M\ddot{u} + C\dot{u} + Ku = B\tau \quad (5)$$

where u is a matrix representing unknown displacements at all discrete points in mesh, M is a mass matrix, C is viscosity matrix assuming Kevin-Voigt viscosity, K is stiffness matrix, B is a matrix representing boundary condition of fault, τ is relative traction.

5 Modeling and Simulation

The Physics-Based Simulation follows a workflow, initially, a region of interest for simulation is selected. The region contains topographical features and fault positioning which is followed by the selection of source rupture parameters that highlights the friction parameters, stress drop in asperity as well as background stresses on the fault. The material properties of the domain like the profile of p, s wave velocities and densities are defined for the domain. These parameters together are called as modeling parameters. Meshing schemes are applied to the domain, the size of the mesh typically depends on the maximum resolving frequency of interest, one can obtain higher resolving frequencies with finer mesh, the mesh size also depends on minimum velocity of the wave, the lower the velocity the finer the mesh should be for particular resolving frequency in the model. The simulation is performed with the help of high-performance computing (HPC).

A three-dimensional vertical strike-slip fault is modeled in a half-space. Earth's surface with a size of 75 by 150 km lies in the XY plane. Fault Plane lies in the XZ plane. Figure 2a and b shows the geometry of the fault plane and earth domain considered for simulations. Rupture is allowed to propagate and is limited to the fault of size 30 km × 15 km. Rupture initiates in a 3 km × 3 km patch (asperity).

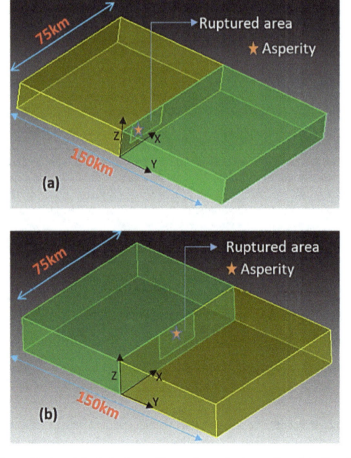

Fig. 2 The positioning of the fault along with asperity to simulate. **a** Case 1 (unilateral rupture) and **b** Case 2 (bilateral rupture)

Figure 2a depicts Case 1 (unilateral rupture) where rupture initiation is at the end and Fig. 2b depicts Case 2 (bilateral) where rupture initiates at the center of the fault. In this nucleation patch, we give high shear stresses to begin fracture in fault plane. A similar description of the cross-section along the fault plane is shown in Fig. 3a and b to enhance understanding of both cases.

Both the simulations are done by using alike dimensions and orientation of fault plane and earth domain. Similar material properties and friction law parameters (critical slip weakening distance, dynamic and static friction coefficients tabularized in Table 1 is engaged in both cases. To learn the directivity effect 4 stations were considered as shown in Fig. 4. The depiction of the 4 stations are categorized as station in front of the fault located at a distance of 2 km from the fault tip (S1-red), station away from the fault located at a distance of 10 km from the fault tip (S2-orange), station located behind the fault (S3-green) and station located perpendicular to the center of the fault (S4-yellow).

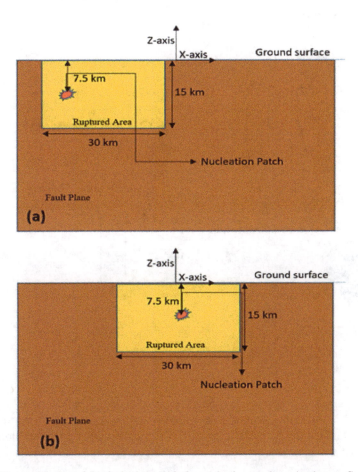

Fig. 3 The positioning of the fault along with the nucleation patch to simulate. **a** Case 1 and **b** Case 2

Table 1 The simulation parameters of the fault in both cases were tabulated

Fault Geometry	
Length of fault	30 km
Width of fault	15 km
Spectral element size	400 m
Density (ρ)	2670 kg/m^3
Shear wave velocity (c_s)	3464 m/s
Friction Parameters in the rupture zone	
Slip weakening distance (D_c)	0.4 m
Dynamic frictional stress (σ_k)	63 MPa
Static Frictional Stress (σ_s)	81.24 MPa
Nucleation patch size	3 km
Static frictional stress ($\sigma_{s,\,nuc}$)	81.24 MPa
Initial shear Stress ($\sigma_{in,\,nuc}$)	81.6 MPa

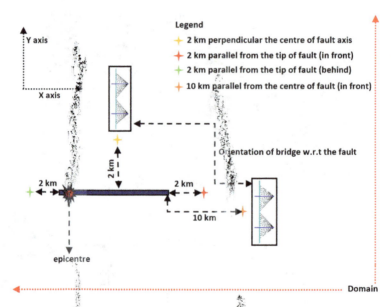

Fig. 4 The orientation of stations around the fault to study Case 1 earthquake, similarities can be drawn for Case 2

The orientation of stations around the fault is similar in both cases taking fault position as a benchmark. Four stations as shown in Fig. 4 were considered to collect the acceleration time history, these accelerations time histories of these stations were considered as inputs to the bridge model to perform the time history analysis.

Table 2 The geometry of the cable-stayed bridge considered

Length of the middle span	200 m
Length of side spans	100 m
Height of the pier above bridge deck	80 m
Height of the pier below bridge deck	60 m
Spacing between cables	10 m

Table 3 Material properties of the bridge model

Structural components	Materials	Yield Stress (MPa)	Density (kg/m^3)
Deck	Concrete	60	2600
	Steel	415	7850
Towers	Concrete	60	2600
	Steel	415	7850
Cables	Steel	1860	7850

The geometry and material properties of the cable-stayed bridge considered are tabularized in Table 2 and Table 3 respectively. Figure 5 shows the 3-D model of the cable-stayed modeled using SAP-2000 software. Table 4 highlights the natural time periods and the corresponding frequencies of the bridge in the 1st and 10th modes.

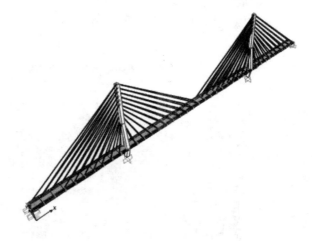

Fig. 5 Cable-stayed bridge modeled in SAP-2000

Table 4 Range of natural frequencies for modeled cable-stayed bridge

Natural time period (s)	Natural Frequency (f)
12.5 (1st mode)	0.08
1.68 (10th mode)	0.59

6 Results and Discussions

A comparative study is performed among the four stations for cases. For the ease of explanation, the nomenclatures of S1, S2, S3 and S4 were adapted for stations in front, far from, behind and perpendicular to the fault respectively. Case 1 and Case 2 nomenclatures were adopted to describe the initiation of rupture propagation at the end and in the middle of the fault correspondingly. Figure 6 project the PGA contour maps for Case 1 and Case 2, it can be observed that the rupture propagation behind the fault plane in both the cases is arrested, which implies that if the asperity is located at the end of the fault, it generates unilateral rupture which is the case in Case 1 and if the asperity is located in the middle of the fault Case 2 exhibits bilateral rupture.

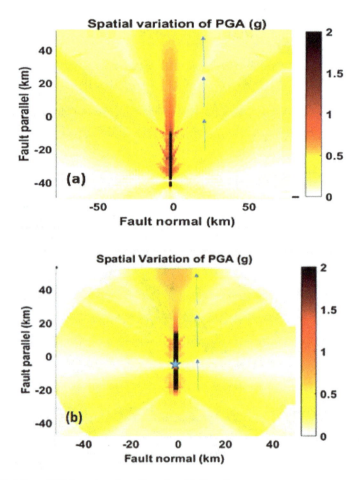

Fig. 6 Variation of PGA contour for **a** Case 1 and **b** Case 2

Figure 7a and b displays the acceleration time history of the 4 stations for both cases, in the instance of Case 1, station S1 experienced maximum PGA of 2.3 g but in Case 2 where rupture begins in the middle, station S2 experienced a higher PGA of 3 g, perceiving the fact that superposition of waves might have occurred at the station S2 for Case 2. Figure 8a and b presents the pseudo acceleration spectrums where station S1 obtained a PSA greater than 4 g for Case 1 event and S2 obtained a PSA above 6 g for Case 2 events. To investigate the response of structures of high time

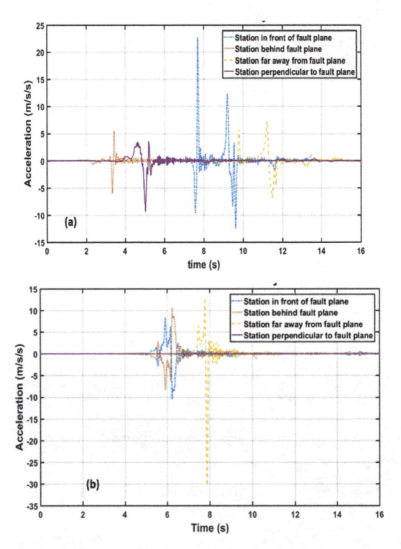

Fig. 7 Acceleration time history of selected 4 stations for **a** Case 1 and **b** Case 2 ruptures

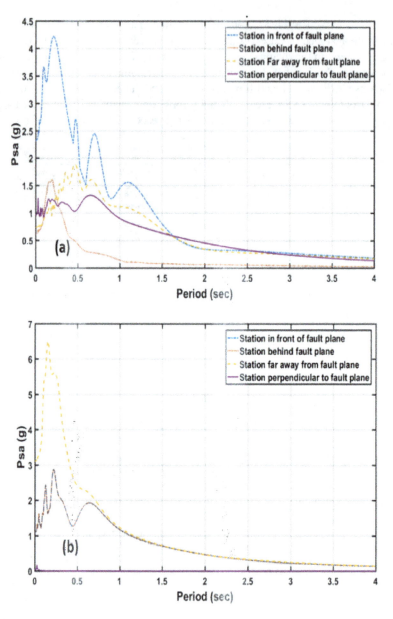

Fig. 8 Pseudo-spectral acceleration of selected 4 stations for **a** Case 1 and **b** Case 2 ruptures

periods a cable-stayed bridge was modeled using SAP-2000 and the time history of the stations for both the cases was given as input.

Figure 9a and b shows the variation of displacements of piers and Fig. 10a and b shows the variation of deck slab displacement for all the stations for both the cases. For Case 1 the side span displacements of the bridge are similar for stations S1 and S2. In the instance of Case 2, the displacements were high for station S1 and S4 citing the fact that for bilateral rupture earthquake i.e. nucleation at the center of the fault, stations in front and behind the fault prone the bridge to similar behavior.

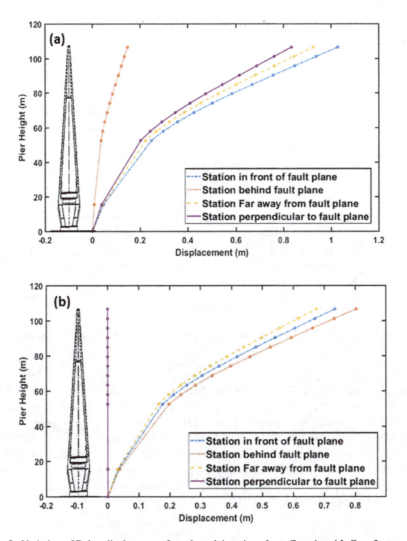

Fig. 9 Variation of Pylon displacement for selected 4 stations for **a** Case 1 and **b** Case 2 ruptures

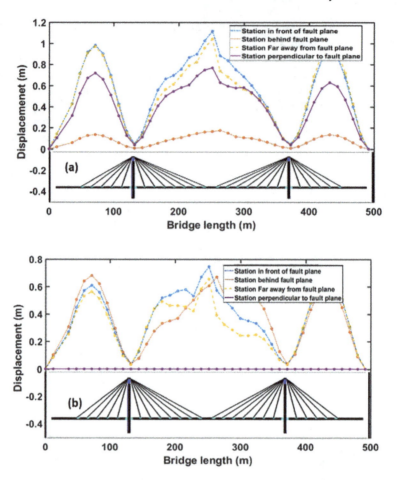

Fig. 10 Variation of bridge deck displacement for selected 4 stations for **a** Case 1 and **b** Case 2 ruptures

Further investigation on why the bridge deck reacted similarly for stations in front and far away from the fault for Case 1 can be argued by the frequency content of the station's velocity time history. Figure 11a and b shows the variation of Fourier amplitude acceleration and Fig. 12a and b displays the variation of Fourier amplitude velocity for the stations in both cases. For Case 1 the Fourier velocity amplitudes of stations S1 and S2 in the resonating frequencies regime of the bridge displayed a decent match. Hence the deck slab might have reacted similarly when rupture propagation initiates at the end of the fault.

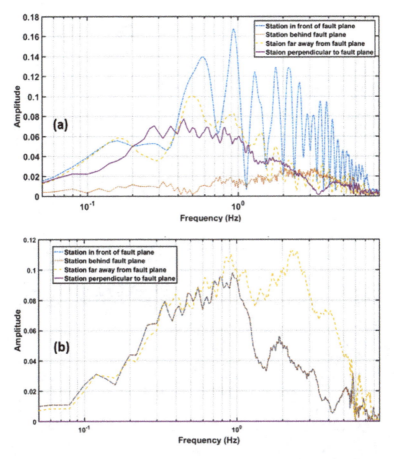

Fig. 11 Fourier amplitude acceleration spectrum for selected 4 stations for **a** Case 1 and **b** Case 2 ruptures

7 Conclusions

- Station in front of the fault plane (S1) experienced a higher PGA than other stations for Case 1 earthquake and in instance of Case 2 earthquake station far from the fault (S2) experienced higher PGA which validates the fact that directivity is governed by the change in position of nucleation patch (asperity).
- Station behind the fault plane (S2) experienced low PGA in Case 1 earthquake event but in Case 2 earthquake stations in front (S1) and behind (S4) experienced similar PGA and spectral content which is a result of the bilateral rupture.

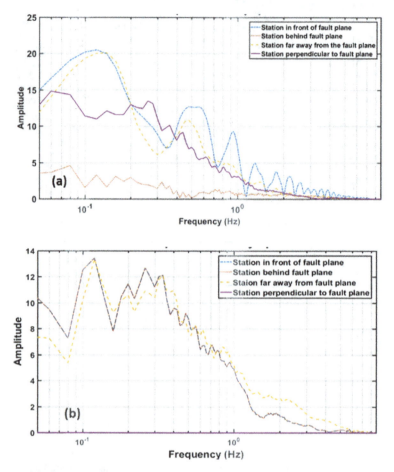

Fig. 12 Fourier amplitude velocity spectrum for selected 4 stations for **a** Case 1 and **b** Case 2 ruptures

- The bridge deck was displaced similarly when it was located in the front (S1) and far away (S2) from the fault plane in event of Case 1 earthquake, it might be because the Fourier velocity amplitudes of these stations are similar in the regime of bridge natural frequencies.
- Fourier velocity amplitudes were able to capture the identical behavior of the stations in front and far away from the fault plane and helped in citing the reason that implies velocity pulses are more decisive for directivity ground motions.

Acknowledgments We thank Indian institute of tropical metrology (IITM) for giving access to HPC Supercomputing resources to run SPECFEM3D. Funding from MoES/P.O(Seismo)/1(304)/2016 is greatly acknowledged.

References

1. Alterman Z, Karal FC Jr (1968) Propagation of elastic waves in layered media by finite difference methods. Bull Seismol Soc Am 58(1):367–398
2. Baker JW, Bozorgnia Y, Di Alessandro C, Chiou B, Erdik M, Somerville P, Silva W (2012) GEM-PEER global GMPEs project guidance for including near-fault effects in ground motion prediction models. 15 WECC, Lisboa (2012)
3. Bray JD, Rodriguez-Marek A (2004) Characterization of forward-directivity ground motions in the near-fault region. Soil dynamics and earthquake engineering 24(11):815–828
4. Frankel A, Vidale J (1992) A three-dimensional simulation of seismic waves in the Santa Clara Valley, California, from a Loma Prieta aftershock. Bull Seismol Soc Am 82(5):2045–2074
5. Li S, Zhang F, Wang JQ, Alam MS, Zhang J (2016) Effects of near-fault motions and artificial pulse-type ground motions on super-span cable-stayed bridge systems. J Bridge Eng 22(3):04016128
6. Shi Z, Day SM (2013) Rupture dynamics and ground motion from 3-D rough-fault simulations. J Geophys Res: Solid Earth 118(3):1122–1141
7. Smith WD (1975) The application of finite element analysis to body wave propagation problems. Geophys J Int 42(2):747–768
8. Somerville PG, Smith NF, Graves RW, Abrahamson NA (1997) Modification of empirical strong ground motion attenuation relations to include the amplitude and duration effects of rupture directivity. Seismol Res Lett 68(1):199–222
9. Soyluk K, Karaca H (2017) Near-fault and far-fault ground motion effects on cable-supported bridges. Procedia Eng 199:3077–3082
10. Taborda R, Bielak J (2013) Ground-motion simulation and validation of the 2008 Chino Hills, California, Earthquake. Bull Seismol Soc Am 103(1):131–156
11. Tabord R, Roten D (2014) Physics-based ground-motion simulation. Encyclopedia Earthq Eng 1–33
12. Udías A, Vallina AU, Madariaga R, Buforn E (2014) Source mechanisms of earthquakes: theory and practice. Cambridge University Press
13. USGS earthquake glossary. https://earthquake.usgs.gov/data/rupture/directivity.php
14. Yi J, Li J (2017) Longitudinal seismic behaviour of a single-tower cable-stayed bridge subjected to near-field earthquakes. Shock and Vibration

Broadband Ground Motion in Indo-Gangetic Basin for Hypothetical Earthquakes in Himalaya

J. Dhanya, S. Jayalakshmi, and S. T. G. Raghukanth

Abstract Indo-Gangetic (IG) Basin, formed between the Indian shield and Himalayas, is the largest sedimentary basin in India. The region also constitutes many metropolitan cities, including the capital city New Delhi. The seismic risk in the region is attributed due to the proximity to seismically active Himalayan faults, the possible seismic wave amplification due to huge sedimentary layers, and the vulnerability due to urban agglomerations. However, the region lacks a dense set of recorded data curbing the direct assessment of ground motion intensities. Hence, seismic hazard needs to be estimated based on synthetic ground motions for possible scenario earthquakes. These ground motion simulations require a proper understanding of the spatial variation of material properties, viz., density and wave velocities in the region of interest. Hence, the present work focuses on developing the 3D regional velocity model for ground motion simulations in IG Basin spanning between longitude 74.5–82.5°E and latitude 24.5–32.5°N. The spatially varying material properties of the region are derived by suitably interpreting the available velocity models constrained according to geological features reported in the literature. The 3D velocity model derived from the study is first employed in a finite element platform to obtain low-frequency ground motion. These ground motions rich in low-frequency content are further combined with the high-frequency ground motion simulated using the Zeng-scattering method on a hybrid broadband ground motion generation platform. Hence, the simulated time histories comprise of energy in the period between 0 and 10 s, thus complying to engineering interest. The 3D velocity model developed from the study is validated using the strong motion data available for an event in the Main Boundary Thrust. The simulated time histories are observed to match the phase and energy content of recorded data. The model is further employed to simulate time his-

J. Dhanya (✉) · S. T. G. Raghukanth
Department of Civil Engineering, Indian Institute of Technology Madras, Chennai, India
e-mail: dhanyaj17@gmail.com

S. Jayalakshmi
Division of Physical Science and Engineering, King Abdullah University of Science and Technology, Thuwal, Saudi Arabia

© The Editor(s) (if applicable) and The Author(s), under exclusive license to Springer Nature Singapore Pte Ltd. 2021
S. K. Saha and M. Mukherjee (eds.), *Recent Advances in Computational Mechanics and Simulations*, Lecture Notes in Civil Engineering 103, https://doi.org/10.1007/978-981-15-8138-0_28

tories for Mw 8.5 hypothetical earthquake in the Himalayas. The obtained response spectra are compared with the IS1893 code recommendations.

Keywords 3D velocity model · Ground motion simulations · Indo-Gangetic Basin

1 Introduction

Seismic waves undergo considerable variations as they travel through medium of variable characteristics. One of the prominent medium characteristics that affect the ground motion at the site is the presence of sedimentary basins. Thus, quantifying the amplification of ground motion in sedimentary basins is one of the key issues faced by seismologists and earthquake engineers. The "basin effect" is manifested in the surface waves that are generated from the interaction of body waves with the boundaries of the basin. The corresponding scattering through the different sediment strata alters the frequency content of the waves [1]. Thus, the waves are either attenuated or amplified by repeated reflections and refractions within the basin. Therefore, the spectral amplitudes at a basin site are functions of the material properties and geometrical features [2]. Furthermore, amplification occurs when this resonant frequency approaches any of the frequencies of earthquake ground motion. Observations from many past earthquakes—1985 Mexico City Earthquake, 1999 Chi-Chi Earthquake, 2010 Darfield Earthquake, 2011 Christchurch Earthquake—have provided insights into seismic wave amplification at basin sites [3, 4].

For engineering application, simple "Ground Motion Prediction Equations" (GMPEs) are proposed to predict seismic wave attenuation. Typically, the site characteristics are accounted in GMPEs only based on shear wave velocity of top 30 m of soil (Vs30) and the spectral decay parameter kappa (κ) [5]. However, for large and deep sedimentary basins extending to several kilometers, the site amplification quantified by Vs30 might not sufficiently account for the attenuation pattern attributed from deeper sediment layers and the geometry of the basin. Hence, suitable models are necessary to combine the effect of both site and basin geometry to capture complete amplification.

Indian subcontinent also constitutes several sedimentary basins. Out of 37 sedimentary basins identified for the country [6], the Indo-Gangetic (IG) Basin is the largest. The IG Basin is formed between the subducting Indian plate and the Eurasian plate. This basin is the depressed part of peninsular India (also known as Peninsular Indian shield) with several hidden faults [7]. The thickness of the sediment fill ranges between few tens of meters and 6 km, increasing progressively from south to north in the form of an asymmetrical wedge. The IG Basin is shown as three distinct regions in Fig. 1—"Indus Plain" in the West, "Ganges Plain" in the center, and "Brahmaputra Plain" in the East. These regions are named after the important rivers that contribute to the alluvial deposition. The active hidden faults embedded within the basin pose a significant seismic hazard to this region.

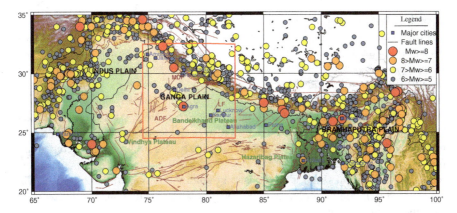

Fig. 1 The tectonic setting of Indo-Gangetic Basin and surrounding region. Seismicity (250-2017) is indicated as circles. [*Note* The study region is the one marked in red box]. Few active faults are marked as MBT (Main Boundary Thrust), MDF (Mahendragarh–Dehradun Fault), ADF (Aravalli–Delhi Fault), LF (Lucknow Fault)

Several investigators attempted to understand basin effects for the IG Basin from many past earthquakes outside the basin. The damaging 1934 M_w 8.2 Bihar–Nepal earthquake and 2001 M_w 7.6 Bhuj earthquake pointed to the ground motion amplification in the IG Basin [8]. Later, Hough and Bilham [9] computed PGA and PGV at hard sites from 1897 M_w 8.1 Assam, 1905 M_w 7.8 Kangra, and 1934 M_w 8.2 Bihar–Nepal earthquakes and inferred basin amplification of up to a factor of 3 near the river banks and flood plains. Furthermore, Srinivas et al. [10] reported the sedimentary thickness and shear wave velocities along an array of cites across Lucknow City from receiver function analysis of broadband seismograms, which aided in deepening the understanding and structural characterization of IG Basin. Later, the network expanded and is currently known as the Central Indo-Gangetic Network (CIGN), consisting of 26 strong motion velocity seismographs. This network recorded PGA and PGV at basin sites 3–4 times higher than the adjacent hard rock sites during 2014 M_w 6.1 Bay of Bengal earthquake and 2015 M_w 7.8 Nepal earthquake [11]. However, the urban agglomerations in our study region still lack dense data to sufficiently constrain the basin structure and to capture seismic wave amplifications for possible large earthquakes in the future. Additionally, the proximity to seismically very active Himalayas faults enhances the vulnerability in the region.

In the absence of recorded data for large events, a suitable alternative to study regional ground motion characteristics is by resorting to analytical or numerical simulations. The studies in this direction include that performed by Jayalakshmi and Raghukanth [12] using the available shear wave structure of the basin to estimate the statistics of expected shaking levels of PGD and PGV on bedrock sites for many large scenario earthquakes. Later, Bagchi and Raghukanth [13] inferred amplifications in ground motions at basin sites due to Himalayan seismic sources, using a 2D finite element model for IG Basin without including the topography of Himalaya. Recently,

Jayalakshmi et al. [14] developed a 3D regional velocity model for Indo-Gangetic Basin to study site amplification. Due to the complexities involved in numerical implementations for the region, the results from the study are valid till 0.5 Hz only. However, for engineering application, there is a need to estimate ground motion for a broader range of frequencies. Hence, the present work aims to implement the 3D velocity model on a hybrid broadband ground motion simulation platform to simulate ground motions for a possible Mw8.5 scenario event in the Himalayan fault. Thus, the model is first employed in a finite element framework to obtain low-frequency ground motion. These ground motions rich in low-frequency content are further combined with the high-frequency ground motion simulated using the Zeng-scattering method on a hybrid broadband ground motion generation platform. The details of simulation algorithm and the corresponding simulations are discussed in the subsequent sections.

2 Broadband Ground Motion Simulation

2.1 Methodology

The near-field ground motions will be strongly influenced by the fault mechanism and direction of rupture propagation. Analytical techniques based on kinematic source models and elastic wave propagation approaches are very efficient in simulating spatial variability of ground motion. These techniques are capable of modeling the source directivity effects on ground motions. In the present study, SPECFEM3D is used to simulate the low-frequency ground motion time histories for study region. The spectral finite element methodology was initially developed by Patera [15] for computational fluid dynamics. Later, Cohen et al. [16], Komatitsch [17], and Komatitsch and Vilotte [18] applied this method for problems related to 2D and 3D seismic wave propagation. The advantage of this method include the ease of implementation of free surface topography, lateral variation in material properties, and the consideration for moment tensor representation of the seismic source. Komatitsch and Tromp [19] have explained in detail the formulations for 3D regional seismic wave propagation. The wave equation of motion in terms of displacement field $\mathbf{u}$ can be expressed as

$$\rho \ddot{\mathbf{u}} = \nabla . \sigma + \mathbf{f} \tag{1}$$

where ρ is the density, σ is the symmetric second-order stress, and $\mathbf{f}$ is the force vector which is considered as a unit impulse force to estimate Green's functions. The integral weak form of Eq. 1 can be obtained by multiplying with an arbitrary test vector w and integrating over the volume Ω imposing boundary conditions expressed as

$$\int_\Omega \rho \mathbf{w}.\ddot{\mathbf{u}}\, d\Omega = -\int_\Omega \nabla \mathbf{w} : \sigma\, d\Omega + \int_\Omega \mathbf{w}.\mathbf{f}\, d\Omega + \int_{\Gamma_{abs}} \mathbf{n}.\sigma.\mathbf{w}\, d\Gamma \tag{2}$$

where **u** is the displacement field, σ is the symmetric second-order stress tensor, and **f** is the external force vector. The stress-free boundary condition is naturally satisfied and hence need not be imposed explicitly. The three terms in the right-hand side of the expression represent stress, source, and traction at absorption boundaries, respectively. Once the weak form is derived, then the region is divided by discretization into non-overlapping hexahedral elements. Further, each element with the assigned material property is mapped to a reference cube using a Jacobian matrix with higher order Lagrange interpolation functions as the basic functions of the displacement field. If the basis function is a polynomial of order n, then SPECFEM requires $(n+1)^3$ control points known as "Gauss–Lobatto–Legendre" (GLL) points inside each element. The GLL quadrature rule is adopted to perform numerical integration over the volume elements. After discretizing the spatial domain, the global system of the equation to be solved from the assembly of individual elements is given by

$$M\ddot{\mathbf{u}} + C\dot{\mathbf{u}} + K\mathbf{u} = F \tag{3}$$

where **u** is the global displacement vector; M, C, and K are the global mass matrix, global absorbing boundary matrix, and global stiffness matrix, respectively; and F is the source term. The efficiency of the scheme in modeling the basin can be analyzed from the ground motion simulation model developed for Los Angeles Basin by Komatitsch et al. [20]. For regions in and around India, Dhanya et al. [21], Sangeetha et al. [22], and Jayalakshmi et al. [14] implemented the methodology to develop regional velocity model and subsequent ground motion simulations.

However, due to the lack of availability of high-resolution images of the earth structure, the above numerical method is efficient in simulating low-frequency ground motion (typically <0.5 Hz). On the other hand, for engineering application, ground motions having a wider range of frequencies are essential. A suitable technique to simulate high-frequency ground motion characteristics is the S-S scattering methodology based on isotropic scattering theory developed by Zeng [23]. This theory allows for the computation of the temporal decay of scattered wave energy, including all multiple-scattering contributions. Zeng [23] showed that the S-to-S-scattered wave field approximates the complete scattered wave field that includes scattering conversion between P and S waves shortly after the S arrival. The energy envelope of the S-to-S-scattered waves $(E(r, t))$ is computed using the scattered wave energy equation from Zeng et al. [24], given by

$$E(r,t) = \frac{\delta\left(t - \frac{r}{\beta}\right) e^{-\eta\beta t}}{4\pi\beta r^2} + \sum_{n=1}^{2} E_n(r,t) +$$

$$\int_{-\infty}^{\infty} \frac{e^{i\Omega}}{2\pi} d\Omega \int_{0}^{\infty} \frac{\left(\frac{\eta_s}{k}\right)^3 \left[tan^{-1}\left(\frac{k}{\eta + i\frac{\Omega}{\beta}}\right)\right]^4}{2\pi^2 \left[1 - \frac{\eta_s}{k} tan^{-1}\left(\frac{k}{\eta + i\frac{\Omega}{\beta}}\right)\right]} dk \tag{4}$$

This is the integral solution for the fundamental wave energy equation ([24] Eq. 4a) computed using Laplace transformation in time and spatial Fourier transformation. $i\Omega$ denotes the Fourier transform solution with respect to time, k is the wave number, η_s is the scattering coefficient of elastic attenuation, η_i is the absorption coefficient for anelastic attenuation, and $\eta = \eta_i + \eta_s$. Anelastic attenuation (η_i) is replaced by $Q(f) = Q_0 f^n$. η denotes the total S-wave attenuation coefficient that mainly controls the exponential decay of the S-wave coda. β is the average S-wave velocity between source and receiver and r is the source–receiver distance. The first term in Eq. 4 denotes the direct arrival, the second term is the singly ($n = 1$) and doubly ($n = 2$) scattered energy, and the third term comprises the multiple-scattering contributions. Green's functions for direct S waves use a $1/R$ geometrical spreading (Eq. 4). Crustal and Moho reflections are not considered. Because the scattered waves arrive at a given site from all directions within the scattering volume surrounding the source and receiver, Zeng et al. [24] assumed isotropic radiation for the direct waves in the derivation of Eq. 4. Effects of the source radiation pattern are therefore averaged out for the high-frequency scattered waves.

In the numerical implementation, the computed coda envelope (Eq. 4) is filled with correspondingly scaled random scattering wavelets and the resulting time series thus represent a scattering Green's function. This theory holds for body waves, while scattered surface waves are not considered. Near-surface effects that control ground shaking at higher frequencies (e.g., site-specific attenuation in the upper layers at frequencies above f_{max}) are modeled with a site-kappa coefficient (κ). The calculation of high-frequency local scattering operators (scatterograms) follows Mena et al. [25]. P-wave and S-wave phase arrivals at each site are computed using a 3D ray-tracing algorithm of Hole [26]. Scattering Green's functions (sgf) are generated for an elementary source (1 cm of slip occurring on a 1 km × 1 km fault patch). The scattering Green's functions are then convolved with the slip-rate function at that point on the fault, forming a scatterogram. The summation of the scattering Green's functions emitted at each subfault is similar to the empirical Green's function method of Irikura and Kamae [27], which adds the contribution of the individual subfaults to obtain the total ground motion at the specified site as follows:

$$A(t) = \sum_{i=1}^{N^2} \frac{r}{r_i} STF(t - t_i) * C(sgf(t)) \tag{5}$$

where r is the hypocentral distance, r_i is the distance from the observation point to the ith subfault, $sgf(t)$ is the scattering Green's function, C is the stress drop ratio between the small and large event, N^2 is the number of subfaults to be summed up to generate the large earthquake ground motion, and the asterisk indicates the convolution. $STF(t)$ is the source time function and $A(t)$ is the resultant high-frequency ground motion.

For computing broadband hybrid seismograms, we adopt a three-stage approach. First, calculate low-frequency ground motions using the SPECFEM methodology as explained earlier. Second, generate the HF scattering contributions for each observer location, considering path-averaged scattering properties and local site conditions based on site kappa (κ) as explained in the previous section. Finally, the two sets of seismograms are reconciled in the frequency domain to form hybrid broadband seismograms. Here, appropriate weighting functions, for low ($W^{LF}(\omega)$) and high ($W^{HF}(\omega)$) frequencies that sum to unity at each frequency, compute the approximate broadband ground motion (u) as

$$u(\omega) = u^{LF}(\omega)W^{LF}(\omega) + u^{HF}(\omega)W^{HF}(\omega) \qquad (6)$$

The inverse Fourier transform of u yields the final broadband ground motion.

3 Implementation of 3D Regional Velocity Profile

We implement the broadband simulation methodology for possible earthquakes in Indo-Gangetic Basin. In this study, the 3D regional velocity model was developed for the region between latitude 24.5–32.5°N and longitude 74.5–82.5°E and reported by Jayalakshmi et al. [14]. The variation of shear wave velocity in the model is shown in Fig. 2. Table 1 lists the material properties (density, Vs and Vp) in the three geological regimes identified for the region. The thickness of sediments in the basin progressively increases from south to reach a total thickness of 6 km at north. A uni-

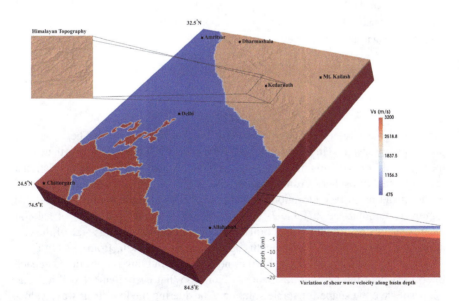

Fig. 2 3D model of IG Basin showing the variation of shear wave velocity in the study region. Bottom Right: A zoomed view of the cross-section of IG Basin, showing the progressively increasing basin depth [14]

Table 1 Material properties for the different regions of the 3D model

Region	Depth (km)	Density (kg/m3)	Vp (km/s)	Vs (km/s)	Quality factor
Indian shield	0–80	2680	5.600	3.200	800
IG Basin	0–0.25	1586	0.740	0.475	150
	0.25–1	1870	1.324	0.850	150
	1–2.25	1950	1.558	1.000	150
	2.25–4.75	2425	3.740	2.400	150
	4.75–6	2560	4.675	3.000	150
Himalaya	0–1	2495	4.200	2.400	250
	1–6	2632	5.200	2.970	250
	6–80	2680	5.600	3.200	800

Note: The IG Basin profile is a wedge shaped which increases progressively from south to north (Fig. 2) and thus the properties describe correspond to the region within this wedge along depth

Table 2 Characteristics of spectral element mesh

Characteristic	Value
Dimension (km × km)	760 × 910
Depth (km)	80
Type of element	Hexahedral
No. of spectral element on each side	1024 × 1024 (top-surface)
Total no. of elements	7.8×10^6
Min size of element (km × km × km)	0.70×0.90×0.50
Max size of element (km × km × km)	2.80×3.60×2.00
No. of GLL points in the entire mesh	9.4 Million
Total degree of freedom	28.3 Million
Time step used for simulation	0.0025 s

form shear wave velocity of 3.2 km/ is considered at bedrock level. These material properties are assigned in all the elements in the spectral element mesh. Furthermore, the high-resolution (30 m interval) topography model available from the US National Oceanic and Atmospheric Administration (NOAA) is smoothened and implemented in the model as surface elevation. The characteristics of finite element discretized region are summarized in Table 2. Thus, the model is divided into 1024×1024×44 elements along x-, y-, and z-directions, respectively. At surface, the size of the element is 0.7 km×0.9 km. Along depth, up to 1 km dz= 0.5 km, from 1–6 km dz = 1 km and beyond 6 km dz=2 km. Furthermore, mesh doubling is considered at each interface where the vertical dimensions change so that each element will have an approximate cube shape for stable solution. Considering the low shear wave velocity of 475 m/s, developed numerical model can accurately resolve up to 0.5 Hz with the computational time step of 0.025 s. The simulations are performed at the Kaust

Supercomputing Laboratory (KSL). Each simulation is performed on 32 compute nodes (1024 processor cores).

The low-frequency ground motion from SPECFEM3D is combined with the Zeng-scattering methodology as explained in the previous section. The matching frequency is taken as 0.5 Hz owing to the numerical resolution of computational mesh. The arrival times of body waves and average velocity along the path are measured according to the 3D velocity model for the region at 0.5 km resolution. The site-specific parameter kappa is taken as 0.03 for those at soil site and 0.01 for those at rock site. The coda wave quality factor (Q_c) for the region is found to be $158 f^{0.97}$. The scattering coefficient is considered as 0.045. The computational effort progressively increases with the increase in number of source points and the number of receivers. In the next step, we validate the 3D model by simulating ground motions for a past earthquake. The corresponding details are discussed further.

3.1 Validation

Indian strong motion network maintained by "PESMOS," comprise about 300 digital accelerographs in the highly seismic zones including northern and northeastern India. We selected an event in the Himalayan MBT fault which has reasonably good coverage of receivers in our study region—Mw 5.4 event at MBT in the Nepal on April

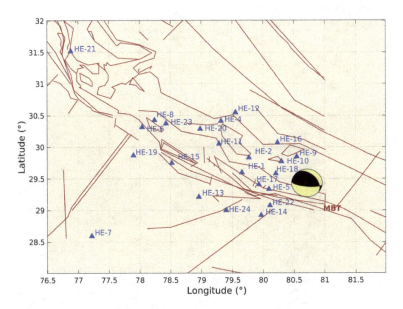

Fig. 3 Focal mechanism of the Mw5.6 Nepal–India Border event on April 4, 2011 used in the study along with the locations of the recording stations

4, 2011. The corresponding moment tensor along with the distribution of receivers is illustrated in Fig. 3. We assume the source characteristics for the event that are reported in USGS (https://earthquake.usgs.gov/earthquakes/eventpage/usp000hz8k/moment-tensor). These include the strike dip and rake for the event, which are taken as 318°, 30°, and 128°, respectively. The epicenter of the event lies in 81.71°E and 29.43°N with a hypocentral depth of 18.8 km. The half duration for the event is 1.2 s.

Ground motions are simulated at the 24 receivers, for the given seismic source and 3D earth model. The simulations are performed by combining SPECFEM methodology for low frequency and Zeng-scattering algorithm for high frequency and the respective time histories are matched at 0.5 Hz to obtain the hybrid broadband ground motions. The comparison of the simulated velocity time histories with the recorded

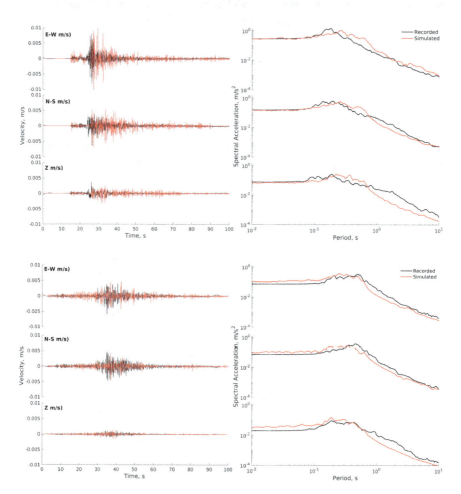

Fig. 4 Left panel: Comparison of velocity time histories at few selected stations for the Mw5.4 event on April 4, 2011 (Event 1: Top panel: HE-5 ($R_{epi} = 60$ km); Bottom panel: HE-8 ($R_{epi} = 260$ km)). Black line indicates instrument data and red line indicates simulated time histories. Right Panel: Comparison of spectral acceleration at the corresponding stations

data at two representative stations, HE-5 at an epicentral distance of 60 km and HE-8 at an epicentral distance of 260 km, is illustrated in Fig. 4. The comparison of waveforms shows that the simulated waveforms are matching in phase as well as amplitude with those of the record data. Furthermore, to understand the frequency dependence of the simulated ground motion intensities, we compute spectral accelerations. We also compare the simulated spectra with the recorded and observe that both the spectra compare well (Fig. 4). Furthermore, the simulations are observed to capture the energy for a broad range of spectra. We repeat the comparison for all the available data and find that our regional model performs accurately in simulating ground motions. Hence, we proceed to predict possible shaking levels for hypothetical Himalayan earthquake.

3.2 Scenario Himalayan Events: Mw 8.5

We simulate ground motions for future great earthquakes (Mw 8.5) in the MBT fault system in Himalaya. Although there have not been any very large earthquakes (>8) in this region, it is important to consider these possibilities in probabilistic seismic hazard analysis. For such great earthquakes, the seismic source plays a major role, especially in the near-field region. The finiteness of the source is therefore important to model in terms of a slip field that has a spatial variation on the fault plane. This spatial slip field is estimated using the methodology proposed by Dhanya et al. [28] for non-tsunamigenic earthquakes. The temporal variation of slip field is expressed

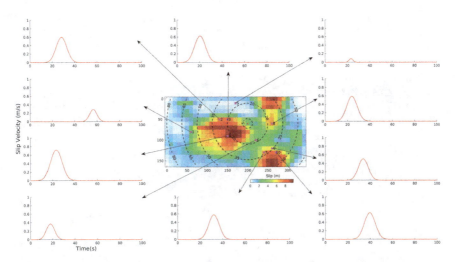

Fig. 5 Slip field (Unit: m) used in the simulation of Mw8.5 Scenario event at MBT. The models are generated following the method of Dhanya et al. [28]. For finite fault dimensions are 340 × 160 km. Hypocenter locations are assumed at the most probable location as per Mai et al. [29] and a constant rupture velocity (0.8 ×Vs) is assumed

in terms of "source time function" for each point on the source. An illustration of source time function for few points on the source for the event is illustrated in Fig. 5. The corresponding length and width of the slip field are 340 km × 160 km with subfault dimensions as 10 km × 10 km. The maximum slip is obtained as 9.76 m. The probability distribution of the slip is taken as truncated exponential with $\mu = 13.9$. The correlation characteristics are considered as stable variogram with sill as 7.51 × 10^5 range along strike as 47.8 km and that along dip as 44.3 km. A constant rupture velocity of 0.8 times shear wave velocity is considered for the model. The hypocenter is estimated using the maximum possible location according to Mai et al. [29], as 200 km along strike and 60 km along dip from top left corner of the slip field. The corresponding location in the region is at 78.99°E and 29.93 °N considering strike as 300° and dip as 16°. The rake angle is taken as 95° corresponding to that of a thrust-type event. This slip field is used in the model to simulate ground motions at some important cites in the region. The trace of the slip field along with the location of the cities is illustrated in Fig. 6. It is noted that stations are chosen such that there are receivers on top of the slip field as well as along different azimuths with respect to the strike of the source. We now simulate broadband ground motion time histories at

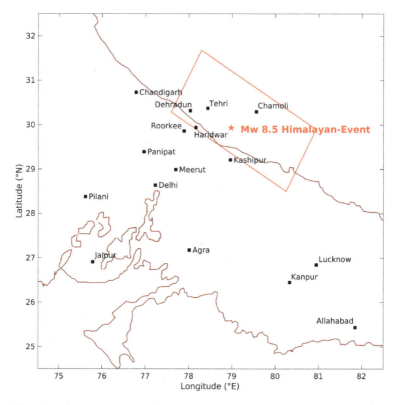

Fig. 6 Location of receivers at important towns and cities in the study region

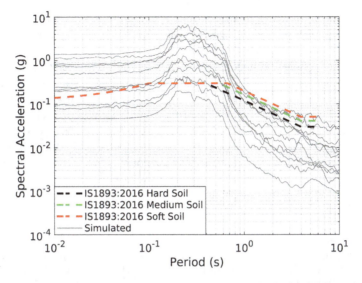

Fig. 7 Comparison of rotd50 spectral acceleration at important cities for Mw8.5 Scenario event at MBT with IS1893:2016 [30] corresponding to Zone IV

the selected cities for the given slip field shown in Fig. 5. The corresponding rotd50 spectral accelerations obtained for all site together are summarized in Fig. 7. It is observed that for near-field stations the PGA exceeded 1 g. The PGA corresponding to cities at large basin depths—Haridwar, Kashipur, and Chandigargh—is observed to be >0.7 g. We also compare the predicted spectral amplitudes with the existing building code provisions IS1893:2016 [30]. Because majority of the region are categorized under seismic Zone IV, according to IS1893:2016 [30], we compute design spectral accelerations for Zone IV. Figure 7 shows that the design spectral amplitudes are underestimated for shorter periods (<1.25 s). This observation warrants that we have to consider these worst-case scenarios that are possibly due to amplifications in sedimentary basins due to large/great earthquakes, and make more efforts to improve design spectra for building safe and resilient structures.

4 Summary and Conclusions

The study aims to implement a 3D regional velocity model for Indo-Gangetic Basin on a hybrid broadband ground motion simulation platform. We adopt the hybrid approach of Mena et al. [25] to simulate broadband ground motions by combining the low-frequency waveforms from spectral element method with the high-frequency waveforms simulated by using Zeng-scattering method. We used a past earthquake (Mw 5.4, Nepal region) to simulate ground motions using our 3D regional model and we observed that the simulated results are in comparison with the recorded data.

We also perform a shake out scenario for a possible Mw8.5 hypothetical event at MBT. We compare the spectral accelerations against the design values reported in the building code IS1893:2016 [30] for cities in seismogenic zone IV. We found that the design values are underestimated for shorter periods and conjecture that these design values have not taken into account the complexities of seismic source, possible amplifications in sedimentary basin, multiple scattering, and presence of topography. However, the results presented in this study are only for one possible scenario earthquake. More suits of events will provide a better understanding of ground motion attenuation characteristics for the region and hence enable us to develop suitable ground motion prediction equations.

References

1. Brad PY, Bouchon M (1985) The two-dimensional resonance of sediment-filled valleys. B Seismol Soc Am 75:519. https://pubs.geoscienceworld.org/bssa/article-pdf/75/2/519/2704980/BSSA0750020519.pdf
2. Brad PY, Campillo M, Chavez-Garcia FJ, Sanchez-Sesma F (1988) The Mexico earthquake of September 19, 1985-a theoretical investigation of large- and small-scale amplification effects in the Mexico City Valley. Earthq Spectra 4:609–633 (1988). https://doi.org/10.1193/1.1585493
3. Fletcher JB, Wen KL (2005) Strong ground motion in the Taipei basin from the 1999 Chi-Chi, Taiwan, earthquake. B Seismol Soc Am 95:1428–1446. https://doi.org/10.1785/0120040022
4. Bradley BA, Cubrinovski M (2011) Near-source strong ground motions observed in the 22 February 2011 Christchurch Earthquake. Seismol Res Lett 82:853 (2011). https://doi.org/10.1785/gssrl.82.6.853
5. Douglas J (2018) Ground motion prediction equations 1964–2018. University of Strathclyde, Glasgow, Review
6. Biswas SK (2012) Status of petroleum exploration in India. Proc Indian Natl Sci Acad 78:475–494
7. Dasgupta S, Narula PL, Acharyya SK, Banerjee J (2000) Seismotectonic atlas of India and its environs. Geological Survey of India
8. Hough SE, Martin S, Bilham R, Atkinson GM (2002) The 26 January 2001 M 7.6 Bhuj, India, earthquake: observed and predicted ground motions. B Seismol Soc Am 92:2061–2079
9. Hough SE, Bilham, R (2008) Site response of the Ganges basin inferred from re-evaluated macroseismic observations from the 1897 Shillong, 1905 Kangra, and 1934 Nepal earthquakes. J Earth Sys Sci 117:773 782. Springer
10. Srinivas D, Srinagesh D, Chadha RK, Ravi Kumar M (2013) Sedimentary thickness variations in the Indo-Gangetic foredeep from inversion of receiver functions. B Seismol Soc Am 103:2257–22652
11. Chadha RK, Srinagesh D, Srinivas D, Suresh G, Sateesh A, Singh SK, Pérez-Campos X, Suresh G, Koketsu K, Masuda T (2015) CIGN, a strong-motion seismic network in central Indo-Gangetic plains, foothills of Himalayas: first results. Seismol Res Lett 87:37–46
12. Jayalakshmi S, Raghukanth STG (2016) Regional ground motion simulation around Delhi due to future large earthquake. Nat Hazards 82:1479–1513. https://doi.org/10.1007/s11069-016-2254-8
13. Bagchi S, Raghukanth STG (2019) Seismic response of the central part of Indo-Gangetic plain. J Earthq Eng 23:183–207. https://doi.org/10.1080/13632469.2017.1323044
14. Jayalakshmi S, Dhanya J, Raghukanth STG, Mai PM (2020) 3D seismic wave amplification in the Indo-Gangetic basin from spectral element simulations. Soil Dyn Earthq Eng 129:105923. https://doi.org/10.1016/j.soildyn.2019.105923

15. Patera AT (2020) A spectral element method for fluid dynamics: laminar flow in a channel expansion. Soil Dyn Earthq Eng 129:105923. https://doi.org/10.1016/j.soildyn.2019.105923
16. Cohen G, Joly P, Tordjman N (1993) Construction and analysis of higher-order finite elements with mass lumping for the wave equation. In: Proceedings of the second international conference on mathematical and numerical aspects of wave propagation, pp. 152–160. SIAM Philadelphia. Pennsylvania, USA
17. Komatitsch D (1997) Spectral and spectral-element methods for the 2D and 3D elastodynamics equations in heterogeneous media. Institut de Physique du Globe, Paris, France
18. Komatitsch D, Vilotte JP (1998) The spectral element method: an efficient tool to simulate the seismic response of 2D and 3D geological structures. B Seismol Soc Am 88:368–392
19. Komatitsch D, Tromp J (1999) Introduction to the spectral element method for three-dimensional seismic wave propagation. Geophys J Int 139:806–822. Blackwell Publishing Ltd., Oxford, UK
20. Komatitsch D, Liu Q, Tromp J, Suss P, Stidham C, Shaw JH (2004) Simulations of ground motion in the Los Angeles basin based upon the spectral-element method. B Seismol Soc Am 94:187–206
21. Dhanya J, Gade M, Raghukanth STG (2017) Ground motion estimation during 25th April 2015 Nepal earthquake. Acta Geod Geophys 52:69–93. https://doi.org/10.1007/s40328-016-0170-8
22. Sangeetha S, Dhanya J, Raghukanth STG (2018) 3D crustal velocity model for ground motion simulations in North-East India. J Earthq Eng 1–37. https://doi.org/10.1080/13632469.2018.1520760
23. Zeng Y (1993) Theory of scattered P-and S-wave energy in a random isotropic scattering medium. B Seismol Soc Am 83:1264–1276
24. Zeng Y, Su F, Aki K (1991) Scattering wave energy propagation in a random isotropic scattering medium: 1. Theory. J Geophys Res: Sol Earth 96:607–619. Wiley Online Library
25. Mena B, Mai PM, Olsen KB, Purvance MD, Brune JN (2010) Hybrid broadband ground-motion simulation using scattering Green's functions: application to large-magnitude events. B Seismol Soc Am 100:2143–2162
26. Hole JA (1992) Nonlinear high-resolution three-dimensional seismic travel time tomography. J Geophys Res: Sol Earth 97:6553–6562. Wiley Online Library
27. Irikura K, Kamae K (1994) Estimation of strong ground motion in broad-frequency band based on a seismic source scaling model and an empirical Green's function technique. J Geophys Res: Sol Earth. Wiley Online Library
28. Dhanya J, Raghukanth STG (2019) A non-Gaussian random field model for earthquake slip. J Seismol 23:889–912. https://doi.org/10.1007/s10950-019-09840-3
29. Mai PM, Spudich P, Boatwright J (2005) Hypocenter locations in finite-source rupture models. B Seismol Soc Am 95:965. https://doi.org/10.1785/0120040111
30. IS:1893-1: (2016) Criteria for earthquake resistant design of structures, Part 1: General provisions and buildings. Bureau of Indian standards (BIS), New Delhi

Earthquake Engineering in Areas Away from Tectonic Plate Boundaries

Nelson Lam

Abstract This article presents findings on many facets of research into engineering for mitigating earthquake hazards in regions remote from tectonic plate boundaries. Topics covered include seismic activities in intraplate regions, ground motion behaviour of intraplate earthquakes, considerations of site modification behaviour, structural design and detailing of reinforced concrete and structural analysis in regions of low-to-moderate seismicity. Much of the materials presented were generated from collaborative research involving the author and numerous esteemed collaborators over many years. The challenges that are being attended to are distinctive to those addressed in mainstream earthquake engineering research which is applicable to areas in proximity to tectonic plate boundaries.

Keywords Earthquake engineering · Intraplate earthquakes · Low-to-moderate seismicity · Seismic design

1 Introduction

Earthquake engineering has been receiving a great deal of attention for more than half a century by researchers from around the world in the discipline of earth sciences and civil engineering. Most active researchers in this field have been working on chal-lenges that are encountered in areas of high seismic activities that are in proximity to tectonic plate boundaries. Regions that are located away from tectonic plate boundaries, which are known as intraplate regions, cover for more than 90‰ of landmass on earth. Intraplate earthquakes are rare. However, once such an earthquake occurs the extent of ground shaking experienced by built facilities located close to the epicentre can be just as life threatening, and destructive. The main challenge with intraplate earthquakes is that the likely locations of future events are very difficult to predict and their occurrences are often not preceded by warnings of any form.

N. Lam (✉)
University of Melbourne, Melbourne, VIC 3010, Australia
e-mail: ntkl@unimelb.edu.au

© The Editor(s) (if applicable) and The Author(s), under exclusive license to Springer Nature Singapore Pte Ltd. 2021
S. K. Saha and M. Mukherjee (eds.), *Recent Advances in Computational Mechanics and Simulations*, Lecture Notes in Civil Engineering 103,
https://doi.org/10.1007/978-981-15-8138-0_29

The added challenge is that built infrastructure in intraplate regions are typically not de-signed to perform satisfactory when subject to moderate ground shaking. Profession-als involved in the design and supervision of construction of buildings and bridges are typically not skilled (nor experienced) in addressing seismic risks. Whilst disturbances generated by an earthquake are not very dependent on whether the earthquake is an intraplate earthquake the nature of the associated engineering challenges is very dif-ferent. The Australian continent is located wholly within an intraplate region and yet the University of Melbourne along with m numerous collaborating institutions from within Australia and internationally have been researching continuously over decades to address the engineering challenges, and with successful outcomes.

In the rest of the paper, different facets of earthquake engineering research and the key findings are covered in the intraplate context under separate headings of seismicity, ground motion behaviour, site effects and design/analysis and detailing considerations.

2 Intraplate Seismicity

A global survey of intraplate earthquakes exceeding magnitude M5 revealed a rate of 5–10 earthquake events with magnitude of at least M5 for a land area of one million square kilometres in a 50-year period [1]. Using M5 as threshold overcomes the issue of completeness of data when the survey covers a period of half a century. This rate of occurrence can be expressed in Eq. 1 in the Gutenberg–Richter form.

$$Log_{10}N = 5.2 - 0.9M \qquad (1)$$

where N is the number of earthquake events in a 50 year for every 1 M sq. km and M is the moment magnitude (Fig. 1).

Unlike a global survey most (local) surveys have to be based on a much lower magnitude threshold which can become problematic. An important finding from the survey was that the recorded rate of earthquake occurrence was consistent across many regions around the globe and hence the figure as quoted is robust. This research finding is consistent with the observation that in Australia (which has a land area of 7.7 million square kilometres) the average rate of occurrence of earthquakes exceeding M5 is about one every year [2].

In comparison, countries like Malaysia (comprising the peninsula, Sarawak and Sabah) seem to be relatively inactive except for the eastern part of Sabah. In Peninsular Malaysia, several earthquake tremors exceeding M4 that once captured attention by the local media have been documented in historical archives but no earthquakes exceeding M5 has ever been recorded [3]. In Singapore, no earthquake exceeding M4 has ever occurred either. It is noted that the area of Peninsular Malaysia is only around 130 k square kilometres which is approximately 1/60th the size of the

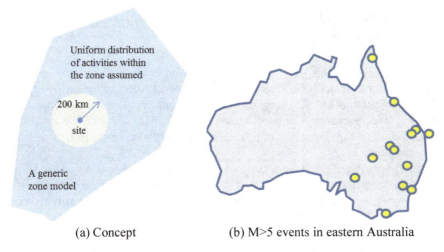

(a) Concept   (b) M>5 events in eastern Australia

Fig. 1 Uniform seismicity model [1]

Australian continent. Thus, there is not enough information of statistical significance to suggest that the inherent rate of earthquake activity in the Australian and Malaysian landmass is any different. This same argument also applies to the British Isles which has a similar land area as peninsula Malaysia. The underlying risk of local earthquakes occurring within the island state of Singapore in the future is also totally unknown and yet there are only long-distance earthquakes provisions in the enacted seismic design code of Sin-gapore.

In other words, making references to statistical observations of seismic activities (or inactivity) of a tiny landmass makes no scientific sense. This same argument can also be used to caution engineers and regulators when interpreting information conveyed on a seismic hazard contour map which typically features areas of relatively very low predicted seismic activities (e.g., Ref. [4]). Recognizing the shortfalls of probabilistic seismic hazard analysis (PHSA), the Australian Standard revised its provisions in 2018 [5] to impose a minimum design seismic hazard factor (Z) of 0.08 across the entire continent irrespective of predictions by PSHA undertaken recently by Geoscience Australia [4]. Note that Z which carries the same meaning as design peak ground acceleration is expressed in units of g's and corresponds to a notional return period of 500 years. The author and co-workers who gave advice to Malaysian regulators over seismic design provisions recommended a minimum design peak ground acceleration of 0.07 g for Malaysia [3].

3 Displacement Demand Behaviour of Intraplate Earthquakes

The displacement behaviour of earthquake ground shaking is best presented in the form of a Displacement Response Spectrum or an Acceleration-Displacement Response Spectrum (ADRS) diagram which is explained in Ref. [6] and Chap. 3 of Ref. [7]. The alternative formats of presenting a response spectrum are also illustrated herein (in Fig. 2) for ease of reference by readers who are not already familiar with the practices.

The saturation of the drift demand of earthquake ground shaking on the structure (i.e. insensitivity of structural drift to change in flexibility of the structure) is a phenomenon which is particular to conditions of low-to-moderate intensity shaking which is generated by a small or medium magnitude earthquake [8–10]. A deterministic second corner period expression has been derived to represent the capping of displacement demand behaviour which is dependent on the moment magnitude of the earthquake [11]. To explain this phenomenon in the layperson language, the average duration of the ground pulse would need to last long enough in order that the structure has the time to drift in one direction. If the ground pulse has limited duration, then the amount of drift that is experienced by the building is accordingly limited. This pulse duration behaviour is characterized by the 'second corner period' parameter (T2) as shown by a displacement response spectrum for ground motions

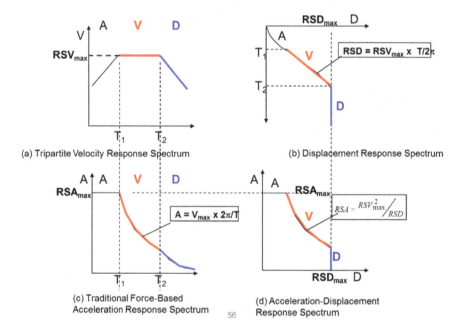

Fig. 2 Alternative response spectrum formats

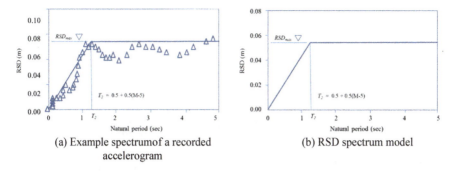

(a) Example spectrum of a recorded accelerogram

(b) RSD spectrum model

Fig. 3 Displacement response spectra on rock (excerpts from Ref. [11])

on rock (Fig. 3). Structures with fundamental natural period of vibration exceeding T2 can be considered as being in displacement-controlled conditions when subject to the designed earthquake.

Displacement-controlled behaviour has strong implications to the risk of overturning [12, 13]. As shown in Fig. 4, the amount of overturning displacement of a free-standing object is well correlated with the amount of displacement of the ground (or value of T2) and is (interestingly) insensitive to the aspect ratio of the object.

4 Torsional Response Behaviour of Intraplate Earthquakes

The underlying concepts of displacement-controlled phenomenon as described in the context of 2D translational motion can have implications to the 3D response behaviour of a torsionally unbalanced building as well. A building featuring asymmetry would respond in both translational and torsional motion. In displacement-controlled conditions (in which the duration behaviour of the ground pulse is limited) the amount of torsional rotation can be capped irrespective of the extent of asymmetry in the building [14]. Thus, the use of static eccentricity to emulate potential torsional behaviour of the building can give misleading results in the sense that the influence of asymmetry in the building could have been exaggerated.

The effects of torsional actions on a building can be expressed in terms of the ratio of the drift demand at the edge of the building and the drift demand in translational motion only (as predicted from 2D analysis of the building). The meaning of the elastic radius parameter 'b' which was first introduced by the author in Ref. [15] can be used to characterize the torsional stiffness behaviour of the building as illustrated in the schematic diagram of Fig. 5. The elastic radius (b) of the frame-pair equivalent model is so selected in order that the two models shown in the figure would experience identical displacement (and rotational) behaviour when subject to the same horizontal force, or when subject to the same ground excitations along the weaker direction of the building.

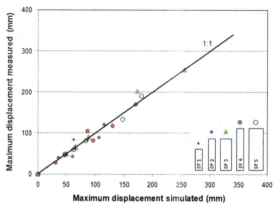
(a) Predicted and recorded peak displacement demand

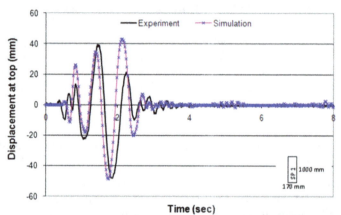
(a) Predicted and recorded displacement time-history

Fig. 4 Prediction of overturning displacement (excerpts from Ref. [12])

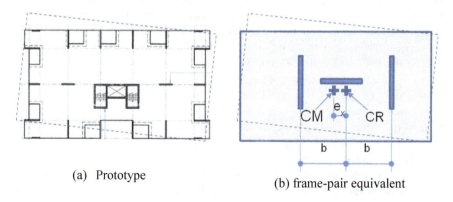

(a) Prototype   (b) frame-pair equivalent

Fig. 5 Introduction to meaning of elastic radius 'b'

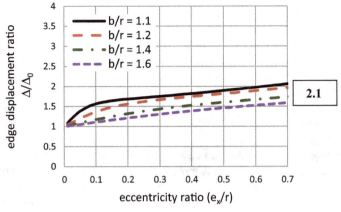

(a) Velocity Controlled Conditions

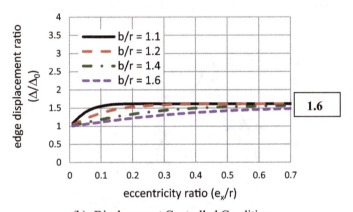

(b) Displacement Controlled Conditions

Fig. 6 Edge displacement ratios in velocity-controlled and displacement-controlled conditions

The drift demand behaviour at the edge of the building is then presented in Fig. 6a and b as function of the eccentricity ratio e_x/r and the elastic radius ratio (b/r) where 'r' is the mass radius of gyration of the building. It is shown in results presented in Refs. [14, 16] that the edge displacement ratio is capped at 2.1 in velocity-controlled conditions (Fig. 6a) in comparison to 1.6 in displacement-controlled conditions (Fig. 6b) provided that the value of b/r is equal to or greater than 1.0.

5 Site Classification and Soil Site Response Spectra

The site modification phenomenon is generally applicable to any region around the globe irrespective of the seismo-tectonic classification. By this logic the code stipulated site classification scheme and the associated site factors (that have been established for many years and adopted in North America, for example) that were originally developed in Western North America, and subsequently adopted across the rest of North America, should be equally applicable to other parts of the world including Australia, India, China and Malaysia. Note that this proposition has only considered the behaviour of the earthquake ground shaking and has neglected the potential response behaviour of a non-ductile, or limited ductile, structure which can be susceptible to very high amplification behaviour because of lack of provisions to dissipate energy in the building.

Non-ductile (or limited ductile) structure is particularly vulnerable to periodic ground shaking as such condition is more prone to high displacement amplification behaviour pertaining to the conditions of resonance (Fig. 7a and b). On soil sites where structures of limited ductility are found upon the likelihood of periodic shaking of the soil sediments is an important design consideration. Such periodic behaviour of the ground motion is associated with multiple reflections of seismic waves that are

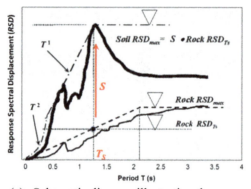

(a) Schematic diagram illustrating the concept

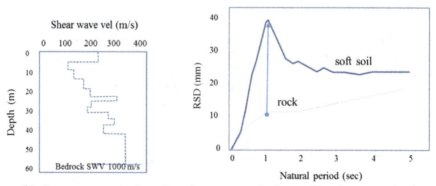

(b) Response spectra based on shear wave velocity profile of an example site

Fig. 7 Response spectra showing high amplification on flexible sites [18]

Table 1 Proposed new site classification scheme for intraplate regions [18]

Site class	Description	Site period T_s (s)
A & B	Rock	$T_s < 0.5$
C	Stiff soil	$0.15 \leq T_s < 0.6$
D	Flexible soil	$0.6 \leq T_s < 0.9$
E	Very flexible soil	$0.9 \leq T_s < 1.2$
S	Special soil	$T_s > 1.2$

trapped between bedrock and the soil surface provided there is a distinct soil–rock interface [17]. Thus, the dominant period of shaking of the soil column (i.e. the site natural period) is dependent on the depth of the soil. For this reason, the conventional format of site classification based on averaging soil shear wave velocities in the upper 30 m of the sedimentary layers has significant shortcomings when adapted for use in regions of low-to-moderate seismicity. The new classification scheme preferred by the author which makes use of the site natural period as the sole classification criterion as shown in Table 1 was proposed in Ref. [18] (and was developed by fundamental research led by Dr Hing Ho Tsang in collaboration with the author and other co-workers over many years, e.g. Refs. [19, 20]). It is shown in Ref. [21] that the shear rigidity of the bedrock as input into a 1D shear wave analysis of the soil column can have significant influence on the site amplification of the response spectrum.

6 Design and Analysis Considerations

In regions of low-to-moderate seismicity, it is seldom the case that seismic performance considerations are taken into account in the conceptual design of the building except for a built facility that requires an exceptionally high level of protection because of its functionality significance or a high consequence of failure. The design hazard factor or design peak ground acceleration is recommended to be based on a return period of 2500 years (and not 500 years). For ordinary construction, a factor of 2/3 may be applied to scale back the intensity of the seismic actions in order to be consistent with international practice (as explained in Chap. 2 of Ref. [7]).

It is recommended that all permanent built facilities have a minimum level of robustness but no well-established guidelines are in place to define the minimum requirements. The lack of guidance is because world research in earthquake engineering is mainly about overcoming challenges in a high seismicity environment whereas this article is aimed at developing minimum requirements for low-to-moderate seismicity environment. This minimum robustness is made up of a limited level of ductility (which is of the order of 2) and a modest amount of over-strength (which is about 1.3) [22, 23]. The quoted ductility ratio and over-strength factor (which are consistent with Australian contemporary code compliance buildings) are sufficiently generic to be applicable internationally in all regions of low-to-moderate seismicity. The terminology: 'limited ductility' used herein is strictly in the context of engineering in regions of low-to-moderate seismicity and is not to be confused with the same terminology used in the New Zealand standard [24]. The so-called 'limited ductile' detailing provisions in the New Zealand standards is referred herein as 'moderate ductile' detailing [23, 24].

Designing a built facility to achieve this minimum standard requires the following list of vulnerable features to be either eliminated or attended to adequately:

(a) buildings which are not adequately braced by use of structural walls, nor braced frames, that have been designed to resist lateral loads;
(b) buildings that are torsionally unstable (i.e. b/r < 1);
(c) use of non-ductile reinforcement in lateral load resisting RC members;
(d) undersized RC columns (which has the smaller dimension to be less than 400 mm) and RC walls of inadequate thickness (of less than 150 mm);
(e) RC walls and columns that have not been adequately reinforced to result in an even distribution of cracks if loaded to the ultimate limit (and this includes walls that are reinforced by a single layer of reinforcement);
(f) inadequately restrained longitudinal reinforcement that is spaced at more than 150 mm apart in a RC column or in a boundary element of a RC wall;
(g) the restraining closed loops or ties are at a vertical spacing exceeding one of the following: 15 times the bar size, smaller dimension of the column (or thickness of the wall) or 200 mm;
(h) RC vertical elements having a compression index exceeding 0.2;
(i) lack of anchorage of RC slabs to the supporting walls or columns, or lack of continuity reinforcement of floor beams going through columns;
(j) squat walls with aspect ratio or less than 2.0 or slender columns or walls with h/t ratio exceeding 16.

Much of the recommendations listed above are based on findings reported in Ref. [23] and stipulations by a recent edition of the Australian Standard for the design of concrete structures [24].

When subjecting the building model to dynamic analysis for checking code compliance it is important to have expertise within the design team to undertake independent checks of results reported from the computer. Performing this type of checks is to ensure that the design has not been compromised by misuse of the computer software or bugs in the software. In undertaking the checks, it is recommended to analyse the building model in 2D for comparison with results obtained from the Generalized Force Method (GFM) which was first introduced in the literature in Ref. [25]. An example of such a comparison is shown in Fig. 8. Once the 2D analyses have been verified the displacement at the edges of the building as obtained from the 2D and 3D analyses can be combined to determine the edge displacement ratio for checking against the limits shown in Fig. 6.

Detailed illustrations of the GFM methodology can be found in Chap. 4 of the textbook [7] for reference by practitioners which can be accessed free using information and the QR code that are shown on the flyer of the book in Fig. 9.

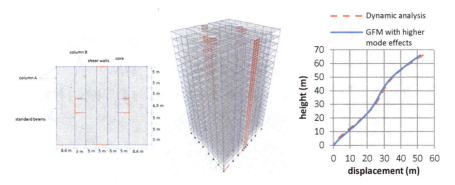

Fig. 8 Use of GFM to check results of dynamic analysis from a commercial software

7 Closing Remarks

Information and recommendations provided in this paper across the area of in-traplate seismicity, ground motion modelling, site effects modelling and structural (and RC) design and detailing, and structural analysis were generated from research undertaken by the author in collaboration with fellow academics and research higher degree candidates over many years. Importantly various common myths are dis-pelled. For example, the notion of certain countries, or areas, that are perceived to be possessing very low rate of seismic activities, and lower than that of Australia, may well have been resulted from the smallness in size of the area of the landmass being surveyed.

In displacement-controlled conditions the overturning behavior of a free-standing object and the torsional response behaviour of an asymmetrical building can be counter intuitive to engineers (who are not familiar to the underlying concepts) as neither the aspect ratio nor the amount of eccentricity controls the outcome of the response. The site factor provisions are not only to address the site hazard but should also to take into considerations the vulnerability of the affected structure.

Design objectives based on limited ductility design principles and the use of Gener-alised Force Method (GFM) and the edge ratio to independently assess results report-ed by a commercial software in the manner as described are also not widely known.

Much of the materials presented herein have been published in peer reviewed journal articles and yet are still not widely known. The underlying objective of writing the article is to facilitate knowledge transfer to frontline practising professionals as well as to regulators (code drafters) who operate in regions of low to moderate seismicity which account for more than 90% of landmass on earth.

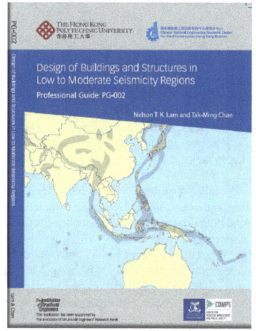

Fig. 9 Flyer containing QR code for textbook available for free-download

Acknowledgements Much of the materials presented in this paper were generated from collaborative research across many institutions including University of Mel-bourne over many years. Contributions from the following persons are specifically acknowledged: Professor John Wilson, Dr Hing Ho Tsang, Dr Daniel Looi, Dr Elisa Lumantarna and Dr Scott Menegon, and along with numerous PhD candidates who were supervised by the author in the past. The assistance given by Rohit Tiwari in relation to my travel to India and the preparation of the keynote presentation is also acknowledged. Finan-cial support of the Commonwealth Government of Australia through the Cooperative Research Centre program is acknowledged.

References

1. Lam NTK, Tsang HH, Lumantarna E, Wilson JL (2016) Minimum loading requirements for areas of low seismicity. Earthq Struct 11(4):539–561. https://doi.org/10.12989/eas.2016.11.4.539
2. McCue K (2004) Australia: historical earthquake studies. Ann Geophys 47(2):387–397
3. Looi DTW, Tsang HH, Hee, MC, Lam NTK (2018) Seismic hazard and response spectrum modelling for Malaysia and Singapore. Earthq Struct 15(1). https://doi.org/10.12989/eas.2018.15.1.067
4. Allen TI, Leonard M, Clark D, Ghasemi H (2018) The 2018 national seismic hazard assessment for Australia-model overview. In: Clark D (ed) Geoscience Australia. Canberra, ACT, Australia
5. Standards Australia AS 1170.4-2007 (R2018) Structural design actions—part 4: earthquake actions in Australia
6. Wilson JL, Lam NTK (2006) Earthquake design of buildings in Australia by velocity and displacement principles. Awarded Warren Medal Aust J Struct Eng Trans Inst Eng Aust 6(2):103–118
7. Lam NTK, Chan TM (2018) Guidelines on design of buildings and structures in low-to-moderate seismicity countries. Professional guide: PG-02, 1st edn. Approved publication by the Research Panel of The Institution of Structural Engineers. The Hong Kong Polytechnic University Press, Hong Kong
8. Lam NTK, Wilson JL, Chandler AM, Hutchinson GL (2000) Response spectral relationships for rock sites derived from the component attenuation model. Earthq Eng Struct Dyn 29(10):1457–1490
9. Lam NTK, Wilson JL, Chandler AM, Hutchinson GL (2000) Response spectrum modelling for rock sites in low and moderate seismicity regions combining velocity, displacement and acceleration predictions. Earthq Eng Struct Dyn 29(10), 1491–1526
10. Lam NTK, Chandler AM (2005) Peak displacement demand in stable continental regions. J Earthq Eng Struct Dyn 34, 1047–1072
11. Lumantarna E, Wilson JL, Lam NTK (2012) Bi-linear displacement response spectrum model for engineering applications in low and moderate seismicity regions. Soil Dyn Earthq Eng 43:85–96
12. Kafle B, Lam NTK, Gad EF, Wilson JL (2011) Displacement controlled rocking behaviour of rigid objects. J Earthq Eng Struct Dyn 40:1653–1669
13. Kafle B, Lam NTK, Lumantarna E, Gad EF, Wilson JL (2015) Overturning of precast RC columns in conditions of moderate ground shaking. Earthq Struct 8(1):1–18
14. Lumantarna E, Lam NTK, Wilson JL (2013) Displacement controlled behaviour of asymmetrical single-storey building models. J Earthq Eng 17, 902–917
15. Lam NTK, Wilson JL, Hutchinson GL (1997) Review of the torsional coupling of asymmetrical wall-frame buildings. Eng Struct 19(3):233–246
16. Lumantarna E, Lam NTK, Wilson JL (2018) Methods of analysis for buildings with uni-axial and bi-axial asymmetry in regions of lower seismicity. Earthq Struct 15(1). https://doi.org/10.12989/eas.2018.15.1.081
17. Lam NTK, Wilson JL, Chandler AM (2001) Seismic displacement response spectrum estimated from the frame analogy soil amplification model. J Eng Struct 23:1437–1452
18. Tsang HH, Wilson JL, Lam NTK, Su RKL (2017) A design spectrum model for flexible soil sites in regions of low-to-moderate seismicity. Soil Dyn Earthq Eng 92:36–45. https://doi.org/10.1016/j.soildyn.2016.09.035
19. Tsang HH, Chandler AM, Lam NTK (2006) Estimating non-linear site response by single period approximation. J Earthq Eng Struct Dyn 35(9):1053–1076
20. Chandler AM, Tsang HH, Lam NTK (2006) Simple models for estimating site period shift and damping in soil. J Earthq Eng Struct Dyn 35(15):1925–1947
21. Tsang HH, Sheikh MN, Lam NTK (2012) Modelling shear rigidity of stratified bedrock in site response analysis. Soil Dyn Earthq Eng 34, 89–98

22. Wilson JL, Lam NTK (Eds) ASS1170.4-2007 (2008) Commentary: structural design actions—part 4: earthquake actions in Australia. Australian Earthquake Engineering Society, Australia
23. Menegon S, Wilson JL, Lam NTK, McBean P (2017) RC walls in Australia: seismic design and detailing to AS 1170.4 and AS 3600. Austr J Struct Eng 18. https://doi.org/10.1080/13287982.2017.1410309
24. Standards New Zealand, NZS 3101 (2006) Part 1: concrete structures standard part 1—The design of concrete structures
25. Standards Australia, AS 3600 (2018) Concrete structures. Standards Australia Limited, Sydney, NSW

Floor Response Spectra Generation Considering Nonlinearity of Reinforced Concrete Shear Walls

Paresh Kothari, Y. M. Parulekar, G. V. Ramarao, and G. V. Shenai

Abstract Many devastating earthquakes exceeding the design basis levels have occurred in the past. Safety assessment of system and component like equipment and piping systems of industrial/nuclear safety-related structures subjected to earthquakes beyond design basis levels necessitates consideration of the structural nonlinearity. The evaluation of the seismic performance of nuclear power plant structures requires the assessment of shear walls, which are its main structural members and for qualification of secondary systems, in-structure or floor response spectra [FRS] is the input required. Hence, it is essential to generate nonlinear FRS considering structural nonlinearity in shear walls. Earlier shake table experiments were performed on two numbers of RC shear walls of aspect ratio 1.98 till failure by Parulekar et al. (Struct Eng Mech 59:291–312, 2016 [1]). This paper addresses the numerical work carried out on generation of FRS for the shear wall considering structural nonlinearity. A stiffness and strength degrading damage-based model given by Sucuoğlu and Erberik (Earthq Eng Struct Dyn 33:69–88, 2004 [2]) is used for modeling strength degradation for concrete hinges in dynamic analysis. The shear wall is modeled with hinge characteristics using pivot hysteretic law. Finally, the floor spectra obtained by numerical simulation of the shear wall are compared with those obtained from the tests.

Keywords Pushover · Nonlinear dynamic analysis · Floor response spectra · Shear wall

---

P. Kothari (✉) · Y. M. Parulekar · G. V. Shenai
Bhabha Atomic Research Centre, Mumbai 400 085, India
e-mail: pareshk@barc.gov.in; lncs@springer.com

G. V. Ramarao
CSIR-Structural Engineering Research Centre, Chennai 600113, India

© The Editor(s) (if applicable) and The Author(s), under exclusive license to Springer Nature Singapore Pte Ltd. 2021
S. K. Saha and M. Mukherjee (eds.), *Recent Advances in Computational Mechanics and Simulations*, Lecture Notes in Civil Engineering 103,
https://doi.org/10.1007/978-981-15-8138-0_30

1 Introduction

The evaluation of the seismic performance of systems and components of a nuclear power plant structures requires the assessment of shear walls and generation of floor response spectra at different heights of such shear walls. During earthquakes, structures amplify the base motion causing floor motions that are typically stronger than base. Hence, it is required to generate floor response spectra (FRS) accurately for design of these components resting on floor.

Response of equipment mounted on structures subjected to dynamic excitation has received much attention in past literature. Villaverde (1997) [3] has given comprehensive details of state of art for linear structures. Effect of structural nonlinearity on FRS was considered way back by Kelly (1978) [4]. Analysis was carried out considering elasto-plastic strain hardening model and it was concluded that for yielding structures higher mode effects are more pronounced. Sewell et al. (1986) [5] carried out investigation of effects of structural nonlinearity on equipment response. It was observed that for SDOF structures inelastic deformations tend to reduce the floor response spectra at and below the fundamental period of vibration. However, analysis of nonlinear MDOF steel framed structures leads to the facts that there were amplifications in equipment response at high frequencies. Later Singh et al. (1996) [6] through a comprehensive parametric study of MDOF system concluded that FRS of yielding structure can be higher in high-frequency range than elastic spectra due to internal resonance if higher mode frequencies are odd multiples of fundamental frequencies. Villaverde (2006) [7] proposed an approximate method to estimate the seismic response of nonlinear non-structural components on nonlinear structures. The method was based on the procedure previously developed for analyzing linear non-structural components attached to a linear primary structure by reduction of stiffness. Later analytical investigations were carried out by Oropeza et al. (2010) [8] to obtain the effect of primary and secondary system frequency and strength reduction factor on FRS of nonlinear structure. Hoffmeister et al. (2011) [9] performed studies using incremental dynamic analyses on simplified models. They evaluated response about floor response spectra for various fundamental periods and ductility ratios achieved during the dynamic response of the supporting structures. They concluded that the floor response spectra shall undergo a systematic revision to enhance both the safety and the effectiveness of the code provisions.

In the literature till date nonlinearity of structure is considered using elastic perfectly plastic models which are applicable to steel structures. The nonlinearity of concrete structures is not used in which stiffness degradation, strength degradation, and pinching effect exist in the hysteretic characteristics. Moreover, there is lot of uncertainty in accurate nonlinear hysteretic modeling of RC elements due to the variation in stiffness degradation and pinching observed with change of axial load and percent reinforcement in the member. Accurate estimates of inelastic behavior may be obtained if structural systems possess stable inelastic behavior. However, a major portion of the existing RC structures subjected to earthquake excitation exhibits capacity loss under strong ground motions, eventually leading to damage accumulation throughout their response duration. Deterioration in the mechanical properties of concrete structures is usually observed under repeated cyclic loading in the inelastic response range. The structure thus represents a stiffness and strength deteriorating (SSD) system and exhibits reduced cyclic energy dissipation capacity even under constant-amplitude cycles. The stiffness degrading model like Takeda model [10] is very common and popular in modeling concrete. However, complex stiffness and strength degrading models [2] must be used for precise evaluation of floor spectra of in-elastically deforming RC structures. Hence, there is a need to perform detailed nonlinear dynamic analysis of structure considering stiffness and strength degradation of concrete and to validate it by experiments. Moreover, a comparative study of analytical and experimental evaluation of the floor spectra for shear wall deforming nonlinearly is not carried out till date. Earlier shake table tests were performed of midrise shear wall structure with increasing base excitation from 0.1 g up to 0.9 g peak till occurrence extensive damage and failure of the structure [1]. These tests lead to damage accumulation on the structure at each level of excitation and hence it was essential to accurately carry out nonlinear modeling of the wall to simulate such damage. In this context, this paper has the aim of discussing analytical evaluation of floor spectra for the midrise shear wall tested on shake table [1] by modeling appropriate concrete hinges considering stiffness and strength degradation. A stiffness and strength degrading damage-based model given by Sucuoğlu and Erberik [2] is used for modeling strength degradation for concrete hinges. The shear wall is modeled with hinge characteristics using pivot hysteretic law. Finally, the floor spectra obtained numerically by performing detailed nonlinear dynamic analysis with hinge characteristics using pivot law are validated by those obtained from tests.

2 Details of the Shear Wall

The RC shear wall model has 3 m height (h), 1.56 m width, 0.2 m thickness (t), reinforcement of 0.4% in vertical direction, and 0.3% in horizontal direction. The foundation is 0.4 m deep and 2 m × 2 m in area and top slab is of dimensions 2.5 m × 2.5 m and 0.5 m thick. The reinforcement details of shear wall are given in Fig. 1. Mass of 8.5 tons in the form of concrete cube of dimension (1650 m × 1650 m × 1420 m) is added on the top slab and the mass of the top slab is 8 tons. Thus, the shear wall will be subjected to total axial load of 16.5 tons and the axial load ratio is 0.03. Two numbers of such shear wall specimens are tested on shake table. Concrete used for the two specimens of shear walls was made from Portland cement, river sand, and crushed gravel. Measured slumps ranged from 60 to 180 mm for all specimens. Compression tests were carried out on six numbers of 150 × 150 mm cubes for each shear wall. Tension tests were carried out on six numbers of HYSD (high yield strength deformed) reinforcement bars for each shear wall. The ultimate strength of rebars was in range of 545–558 MPa. The details are mentioned in Table 1.

Fig. 1 Shear wall details

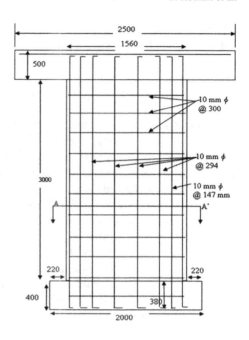

Table 1 Shear wall specimen details [1]

Shear wall no	Aspect ratio	Cube compressive strength (Mpa) 28 days	Yield stress of rebars (MPa)	Loading condition
SW1	1.92	46.5	410.56	Shake table
SW2	1.92	45.4	429.85	Shake table

3 Shake Table Tests

Two identical specimens were tested on shake table in the test setup as shown in Fig. 2. The out-of-plane deformation of the shear walls is restrained by setting steel frame as shown in Fig. 2. The steel frame was rigid with frequency of 40 Hz. The shake table was having capacity of 50 Tons with maximum acceleration of 1 g. Accelerometers and laser displacement sensors were attached at the top slab and at the middle of shear wall to measure the acceleration and displacement of the shear wall at the top and middle. Initially, sine sweep tests were carried out on the shear wall to evaluate the frequency and damping. Sine sweep excitation test was carried out and it gave a frequency of 11.5 Hz. Also some amplification was observed at 3.2 Hz. The global mode of the shear wall is 11.5 Hz and it has local frequency of 3.2 Hz which is the frequency of the mass attached to the top slab of the wall. The wall frequency through analysis is obtained as 12 Hz. Damping ratio obtained from tests is found to be 0.06 which is higher and can be attributed to friction between the mass and top slab. Two specimens of shear walls were subjected to same series of spectrum

Fig. 2 Test setup (shake table test)

compatible time histories with increasing excitation from 0.1 g peak acceleration to 0.9 g peak acceleration. The input time history for 0.2 g peak acceleration and the compatible test spectrum along with required response spectrum is shown in Fig. 3.

At 0.9 g PGA, the walls failed due to sliding shear failure and both the shear walls slide at the base. The maximum displacements of the walls SW1 and SW2 before failure are 35 mm and 42 mm, respectively, relative to base in the shake table test for excitation of 0.8 g PGA. Tests were also carried out for the shear wall of same details and configuration for quasi-static monotonic loads and cyclic loads [1]. The comparison of the experimental F-δ diagram obtained from tests is shown in Fig. 4.

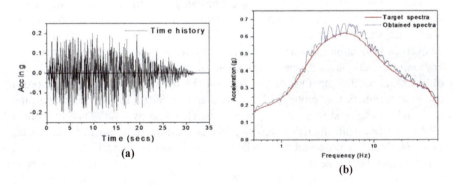

Fig. 3 a Spectrum compatible base time history. **b** Test response spectra

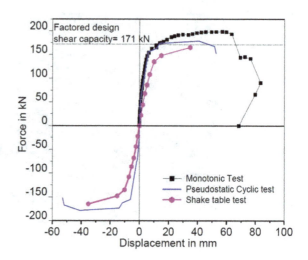

Fig. 4 Comparison of experimental F-δ diagram of the wall from tests

4 Theoretical Study

4.1 Modeling of Hysteretic Behavior of Structural Elements

Concrete is known to be a highly nonlinear material which tends to undergo severe hysteresis beyond yield. In case of reinforced concrete, the hysteretic behavior becomes more complex due to the effect of reinforcing steel and the behavior of bond between concrete and reinforcement under cyclic loads. Over the years, various researchers have proposed various models to predict the hysteretic behavior of concrete and reinforced concrete structural elements. In the past, elasto-plastic hysteretic rules that idealize the hysteretic loops in bilinear format were frequently used. Such an idealization, though reasonable for steel members, was found to be an over-simplification for concrete structures. Some other models give more consideration to effects like stiffness degradation, pinching due to opening and closing of cracks, bond slip, etc. One of the most popular hysteretic models for reinforced concrete members was proposed by Takeda et al. (1970) [10]. In this model, the constantly changing stiffness of reinforced concrete was considered. This contributed less energy dissipation than that predicted by elasto-plastic response. In 1998, Dowell et al. (1998) [11] proposed a so-called "Pivot hysteretic model" for reinforced concrete members. The model was basically developed for circular columns with the aim of serving the need of inelastic analysis of bridge pier columns. The model could consider the effect of axial load, lack of section symmetry, stiffness degradation, and pinching effect. This model is quite simple and effective in modeling the force-deformation or moment–rotation response for the members and has been used is the present analysis. The advantage of using the Pivot model is essentially due to the fact that this model is based mainly on geometrical rules that define

Fig. 5 Pivot hysteretic model

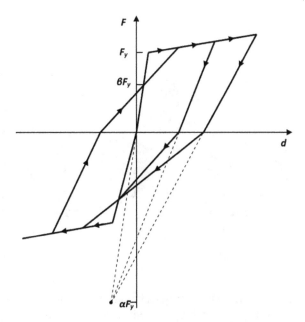

loading and unloading branches rather than analytical laws. This reduces not only the computational effort but also the number of hysteretic parameters involved.

The pivot model is governed by a set of rules that depend on the properties of the member and loading history. The two basic and most important parameters that define the pivot hysteretic model are "α" parameter and "β" parameter. As shown in Fig. 5, in pivot model, it is considered that while unloading, the load deflection path is guided toward a common point known as primary pivot point (αFy). Also, in concrete elements, the force-displacement paths tend to cross the elastic loading line at approximately the same point called pinching pivot point (βFy). Thus, these two parameters, namely, "α" and "β" are to be accurately assigned to the nonlinear springs for the members the hysteretic rules are set. The parameters "α" and "β" were provided as a function of axial load ratio and longitudinal reinforcement ratio, respectively, by Dowell et al. (1998) [11].

4.2 Definition of Pivot Model for RC Shear Wall

The structural model (Fig. 9) was formulated using finite element software. Beam element was used for modeling of shear wall. As the shear wall is having aspect ratio about 2, the peak ultimate load is governed by bending failure and hence mechanical nonlinearities of shear wall were introduced by means of interacting axial load-moment plastic hinges placed at the ends of the shear wall. Referring to

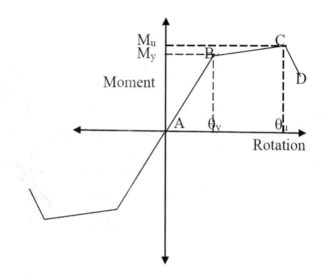

Fig. 6 Typical moment–rotation relation

plastic hinges, a multi-linear moment–curvature law was obtained. Moment rotation was obtained from moment–curvature relationship by considering plastic hinge length by Mattock et al. [12] Moment–rotation relation as shown in Fig. 6 with yield moment, yield rotation (point B), ultimate moment and ultimate rotation (point C), and failure moment and failure rotation (point D) obtained from the reinforcement and concrete characteristics. During the shake table tests of various base excitation from 0.1 to 0.9 g, the shear wall is deteriorated after few base excitations eventually leading to damage accumulation throughout its response duration. Deterioration in the mechanical properties of concrete structures is usually observed under repeated cyclic loading in the inelastic response range (0.8 g base excitation). Hence, for such type of structures, stiffness and strength degrading hysteretic models (SSD hysteretic model) should be appropriately used for calculation of accurate analytical response. Energy-based strength degrading model was developed by Sucuoglu and Erberick (2004) [2]. This model is given by Eq. (1), where $E_{h,n}$ is the normalized hysteretic energy dissipated at nth cycle for deteriorating systems and a, b are the fatigue parameters.

$$E_{h,n} = a + (1-a)e^{b(1-n)} \quad (1)$$

According to the above model, the energy dissipation at each cycle goes on decreasing as the number of cycles increases due to degradation of strength and stiffness of the system. This energy dissipation is a function of yield moment/yield strength of the SSD system. Sucuoglu and Erberick (2004) [2] have given parameter a = 1 and b = 0 for non-deteriorating system, a = 0.5 and b = 1 for moderately deteriorating system, and a = 0.22 and b = 0.82 for severely deteriorating system. The graphical representation of the fatigue model for three systems is shown in Fig. 7. Thus, for non-deteriorating system, which is represented by elastic–plastic hysteretic

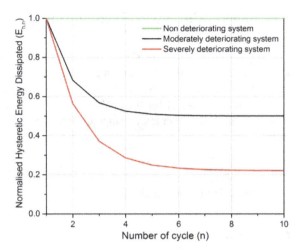

Fig. 7 Fatigue model for three systems

relation, the normalized hysteretic energy value will be 1 and there is no decrease in yield strength of the element with number of cycles. For the RC shear wall in the present work subjected to shake table tests with increasing level of excitation from 0.1 g PGA to 0.9 g PGA, the yield moment of the multi-linear hinge characteristics of members shall be decreased as per the damage level of the structure. It is observed from Fig. 4 that the ultimate load of the shear wall is decreased by factor 0.85 for shake table tests for 0.7 and 0.8 g peak base excitation levels with respect to static monotonic tests. Hence, three cases of shake table tests peak base excitation are studied in this paper, viz., 0.1, 0.2, and 0.8 g. Among these cases, 0.1 and 0.2 g base excitations represent the non-deteriorating system, where shear wall is not yielded but concrete is getting cracked. However, 0.8 g base excitations represent the system between non-deteriorating and moderately deteriorating system, where shear wall is completely yielded. The maximum displacement of 0.1 g, 0.2 g, and 0.8 g are, respectively, 2.11 mm, 3.74 mm, and 31.5 mm. Thus, the normalized hysteretic energy value for slightly deteriorating structure with a = 0.85 and b = 0.82 is taken as 0.85 corresponding to ten number of cycles as per Sucuoglu and Erberick (2004) [2] for modeling hinge characteristics of shear wall subjected to 0.8 g peak base excitation case. In the present work, the yield moment of the multi-linear hinge characteristics input to the finite element program is decreased by the same factor of 0.85 by which the normalized hysteretic energy is decreased for generating response spectra of nonlinear structure.

5 Moment–Curvature Relationship

The shear wall is a midrise shear wall with small axial compression and thus the failure of the shear wall is based on flexural yielding and the evaluation of flexural strength of the shear wall will be based on the basic principles.

Mathematical model for the analysis of curvature and deflections of reinforced concrete shear wall in cracking stage as described by Kent and Park (1971) [13] is used.

6 Analytical Simulation

6.1 Pushover Analysis

In order to get the peak lateral load taken by the shear wall, nonlinear static pushover analysis (NLSPA) was carried out for the shear wall. Modeling of shear wall was done using beam element of representative size of shear wall and additional mass was lumped at the top node of shear wall. Fixed boundary condition was assigned to the base of shear wall (Fig. 8). Moment–curvature characteristics were generated using the sectional details of the shear wall (Fig. 9). Nonlinear static hinge was provided at the base of shear wall. Targeted lateral displacement was assigned at the top of shear wall such that it simulates the experimental boundary conditions. In the experiments, the top slab of shear wall was restrained by rollers such that out-of-plane movement of the shear wall was prevented. Hence, in the analytical simulation, the same boundary condition was simulated using set of springs. NLSPA was performed and capacity curve of shear wall was obtained and compared with tests as shown in Fig. 10. It is observed that the analytically obtained pushover curve is in good agreement with experiments.

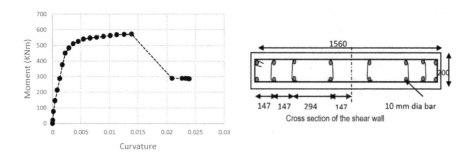

Fig. 8 Moment–curvature relationship and cross section of test shear wall

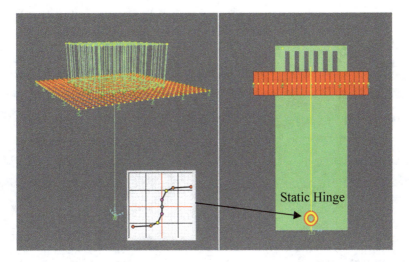

Fig. 9 Analytical model for pushover analysis

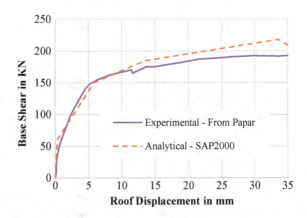

Fig. 10 Analytical pushover curve comparison with experimental [1]

6.2 Nonlinear Dynamic Analysis

Evaluation of the shear wall subjected to earthquake loading needs detailed nonlinear dynamic analysis of shear wall subjected to spectrum compatible time history.

Experiment was performed for various base excitations 0.1 g to 0.9 g and shear wall failed at 0.9 g base excitation. In the present study, 0.1, 0.2, and 0.8 g base excitations are simulated.

The same finite element model considered in nonlinear static pushover analysis model was used for performing nonlinear dynamic analysis except that the hinge was replaced by spring with cyclic characteristics. Cyclic hinges were modeled using multi-linear plastic-link element. The length of link element is kept approximately equal to plastic hinge length. Pivot hysteretic model parameters α and β are calculated

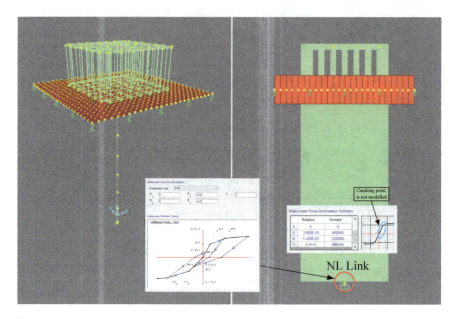

Fig. 11 Analytical model for nonlinear dynamic analysis

as 6 and 0.55, respectively, for shear wall considering the axial load on the wall and the provided shear reinforcement. The calculated pivot hysteretic model parameters α and β were assigned for shear wall bottom hinge and the hinge moment–rotation characteristics were input for each case of base excitation level. For 0.1 and 0.2 g level, the yield moment was not reduced as strength of shear wall was intact at this stage but cracking point was not given in moment–rotation curve and directly yielding point was modeled as shown in Fig. 11. And for 0.8 g base excitation, the yield and the ultimate moment values were reduced to 0.85 times yield and ultimate moment of moment–rotation relationship as per the damage state of shear wall between non-deteriorating and moderately deteriorating system. Rayleigh damping model is used in nonlinear dynamic analysis using direct time integration technique. Damping ratio was taken as 6% as per experiment data for pre-yield stage of shear wall that is 0.1 and 0.2 g base excitation [1]. It is observed by researchers that the stiffness and mass proportional Raleigh's damping following the changes in stiffness will drop after yield, practically resulting in insignificant elastic damping [14]. The common modeling choice of Rayleigh damping may lead to over-damped higher modes or to an overestimation of the contribution of the second mode, depending on to which modes the damping ratio is allocated. Therefore, for inelastic analysis, it was proposed by researchers [14] a user-specified modal damping model with a constant value of 5% in all modes, apart from the first mode, where an artificially lower value is set, as a more suitable modeling solution. This was aimed to avoid excessive damping in the post-yield phase and counter-balance the additional damping introduced due to the use of a constant value of damping ratio. In the present work, the structure acts

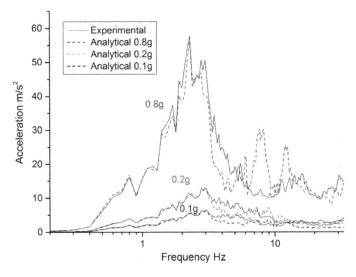

Fig. 12 Experiment and analytical FRS for 2% damping for 0.1, 0.2, and 0.8 g base excitations

like as single degree-of-freedom system and hence only first mode is most effective. So, it will be appropriate to consider a lower value of damping in Raleigh damping for the nonlinear dynamic analysis of shear wall at 0.8 g PGA for considering the stiffness reduction due to yielding effect. The stiffness reduction from elastic stiffness is obtained from the peak force taken by the wall and failure displacement of 35 mm which is achieved at 0.8 g. Using this modified stiffness 2% damping ratio is evaluated as the appropriate damping to be input for nonlinear dynamic analysis at 0.8 g peak base excitation level. Hence, lesser damping ratio of 2% was taken for post-yield stage (0.8 g base excitation). Nonlinear time history analysis was performed for 0.1, 0.2, and 0.8 g base excitation and results were compared with experiment. Experimental and analytical floor spectra are shown in Fig. 12, for 0.1, 0.2, and 0.8 g base excitations for first floor, second floor, and roof for 2% damping. There is good agreement with the FRS simulated by nonlinear dynamic analysis and with that obtained from experiments. It is observed that the stiffness and strength degradation of the structure leads to decrease in fundamental frequency of the structure from 12 Hz at 0.1 g peak base excitation to 8 Hz at 0.8 g peak base excitation which is about 33% decrease in frequency. The combined frequency of shear wall with local frequency of top mass attached to shear wall is reduced from 3.2 Hz at 0.1 g peak base excitation to 2.5 Hz due to stiffness and strength degradation at 0.8 g peak base excitation. Hinge rotations at the shear wall bottom at different levels of excitation are presented in Figs. 13, 14 and 15.

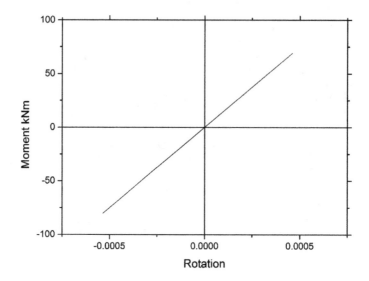

Fig. 13 Shear wall bottom hinge characteristic for 0.1 g base excitation

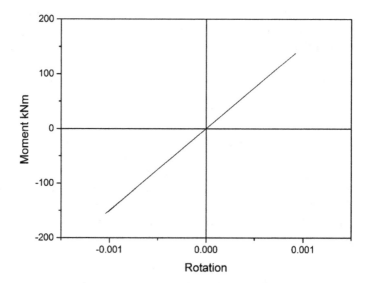

Fig. 14 Shear wall bottom hinge characteristic for 0.2 g base excitation

7 Conclusion

Detailed nonlinear dynamic analytical simulation of shake table experiments of RC shear wall is attempted in the present work. Appropriate pivot hysteretic model with strength and stiffness degradation is employed with accurate damping in order to

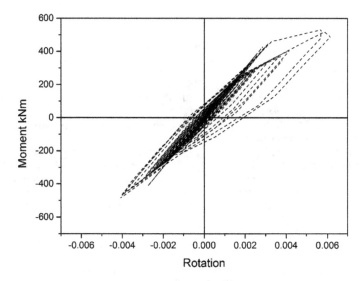

Fig. 15 Shear wall bottom hinge characteristic for 0.8 g base excitation

evaluate the floor response spectra at the top of the RC shear wall. The following conclusions are drawn from the work carried out:

- The floor response spectra obtained on RC shear wall structures considering degradation of the strength and the stiffness and its ultimate deformability into nonlinear domain is entirely different from that obtained in linear structures.
- Rayleigh's damping to be considered for nonlinear dynamic analysis of the structure will be lesser than that of the linear structure due to the reduction in stiffness which will drop after yield, practically resulting in insignificant elastic damping.
- The frequency of nonlinear shear wall structure is reduced to about 8 from 12 Hz for a linear structure. The stiffness degradation leads to decrease in fundamental frequency by about 33% than that of linear RC shear wall structure.
- This reveals that the realistic evaluation of FRS is essential for structures subjected to beyond design basis earthquakes for safety of the equipment/piping systems.

References

1. Parulekar YM, Reddy GR, Singh RK, Gopalkrishnan N, Ramarao GV (2016) Seismic performance evaluation of mid-rise shear walls: experiments and analysis. Struct Eng Mech 59(2):291–312
2. Sucuoglu H, Erberik A (2004) Energy-based hysteresis and damage models for deteriorating systems. Earthq Eng Struct Dyn 33:69–88
3. Villaverde R (1997) Seismic design of secondary structures: state of the art. J Struct Eng 123(8):1011–1019

4. Kelly TE (1978) Floor response of yielding structures. Bull NZ Nat Soc Earthq Eng 11(4):255–272
5. Sewell RT, Cornell CA, Toro GR, McGuire RK (1986) A study of factors influencing floor response spectra in nonlinear multi-degree of freedom structures, Report No. 82, The John Blumo Earthquake Engineering Center, Stanford University
6. Singh MP, Chang TS, Suarez LE (1996) Floor response spectrum amplification due to yielding of supporting structure. In: Proceedings 11th World conference on earthquake engineering, Acapulco, Mexico, Paper No 1444
7. Villaverde R (2006) Simple method to estimate the seismic nonlinear response of non-structural components in buildings. Eng Struct 28(8):1209–1221
8. Oropeza M, Favez P, Lestuzzi P (2010) Seismic response of non-structural components in case of nonlinear structures based on floor response spectra method. Bull Earthq Eng 8:387–400
9. Hoffmeister B, Gündel M, Feldmann M (2011) Floor response spectra for dissipative steel supports of industrial equipment. In: Thematic conference on computational methods in structural dynamics and earthquake engineering, COMPDYN-2011
10. Takeda T, Sozen MA, Nielsen NN (1970) Reinforced concrete response to simulated earthquakes. J Struct Div (ASCE) 96:2557–2573
11. Dowell RK, Seible F, Wilson EW (1998) Pivot hysteresis model for reinforced concrete members. ACI Struct J 95:607–617
12. Mattock AH (1967) Discussion of Rotational capacity of reinforced concrete beams by W.G. Corley. J Struct Div ASCE, 93, 519–522
13. Kent DC, Park R (1971) Flexural members with confined concrete. J Struct Div ASCE, 97(7), 1969–1990
14. Smyrou E, Priestley MN, Carr AJ (2011) Modelling of elastic damping in nonlinear time-history analyses of cantilever RC walls. Bull Earthq Eng 9(5):1559–1578

Efficient Arrangement of Friction Damped Bracing System (FDBS) for Multi-storey Steel Frame

Saikat Bagchi, Avirup Sarkar, and Ashutosh Bagchi

Abstract The Friction Damped Bracing System (FDBS) is able to significantly control the vibration of framed structure without dissipating energy through the inelastic yielding of its structural components. Therefore, it is a useful tool to design the structural system by isolating the energy dissipation components at some specific as well as desired locations. This purposeful isolation of the critical components helps, in turn, to monitor the health of the system efficiently, especially for the large and complicated systems such as process plant structures, offshore structure, etc. Therefore, effective placement of the energy dissipation devices in terms of their numbers as well as locations is essential to meet the optimum requirement of serviceability, safety, and stability. In this article, FDBS is modelled numerically following standard Friction Damper guideline. The 2D building frames with FDBS at various locations are used to study the responses of multi-storey building frames having different vertical bracing configurations. Locations of the energy dissipation devices are altered for each of the structures to study the effect of load flow through the desired load path. It is intended to isolate the FDBS in such a way so that the operational constraints do not interfere with the monitoring and maintenance of the critical dissipating system, which is the lifeline for the structural stability. Nonlinear time history analysis is performed for each of the frames for a scaled ground motion obtained using Conditional Mean Spectra for the city of Vancouver. Energy dissipation behavior of the structures is compared in order to comprehend the effect of damper arrangement. Load versus deflection behaviour of the structures at different levels indicate that structures with regular configurations show better behaviour in comparison to the customized structures with special configurations. Therefore, it is concluded that FDBS enabled structural systems are suitable as well as necessary for the complicated structures where the horizontal load transfer system is expected to be flexible to meet the process requirements.

S. Bagchi (✉) · A. Sarkar · A. Bagchi
Concordia University, Montreal, QC H3G 1M8, Canada
e-mail: er.saikat.ac@gmail.com

Keywords Friction damped bracing system (FDBS) · Conditional mean spectra ·
Nonlinear time history analysis · Storey drift · Base shear

1 Introduction

Aseismic design of a multistory steel frame especially in case of an industrial facility like steel or power plant is very much complicated because of its process-specific requirement. In those cases, the design engineer needs to take care of not only the structural constrains but also must accommodate the equipment or facilities for which the structure is to be constructed. Many a time process requirement is such that the structural stability system cannot be planned in the desired way. For example, in case of a silo or water tank supporting Junction House, where in addition to the huge overburden load belt conveyor enters or exits the building at an angle and an elevation-difference to each other, usual bracing planning is not possible. Aseismic design of such structure may call for some amount of energy dissipation using the ductility of the constituting members. As per the philosophy of capacity design, predetermined members of the structure are expected to perform in the zone of plastic deformation while others continue to remain elastic. Energy dissipation through this process has a serious limitation because in such a situation the weak link of the structural lattice, i.e., the ones which are designed to operate in the nonlinear range undergo permanent deformation. In the post-earthquake situation, repair and rehabilitation of such structures become a difficult task because of those partially damaged members. Therefore, it is imperative from the perspective of the seismic resilience that the energy dissipation system be planned and designed in such a way so that it doesn't impair the major load-bearing system and can perform uninterruptedly even after the seismic event. Friction damped bracing system (FDBS) is one of such solutions which has been popular for the last three to four decades. Although the usage of FDBS is not new, as of now the existing international design standards doesn't provide with an integrated design approach where the influence of the passive energy dissipation system, i.e., FDBS in this case, is considered. One of the major reasons behind this incompleteness is the insufficient study of the structural behaviour with and without FDBS. The current article reports a numerical study of a five-storey five-bay 2D frame for different bracing configurations and compares the behaviour of FDBS with conventional braced frame and moment-resisting frames. The study also tries to identify the basic scheme for bracing placement which facilitates the maintenance without hindering the process requirements. Before the reader is introduced with the detailed working methodology a brief review of the existing state of art friction damper enabled structural analysis and design practices are presented in the next section. Subsequently, the models, ground motion input, its selection procedure and the results are discussed. The article concludes by summarizing the findings of the study and the scope of further research.

2 Friction Damped Bracing System (FDBS): Brief History and Questions Pertaining to the Design and Application

Passive energy dissipation system works as a response to the external force that is applied to the structure and is intended to be resisted [1]. Friction damper is one of the force-based passive energy dissipation devices which uses friction between two metallic and/or non-metallic surfaces to dissipate energy. The device is widely used in the automobile industry as a component of the automotive brake to dissipate kinetic energy. Avtar Pall introduced the mechanism to the structural requirements with the innovative design of Pall damper applied to the panelled structures [2]. After that, through many studies, the energy dissipation behaviour of this kind of damper including its surface attributes are studied thoroughly [3–5]. Over the time, different types of friction devices are invented and introduced to the structure such as Sumitomo friction damper [6], Energy dissipating restraint [7], the energy dissipator that uses slotted-bolted connection [8, 9], cylindrical friction damper [10], etc. Recently, Quaketek Inc. [11] has developed a friction damper which uses a similar principle that of the Pall damper. In terms of quantifying the behaviour of these devices, two of the important parameters are the initial slip force to storey yield force ratio and the ratio of bracing stiffness to the corresponding storey stiffness [12]. Although several attempts have been made to practice the structural design involving FDBS [13–15], till date no international design standard prescribes designing the structure considering active participation of the passive energy dissipation device [16]. As per the National Earthquake Hazards Reduction Program (NEHRP) recommended provisions (FEMA P-750, 2009) the design load applicable for the seismic force-resisting system without the damper can be reduced up to 75 per cent if the damper is installed [17]. Additionally, in terms of damper arrangement also available guideline is not sufficient. FEMA-356 (2000) stipulates that there should be at least four displacement activated dampers in a storey along any principal direction [18], whereas usually an undamped bracing system may comprise of only one set of vertical bracing in the storey. Therefore, the question is how to use FDBS as a replacement of conventional bracing system (CBS). Moreover, in a horizontally spread building location of the horizontal load transfer system is very important. With the understanding of these requirements in the next section working methodology of this study is described.

3 Working Methodology

In order to understand the behaviour of FDBS over the moment-resisting frame and the conventional braced frame, two sets of frames with both types of bracing configuration are studied through nonlinear time history analysis along with additional two moment-resisting frames. Frames of various bracing configurations (both with FDBS and CBS) are studied and compared to achieve a quantitative inference.

3.1 The Numerical Model of the Structure

The structure is modelled numerically using ETABS commercial software [19]. In one of the sets, the vertical bracing system (with and without FD) is located at the central bay of the 5 bay 2D frame and in the other sets the same is located at the corner bay. Apart from these, another two structures with moment-resisting frame (MRF) are studied. In these cases, also the location of the MRF is either at the central bay or the corner bay. Frames with the lateral load transfer system located at the central bay and the corner bay are presented in the Figs. 1 and 2, respectively. Figure 3 shows a typical frame with a soft storey; in the presented case it is shown to be in the 2nd storey. Loads are applied as per the recommendations of the National Building Code of Canada, 2015 (NBCC 2015) [20].

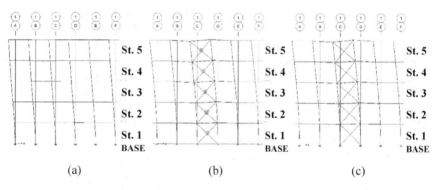

Fig. 1 Frames with the lateral load transfer system located at the central bay; **a** Moment-resisting frame, **b** Frame with FDBS, **c** Frame with CBS

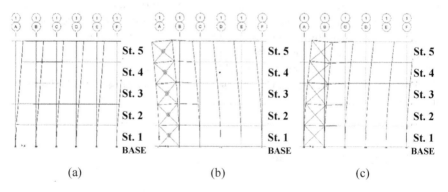

Fig. 2 Frames with the lateral load transfer system located at the corner bay; **a** Moment-resisting frame, **b** Frame with FDBS, **c** Frame with CBS

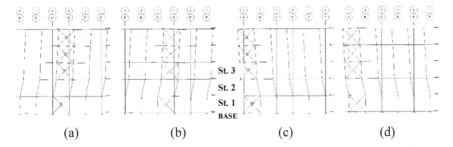

Fig. 3 Typical frames with a soft storey (at the 2nd level); **a** Frame with FDBS at the central bay, **b** Frame with CBS at the central bay, **c** Frame with FDBS at the corner bay, **d** Frame with CBS at the corner bay

3.2 Modelling the Friction Damped Bracing System (FDBS)

Passive energy dissipation using friction damper (FD) involves nonlinear hysteresis energy dissipation technique where nearly rectangular hysteresis loop of the FD characterizes the damper behaviour. In this case, the slip load of a specific damper is considered as the yield load of the member which connects the device. The current numerical study considers Pall type friction damper in the model [21]. Here the FD modelling guidelines provided by Pall Dynamics Inc. and Quaketek Inc. [11] are considered. As stipulated in by the Pall Dynamics, for tension only cross bracing when the damper in the tension brace slips, the compression brace is shortened suitably to avoid the buckling. This feature enables the compression brace to remain perfectly elastic and therefore, during the next cycle of loading when this brace takes the load in tension, it remains ready to take tension in the similar way. The friction damper is idealized as an ideal elasto-plastic link (Wen) as per the procedure provided by Quaketek Inc. The mass of the damper enabled bracing is taken as the sum of the bracing mass and the mass of the damper provided in the catalogue for 450 kN friction damper. The rotational inertia of the damper is not considered because of its strictly axial behaviour.

3.3 Ground Motion Selection

The Selection and scaling of time history records is an important part of the procedure of performance evaluation of structures by time history analysis. There are several methods used in the scaling of earthquake time histories. Scaling makes the selected time histories compatible with a target design spectrum at the fundamental period of the structure so that the results of time history analysis are comparable to the results of response spectrum analysis. In general, the target design spectra for scaling is the Uniform Hazard Spectra (UHS) provided in the design codes for a particular country. However, research has suggested that the UHS is an envelope curve and not suitable as a target spectrum for scaling procedures. Baker developed the concept of Conditional Mean Spectra (CMS) to be used as a suitable response spectrum for scaling procedures, which is not an envelope curve and matches the spectral amplitude of the UHS at the period of interest, i.e., the fundamental period of the structure but has much lower spectral ordinates at other time period values, using the concept of conditional probability [22]. In this article, scaling has been done using the CMS method provided by Baker. Selection of time history is also important; the chosen time history records should satisfy the seismic parameters of the site for which the structure is designed. In this case, as the building is designed for the city of Vancouver, tentime history records are selected which has a ratio of peak ground acceleration and peak ground velocity as 1, which matches the seismic parameter of Vancouver [23]. For all these ten records, the corresponding response spectrum is developed by MATLAB [24] coding. For the CMS method, the sum of squared errors, over the entire period range, is calculated between the developed CMS for Vancouver and the spectra developed by MATLAB codes corresponding to the time histories selected. Only those time histories which give the minimum values of the sum of squared errors are chosen for scaling by CMS method. Finally, the ratio of spectral ordinates of the CMS and the response spectra of the chosen time histories are calculated to be the scale factor and the time histories are multiplied by that factor to obtain the scaled records, used as the input motions for time history analysis. The input acceleration time histories are shown in Fig. 4.

4 Results and Discussion

As elaborated in the previous section, 2D numerical models of four sets of braced frame steel structures and two moment-resisting frame structures are analyzed against three scaled realistic ground motion acceleration time histories for the site at Vancouver city to comprehend the influence of FDBS and its arrangement in the structural response. In this section, the hysteresis plot of the FDs and different response parameters of the frames such as Base shear, Horizontal displacement, Pseudo spectral acceleration (PSA), etc. are presented.

4.1 Hysteresis Behavior

The Hysteresis behaviour shown in Fig. 5 represents the nonlinear behaviour of the friction dampers for the five storeys of the central FDBS against the GMR—1. It is evident that the damper in the top storey doesn't operate under enough load to be predominantly in the nonlinear range. Thus, the hysteresis loop for the fifth storey is not that prominent.

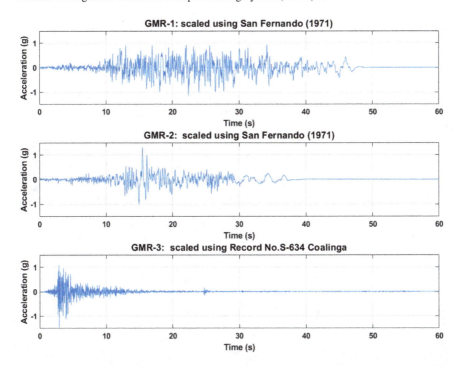

Fig. 4 Scaled ground motion records for Vancouver obtained using conditional mean spectra [25, 26]

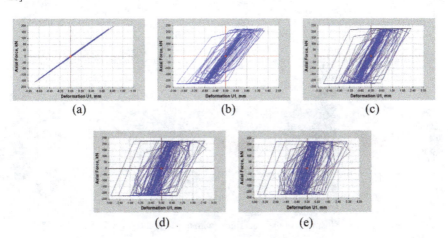

Fig. 5 Hysteresis behaviour of the plastic links at different storey levels; **a** Fifth (Top) storey, **b** Fourth storey, **c** Third storey, **d** Second storey, **e** First (Ground) storey

4.2 Base Shear (V_b)

Base shear time histories for the three types of frames with no soft storey are presented in Figs. 6, 7, 8, 9, 10, and 11. It is evident that for all the cases the base shear is lesser with FDBS in comparison to the conventional frame and in some cases, it is lesser than the rigid frame system. Moreover, FDBS slightly reduces the shake duration of the frame.

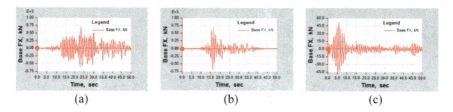

Fig. 6 Base Shear time histories of the steel frame with the central rigid frame system, **a** GMR—1, **b** GMR—2, **c** GMR—3

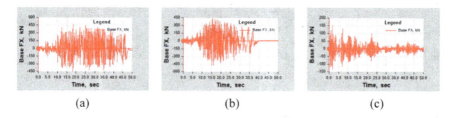

Fig. 7 Base Shear time histories of the steel frame with the central FDBS, **a** GMR—1, **b** GMR—2, **c** GMR—3

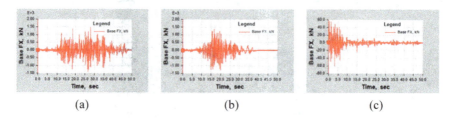

Fig. 8 Base Shear time histories of the steel frame with the central CBS, **a** GMR—1, **b** GMR—2, **c** GMR—3

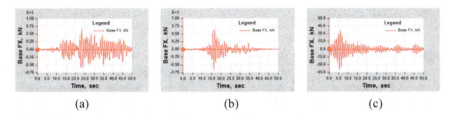

Fig. 9 Base Shear time histories of the steel frame with corner rigid frame system, **a** GMR—1, **b** GMR—2, **c** GMR—3

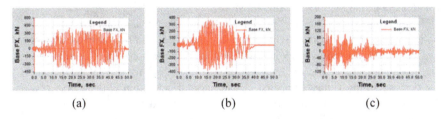

Fig. 10 Base Shear time histories of the steel frame with corner FD brace system, **a** GMR—1, **b** GMR—2, **c** GMR—3

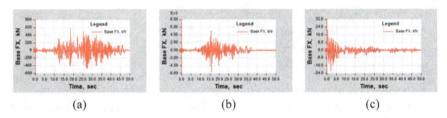

Fig. 11 Base Shear time histories of the steel frame with the corner brace system, **a** GMR—1, **b** GMR—2, **c** GMR—3

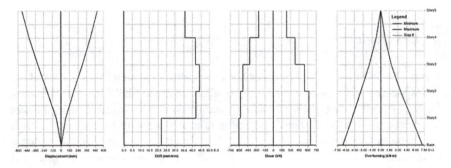

Fig. 12 Envelope of the storey responses of the steel frame with the rigid frame system located at the corner bay of the frame

4.3 Storey Response

Figure 12 shows the storey responses for the moment-resisting rigid frame system located at one corner bay of the frame. Displacement mode shape for both the types of bracing systems are found to be similar but the magnitude of the displacement usually increases significantly if FDBS is used. However, corresponding storey shear decreases. Figures 13 and 14 show storey responses for different soft storey locations in the frames with FDBS and CBS, respectively. As expected, with the upward shift of the soft storey deflection increases and base shear decreases but the trend is not true for the ground storey and the 4th storey.

It is observed that if the bays adjacent to the soft storey are braced using FDBS an optimized combination of the deflection and base hear can be achieved. Figures 15 and 16 present the storey responses for such an arrangement where both base shear and horizontal deflection of the top storey (5th storey) for FDBS are lesser than the corresponding values for CBS. The horizontal deflection at the top storey, i.e., cumulative elastic maximum drift at the top of the building is 65 mm which is well within the prescribed limit of 2.5 per cent of the height. Corresponding base shear is ~457.34 kN which is less than the corresponding base hear value for the CBS system (~777.83 kN).

4.4 Pseudo Spectral Acceleration (PSA)

Pseudo spectral acceleration (PSA) noted at the top floor for different locations of the soft storey in the FDBS are presented in Fig. 17. It shows that the presence of the soft storey increases the value of PSA at the top floor. As the location of the soft storey moves upward, there is a slight reduction in the spectral quantity.

5 Conclusions

The prime objective of the current study is to identify the suitable scheme and the corresponding parameters to quantify the influence of the FDBS in a steel building frame. It is well known that for the same ground motion force in a particular structure is dependent on its fundamental natural period of vibration (T_f), which in turn is a signature of its flexibility. Hence, in order to capture the actual frame response depending on the structure-specific T_f, period-specific conditional mean spectra is used in this study, i.e., eventually, each of the structures is analyzed against a unique input ground motion tailor-made for that specific frame.

It is observed that the frames with corner bracing system are flexible in comparison to the frames with the central bracing system. However, as expected, the deference is comparatively less in the case of FDBS. Another important observation is

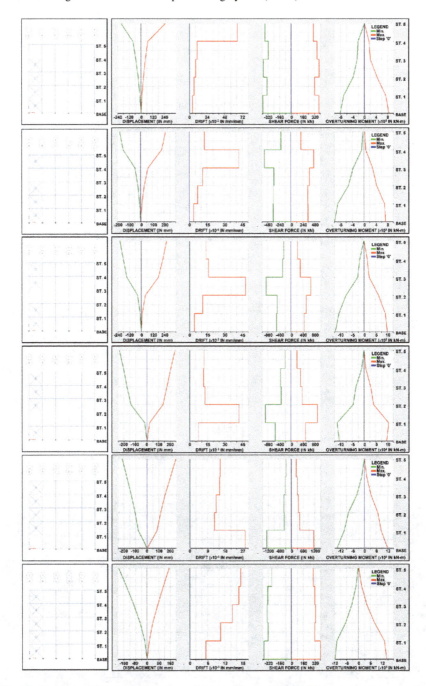

Fig. 13 Envelope of the storey responses of the steel frame with FDBS located at the corner bay of the frame for different soft storey locations

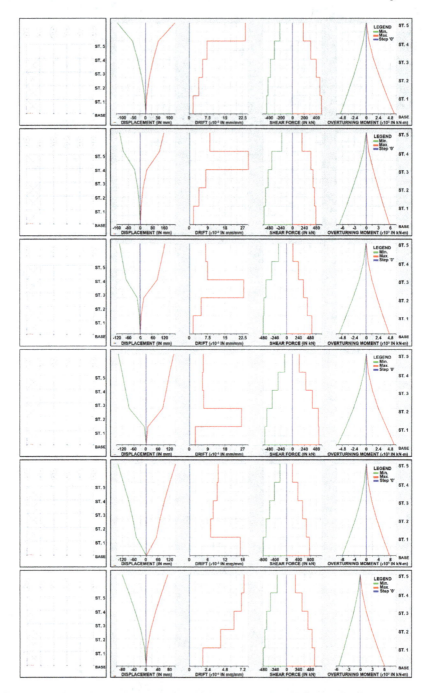

Fig. 14 Envelope of the storey responses of the steel frame with CBS located at the corner bay of the frame for different soft storey locations

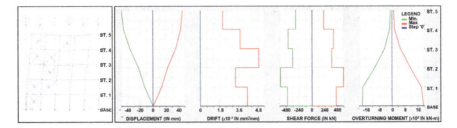

Fig. 15 Storey response of the frame with FDBS and a soft storey at the 3rd level

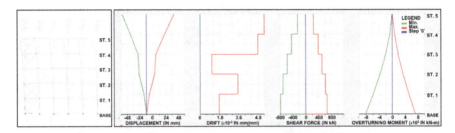

Fig. 16 Storey response of the frame with CBS and a soft storey at the 3rd level

the arrangement of the FDBS has very minimum influence on the variation of T_f values.

Usage of FDBS reduces the base shear for all the cases in comparison to CBS. Usually, as the flexibility of the frame increases, base shear reduces. On the other hand, as the location of the soft storey elevates, base shear usually increases but for a rigid base frame with a soft storey at the 2nd level also increases the base shear significantly. The trend is true for both FDBS and CBS. Additionally, FDBS slightly reduces the shake duration also. The study concludes that it is possible to optimize the base shear value and the corresponding elastic cumulative maximum top storey drift using FDBS.

PSA values are found to be more if the soft storey is introduced in the structure. The important inference is that the spectral location of the amplification has a weak but direct relation with the T_f value of the structure. For the frames with a soft storey at the 4th or 5th storey, reduction in the spectral quantity may be attributed to the reduction in the overburden load, not to any structural attribute.

5.1 Contribution, Limitation, and Further Scope

Summarizing the inferences drawn in this article it is possible to say that FDBS can emerge as a reliable solution to handle the dynamic response of the framed structure. The major take-aways from the current study are as follows.

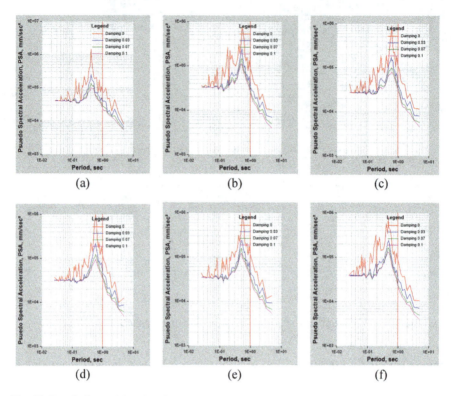

Fig. 17 Pseudo Spectral Acceleration (PSA) at the 5th storey (St.) of the steel frame with the corner FDBS, **a** No soft storey, **b** soft storey at St. 1, **c** soft storey at St. 2, **d** soft storey at St. 3, **e** soft storey at St. 4, **f** soft storey at St. 5

Contribution: The article reports the first-ever quantitative study of steel structural frame with FDBS against T_f specific acceleration time history. It tries to identify the possible parameters, necessary to formulate the basic member arrangement scheme, which may be important for a process plant structure where bracing arrangement may be random to meet the process requirements. Once identified, the exercise is helpful for structural maintenance also. The learnings can as well be applied for the regular frames. Nevertheless, the study confirms that if possible, the regular arrangement of bracing is beneficial.

Limitations: Two-dimensional model of the study is its major limitation. A 3D model will be able to capture the real behaviour of the structure including the influence of the torsional modes. Additionally, the current study doesn't report the velocity response.

Further scope: The preliminary results obtained in this study opens the scope for further detail study of the frames with regular as well as irregular configurations in the temporal and spectral domains. Storey wise transfer functions for displacement, velocity, and acceleration need to be studied for 3D frames with various configurations.

Acknowledgements The support of the Natural Sciences and Engineering Research Council (NSERC), Ottawa, Canada and Mr Avik Dhar are gratefully acknowledged.

References

1. Housner GW, Bergman LA, Caughey TK, Chassiakos AG, Claus RO, Masri SF, Skelton RE, Soong TT, Spencer BF, Yao JTP (1997) Structural control: past, present, and future. J Eng Mech 123(9):897–971
2. Pall AS (1979) Limited slip bolted joints: a device to control the seismic response of large panel structures. Doctoral dissertation, Concordia University
3. Filiatrault A, Cherry S (1987) Performance evaluation of friction damped braced steel frames under simulated earthquake loads. Earthq Spectra 3(1):57–78
4. Filiatrault A, Cherry S (1990) Seismic design spectra for friction-damped structures. J Struct Eng 116(5):1334–1355
5. Cherry S, Filiatrault A (1993) Seismic response control of buildings using friction dampers. Earthq Spectra 9(3):447–466
6. Aiken IA, Kelly JM (1990) Earthquake simulator testing and analytical studies of two energy-absorbing systems for multistory buildings. Report No.: CB/EERC, 90(03)
7. Nims DK, Richter PJ, Bachman RE (1993) The use of the energy dissipating restraint for seismic hazard mitigation. Earthq Spectra 9(3):467–490
8. Fitzgerald TF, Anagnos T, Goodson M, Zsutty T (1989) Slotted bolted connections in aseismic design for concentrically braced connections. Earthq Spectra 5(2):383–391
9. Grigorian CE, Yang TS, Popov EP (1993) Slotted bolted connection energy dissipators. Earthq Spectra 9(3):491–504
10. Mirtaheri M, Zandi AP, Samadi SS, Samani HR (2011) Numerical and experimental study of hysteretic behavior of cylindrical friction dampers. Eng Struct 33(12):3647–3656
11. https://www.quaketek.com/
12. Scholl RE (1993) Design criteria for yielding and friction energy dissipators. In: Proceedings of ATC 17-1 on seismic isolation, energy dissipation and active control, vol. 2, pp 485–495
13. Moreschi LM (2000) Seismic design of energy dissipation systems for optimal structural performance. Doctoral dissertation, Virginia Tech
14. Ciampi V, De Angelis M, Paolacci F (1995) Design of yielding or friction-based dissipative bracings for seismic protection of buildings. Eng Struct 17(5):381–391
15. Moreschi LM, Singh MP (2003) Design of yielding metallic and friction dampers for optimal seismic performance. Earthq Eng Struct Dyn 32(8):1291–1311
16. Christopoulos C, Filiatrault A (2006) Principles of passive supplemental damping and seismic isolation, 1st edn. IUSS Press, Pavia, Italy
17. NEHRP Recommended Seismic Provisions for New Buildings and Other Structures (FEMA P-750) (2009) Report Prepared for the Federal Emergency Management Agency (FEMA), Building Seismic Safety Council (BSSC), Washington D. C., USA
18. FEMA (2000) Prestandard and commentary for the seismic rehabilitation of buildings (FEMA-356) Federal Emergency Management Agency, American Society of Civil Engineers (ASCE), Reston, USA
19. ETABS®, Computers & Structures, Inc., 1646 N. California Blvd., Suite 600 Walnut Creek, CA 94596 USA. Retrieved from https://www.csiamerica.com/products/etabs)
20. National Building Code of Canada (2015) NRCC, Ottawa, Canada (2015)
21. http://www.palldynamics.com/
22. Baker JW (2011) Conditional mean spectrum: tool for ground motion selection. J Struct Eng 137(3):322–331

23. Naumoski N, Saatcioglu M, Amiri-Hormozaki K (2004) Effects of scaling of earthquake excitations on the dynamic response of reinforced concrete frame buildings. In: Proceedings of 13th World conference on earthquake engineering, Vancouver BC, Canada, Paper No. 2917
24. The MathWorks Inc., Natick, A. H. Campus, 1 Apple Hill Drive, Natick, MA 01760-2098
25. Ram SRKMM, Bagchi A (2015) Seismic performance evaluation of steel moment resisting frames. In: 11th Canadian conference on earthquake engineering
26. PEER Ground Motion Database (Web source: https://ngawest2.berkeley.edu/).

Bidirectional Pushover Analysis considering the Effect of Angle of Seismic Incidence

Prabakaran Kesavan and Arun Menon

Abstract The general practice in structural assessment or design using 3D seismic analysis of inelastic structures is to obtain the responses (such as storey displacement, inter-storey drift, etc.) by applying the bidirectional components along any arbitrarily chosen structural axes. However, such response quantities may not remain the same when the bidirectional components are rotated with respect to the structural axes. It becomes relevant to identify the angle of seismic loading which maximizes the response quantity in the structure, and that angle would be called as critical angle of incidence (CAI), and the associated response as critical response. Nevertheless, a single CAI which can simultaneously maximize all the responses of the structure does not exist, rather the critical angle varies according to the type of response and earthquake loading. Therefore, any structural assessment or design of buildings based on the responses obtained from applying the earthquake loading along user defined structural axes may not be always conservative. To avoid such scenarios the critical responses should be estimated considering the effect of angle of seismic incidence. To address such an issue, pertaining to inelastic structures, this study presents a new approach which essentially combines bidirectional pushover analysis with an already existing non-linear static procedure called the Extended-N2 method.

Keywords Bidirectional pushover analysis · Seismic incidence angle · Multidirectional time history analysis

P. Kesavan (✉) · A. Menon
Indian Institute of Technology Madras, Chennai 600 036, India
e-mail: kesavapraba@gmail.com

A. Menon
e-mail: arunmenon@iitm.ac.in

© The Editor(s) (if applicable) and The Author(s), under exclusive license to Springer Nature Singapore Pte Ltd. 2021
S. K. Saha and M. Mukherjee (eds.), *Recent Advances in Computational Mechanics and Simulations*, Lecture Notes in Civil Engineering 103,
https://doi.org/10.1007/978-981-15-8138-0_32

1 Introduction

For frame structures, such as Reinforced Concrete (RC) buildings, Non-linear Time History Analysis (NLTHA) with bidirectional ground motion inputs is considered to be more accurate and realistic, even such an analysis is computationally demanding. Alternatively, Non-linear Static Procedure (NSP) or Pushover analysis is a popular tool among practitioners, which essentially captures the failure mechanism and is computationally cost and time effective. It is important to note that there are limitations in non-linear static procedures such as absence of effects of the dynamic characteristics of ground motion on structural response, which can be accurately estimated by non-linear time history analysis only. Nevertheless, NSP is recognized as an important tool to understand the failure mechanism and damage pattern. However, most of the codal provisions and guidelines, though mention about NLTHA, do not specify the principal directions along which the bidirectional ground motion records (i.e., assumed angle seismic incidence) should be applied during the analysis. It is left to the choice of the analyst to decide on the directions, which will lead to different choices if the building plan has non-orthogonal frames. Accordingly, the estimated responses will be different for each choice of the direction, and end up with underestimation of structural responses in some cases. Hence, the angle of seismic incidence at which an EDP (Engineering Demand Parameter, such as storey displacement etc.,) attains its maximum value (critical angle) and then the estimation of corresponding EDP (critical response) which is independent of angle of seismic incidence is recognized as an important issue in seismic analysis of structures.

It is in the authors' opinion that the discussion about critical angle of seismic incidence started from the work done by Penzien and Watabe in [1], which proposed a method to obtain uncorrelated orthogonal principal components of a given earthquake based on stochastic approach, and gained attention of many researchers afterwards. For linear-elastic analysis, Wilson and Button in [2] proposed a method under Response Spectrum Analysis (RSA) approach, where a single analysis produces, for every point in a 3D structure, the mean maximum response and the corresponding critical angle of seismic incidence. Maximum values of forces in all members were calculated from one computer run in which two global dynamic loadings were applied. An important point brought out of the study was that each internal response quantity can have a different critical angle, and a single critical angle which can maximize all the responses in a structure is not possible. However, the correlation of input ground motion records was not accounted for in the work, which may result in unrealistic responses. Determination of critical response and the corresponding angle based on stationary random vibration theory was attempted by Smeby and Der Kiureghian in [3] in order to provide practical solution involving response spectra. Later, Lopez et al. [4] critically questioned the validity of the formula recommended by Wilson in [5] to calculate the critical responses and angles. It was proved that the formula is strictly valid only when the dynamic response takes place in one vibrational mode with fully correlated spectra, and later in [4] presented a correct formula by addressing those shortcomings. A spatial combination rule which considers the correlation among ground motion components and also critical response aspect, called CQC3, was introduced as a replacement for the percentage (30% & 40%) [6, 7] and SRSS rules by Menun and Der Kiureghian in [8]. An explicit formula was derived for the calculation of critical value of a response, without a need to determine a-priori the critical angle, when the two horizontal principal components were applied at any angle with respect to structural axes along with vertical component of ground motion in [9]. Apart from the previously discussed works one can find other interesting studies related to response spectrum and equivalent lateral static analysis, which were carried out in later years, and confirmed the similar conclusions drawn earlier [10–14].

The literature on estimation of critical responses for non-linear analysis of structures were, however, limited to only time history analysis. The general reason stated by the works of [15–18] was that a wide statistical dispersion of results was observed from similar studies involving non-linear time history analysis. It is interesting to note that for a single EDP, its critical angle would never be the same for two different ground motions, and for the same ground motion two different EDPs (e.g., storey displacement and storey drift) can have different critical angles as well. Notwithstanding, the studies proved that the fundamental period, type of model and level of inelastic behaviour can have influences on critical angle of seismic incidence and states that it is difficult to determine the angle a-priori like that of a linear-elastic structure. An easy and immediate alternative to NLTHA is NSP, or otherwise known as pushover analysis which is actually a monotonic non-linear static analysis. The monotonic load (force or displacement) is incrementally applied, either as invariant or variant load pattern on the structure to represent the equivalent seismic action. Even though different NSPs are available to address torsional [19], higher mode [20] and bidirectional effects [21], clearly no work has been done till now to bring in the aspect of critical angle of seismic incidence into NSPs. It can be stated from the reviewed literature that: (i) compared to other types of analysis, many researchers have contributed to a well-established study of critical angle of seismic incidence and the estimation of corresponding critical responses, particularly with closed-form solution, under the ambit of response spectrum analysis, (ii) traditional SRSS method is not meant for estimating critical responses, (iii) the estimation of critical responses for inelastic structures using simplified approach, such as pushover analysis, instead of time history analysis is required, and (iv) till now, no study has attempted to address the critical response issue in the context of NSP, which is simple and consumes less computational time.

Therefore, the objective of the present study is to propose and demonstrate a new NSP (named as Critical-N2) based on the Extended-N2 method in [19] and the bidirectional pushover analysis in [21]. Extended-N2 is an established NSP which essentially combines Responses Spectrum Analysis with original-N2 method [22] to address torsional effects in asymmetric reinforced concrete (RC) frame buildings. The proposed NSP takes into account the critical angle of seismic incidence along with the consideration of bidirectional ground motion input and torsional effects due to asymmetry of the structures. All types of non-linear analysis (pushover and time history) for the study were programmed in an open-source simulation platform called OpenSEES [23].

2 Proposed NSP for Estimating Critical Responses

This section explains the methodology followed in the present study to develop a simple and practical Non-linear Static Procedure (NSP) to account for the effect of critical angle of seismic incidence. The methodology derives its base from the Extended-N2 method which was developed by Fajfar et al. in [19].

It is required to perform two analysis before obtaining the final non-linear critical responses; one from linear dynamic analysis (RSA), and other from bidirectional pushover analysis (BPA). Accordingly, two separate numerical models are developed for the frame structure which is to be assessed/designed. Seismic demand for the structure can be considered according to any specified codal provisions/guidelines in the form of response spectrum. Elastic critical responses are calculated from the results of RSA analysis following the work in [8]. In the next step, four bidirectional pushover analysis along the principal axes $(+X' + Z', +X' - Z', -X' + Z', -X' - Z')$, with invariant load patterns proportional to the first two translational mode shapes (applied simultaneously in two orthogonal directions) are carried out to obtain non-linear responses of the structure. It is also assumed that the principal axes are same for both linear and non-linear pushover analyses, hence all the pushover analyses were performed along the principal axes suggested in [24]. Though the loads are applied along the two principal axes, the capacity curve in the major principal axis, which has the least stiffness and so the largest roof displacement, was considered to obtain target displacement. The capacity curve which has the least capacity out of four curves obtained from BPA is chosen for further processing to fetch non-linear responses. The responses from the RSA and BPA are compared, and the correction factors are applied to the non-linear responses. Thus, the corrected responses are called as the critical responses of Critical-N2 method, which are actually approximate estimations. The flow chart in Fig. 1 explains the steps involved in the proposed NSP.

The Critical-N2 method was validated with the actual responses from Multidirectional Time History Analysis (MDTHA). MDTHA is a non-linear dynamic analysis, which is conducted with a set of ground motion records with each record having two components, at a desired interval of angle (10°). It involves proper selection of ensemble of ground motion records, and programming an automated multidirectional analysis without any convergence issues in OpenSEES.

3 Numerical Modelling of the Study Building

A hypothetical RC frame building was conceived for the study purpose as shown in Fig. 2. The structure is a three-storeyed building with the outer dimension of 9 m × 11 m and same floor height of 3 m. It has totally nine columns with two different sizes, viz., 300 mm × 300 mm (C1–C7, C9), and 750 mm × 300 mm (C8). All the beams in every floor has the same dimension, i.e., 250 mm × 400 mm. The arrangements of columns in the building make it asymmetric in both ways (the centre

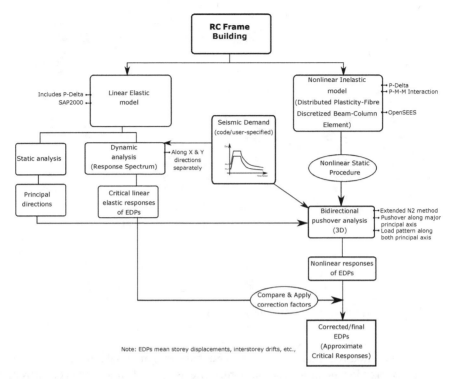

Fig. 1 Methodology for developing the Critical-N2 method to estimate critical responses

of stiffness (CR) and the centre of mass (CM) are not coincidental), which additionally induces the torsional effect during any lateral loading. Eccentricity ratio in percentage with respect to the corresponding plan dimension at floors are: third ($e_{x3} = 4$, $e_{z3} = 1.6$), second ($e_{x2} = 5.1$, $e_{z2} = 1.9$), and first ($e_{x1} = 6.9$, $e_{z1} = 3.3$). Two numerical models of the same building were developed: linear-elastic model using SAP2000 [25] and non-linear-inelastic model using OpenSEES [23]. The building was a-priori designed according to the design criteria laid out by IS 1893-2002 (Part-1) [26] in Zone-IV in Rocky soil site. In linear-elastic model, the stiffnesses of columns and beams were reduced by applying the factors 0.5 and 0.7, respectively, to account for the effective sectional properties of RC section, along with ($P - \Delta$) effect, which also nearly matches the time period of the building in both linear and non-linear models. The mean strength of concrete and yield strength of rebars used are 25 (M25) & 415 N/mm^2 (Fe415), respectively. In the non-linear model, all the columns were modelled with distributed plasticity approach employing force-based beam column element with fiber discretized cross section, while beams with elastic beam element with concentrated plasticity at its ends. A fiber section was assigned with uniaxial constitutive laws available in [23]; Concrete01 material for confined and unconfined concrete, and Giuffre–Menegotto–Pinto steel material (Steel02) for

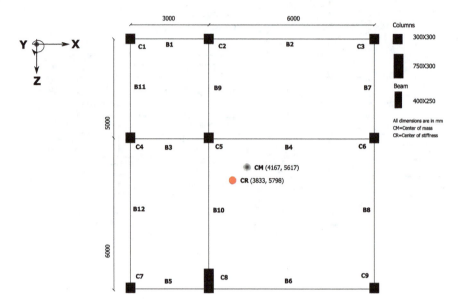

Fig. 2 Structural plan of the study building at top floor

rebars. Slab action was modelled with rigid floor diaphragm. Axial force and bending moment interaction along with second-order effects ($P - \Delta$) have been considered in all analysis.

4 Bidirectional Pushover Analysis

General practice in seismic analysis is to apply same or separate response spectra in two mutually orthogonal directions of the building, and combining the results by SRSS or CQC3 rules. However, the same cannot be applicable in non-linear static analysis, as noted by Cimellaro et al. in [21] which proposes a bidirectional pushover analysis to overcome such problem. Current study follows the work in [19, 21], but with a different approach to calculate the principal directions for static analysis. The steps involved in the present bidirectional pushover analysis are as follows:

- A multi-degree of freedom (MDOF) model of the study building is created with required non-linear and inelastic structural properties. The first natural frequencies (ω_n) and mode shapes (ϕ_n) in respective orthogonal directions are obtained from Eigenvalue analysis. Considering only the degrees of freedom associated with the lateral translation in the direction of study, the mathematical model of MDOF system is written as

$$\mathbf{M\ddot{U}} + \mathbf{R} = -\mathbf{M}1 a_g, \qquad (1)$$

where **U** is displacement vector, **R** is the internal resistance force vector which is equal to **F** in Eq. 2 from statics theory, and a_g is the ground acceleration as function of time. For the purpose of simplicity, the damping in the structure is not directly included in the equation, but in the demand response spectrum. If the displacement vector is assumed as $\mathbf{U} = \mathbf{\Phi} d_t$, where d_t is the time-dependent roof displacement and **Φ** is the normalized first mode shape in order to have its top component equal to 1, the equivalent SDOF system can be written in the following format:

$$m^* \ddot{d}^* + F^* = -m^* a_g \tag{2}$$

where m^* is the equivalent mass and equal to $\mathbf{\Phi}^T \mathbf{M} \mathbf{1}$ and F^* is the base shear of the SDOF system. The roof displacement and base shear of the SDOF system are obtained as in Eqs. 3 and 4 below:

$$d^* = \frac{d_t}{\Gamma} \tag{3}$$

$$F^* = \frac{F}{\Gamma} \tag{4}$$

Then transformation/modal participation factor (Γ) is calculated as follows:

$$\Gamma = \frac{\mathbf{\Phi}^T \mathbf{M} \mathbf{1}}{\mathbf{\Phi}^T \mathbf{M} \mathbf{\Phi}} = \frac{\sum_i m_i \phi_i}{\sum_i m_i \phi_i^2} = \frac{m^*}{\sum_i m_i \phi_i^2} \tag{5}$$

where Φ = roof displacement normalized mode shape; **M** = diagonal mass matrix; ϕ_i = component of mode shape at ith storey; m_i = mass at ith storey.
- Seismic demand for the equivalent SDOF system is determined in the form of inelastic response spectra using Acceleration-Displacement Response Spectrum (ADRS) format by applying a reduction factor (R_μ) on elastic acceleration response spectra as given in [27].
- Since the pushover analysis is a static analysis, principal directions/axes of the building plan are calculated following the approach in [24]. Subsequently, the three-dimensional MDOF model is subjected to invariant and monotonically increasing lateral load pattern. The loads are simultaneously applied along the principal directions (X' and Z'), which are proportional to the first two translational modes with higher mass participation ratio. The lateral load patterns along X' and Z' are $\mathbf{F}_{X'} = \mathbf{m}.\mathbf{\Phi}'_X$ and $\mathbf{F}_{Z'} = \gamma.\mathbf{m}.\mathbf{\Phi}'_Z$, respectively. Where Φ'_X and Φ'_Z are the vectors of first two translational mode shapes; **m** = vector of floor masses; γ = a scale factor obtained from peak-ground acceleration (PGA) values of the acceleration response spectra along the principal directions as defined below:

$$\gamma = \frac{(PGA)_{X'}}{(PGA)_{Z'}}, \quad 0 < \gamma \leq 1 \tag{6}$$

Though the bidirectional pushover analysis is done in two orthogonal principal directions simultaneously, the capacity curve from the major principal axis ($\mathbf{X'}$) is considered as the characteristic capacity curve of the MDOF system. Further, the Eqs. 3 and 4 are used to transform the MDOF system into an equivalent SDOF system with bilinear force-displacement ($F^* - d^*$) relationship.

- Performance point of the SDOF system, in terms of target displacement (d_t^*), is evaluated by comparing its bilinearized capacity curve with the demand curve using ADRS format. Applying the transformation factor on d_t^*, the target displacement (d_t) for the MDOF system is obtained. Thus, the non-linear responses from bidirectional pushover analysis correspond to the state of the analysis where the roof displacement reaches d_t value along X'.
- The non-linear responses would be further corrected by the factors obtained using critical elastic responses from RSA. The procedure is same as mentioned in [19], except using the critical responses which is the main difference between the present study and the works by [19, 21]. By incorporating the critical responses in the Extended-N2 method and BPA, the study addresses the need to calculate the non-linear critical responses. Thus, the present study explicitly considers the elastic critical responses from RSA as per [28] and estimates the correction factors as per [19], and checks if the methodology provides acceptable results compared to MDTHA.
- Termination criteria for the pushover analysis was set as either the base-shear capacity reaching 80% of the ultimate base-shear capacity in the post-peak, or the compressive strain in the core concrete reaching its ultimate strain.

5 Multidirectional Time History Analysis

A multidirectional time history analysis strategy was formulated to obtain actual critical responses of all EDPs for the validation of the non-linear responses obtained by the Critical-N2 method. The non-linear model created in OpenSEES platform was subjected to a set of nine bidirectional ground motion records with rotating angle interval of 10° each time, thus requiring 35 such analysis for each ground motion (totally, $9 \times 35 = 315$) to complete the entire 360°. As each analysis is independent, a parallel version of OpenSEES, called OpenSeesMP, was used to minimize the simulation duration. Damping matrix was constructed to be proportional to the last-committed stiffness matrix and mass matrix. Newmark's two-parameter ($\gamma_N = 0.5$, $\beta_N = 0.25$) time-stepping scheme was used in time integration process.

5.1 Ground Motion Records

Earthquake strong ground motion records are required for conducting non-linear time history analysis of the structural model. Ground motion selection was performed by a standard protocol using PEER ground motion database. Although definitive standards are not available for selection of the records, few guidelines [29, 30] and literature accounts [31, 32] exist. Totally nine pairs of records (Table 1) were selected from PEER NGA-2 database [33] based on the criteria given in the Table 2. It was ensured that all the records in the ensemble are from the distinct event to introduce the variability in the nature of ground motion, and there is no pulse-like record as to avoid any near-field effects. The design response spectrum of IS 1893–2002 (Part-1) for rocky soil condition in seismic Zone-IV was used as target spectrum. Scaling factors are calculated on a basis that: the geometric mean spectrum of the selected accelerograms was compared with the target spectrum to minimize the deviation (difference between the SRSS spectral ordinates of the pairs and the ordinate of the target spectrum) in the time period range $0.1 - 2.0$ T_1 (significant time periods), where T_1 is the fundamental time period of the building. Same scaling factor was applied to both the horizontal components of a record. Figure 3a, b shows the psuedo-spectral acceleration and displacement response spectrum, respectively. It is evident from the figures that acceptable level of spectral values matching exist in the range of significant time periods.

Table 1 Details of selected earthquake ground motion records

Sl No	Earthquake name	Station name	Year	Magnitude (M_w)	Closest distance to rupture plane (R_{rup}, km)	Scale factor
1	Lytle Creek	Cedar Springs Allen Ranch	1970	5.3	19.4	6.69
2	Duzce_Turkey	Lamont 1060	1999	7.1	25.9	2.05
3	Chi-Chi_Taiwan-04	CHY102	1999	6.2	39.3	2.69
4	Hector Mine	LA-Griffith Park Observatory	1999	7.1	185.9	5.14
5	Tottori_Japan	HYG004	2000	6.6	108.3	5.92
6	Chuetsu-oki_Japan	NGN013	2007	6.8	142.0	5.98
7	Iwate_Japan	MYG011	2008	6.9	82.9	3.02
8	El Mayor Cucapah Mexico	San Diego Road Dept	2010	7.2	111.0	3.79
9	San Simeon CA	Diablo Canyon Power Plant	2003	6.5	38.0	1.86

Table 2 Summary of PEER ground motion database search criteria

Magnitude min	5
Magnitude max	7.5
Rrup min (km)	15
Rrup max (km)	300
Rjb min (km)	15
Rjb max (km)	300
Vs30 min (m/s)	760
Vs30 max (m/s)	2000
D9-95 min (s)	15
D9-95 max (s)	60
Scale factor min	0.1
Scale factor max	10
User defined maximum number of records	100
Pulse-like records	No
Damping ratio	0.05
Scaling method	Minimize MSE
Suite average	Geometric
Component	SRSS
Fault type	All types

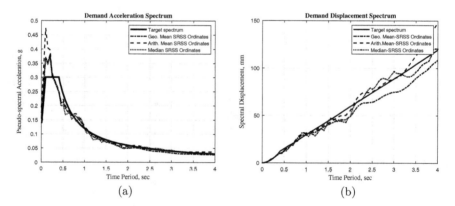

Fig. 3 Comparison of (**a**) acceleration and (**b**) displacement demand spectra with target spectrum

5.2 Actual Critical Responses from MDTHA

The actual critical responses are extracted from the MDTHA analysis after processing the results using MATLAB programs. The polar plots in Figs. 4 and 5 show the statistical variations of maximum storey displacement and storey drift, respectively, at a typical node (at C9-roof location) along with the critical angles. Storey displacement and drift at all the nodes are considered to obtain their critical values from the polar plots. It is interesting to observe from the plots that the conclusions of [15] that critical angles vary according to the type of response is proved again.

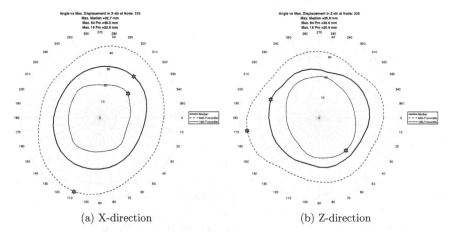

(a) X-direction (b) Z-direction

Fig. 4 Angle versus maximum displacement of a typical node from MDTHA

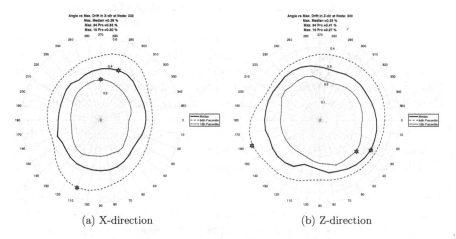

(a) X-direction (b) Z-direction

Fig. 5 Angle versus maximum drift of a typical node from MDTHA

6 Validation of the Critical-N2 Method

Lateral displacement, storey drift, Top Displacement Ratio (TDR), and Normalized Top Displacement (NTD) are the different measures used for validating Critical-N2 method with the actual responses from MDTHA. TDR gives an estimate of bias by NSP in predicting a particular response, which is defined as the ratio between the top displacement from NSP and the median top displacement from MDTHA. However, the NTD is used to understand the torsional amplification in asymmetric buildings and thus captures how well an NSP predicts the torsional effects. Hence it is obtained by normalizing the edge displacement values with respect to that of the centre of mass.

6.1 Results and Discussion

Three column locations were considered for all comparison purposes: one on flexible edge (C3), one on stiff edge (C7), and the other near the centre of mass (C5). Flexible and stiff edges were decided from the details of CM and CR locations. Responses of Critical-N2 method are compared with that of RSA and MDTHA (median, 84 & 16 percentiles). It is emphasized that the responses from MDTHA are envelope of maximum of maximum values which do not occur at the same time during the analysis (i.e., asynchronous). Hence, the results from MDTHA might be used to observe a general rather than specific trend.

Figure 6 shows the lateral displacement profiles at the considered column locations. It is clear from the figure that the response estimates by Critical-N2 method is consistently conservative in all the locations as the results lie within mean $\pm \sigma$ range. In the lower floors, the displacement predictions along both X and Z directions are closer to the median results than that of the roof at all locations, even though the eccentricity ratio increases from roof to first floor. This indicates that the Critical-N2 method performs well even if plan eccentricity increases. Lesser underestimation happened at the location C3 on the flexible edge along X-direction. The location of C3 on the plan and the level of inelastic torsional demand from bidirectional ground motions are the reasons attributed for such response. Storey drift profile shown in Fig. 7 follows a different trend when compared to lateral displacement profile. Predictions are better at the roof floor level though there is no trend in the lower floors. Nonetheless, the results are within the mean$\pm\sigma$ range. The prediction at C5 is better than the other two locations as its location is near the CM. TDR values are compared at different locations in Fig. 8 which shows that conservative predictions are achieved except at C3 location where Critical-N2 method marginally underestimates the response along X-direction. However, along Z-direction the NSP performed well. The torsional amplification is well captured by the Critical-N2 method along Z-direction at all locations as evident from the Normalized Top Displacement (NTD) plot in Fig. 9. Although the method predicts the torsional amplification at stiff edge

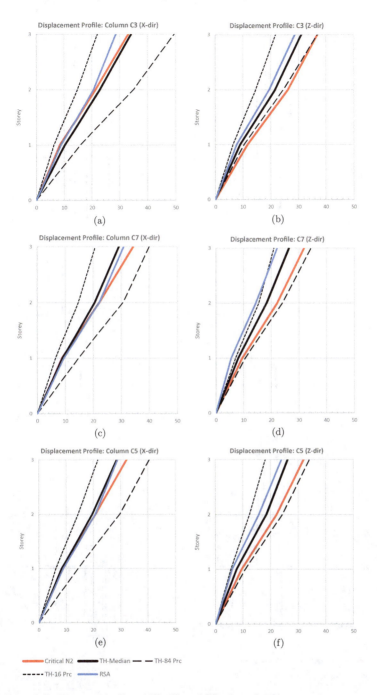

Fig. 6 Lateral displacement profiles at C3, C7, and C5 column locations

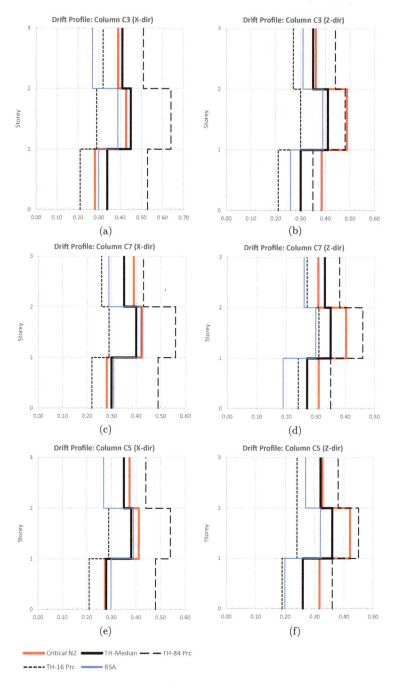

Fig. 7 Storey drift profiles at C3, C7, and C5 column locations

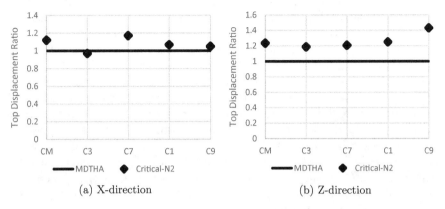

(a) X-direction (b) Z-direction

Fig. 8 Comparison of Top Displacement Ratio

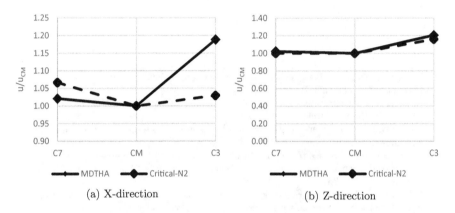

(a) X-direction (b) Z-direction

Fig. 9 Comparisons of Normalized Top Displacement values for torsional amplification effects

location (C7) well, it misses out on the flexible edge location (C3). It should be pointed out that the calculated NTD is based on the median results, not on the actual synchronous NTD values. Moreover, this is not due to inherent characteristics of BPA. From the above discussion, it is evident that the Critical-N2 method performed well for the study building by estimating the non-linear critical responses which are within the acceptable statistical range of responses from MDTHA.

7 Conclusions

This paper presented a new approach to address the problem of estimating critical responses for inelastic RC frame buildings due to the effect of angle of seismic incidence. The methodology demonstrated the proposed Critical-N2 method, which is based on a well-established Non-linear Static Procedure called the Extended-N2 method that essentially applies correction factors on non-linear responses from linear dynamic analysis. As part of the Critical-N2 method, a bidirectional pushover procedure, which simultaneously applies the loading in two mutually orthogonal directions (principal axes) was formulated to obtain the capacity curve of the study building. Critical-N2 method was validated with the responses obtained from multidirectional time history analysis using four different measures. The Critical-N2 method almost estimated all the critical responses with the acceptable level of accuracy, mostly on conservative side. Hence, it is concluded that the Critical-N2 method can be used as an approximate non-linear analysis to obtain quick estimates of critical responses at design offices for seismic structural assessment. As the proposed method is validated with only one RC building model, more number of structural models with different asymmetric properties are required to be validated and the same is being explored as an extension of the current work.

References

1. Penzien J, Watabe M (1974) Characteristics of 3-dimensional earthquake ground motions. Earthq Eng Struct Dyn 3(4):365–373
2. Wilson EL, Der Kiureghian A, Bayo EP (1981) A replacement for the SRSS method in seismic analysis. Earthq Eng Struct Dyn 9(2):187–192
3. Smeby W, Der Kiureghian A (1985) Modal combination rules for multicomponent earthquake excitation. Earthq Eng Struct Dyn 13(1):1–12
4. López OA, Torres R (1996) Discussion of "a clarification of orthogonal effects in a three-dimensional seismic analysis by E.L. Wilson, I. Suharwardy, and A. Habibullah". Earthq Spectra 12(2):357–361
5. Wilson EL, Suharwardy I, Habibullah A (1995) A clarification of the orthogonal effects in a three-dimensional seismic analysis. Earthq Spectra 11(4):659–666
6. Newmark NM (1975) Seismic design criteria for structures and facilities, Trans-Alaska pipeline system. In: Proceedings of the US national conference on earthquake engineering. Earthquake Engineering Research Institute, pp 94–103
7. Rosenblueth E, Contreras H (1977) Approximate design for multicomponent earthquakes. J Eng Mech Div (ASCE) 103(5):881–893
8. Menun C, Der Kiureghian A (1998) A replacement for the 30%, 40%, and SRSS rules for multicomponent seismic analysis. Earthq Spectra 14(1):153–163
9. López OA, Chopra AK, Hernandez JJ (2000) Critical response of structures to multicomponent earthquake excitation. Earthq Eng Struct Dyn 29(12):1759–1778
10. Nie J, Morante R, Miranda M, Braverman J (2010) On the correct application of the 100-40-40 rule for combining responses due to three directions of earthquake loading. Brookhaven National Laboratory (BNL)
11. Gao X-A, Zhou X-Y, Wang L (2004) Multi-component seismic analysis for irregular structures. In: 13th world conference on earthquake engineering, Vancouver, BC. Canadian Association of Earthquake Engineering, Paper No. 1156
12. Khoshnoudian F, Poursha M (2004) Responses of three dimensional buildings under bi-directional and unidirectional seismic excitations. In: Proceedings of the 13th world conference on earthquake engineering, pp 1–6
13. Morfidis KE, Athanatopoulou AM, Avramidis IE (2008) Effects of seismic directivity within the framework of the lateral force procedure. In: 14th world conference on earthquake engineering
14. Skoulidou D, Romao X (2017) Critical orientation of earthquake loading for building performance assessment using lateral force analysis. Bull Earthq Eng 15(12):5217–5246

15. Rigato AB, Medina RA (2007) Influence of angle of incidence on seismic demands for inelastic single-storey structures subjected to bi-directional ground motions. Eng Struct 29(10):2593–2601
16. Magliulo G, Maddaloni G, Petrone C (2014) Influence of earthquake direction on the seismic response of irregular plan RC frame buildings. Earthq Eng Eng Vib 13(2):243–256
17. Emami AR, Halabian AM (2015) Spatial distribution of ductility demand and damage index in 3D RC frame structures considering directionality effects. Struct Des Tall Spec Build 24(16):941–961
18. Kostinakis KG, Manoukas GE, Athanatopoulou AM (2018) Influence of seismic incident angle on response of symmetric in plan buildings. KSCE J Civ Eng 22(2):725–735
19. Fajfar P, Kilar V, Marusic D, Perus I, Magliulo G (2005) The extension of the N2 method to asymmetric buildings. In: Proceedings of the 4th European workshop on the seismic behaviour of irregular and complex structures, No. 41
20. Chopra AK, Goel RK (2004) A modal pushover analysis procedure to estimate seismic demands for unsymmetric-plan buildings. Earthq Eng Struct Dyn 33(8):903–927
21. Cimellaro GP, Giovine T, Lopez-Garcia D (2014) Bidirectional pushover analysis of irregular structures. J Struct Eng (ASCE) 140(9):04014059
22. Fajfar P, Gašperšič P (1996) The N2 method for the seismic damage analysis of RC buildings. Earthq Eng Struct Dyn 25(1):31–46
23. McKenna F (2011) OpenSees: a framework for earthquake engineering simulation. Comput Sci Eng 13(4):58–66
24. Athanatopoulou AM, Doudoumis IN (2008) Principal directions under lateral loading in multistorey asymmetric buildings. Struct Des Tall Spec Build 17(4):773–794
25. Computers and Structures Inc (2017) SAP2000 Version 17
26. Bureau of Indian Standards (2002) IS:1893-2002 (Part-1), Criteria for earthquake resistant design of structures, New Delhi, India
27. Fajfar P (1999) Capacity spectrum method based on inelastic demand spectra. Earthq Eng Struct Dyn 28(9):979–993
28. López OA, Torres R (1997) The critical angle of seismic incidence and the maximum structural response. Earthq Eng Struct Dyn 26(9):881–894
29. NIST: National Institute of Standards and Technology (2011) Selecting and scaling earthquake ground motions for performing response-history analyses
30. ASCE/SEI 7-05 (2005) Minimum design loads for buildings and other structures
31. Bommer JJ, Acevedo AB (2004) The use of real earthquake accelerograms as input to dynamic analysis. J Earthq Eng 8(spec01):43–91
32. Beyer K, Bommer JJ (2007) Selection and scaling of real accelerograms for bi-directional loading: a review of current practice and code provisions. J Earthq Eng 11(S1):13–45
33. PEER: Pacific Earthquake Engineering Research Center. University of California, Berkeley (2018)

Seismic Energy Loss in Semi-rigid Steel Frames Under Near-Field Earthquakes

Vijay Sharma, M. K. Shrimali, S. D. Bharti, and T. K. Datta

Abstract A comparative study is carried out to estimate the seismic energy losses between the semi-rigid steel frames, modeled in two different approaches and rigid frames. For this purpose, three variant of earthquakes is considered, namely, far-field and near-field with forward directivity and fling step effect. These earthquakes are scaled to a peak ground acceleration (PGA) level of 0.4 and 0.6 g. The seismic energy loss is evaluated along with other seismic response parameters. The responses parameters of interest are maximum roof displacement, base shear, the total number of formation of plastic hinges with their square root of the sum of square (SRSS) values of maximum hinge rotations, and the energy dissipation in the form of modal damping and link hysteretic energy. For this numerical simulation study, a five-story rigid frame is designed as per Indian standard provisions as an illustrative problem. A nonlinear response history analysis is performed using the SAP2000 platform to evaluate the desired responses. The results of present work reveal that (i) the seismic energy dissipation significantly more in semi-rigid connected frame with plastic link as compared to elastic link; (ii) the energy dissipation in the form of plastic hinges are substantial in rigid frames as compared to semi-rigid frames with plastic and elastic link, plastic link model provides comparable loss in seismic energy with rigid frames; and (iii) the significance of seismic energy loss depends on earthquakes type, PGA level, degree of semi-rigidity and connection type.

Keywords Semi-rigid · Near-field · Far-field · Energy dissipation

1 Introduction

The efficiency of steel moment frames (SMF) is considerably better as compared to other civil engineering structures, especially in major seismic prone areas due to their strength and high ductility. The efficacy of moment frames is dependent significantly

V. Sharma (✉) · M. K. Shrimali · S. D. Bharti · T. K. Datta
Malaviya National Institute of Technology Jaipur, Jaipur 302017, India
e-mail: vrsgec2011@gmail.com

on beam-column connections. The beam-column connections are crucial in seismic events due to the cyclic nature of ground motions. Conventionally, the steel beam-column connections are designed as rigid with extremely high stiffness to compel the safety consideration of high stiffness as well as sufficient over-strength irrespective of the economy of construction. The 1994 Northridge and 1995 Kobe earthquake events were significantly damaged the several SMF buildings, and the beam-column welded connections were severely affected. These events diverted the attention of seismic analysts toward rigid to semi-rigid (SR) connections. During the seismic event, the huge amount of stocked energy dissipation is carried out in the form of plastic hinges, mainly in flexural members. Depending upon the stocked seismic energy, the number and rotations of the plastic hinges could be significant, resulting in a high level of damages. The application of SR connections considerably reduces the plastic hinge formation and enhances the performance level (lower down the collapse prevention 'CP' to life safety 'LS' or immediate occupancy 'IO' as described in ASCE 41-17 [1]. Thus, most of the current standards, like Indian, American and European codes incorporated the three types of the beam-column connections, i.e., fully restrained/ rigid, partially restrained/ semi-rigid and flexible/ pinned connection [2–4].

Diaz et al. [5] reviewed comprehensively on the development of SR connections and classified the SR connections based on moment-rotation behavior in different categories such as experimental, analytical, numerical, mechanical, and informational. Enhancing the global hysteretic energy can be attained either by improving the hysteretic path of structural components (member and connection) as provided in FEMA 355D [6] or actuating the enhanced number of locations for plastification before the structural collapse. The second measure is implied in the current practices in AISC 341-16 [4] and Eurocode 8 [7]. The seismic performance of SMF with SR connections was investigated by various authors in the past [8, 9]. Aksoylar et al. [10] investigated the hysteretic moment-rotation behavior of low-rise semi-rigid frames. Abolmaali et al. [11] experimentally investigated the energy dissipation in welded/ bolted connections. Sekulovic and Nefovska-Danilovic [12] investigated the seismic performance of the multistory SR frames under different peak ground acceleration (PGA) levels and observed that the earthquake energy is primarily dissipated in beam-column connections and flexural plastic hinges. The efficacy of tuned mass dampers for multi-mode seismic response control and energy dissipation for different structures was investigated by various researchers [13–16]. Recently, Lemonis [17] analyzed the seismic performance of SMF, especially in this regard to energy dissipation in joints and beams.

The previous studies focused on far-field seismic excitations to assess the structural seismic performance. Fewer studies were carried out considering the near-field ground motions, and primarily these were constrained to rigid frames only [18]. The seismic behavior of SR frames and energy dissipation characteristics in these studies are not comprehensively studied for near-field seismic excitations. This paper aims to investigate and compare the seismic behavior of the 5-Story SMF SR frame with rigid frames. The beam-column connections in SR frames are modeled in two ways, namely, the multi-linear elastic link and multi-linear plastic link element. The seismic demand parameters and the energy dissipation are evaluated by a nonlinear response history analysis (NRHA) for three variety of ground motions, viz., the far-field and the near-field with forward directivity and fling effects. For each earthquake, two PGA levels (design and high level) are considered. The seismic demand parameters included the roof displacement, maximum base shear, formation of a total number of plastic hinges, the SRSS of maximum plastic hinge rotations and the energy dissipation in the form of connection/link hysteretic energy and modal damping energy.

2 Theory

The standard software SAP2000 [19] platform is used to perform the nonlinear response history analysis (NRHA) to compare the responses in rigid and semi-rigid frames. The modelings of frames are explained in subsection with some attentive measures.

2.1 Implementation of Semi-rigid Connection Link Element in SAP2000

The semi-rigid beam-column connections in this study are modeled in two different approaches. The two-jointed zero-length link element is used to represent the two types of SR connections with different hysteretic behaviors. The SR connection with a multi-linear elastic link element (MLE) exhibits the isotropic hysteretic behavior, whereas the kinematic hysteretic behavior is exhibited by SR connections with multi-linear plastic link (MLP) elements (see Fig. 1). The considerable amount of seismic

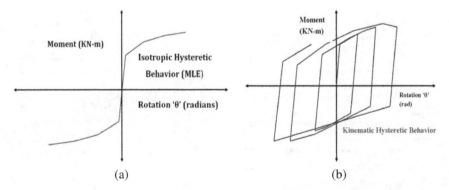

Fig. 1 Semi-Rigid Connection Behavior. **a** Isotropic Hysteretic Behavior for Elastic Link. **b** Kinematic Hysteretic Behavior for Plastic Link

energy is dissipated in cyclic loading in the MLP link element. The MLP shows the kinematic hardening behavior, and it is pertinent for ductility in connections.

Figure 2a explains the typical generic moment-rotation (M-θ) curve adopted for SR connections. The three parameters, namely, stiffness parameter (k), flexural strength parameter (s), and ductility parameter (μ), decide the shape and values of the M-θ curve. The acceptance criteria for these parameters are based on AISC 341-16 for seismic strengthening of beam-column SR connections. The flexural strength of the connection is chosen in such a way that the ratio of yield moment capacity ($M_{y,c}$) of connection to the plastic moment capacity ($M_{p,c}$) of connections is maintained at 0.67. The flexural strength of connection at the column end should be 0.8 multipliers of plastic moment capacity of the adjoining beam ($M_{p,b}$) to satisfy the story drift limit (greater than 0.04 rad), recommended in ASCE 341-16. The degree of semi-rigidity in connection is explained by parameters 'k' and 's' shown in Eq. 1 (a, b). Chan

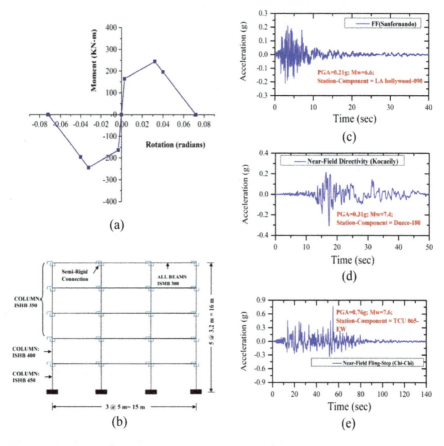

Fig. 2 **a** Typical moment Rotation Curve for Semi-rigid connection for k = 15, s = 1.5; **b** 5-Story Semi-rigid frame, and Ground Motion Records: **c** Far-Field, **d** Near-Field Directivity and (**e**) Near-Field Fling Step

et al. [20] prescribed the ductility limit to a minimum value of 0.04 rad for partially restrained connections for seismic strengthening. The connection parameters are

$$k = \frac{S_i}{\frac{EI_b}{L_b}}; \; s = \frac{M_{p,c}}{M_{p,b}} \quad (1)$$

where S_i is initial connection stiffness; EI_b/L_b is the flexural strength of the adjoining beam.

Three types of nonlinearity are considered in modeling, namely, connection nonlinearity in R3 direction in both types of link connection, material nonlinearity for flexural plastification in the form of plastic hinges as per ASCE 41-17 criterion and geometric nonlinearity considering second-order P-Δ effects. The panel zone behavior of joints is excluded in this work.

2.2 Nonlinear Response History Analysis (NRHA)

Nonlinear response history analysis (NRHA) is carried out for numerical simulation using the Hilber-Hughes-Taylor time integration approach with default values (gamma = 0.5, and beta-0.25) and second-order P-Δ effects are also considered. The 5% proportional Rayleigh damping is accounted for considering the first and second modes of vibration. In all, 30 NRHA simulations were performed for the 5-Story frame with different degrees of semi-rigidity.

3 Numerical Simulation

In this study, the numerical simulations are performed on a 5-Story frame with different degrees of semi-rigidity and link properties. The 5-Story steel frame has a uniform height of 3.2 m in each floor and 5 m bay width in both directions (see Fig. 2b). The building consists of special moment frames (SMF) with rigid beam-column connections, satisfying the primary requirements of the capacity design concept. The column sections are selected in such a way that the ratios of plastic moment capacity of columns are maintained 1.2 times to the plastic moment capacity of the adjoining beam (Strong Column-Weak Beam SCWB). The gravity loads on the particular beams consist of 20KN/m floor dead load, 15 KN/m roof dead load and 4 KN/m as live load, uniformly distributed on all floors. The SMF is designed for Indian seismic code requirements as per IS 1893-2016, IS 800-2007, and IS 875-1987 [2, 21, 22]. The seismic design parameters comprised of zone factor (Z = 0.36; Zone V), medium soil condition, importance factor (I = 1 for multistory frames) and response reduction (R = 5 for SMF) factor. Three types of earthquakes are selected as shown in Fig. 2 (c–e) for simulation, namely, San Fernando as Far-field (FF), Kocaeli as Near-field with the forward directivity (NFD) and Chi-Chi TCU065 as near-field with fling step (NFF) effects. A typical internal frame is chosen for analysis. Two types of beam-column connections, rigid and two sets of semi-rigid with a multi-linear elastic link and plastic link elements are considered for numerical simulation. The degree of semi-rigidity is described by two parameters: k and s (see Eq. 1). The semi-rigid frames are entitled as (i) SR-1E and SR-1P (k = 5; s = 1.2); (ii) SR-2E and SR-2P (k = 15; s = 1.5) in which 'E' stands for multi-linear elastic link whereas 'P' stands for multi-linear plastic link elements. The responses from SR frames are compared with rigid frames. The material nonlinearity in frames are modeled as concentrated default plastic hinge, defined in ASCE 41-17 at both ends of flexural members (M3 hinges for Beams and P-M3 hinges for columns in SAP2000).

4 Numerical Results and Discussions

The performance of 5-Story steel frames with different degrees of semi-rigidity, including rigid frames, is investigated for three types of ground motions at design and high PGA (i.e., 0.4 and 0.6 g) level. The seismic energy dissipation in SR frames with multi-linear elastic and plastic links are compared with rigid frames. The seismic demand parameters for comparison included maximum roof displacement, the maximum base shear, the total number of plastic hinges formed (NH), square root of sum of square values of maximum plastic hinge rotations (SRSS) and energy dissipation in link hysteretic energy and modal damping.

4.1 Energy Dissipation in Rigid and Semi-rigid Steel Frames

Figure 3 represents the variation of link hysteretic and modal damping energy in the SR frames for three types of seismic excitations at a high PGA level of 0.6 g. It is apparently visualized from the figure that the modal damping energy is more in all types of earthquakes as compared to the link hysteretic energy. Further, modal damping energy is higher in near-field fling step earthquakes. It is also noticed that the elastic link frames exhibited less link hysteretic energies as compared to plastic link connected SR frames in all cases. The link energy is decreased with an increase in the degree of semi-rigidity (k = 15 to 5). The reverse pattern is observed in modal damping energy.

Figure 4 shows the distribution of different types of energy in two types of SR connection modeling (multi-linear elastic and plastic) for the NFF type earthquake at the PGA level of 0.6 g. It is seen from the figure that, the multi-linear elastic link model, the model energy stored most of the input energy, very less amount of energy is dissipated in the form of other energies. As a result, the energy dissipation by way of the formation of plastic hinges is small in case of multi-linear elastic link SR frames with less number of plastic hinges (see Table 1). It is also observed that in

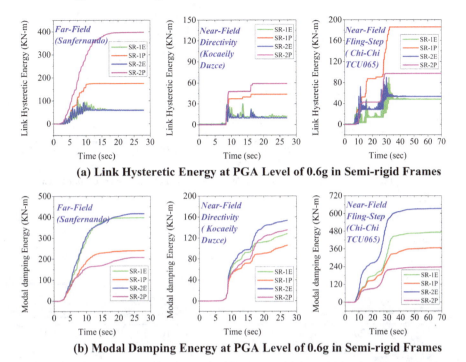

Fig. 3 Variation of **a** Link Hysteretic Energy, and **b** Modal Damping Energy under FF, NFD and NFF ground motion at a PGA level of 0.6 g for 5-Story Semi-rigid frames

the case of multi-linear plastic link SR connection, the most of energy is consumed in both modal damping and link hysteretic energy, not much energy is available to be dissipated by way of formation of plastic hinges. As a result, the formation of a number of hinges in SR connection with the plastic link element, there is a significant reduction.

Table 1 shows that the number of plastic hinges increased with an increase in the strength factor 's' (1.2–1.5). Further, in fully rigid (FR) frames, the near-field earthquakes stocked more energy as compared to the far-field earthquake at the same PGA level (see Fig. 5). It is clearly seen from Fig. 5 that modal energy together with other energies consumes only a part of total energy; much energy is left, which gets dissipated in the formation of plastic hinges. Thus, a large number of plastic hinges formed in the rigid frame (see Table 1).

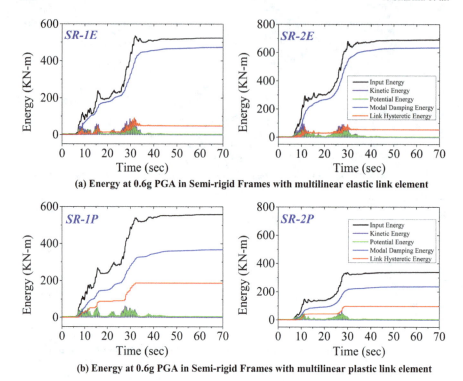

Fig. 4 Variation of Energy in SR frames with Multi-linear **a** Elastic Link, and **b** Plastic Link under the Chi-Chi TCU065 (NFF) earthquake at PGA level of 0.6 g

Table 1 Variation of total number of plastic hinges

Frame id	Total number of plastic hinges (NH)						
	FF		NFD		NFF		
	0.4 g	0.6 g	0.4 g	0.6 g	0.4 g	0.6 g	
SR-1E	4	8	3	4	2	6	
SR-1P	4	6	0	4	0	4	
% Difference^	0.00	25.00	100.00	0.00	100.00	33.33	
SR-2E	0	18	0	9	4	18	
SR-2P	0	4	0	4	4	6	
% Difference^	0.00	77.78	0.00	55.56	0.00	66.67	
FR	16	27	21	30	26	31	

^Percentage Difference = $((NH_{SR\text{-}Elastic} - NH_{SR\text{-}Plastic})/NH_{SR\text{-}Elastic}) \times 100$

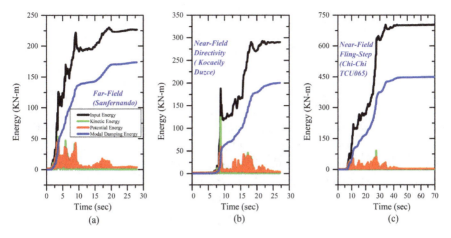

Fig. 5 Variation of Energy in Rigid frame at PGA 0.6 g for **a** FF, **b** NFD, and **c** NFF

4.2 Inelastic Excursion in the Semi-rigid and the Rigid Frames

Earlier section explained that the less number of plastic hinges formed in SR frames as compared to FR. Thus, the damages in terms of plastic hinges incurred at flexural member ends are less in the SR frames. However, considerable inelastic excursion occurs in plastic link connections dissipating energy through a hysteretic cycle in SR frames as discussed in earlier subsection. Table 1 shows a large number of plastic hinges formed in rigid frames, and it is increased with PGA level and type of earthquakes, maximum in near-field with fling step earthquakes. This number is considerably reduced in SR frames as energy is dissipated more in link hysteretic loops as compared to SR frames.

Table 2 shows the SRSS of maximum plastic hinge rotations in the rigid and semi-rigid frames. The SRSS values are considerably higher in FR as compared to SR frames. It is again observed that more seismic energy is dissipated in the plastic hinges in FR frames as compared to SR frames, as it would be expected.

4.3 Variation of Maximum Roof Displacements in Semi-rigid and Rigid Frames

Table 3 describes the variation of roof displacement in rigid and semi-rigid frames. It is observed from the table that the way of SR modeling affects significantly on roof displacement. The plastic link SR connection dissipates seismic energy considerably, so the maximum roof displacement in all cases is less as compared to elastic SR connection and reduced with decrease in the degree of semi-rigidity which means

Table 2 Variation of SRSS of maximum plastic hinge rotations in radians

Frame id	SRSS (radians)					
	FF		NFD		NFF	
	0.4 g	0.6 g	0.4 g	0.6 g	0.4 g	0.6 g
FR	0.01058	0.03175	0.01771	0.03854	0.03389	0.05000
SR-1E	0.00292	0.01232	0.00032	0.00346	0.00007	0.00700
SR-1P	0.00059	0.00397	0.00000	0.00304	0.00000	0.00481
% Difference^^	79.67	67.76	100.00	12.13	100.00	31.30
SR-2E	0.00000	0.00772	0.00000	0.00356	0.00360	0.01363
SR-2P	0.00000	0.00216	0.00000	0.00356	0.00129	0.00756
% Difference^^	0.00	71.98	0.00	0	64.33	44.49

^^Percentage Difference = $((SRSS_{SR\text{-}Elastic} - SRSS_{SR\text{-}Plastic})/SRSS_{SR\text{-}Elastic}) \times 100$

Table 3 Variation of roof displacement in semi-rigid frames

Frame id	Roof displacement (mm)					
	FF		NFD		NFF	
	0.4 g	0.6 g	0.4 g	0.6 g	0.4 g	0.6 g
SR-2E	183.63	313.05	183.63	241.89	270.09	354.28
SR-2P	179.56	228.90	170.15	239.04	220.59	291.27
SR-1E	325.83	463.03	173.59	270.71	254.56	358.48
SR-1P	264.56	344.66	173.62	271.74	230.24	291.88
FR	160.75	198.87	190.30	224.96	207.73	272.93

lowest in FR and highest in SR-1E case. The % change in maximum values are increased with PGA values from the design level to high PGA level for all cases.

4.4 Variation of Maximum Base Shear in Semi-rigid Frames

Table 4 represents the percentage variation in the maximum value of base shear in SR elastic and plastic frames for three types of earthquakes. It is clearly observed that the maximum value of base shear reduced in plastic SR frames, and it is increased with the PGA level in all cases. As the PGA increased, more energy is attracted and the more amount of energy is dissipated in the multi-linear plastic links (due to kinematic hysteretic behavior) as compared to multi-linear elastic links (due to isotropic hysteretic behavior).

Table 4 Variation of maximum base shear in semi-rigid frames

Frame id	Percentage difference in maximum base shear (KN)					
	FF		NFD		NFF	
	0.4 g	0.6 g	0.4 g	0.6 g	0.4 g	0.6 g
SR-2E	615.67	792.82	615.92	781.35	680.46	788.29
SR-2P	538.98	665.51	582.74	702.79	654.03	754.81
% Difference#	12.46	16.06	5.39	10.05	3.89	4.25
SR-1E	608.77	755.07	603.79	757.59	496.10	630.55
SR-1P	589.72	707.45	581.93	676.69	490.86	612.64
% Difference#	3.13	6.31	3.62	10.68	1.06	2.84

 # Percentage Difference = ((Base Shear$_{SR-Elastic}$ - Base Shear$_{SR-Plastic}$)/Base Shear$_{SR-Elastic}$) X 100

5 Conclusions

The seismic energy dissipations and seismic demand parameters of 5-Story steel rigid and semi-rigid (SR) frames are examined at two peak ground acceleration (PGA) levels (designated as design level and high level) for three variants of earthquakes, viz., the far-field and near-field with the forward directivity effect and fling step effects. For the simulation study, a nonlinear response history analysis is carried out to obtain the seismic energy along with other responses. The semi-rigid frames are simulated in two different ways, namely, semi-rigid connection with a multi-linear elastic link and plastic link connections. The degree of semi-rigidity is taken as a prime parameter. The results of numerical simulation achieve the following outcomes:

i. The way of the dissipation of seismic energy in the rigid and semi-rigid frames are in different ways. The modal energy in the plastic hinges shared the maximum of input energy in rigid frames. Whereas for semi-rigid frames, the seismic energy is dissipated in the modal damping energy along with link hysteretic energy in plastic hinges.
ii. For semi-rigid frames, the modal energy consumes the maximum share of input seismic energy in connection with an elastic link, and the link hysteretic energy is considerably less as compared to modal energy. On the other end, in plastic semi-rigid connections, most of the input energy is dissipated in the form of link hysteretic and modal energy along with a very little share of other energies.
iii. The number of plastic hinges formed and their SRSS of maximum plastic hinge rotations are significantly higher in the rigid frames and considerably less in semi-rigid frames.
iv. The maximum values of roof displacement are observed in semi-rigid frames due to less strength and stiffness values as compared to the rigid frames.
v. The base shear is less in semi-rigid frames as compared to rigid frames. Especially, the semi-rigid frames with plastic link provide less base shear than elastic

semi-rigid frames, and these difference decreases with increase in the degree of semi-rigidity.

vi. The multi-linear plastic link semi-rigid frames are performed considerable better during earthquakes as compared to rigid frames and semi-rigid frames with multi-linear elastic links.

References

1. ASCE-41 (2017) Seismic evaluation and retrofit of existing buildings. ed. American Society of Civil Engineers, Reston, Virginia p 518
2. IS-800 (2007) General construction in steel-code of practice. vol. 3rd Revision, ed. Beureau of Indian Standards, New Delhi
3. Eurocode-3 (2005) Design of steel structures (part 1–8: design of joints). vol. 1, ed. European Committee for Standardization, Brussels
4. ANSI/AISC-341 (2016) Seismic provision for structural steel buildings. American Institute of Steel Construction, Chicago
5. Díaz C, Martí P, Victoria M, Querin OM (2011) Review on the modelling of joint behaviour in steel frames. J Constr Steel Res 67:741–758
6. FEMA (2000) State of the art report on connection performance. FEMA-355D
7. Code P (2005) Eurocode 8: design of structures for earthquake resistance-part 1: general rules, seismic actions and rules for buildings. European Committee for Standardization, Brussels
8. Nader M, Astaneh A (1991) Dynamic behavior of flexible, semirigid and rigid steel frames. J Constr Steel Res 18:179–192
9. Elnashai A, Elghazouli A (1994) Seismic behaviour of semi-rigid steel frames. J Constr Steel Res 29:149–174
10. Aksoylar ND, Elnashai AS, Mahmoud H (2011) The design and seismic performance of low-rise long-span frames with semi-rigid connections. J Constr Steel Res 67:114–126
11. Abolmaali A, Matthys JH, Farooqi M, Choi Y (2005) Development of moment–rotation model equations for flush end-plate connections. J Constr Steel Res 61:1595–1612
12. Sekulovic M, Nefovska-Danilovic M (2008) Contribution to transient analysis of inelastic steel frames with semi-rigid connections. Eng Struct 30:976–989
13. Elias S, Matsagar V, Datta T (2016) Effectiveness of distributed tuned mass dampers for multi-mode control of chimney under earthquakes. Eng Struct 124:1–16
14. Elias S, Matsagar V, Datta T (2017) Distributed tuned mass dampers for multi-mode control of benchmark building under seismic excitations. J Earthq Eng 23:1137–1172
15. Elias S (2018) Seismic energy assessment of buildings with tuned vibration absorbers. Shock Vibr, 2018
16. Elias S, Matsagar V (2019) Seismic vulnerability of a nonlinear building with distributed multiple tuned vibration absorbers. Struct Infrastruct Eng, 1–16
17. Lemonis M (2018) Steel moment resisting frames with both joint and beam dissipation zones. J Constr Steel Res 147:224–235
18. Kunnath SK, Kalkan E (2004) Evaluation of seismic deformation demands using nonlinear procedures in multistory steel and concrete moment frames. ISET J Earthq Technol 41:159–181
19. SAP2000 (2017) SAP 2000 v19: integrated software for structural analysis and design. Computers and Structures, Inc., Berkeley, California
20. Chan SL, Chui PT (2000) Non-linear static and cyclic analysis of steel frames with semi-rigid connections. Elsevier Science Ltd, Oxford, UK

21. IS-1893 (2016) Criteria for earthquake resistant design of structures in part 1 general provisions and buildings (6th Revision), vol. 6th (ed.) Bureau of Indian Standards, New Delhi
22. IS-875 (1987) Code of practice for design loads (other than earthquake) for buildings and structures. in part 1 dead loads-unit weights of building materials and stored materials (ed.) Bureau of Indian Standards, New Delhi, India

Seismic Behavior of Baffled Liquid Storage Tank Under Far-Field and Near-Field Earthquake

Sourabh Vern, M. K. Shrimali, S. D. Bharti, and T. K. Datta

Abstract The liquid storage tanks (LST) are one of the essential civil structures. The inter-action of LSTs with an earthquake during its service life is a critical and crucial factor that cannot be ignored. The present study describes effect of the earthquake on the response parameters of LSTs. Depending on the nature of the earthquake these responses can increase or decrease. The study is further extended by implementing the passive control device. A vertical baffle plate is placed at the base of the tank. The non-linear dynamic analysis of the LST is performed on the ABAQUS. The response quantities of interest include top board displacement, Von-Mises stress, shear force, overturning moment, wave height, and hydrodynamic pressure in the tank. The results show that wave height for the different earthquakes can yield a difference of about four times. The baffle plate can significantly impact the sloshing height. The maximum reduction of 77% in wave height for far-field earthquake is present. However, the presence of the baffle increases the shear force response of the tank.

Keywords Fluid-Structure interaction · Liquid storage tanks · Far-Field and Near-Field earthquakes · Abaqus

1 Introduction

Storage Vessels and Containers have become integral Lifeline Structures of many cities. The proclivity for concrete ground supporting liquid storage tanks in various industries such as for nuclear power plants, chemical treatment plants, food storage facilities can be seen because of their maintenance and construction. However, the risk of these storage structures from seismic hazards in their service life is always high, and damages in them possess an even greater threat to its surroundings. The failed epitomes of liquid storage tanks in past are evidence that these structures

---

S. Vern (✉) · M. K. Shrimali · S. D. Bharti · T. K. Datta
Malaviya National Institute of Technology, Jaipur 302017, India
e-mail: sourabh.vern@gmail.com

© The Editor(s) (if applicable) and The Author(s), under exclusive license to Springer Nature Singapore Pte Ltd. 2021
S. K. Saha and M. Mukherjee (eds.), *Recent Advances in Computational Mechanics and Simulations*, Lecture Notes in Civil Engineering 103,
https://doi.org/10.1007/978-981-15-8138-0_34

require constantly developing design implementations and recommendations. Huge Losses were reported in Chile earthquake (1960), Eureka earthquake, California, (1980), Great East Japan earthquake, Japan (2011), etc. Among these failed LSTs major failures were elephant foot buckling, damage of the tank roof, and uplifting of the base.

The development of the various control techniques to suppresses the formation of the wave in the liquid storage tanks is still in the nascent stage. Currently there is mainly a couple of ways to control the responses in LSTs, firstly by employing the base isolations and second is by introducing the inner obstructions in the fluid media. By the application of base isolation, the fundamental natural period of the system is elongated which helps in slow dissipation of the inertial energy. The shortcoming of the base isolation control appears in terms of the sloshing height as the two different parts of the LSTs have two different fundamental periods. As the base isolation extends the impulsive period of the system, the time period values shift toward the natural period of the convective period. As the fundamental period approaches the convective period, the amplification of the wave height is reported [1–3]. However, the baffles inside the fluid media can subsequently decrease the sloshing height which also aids in mitigating the overturning moments and hydrodynamic pressure on the tank walls.

The responses of the LSTs can be simplified into the two parts, namely, impulsive and convective. The impulsive response of the tank is responsible for the base shear and overturning moment. Due to the higher magnitude of the overturning moment at the bottom the compressive stresses are formed tank walls which are responsible for the buckling of the tank walls. The remaining response is created by the convective part. The wave height development in the LST because of the earthquake motion can be contributed to convective part. Not only the formation of wave height is related to this response quantity, but also convective response makes its contribution to the resultant overturning moment and shear forces of the total system. The behavior of the LSTs involves the difficult fluid-structure interaction that makes the analysis cumbersome. Housner [4] carried out more defined and exact analyses in which pressure can be subdivided into two major parts, Convective and Impulsive. Impulsive pressure is exerted by the inertial response of the fluid which is because of the inertial response of the tank walls. Wozniak et al. [5] studied a simplified procedure which has been prescribed by API standard 650. Liu [6] examined the fluid-structure interaction inside a liquid storage tank by finite-element method with emphasis on dynamic and buckling analyses. Lagrangian-Eulerian kinematical description for modeling fluid subdomains in fluid-structure interaction problems. Evans and Mciver [7] studied the interaction between the vertical baffle with the resonant frequencies of the fluid storage tank by linearized wave theory of water waves. The validation is done by an accurate solution of eigenfunction expansions. The surface piercing barrier can develop a significant change in the tank resonant frequencies. Malhotra [8] explained the design criteria for the simplified cylindrical ground support liquid storage tank. While examining the design criteria, impulsive and convective part of the liquid in flexible steel or concrete tanks fixed to rigid foundations is considered. Fischer and Seeber [9] presented a dynamic response analysis of vertically excited liquid storage tanks including both liquid-tank and liquid-soil interaction. Virella et al. [10] studied the sloshing natural periods and their modal pressure distributions by the influence of nonlinear wave theory for two-dimensional behavior rectangular tanks. Panigrahy et al. [11] conducted a series of experiments for studying the impact of various sloshing frequencies with baffles and without baffles. Due to the presence of baffles sharp edges the turbulence in flow field is developed which dissipates maximum kinetic energy to the tank walls. Sygulski [12] evaluated the natural frequencies and the modal shapes of the liquid sloshing profile for a three-dimensional arbitrary geometry with baffles. The tank base and the baffle were taken as rigid components boundary element method with 6 node triangular curvilinear boundary elements was taken up for the solution. Jung et al. [13] studied impact of the baffle height on the sloshing for a laterally moving three-dimensional (3D) rectangular tank. The volume of the fluid method which was based on the finite volume method was used for the analysis. As expected, the presence of the baffle in the liquid storage tank initiated a vortex due to the flow separation at the baffle tip. As a result of the damping effect of the baffle can be successfully recorded. Cao et al. [14] developed the paralleled SPH codes are programmed to study the liquid sloshing in both two-dimensional and three-dimensional tanks of single and multi-degrees of freedom. Wang et al. [15] studied the semi-analytical scaled boundary finite-element method to evaluate the effects of the T-shaped baffle on liquid sloshing for an elliptical tank. The influences of the wide range parameters including liquid fill level, baffled arrangement, and length of those baffles, the ratio of the major and minor semi-axes being equal of elliptical tank and angle of the Y-shaped baffle are studied in details and the results in terms of the sloshing frequencies, sloshing mode shapes and water surface elevation are presented. Xue et al. [16] presented an experimental setup to determine the pressure distribution of the displaced liquid profile of a baffled tank, the vertical baffle flushing with free surface and the perforated vertical baffle are two more effective tools in reducing the dynamic impact pressure. Cho et al. [17] examined sloshing suppression capability of the porous baffles by the application of the boundary element method.

In the present study, the impact of the vertical baffle on the suppression of sloshing and eventual decrease of the responses of interests are critically investigated for different types of earthquakes. The responses of interests include the top board displacement, base shear, overturning moment in the tank walls and the fluid surcharge at the free surface. A finite-element analysis is carried out by simulating the fluid and tank media as finite elements. For the study, ABAQUS software is used. The nonlinear dynamic analysis of the rectangular liquid storage tank is performed under various seismic ground motions, which includes both near field and far-field. The effect of the vertical baffle on the responses is also critically examined.

2 Methodology

The motion of the LSTs due to seismic excitation is classified as a nonlinear analysis of the liquid-tank system. The movement at the base of the tank creates inertial resistive forces which get amplified further due to the hydrodynamic motion of the fluid stored inside the LST. The increase in the slosh forces at the walls of the tank further pushes the material of the tank walls into the nonlinear range. The complexity of the material nonlinearity is handled by the Abaqus FEM platform. The dynamic explicit time step integration techniques involve an explicit central-difference integration rule which inputs the diagonal mass matrix at initial of the time step increment results as given in Eq. (1).

$$\ddot{u}_{(i)}^N = (M^{NJ})^{-1}(P_{(i)}^J - I_{(i)}^J) \qquad (1)$$

where P_J, I_J, and M_{NJ} are load vector, force vector, and mass matrix.

$$\dot{u}^{(i+\frac{1}{2})} = \dot{u}^{(i-\frac{1}{2})} + \frac{\Delta t^{(i+1)} + \Delta t^{(i)}}{2} \ddot{u}^{(i)} \qquad (2)$$

$$u^{(i+1)} = u^{(i)} + \Delta t^{(i+1)} \dot{u}^{(i+\frac{1}{2})} \qquad (3)$$

The complexity of the analysis is made simple to the solution by the equation of motion and it is solved by using the central difference integration rule as stated in Eqs. (2) and (3), where $\dot{u}$ and $\ddot{u}$ are the velocity and the acceleration values, respectively. The superscript (i) shows the incremental series number; ($i-1/2$) and ($i+1/2$) refer to the mid incremental series numbers. As the condition stability of the central difference operator remains available for the solution the stable time increment is then calculated by Eq. (4). The presence of high-frequency oscillations pushes the stable time increment of the solution to the higher range, to suffice this problem a low level of damping is added into the operator for compensating the requirement of time increment.

$$\Delta t \leq \frac{2}{\omega_{max}} \left(\sqrt{1+\xi^2} - \xi \right) \qquad (4)$$

The maximum value of the frequency of the whole system is extracted by the explicit module. In the present study, to perform the complex FSI the Mie-Gruneisen equation of state is used for the analysis. The Mie-Gruneisen equation takes the velocity of the particle and velocity of shock in the Hugoniot form which represented as a linear form in Eq. 5.

$$U_s = c_0 + sU_p \qquad (5)$$

The final form of the equation available for the analysis as a Hugoniot form is given in Eq. 6

$$p_H = \frac{\rho_0 c_0^2 \eta}{(1-s\eta)^2} \left(1 - \frac{\Gamma_o \eta}{2}\right) + \Gamma_o \rho_0 E_m \qquad (6)$$

where p_H and E_m are Hugoniot pressure and specific energy, respectively, in the form of density function, whereas Γ is known as Gruneisen ratio [18]. The impulsive and convective components of weights in the accelerating liquid are calculated by the guidelines of ACI 350,2006 [19].

$$\omega_C = \sqrt{(n\pi g/L)tanh((n\pi H_L/L))} \tag{7}$$

Equation 7 links the fundamental circular frequencies in the convective mode of the liquid inside the tank which is responsible for the sloshing phenomena whereas the impulsive frequency can be contributed to the development of the stresses in the tank walls. As the convective mode is not dependent on the motion of the tank, the sloshing response is independent of the material property of the tank.

2.1 Finite-Element Modeling Details

For FEM, the tank is considered flexible and surface to surface interaction is used between the fluid and the wall surfaces. The tank is modeled by an eight-node linear brick element with reduced integration and hourglass control. The fluid is also modeled by the same solid element with combined hourglass control and providing the stiffness-viscous weight factor of 1. A mesh convergence study is conducted to get a stable mesh with sound results. For the tank, a mesh size of 0.1 m is taken and for the fluid media mesh of size 0.08 m is taken. Arbitrary Lagrangian and Eulerian (ALE) approach is used, which maintains the mesh and results in faster and higher accuracy outputs, as compared to the pure Lagrangian approach. The ALE approach gives the advantage in analysis by maintaining aspect ratio of the original mesh. As a result of the implementation of the ALE step stable fluid motion is obtained. As the ALE step is selected for the fluid mesh, it provides the independent motion of the fluid mesh nodes, therefore in case of the high scale deformation, the program doesn't lose its stability.

3 Numerical Study

For the present study, a flexible rectangular tank is taken with ground support. The nature of the earthquakes can significantly alter the natural responses of the LST. To study the influence of the nature of the earthquake on the LSTs, different types of earthquakes are taken up for the analysis. These earthquakes are primarily divided into two groups, Far-Field, and Near-Field Earthquakes. Further the Near-Field earthquake can be divided into three parts, Fling Step, High, and Low directivity. The high directivity is defined when the rupture front propagates toward the site and the direction of slip fault is along the structure site. The details of the various earthquakes taken up for the analysis are given (see Fig. 2). The duration of the time history of the earthquakes for the analysis is 23 s. The peak ground acceleration of the earthquakes is 0.25 g. For a better understanding of earthquakes taken up for the analysis their response spectrum and the time history plot are given (see Figs. 1 and 2).

The tank is taken as a concrete tank with a thickness of 300 mm and with a cross-sectional size of 6 m X 6 m. The height of the fluid stored in the tank is of 4.8 m (Tables 1 and 2). The tank is taken up in the present study is a chemical industry instrument [20]. The properties of the tank and fluid media are given *in* Table 1 for reference.

The various fundamental frequencies of the LST is given in Table 2. The effect of the natural frequency in base shear and overturning moment is crucial.

The study of comparison between the uncontrolled responses and the various controlled responses of the LST is done. For the controlled responses, a vertical baffle of height 2.4 m and width 0.3 m is taken. The position of the baffle is critical; in the present study the baffle is placed at the center of the tank base and is perpendicular to the direction of the earthquake motion. The response quantities of the interests are free surface charge in fluid, top board displacement (TBD), hydrodynamic pressure in fluid, Von-mises stress, base shear, and overturning moment. The TBD and wave height are calculated at the center of the end of right side facing wall, whereas Mises stress is calculated at the bottom section of the right side facing wall. The contour diagrams of the mises stress are given (see Fig. 3). High intensity of the stresses is seen in the tank walls. However, due to the presence of the baffle in the tank, its stress contour diagram displays lower value. The magnitude of the stress in both cases is highest at the bottom of the base, and it can be related to the elephant footing failure that can be seen during earthquakes.

The flexibility of the tank wall can significantly affect the overall response of the tank responses. The plot of the far-field and near-field with high directivity is

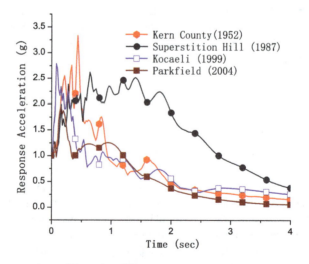

Fig. 1 Response spectrum of the various EQs

Seismic Behavior of Baffled Liquid Storage ...

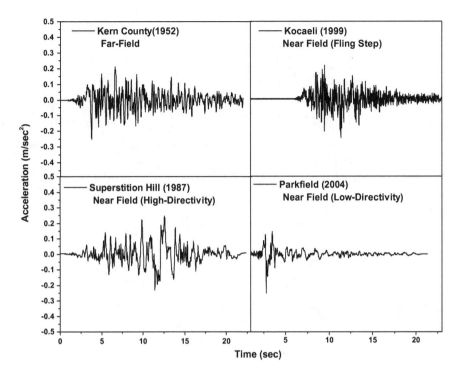

Fig. 2 Time history plot of the various earthquake

Table 1 Material properties for the tank and Fluid media

Concrete	Fluid
Modulus of Elasticity, E_c = 30GPa	Density, ρ_w = 983.204 kg/m3
Density, ρ_c = 2490 kg/m^3	Equation of state: c_0 = 1450, s = 0, γ_0 = 0
Poisson's ratio, υ = 0.2	Dynamic Viscosity = 0.001 N-sec/m^2

Table 2 Natural Frequency for Different Modes

Type of mode	Frequency (Hz)
Impulsive	3.125
Convective	0.278

shown (see Fig. 4), it can be seen that the effect of the baffle has little effect on the curtailment of the top board displacement. The time history responses reach peak in the middle section of the time history, whereas the far-field earthquake response shows peak value at the initial section of the time history plot. It can be inferred

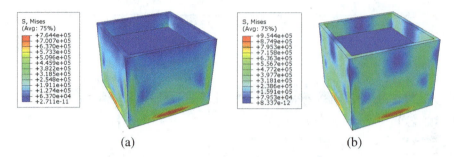

Fig. 3 Contour plots of mises stress for **a** Near-field low directivity **b** Near-field high directivity

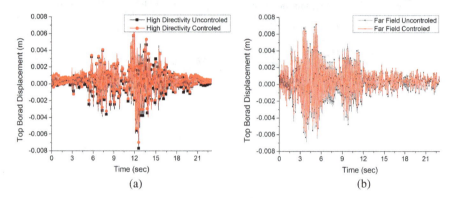

Fig. 4 Time history plots of uncontrolled and controlled Top Board displacement for **a** Near-Field High directivity **b** Far-Field earthquake

that the position of the baffle and height of the baffle plate can significantly alter the nature of responses.

In the present case, the wave height of the fluid can be seen to follow a sinusoidal flow pattern (see Fig. 4). The time history pattern of the near-field earthquake starts with a small height which eventually achieves its peak at the end segment of the time history plot. The different patterns of the wave height can be observed for the far-field earthquake; the fluid response is maximum in the middle segment and then starts to degrade (Fig. 5).

There is a significant decrease in the wave height due to the application of the baffle plate. This may be due to the formations of the eddies at the baffle surface, which leads to the turbulent flow near the baffle interface. Thus, a change in the response can be noted. The base shear plot history of the near-field high directivity and far-field of both, uncontrolled and controlled are given (see Fig. 6). The time history shows that there is a considerable difference between the 2 plot histories; thus it can be concluded that nature of the earthquake can significantly change the response of the LSTs. The base shear requirement of the near-field earthquake is significantly much more than

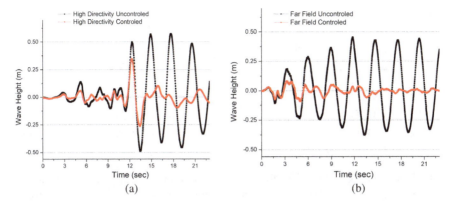

Fig. 5 Time history plots of uncontrolled and controlled wave height for **a** Near-Field High directivity **b** Far-Field earthquake

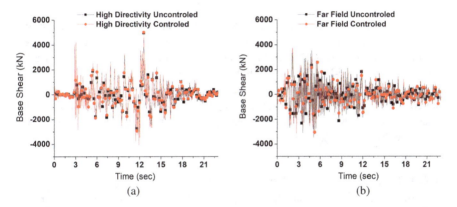

Fig. 6 Time history plots of uncontrolled and controlled Base Shear for **a** Near-Field High directivity **b** Far-Field earthquake

that of the far-field earthquake. This may be due to the more energy content in near-field earthquake, as the distance between the structure and origin of the earthquake is lesser and lesser wave is dissipated. The controlled response due to the baffle shows an increased value of the base shear. This increase in the response might be due to the additional rigidity provided by the baffle section at the base of the tank. The optional benefit of the baffle can be achieved in tanks in which the stiffeners are required for the structural strength, thus serving dual purpose of the sloshing damper and a stiffener. However, a decrease in the overturning moment. The overturning moment of tank can lead to the base uplifting failure, thus deploying baffles inside the tank should be considered. The absolute peak values of both the uncontrolled and controlled responses are obtained. It is quite interesting to note the effect of the external solid geometry in altering the natural responses of the tank and the fluid. Not only the complex problems of fluid-structure interaction (FSI) requires additional efforts but

Table 3 The absolute maximum values of uncontrolled responses of interests

Type of earthquake	Von-Mises stress (kPa)	Hydro-dynamic pressure (kPa)	Wave height (m)	Base shear (kN)	Overturning moment (kN-m)	Top board displacement (m)
FF	5270.3	43.4	0.45	3575.7	9817.87	0.0072
NF-FS	6758.4	44.1	0.29	5073.9	17030.5	0.0111
NF-HD	4523.4	41.6	0.16	3884.5	8763.53	0.0070
NF-LD	4363	40.5	0.57	4967.4	13502.8	0.0064

also their behavior under the different type of earthquakes can give unconventional results. The following Table 3 gives the absolute maximum values of uncontrolled responses of interests. The maximum value of the sloshing height for uncontrolled response is reported in the case of the near field with low directivity(NF-LD). This value of fluid surcharge is almost four times that of near-field with high directivity (NF-HD).

Similarly, the overturning moment of the NF-LD is almost double in the magnitude with respect to the NF-HD. The maximum value of the peak tank wall displacement is reported in the near-field with the fling step (NF-FS) case, which is in itself is almost double of NF-LD. The following Tables 4 and 5 give the absolute maximum values of responses of interests and the percentage change. It can be seen that NF-FS and NF-HD show the increased values of the hydrodynamic pressure for controlled cases, whereas both the far-field and near field low directivity earthquakes showed almost negligible change in the pressure.

Table 4 The absolute maximum values of controlled responses of interests

Type of earthquake	Von-Mises stress (kPa)	Hydro-dynamic pressure (kPa)	Wave height (m)	Base shear (kN)	Overturning moment (kN-m)	Top board displacement (m)
FF	4951.83	44.3	0.10	4758.4	9243.9	0.006229
NF-FS	6242.29	64.8	0.20	6242.1	14341.3	0.009926
NF-HD	4687.32	60.6	0.12	4233.6	7212.2	0.006558
NF-LD	4421.69	41.9	0.35	5223.6	10757.9	0.005572

Table 5 The percentage change in responses with respect to uncontrolled responses

Type of earthquake	Von-Mises stress (%)	Hydro-dynamic pressure (%)	Wave height (%)	Base shear (%)	Overturning moment (%)	Top board displacement (%)
FF	−6	02	−77	33	−06	−14
NF-FS	−8	47	−31	23	−16	−11
NF-HD	4	46	−23	09	−18	−06
NF-LD	1	03	−39	05	−20	−13

The following Table 5 gives the absolute maximum values of responses of interests. It can be seen that NF-FS and NF-HD show the increased values of the hydrodynamic pressure for controlled cases, whereas both the far-field and near field low directivity earthquakes showed almost negligible change in the pressure. There is a notable decrease in the TBD for the case of the controlled response thus it can be inferred that the rigidity offered by the baffle to the tank wall can provide it with more resistance to the motion. They are thus making the overall responses of the tank walls rigid. The maximum decrease in the TBD is for the FF ground motion of almost 15 %. The cause of the base uplift can be contributed to the large overturning moment in the base of the tank. The NF-LD shows a good reduction in the overturning moment of the tank. Never the less the least reduction is for the FF earthquake.

4 Conclusions

The behavior of the LST under the uni-directional earthquake motion is studied and then the controlled effect of the vertical baffle plate is critically examined in the present study. Four various types of earthquakes are taken, namely, Far-Field and Near-field. A square tank with fixed base is taken up for the analysis. The modeling and non-linear analysis of liquid and tank are done in ABAQUS software. The following conclusions can be drawn from the numerical study:

1. The time history responses of the various earthquakes are different. The peak of the responses is developed at a different time interval. Thus, the nature of the earthquake can play significant role in the response. The pattern of the uncontrolled and controlled responses also showed the deviations.
2. The reduction in the sloshing height is more for the far-field earthquake whereas the reduction for the near-field earthquake is almost half of the far-field earthquake.
3. The shear force response displayed an increase in the magnitude of the controlled response. This may be due to the additional rigidity is provided by the vertical baffle in the LST. Whereas the overturning moment of the LST is reduced to an order of 20% in case of the near-field earthquake with low directivity.
4. The top board displacement showed a uniform decrease of 10% in the magnitude. The minimum reduction is for the near-field earthquake with high directivity.

References

1. Shrimali MK, Jangid RS (2002) Seismic response of liquid storage tanks isolated by sliding bearings. Eng Struct 24:909–921
2. Cho KH, Kim MK, Lim YM, Cho SY (2004) Seismic response of base-isolated liquid storage tanks considering fluid-structure-soil interaction in time domain. Soil Dyn Earthq Eng 24:839–852

3. Hashemi S, Aghashiri MH, Kianoush MR (2017) Behavior of concrete rectangular containers isolated using different isolation systems subjected to bi-directional excitation, pp 3350
4. Housner GW (1957) Dynamic pressures on accelerated fluid containers. Bull Seismol Soc Am 47:15–35
5. Wozniak RS, Mitchell W (1978) Basis of seismic design provisions for welded steel oil storage tanks. In: Advances in storage tank design. American Petroleum Institute, Washington, USA
6. Liu WK (1981) Finite element procedures for fluid-structure interactions and application to liquid storage tanks. Nucl Eng Des 65:221–238
7. Evans DV, Mciver P (1987) Resonant frequencies in a container with a vertical baffle. J Fluid Mech 175:295–307
8. Malhotra PK (1997) New Method for Seismic Isolation of Liquid-Storage Tanks. J Earthq Eng Struct Dyn 26:839–847
9. Fischer FD, Seeber R (1988) Dynamic response of vertically excited liquid storage tanks considering liquid-soil interaction. J Earthq Eng Struct Dyn 16:329–342
10. Virella JC, Prato CA, Godoy LA (2008) Linear and nonlinear 2D finite element analysis of sloshing modes and pressures in rectangular tanks subject to horizontal harmonic motions. J Sound Vib 312:442–460
11. Panigrahy PK, Saha UK, Maity D (2009) Experimental studies on sloshing behavior due to horizontal movement of liquids in baffled tanks. Ocean Eng 36:213–222
12. Sygulski R (2011) Boundary element analysis of liquid sloshing in baffled tanks. Eng Anal Bound Elem 35:978–983
13. Jung JH, Yoon HS, Lee CY, Shin SC (2012) Effect of the vertical baffle height on the liquid sloshing in a three-dimensional rectangular tank. Ocean Eng 44:79–89
14. Cao XY, Ming FR, Zhang AM (2014) Sloshing in a rectangular tank based on SPH simulation. Appl Ocean Res 47:241–254
15. Wang W, Guo Z, Peng Y, Zhang Q (2016) A numerical study of the effects of the T-shaped baffles on liquid sloshing in horizontal elliptical tanks. Ocean Eng 111:543–568
16. Xue MA, Zheng J, Lin P, Yuan X (2017) Experimental study on vertical baffles of different configurations in suppressing sloshing pressure. Ocean Eng 136:178–189
17. Cho IH, Choi JS, Kim MH (2017) Sloshing reduction in a swaying rectangular tank by an horizontal porous baffle. Ocean Eng 138:23–34
18. Dassault Systèmes Simulia (2012) Analysis user's manual volume 2: analysis. Abaqus 6.12 II
19. ACI 350 (2007) Seismic design of liquid-containing concrete structures and commentary (ACI 350.3-06)
20. Cheng X, Jing W, Gong L (2017) Simplified model and energy dissipation characteristics of a rectangular liquid-storage structure controlled with sliding base isolation and displacement-limiting devices. J Perform Constr Facil 31:04017071

Seismic Response of Asymmetric Structure with Soil Structure Interaction Using Semi-Active MR Damper

Shuvadeep Panchanan, Praveen Kumar, Swagata Basu, and R. S. Jangid

Abstract The seismic response of a four-storey asymmetric building with SSI using MR damper with semi-active damping with friction type damping algorithm is studied. The effect of SSI is assessed by comparing the seismic response of a four-storey one-way asymmetric building resting on stiff soil and soft soil with that of the same building considering fixed base case. The optimal damper parameters considering four different ground motions are obtained. The controlled response of the building resting on soft soil is also compared with that of the corresponding uncontrolled symmetric structure to assess the effects of lateral torsional coupling. It is observed from the present study that MR damper with friction type damping scheme could significantly reduce the detrimental effects of SSI and lateral torsional coupling but the reduction in response varies widely with the ground motion considered.

Keywords Seismic response · Asymmetric · Soil structure interaction · Semi-active damper

1 Introduction

Post-earthquake reconnaissance studies conducted in the past show that asymmetric buildings are more vulnerable to earthquake-induced vibrations in comparison to symmetric buildings. In structures having an asymmetric distribution of mass and/or stiffness, torsional motions are coupled with lateral translational motions. Such coupling of motion changes the distribution of shear force in building columns resulting in higher shear force in outer columns than that in inner columns and also leads to increased floor displacements of the building. One of the approaches to reduce the above-mentioned detrimental effects of lateral torsional coupling is

---

S. Panchanan (✉) · P. Kumar
Bhabha Atomic Research Centre, Mumbai 400085, India
e-mail: sdeep@barc.gov.in

S. Basu · R. S. Jangid
Indian Institute of Technology Bombay, Mumbai 400076, India

© The Editor(s) (if applicable) and The Author(s), under exclusive license to Springer Nature Singapore Pte Ltd. 2021
S. K. Saha and M. Mukherjee (eds.), *Recent Advances in Computational Mechanics and Simulations*, Lecture Notes in Civil Engineering 103,
https://doi.org/10.1007/978-981-15-8138-0_35

the use of structural control devices that dissipate the energy of structures during earthquakes. Significant number of studies have been conducted in the past such as [3, 8, 15, 22] which have demonstrated the use of dampers. There has been significant application of dampers in India and abroad to control the detrimental effects associated with torsional motion of buildings. Friction type dampers are installed to control the torsional response of La Gardenia housing complex in Gurgaon, India; the complex consists of 7 towers, each having 18 storeys [4]. Other applications of friction dampers include the Sonic City Office Tower and Asahi Beer Tower in Japan [11]. A number of medium to high-rise buildings in Japan like Herbis Osaka, Hanku Chayamachi, Sendagaya INTES, and Kyobashi Seiwa Buildings are equipped with Active Mass Dampers (AMD) to reduce torsional motions arising due to structural eccentricity in these buildings [11]. Thus the efficiency of dampers to control torsional motion has been well demonstrated over the period of many years.

Setting aside asymmetry, the soil which is surrounding the building along with its foundation may also play an important role while determining the response of building subjected to ground motions. Depending on the characteristics of the soil beneath the building, the foundation of buildings can have translational and rotational motions during earthquakes. This interaction between structure and soil through foundation is commonly called as Soil Structure Interaction (SSI). A detailed analysis of SSI by generating an FEM-based continuum model of the soil surrounding the structure can be too complicated and computationally time-consuming for simple structures. But, totally neglecting SSI by considering full fixity at foundation level can also be very simplistic and may not paint the true picture even for simple buildings. Previous studies such as the ones conducted by Mylonakis and Gazetas [16], Dutta et al. [5], Raychowdhury and Singh [18] and Jayalekshmi and Chinmayi [10] have demonstrated that SSI can greatly change the response of structures than that when ground storey columns are considered as fixed to their bases. Barring a few studies such as Mylonakis and Gazetas [16], it has generally been observed that the consideration of SSI may be beneficial for reducing response demands of structures such as base shear; however, floor displacements of these structures may increase due to their increased flexibility in comparison to structures which are considered to be fixed at the foundation level [10, 18].

Past research has evaluated the effect of SSI on asymmetric buildings without control devices [2, 12, 19], explored the possibility of using structural control devices to reduce detrimental effects of SSI in symmetric buildings [1, 20, 23], use of dampers to reduce the effect of lateral torsional coupling in asymmetric buildings without SSI [3, 8, 15, 22], use of dampers to reduce the effect of lateral torsional coupling and SSI in asymmetric buildings [17, 21]. So, the past three to four decades has witnessed significant research on the effect of SSI, asymmetricity on buildings and reduction of their ill effects by making use of control devices.

Therefore, adequate knowledge on individual effects of lateral torsional coupling and SSI on response of buildings equipped with dampers as well as combined effect of lateral torsional coupling and SSI on buildings with active and passive dampers are gathered from the afore mentioned past studies. However, the integrated effect of lateral torsional motion and SSI on edge deformations of asymmetric buildings equipped with semi-active dampers is yet to be fully explored. Moreover, it is desirable to observe the change in behaviour of a controlled asymmetric structure with SSI from that of a corresponding uncontrolled symmetric structure with SSI.

With an objective of exploring the same, the current article analyses the seismic response of a four-storey, one-way asymmetric building with SSI equipped with semi-active MR dampers. In this study, Magneto Rheological (MR) damper with friction type damping scheme is used as a semi-active method of structural control. The benefit of these devices in comparison to active control devices is that semi-active devices require a relatively small amount of power which can be provided with the help of batteries and do not produce instability in the structural system. Thus these devices can be used in the event of a power outage which may be expected at the time of strong earthquakes which might render active control devices ineffective. Response of the controlled building with SSI is measured in terms of displacements and rotations at the fourth storey of the building. Objectives of present study are (i) to evaluate the effect of SSI on the asymmetric building, (ii) to evaluate the optimal control parameters of dampers, (iii) to find the response of asymmetric building with SSI, without and with semi-active control devices and (iv) to evaluate the effects of lateral torsional coupling through a comparison between controlled asymmetric building with corresponding uncontrolled symmetric building.

2 Structural Model

The four-storey asymmetric shear building used in this study is developed based on a doubly symmetric building with fixed bases from Mariani et al. [14]. Relevant model modifications are made in order to meet the requirements of this ensuing study and the current asymmetric model is considered to be rested on an elastic half space. This reinforced concrete (RC) frame building is consisting of five 3.5 m bays and two 4.5 m bays, in the longitudinal and transverse directions, respectively. The building is symmetric in the transverse direction and asymmetric in the longitudinal direction with an eccentricity of 10% of the plan dimension along the transverse direction in each storey. Hence, the eccentricity, e is equal to 0.16r where r is the radius of gyration about the center of mass at each floor. All columns are uniform with cross-section dimensions of 30 cm × 30 cm and beams are assumed to be infinitely rigid in flexure. Figure 1 represents a schematic plan view of the building. Height of each storey is taken to be equal to 3 m with a uniform floor mass of 6 kN/m^2 in each storey. The grade of concrete in all structural members is considered as M-25. The short-term elastic modulus of concrete is taken as $5000\sqrt{fck}$ following the guideline of IS 456 [9] in which fck represents the characteristic cube strength of concrete.

The building is assumed to have a mat foundation resting on soil surface having dimensions same as that of building floors (17.5 m × 9 m) and the foundation to floor mass ratio is taken as 3 as mentioned in Lin et al. [12]. The damping matrix of the building is assumed to be Rayleigh damping with 5% damping in first two modes of vibration. The plan and lumped mass model of the building is shown in

Fig. 1 **i** Plan of building.
ii Lumped mass model

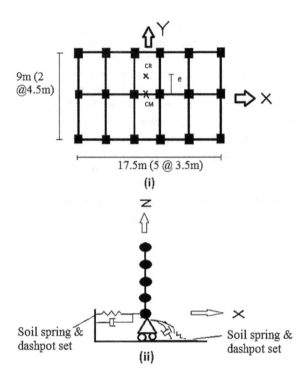

Fig. 1 (i) and (ii), respectively. The original building is symmetric in Y-axis; hence, two degrees of freedom, namely, horizontal displacement along X-axis and torsional rotation about Z-axis are selected for each storey. The foundation has three degrees of freedom, i.e. horizontal displacement, rocking rotation and torsional rotation. During analysis, ground motions are applied along the longitudinal direction (i.e., X-axis) of the building.

First Section The governing differential equation for an asymmetric system with SSI having semi-active dampers is given in Eq. (1) [2, 15, 19].

$$M\ddot{x} + Kx + C\dot{x} = -M\Gamma\ddot{u}_g + \Lambda F \qquad (1)$$

in which M, K, C are mass, stiffness and damping matrices, respectively, Γ is the earthquake influence coefficient vector, matrix Λ defines how control forces affect the structure, and F is a vector of control forces. Functional forms of M, K, C matrices are given below in Eqs. (2a, 2b and 2c).

$$M = \begin{bmatrix} m & & & & \\ & mr^2 & & & \\ & & m_o & & \\ & & & m_o r^2 & \\ & & & & I_y \end{bmatrix} \qquad (2a)$$

$$K = \begin{bmatrix} k & ek & -k & -ek & -kh \\ & k_\theta & -ek & -k_\theta & -ekh \\ & & k_T+k & ek & kh \\ & symm. & & k_Z+k_\theta & ekh \\ & & & & k_r+h^2k \end{bmatrix} \quad (2b)$$

$$C = \begin{bmatrix} c & ec & -c & -ec & -ch \\ & c_\theta & -ec & -c_\theta & -ech \\ & & c_T+c & ec & ch \\ & symm. & & c_Z+c_\theta & ech \\ & & & & c_r+h^2c \end{bmatrix} \quad (2c)$$

where, m, k, mo, r, Iy denote mass of the superstructure, stiffness of the superstructure, mass of the foundation, radius of gyration and moment of inertia about y-axis, respectively. kT, kZ, kr, denote the stiffness of soil spring in translation, rotation and rocking motions while cT, cZ, cr denote the damping coefficient of dashpot in translation, rotation and rocking motions and c, e, h denote damping coefficient of superstructure, eccentricity and height of storey respectively. The interaction forces for SSI can be modelled through frequency dependent spring with dashpot systems.

The coefficients for equivalent spring and damper of SSI interaction forces as given by Gazetas [7] are utilized. Table 1 shows the expressions used for obtaining static and dynamic stiffness and dashpot coefficient. Here K stands for static stiffness coefficient and ko stands for dynamic stiffness coefficient of soil springs. Similarly

Table 1 Dynamic stiffness and dashpot coefficients for foundation resting on homogeneous half space

Vibration mode	Static stiffness, K	Dynamic stiffness, k_o	Dashpot coefficient, C
Horizontal (y)	$K_y = \left[\frac{2GL}{2-\upsilon}\right]$ $(2+2.5\chi^{0.85})$	$k_{oy} = k_{oy}(L/B; a_0)$	$C_y = (\rho v_s A_b) \times \widetilde{c}_y$ where $\widetilde{c}_y = \widetilde{c}_y(L/B; a_0)$
Horizontal (x)	$K_x = K_y - \left[\frac{0.2}{0.75-\upsilon}\right]$ $GL[1-(B/L)]$	$k_{ox} \cong 1$	$C_x = \rho v_s A_b$
Rocking about y axis	$K_{ry} = \left[\frac{3G}{1-\upsilon}\right] I_{by}^{0.75}$ $(L/B)^{0.15}$	$k_{ory} \cong 1-0.26a_o$	$C_{ry} = (\rho v_{La} I_{by}) \times \widetilde{c}_{ry}$ where $\widetilde{c}_{ry} = \widetilde{c}_{ry}(L/B; a_0)$
Torsion	$K_t = 3.5 G I_{bz}^{0.75} \left(\frac{B}{L}\right)^{0.4}$ $(I_{bz}/B^4)^{0.2}$	$k_{ot} = 1-0.14a_o$	$C_t = (\rho v_s I_{bz}) \times \widetilde{c}_t$ where $\widetilde{c}_t = \widetilde{c}_t(L/B; a_0)$

C and $\bar{c}$ stands for damping coefficient and dynamic damping coefficient of soil dashpots, respectively. Here $\chi = A_b/4L^2$, $a_o = \frac{\omega B}{v_s}$, A_b = area of foundation, ω denotes dominant frequency of excitation, v_s denotes shear wave velocity in soil, I_{by}, I_{bz} are the area moment of inertia of foundation about y- and z-axis.

3 Damping Force Algorithm

In this study, one semi-active damping algorithm is used, namely, the Friction type damping scheme. The MR damper can be configured to generate this pre-set control force scheme as mentioned in Madhekar and Jangid [13].

In the case of a conventional viscous damper, the damping force is proportional to velocity of the damper, as a result of which maximum damping force is obtained only when the velocity of damper is the highest, whereas in the friction type damping scheme the maximum damping force, F_d, can be provided for the entire range of the stoke of damper. It can be represented as shown in Eq. (3) [13]

$$F_d = \begin{cases} -f_d; \dot{u} \leq -\dot{u}_t \\ \left(\frac{f_d}{\dot{u}_t}\right)\dot{u}; -\dot{u}_t < \dot{u} \leq \dot{u}_t \\ f_d; \dot{u} > \dot{u}_t \end{cases} \quad (3)$$

where f_d is the frictional force, $\dot{u}$ is the velocity of the damper and $\dot{u}_t$ is the transition velocity.

4 Numerical Study

The modified version of the four-storey building from Mariani et al. [14] as described earlier is considered. Density of the soil medium is taken as 1922 kg/m^3 with a Poisson's ratio of 0.33 as taken by Lin et al. [19]. In this study two soil types with different shear wave velocities are taken. The shear wave velocity in stiff soil is assumed to be 300 m/s and the same in soft soil case is taken as 100 m/s corresponding to site class D and E respectively from FEMA 450 [6]. The SSI is modelled by frequency dependent spring and dashpot set. To compare the effects of SSI in both the soft soil case and the stiff soil case, the same structure is also considered to have full fixity at the foundation level (i.e., columns fixed at the base). The response quantities of interest here are the displacement and rotation of the fourth storey. The displacement and rotation time history of the fourth storey for the four-storey building resting on soft soil, stiff soil and fixed base condition, subjected to N-S component of El Centro (1940) earthquake is shown in Fig. 2.

It is observed from Fig. 2 that there is not much difference between the responses in case of hard soil and the fixed base case due to significant flexibility possessed by

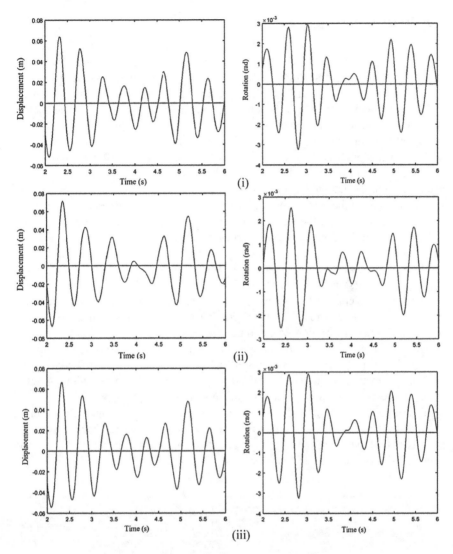

Fig. 2 Displacement and rotational time history of top storey in **i** fixed base case **ii** soft soil **iii** stiff soil

the building. Further the displacement response in soft soil case is higher than both hard soil and fixed base cases but the trend is reversed in case of torsional rotation. This finding is consistent with the findings of Lin et al. [19] and Balendra et al. [3].

4.1 Optimum Parameters of Dampers

In order to obtain the optimum damper parameters, two dampers of 1000 kN capacity [13] are placed at the first storey of the four-storeyed asymmetric building with SSI resting on soft soil as described in the previous section, with one damper on the flexible edge and the other on the stiff edge and a parametric study of the friction type damping scheme is carried out for four different ground motions. The different earthquake ground motions used are N-S component of El Centro (1940) with PGA 0.35 g, Imperial Valley (1979) with PGA 0.45 g, Loma Prieta (1989) with PGA 0.47 g and Northridge (1994) with PGA 0.5 g. The chosen earthquakes are deliberately selected to ensure that the PGA is not very high and the structure can be realistically assumed to stay in the elastic range. As there was not a large difference between soft soil and stiff soil cases, the optimum damper parameters are obtained for the soft soil case only. In friction type damping scheme the transition velocity is varied from 0.05 to 0.5 m/s.

For the friction type damping scheme the ratio of peak displacement after application of dampers and peak displacement before application of dampers for stiff edge and flexible edge of top storey, for four different ground motions as well as the average for the four cases is plotted in Fig. 3. It is visible from Fig. 3 that the optimum transition velocity in friction type damping scheme is obtained as 0.25 m/s and the response reduction varies significantly with the ground motion considered. The efficiency (%) of friction type damping scheme for flexible edge and stiff edge taking optimum transition velocity of 0.25 m/s for various ground motions with respect to peak displacement of the top storey is shown in Table 2.

The variation of mechanical energy of the structure with time subjected to different ground motions before and after application of dampers utilizing the optimum damper properties is shown in Fig. 4. It can be seen from Fig. 4 that the chosen damper can significantly reduce the mechanical energy of the structure.

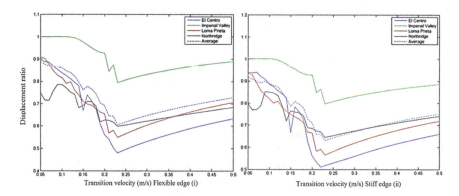

Fig. 3 Variation of damping parameters for Friction type damping scheme (i) Flexible edge (ii) Stiff edge

Table 2 Damper efficiencies for different ground motions

	El Centro		Loma Prieta		Imperial Valley		Northridge	
	Flex. Edge	Stiff Edge	Flex. Edge	Stiff Edge	Flex. Edge	Stiff Edge	Flex. Edge	Stiff Edge
Efficiency	52%	46%	45%	43%	20%	20%	38%	35%

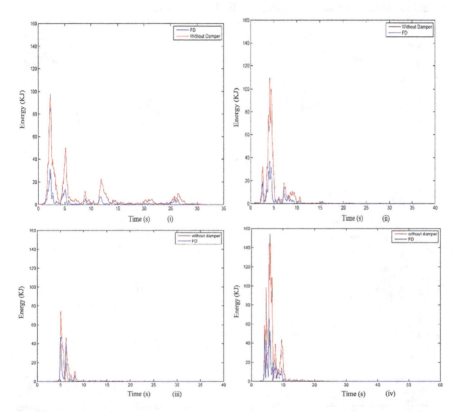

Fig. 4 Comparison of mechanical energy for (i) El Centro (ii) Loma Prieta (iii) Imperial Valley (iv) Northridge

4.2 Comparison of Asymmetric Structure with Symmetric Structure

The controlled response of asymmetric structure in soft soil is compared with the corresponding uncontrolled symmetric structure in order to assess the effects of lateral torsional coupling. For carrying out this comparison three different eccentricity values are utilized namely $0.16r, 0.2r, 0.3r$ where r is the radius of gyration. The structures are subjected to N-S component of El Centro (1940) earthquake and the response quantities of interest are displacements of stiff and flexible edge of the top storey in the soft soil case, utilizing the optimum damper parameters for friction type damping scheme (FD) obtained in the previous section. Displacement time history of fourth storey for different eccentricity values of the controlled asymmetric structure, the uncontrolled symmetric structure and the uncontrolled asymmetric structure is shown in Fig. 5. It can be seen from Fig. 5 that as the eccentricity increases the displacement of flexible edge increases while that of stiff edge decreases due to lateral torsional coupling in comparison to the symmetric case.

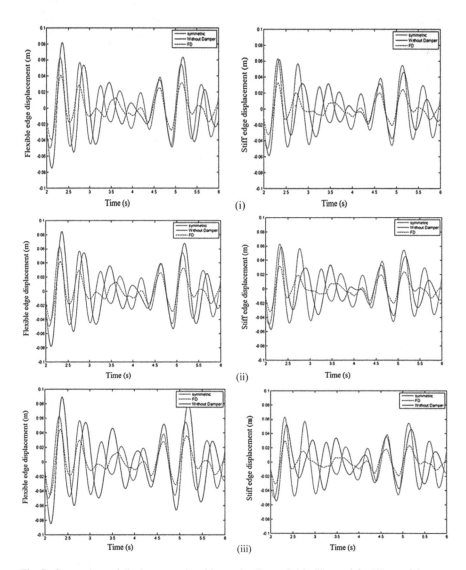

Fig. 5 Comparison of displacement time history for (i) $e = 0.16r$ (ii) $e = 0.2r$ (iii) $e = 0.3r$

The friction type damping scheme could reduce the displacements of flexible edge and stiff edge significantly which is even lesser than the response in the uncontrolled symmetric structure. Thus, the ill effects of lateral torsional coupling could be mitigated.

5 Conclusions

The seismic response of linear elastic four-storey one-way asymmetric building with SSI equipped with semi-active MR dampers excited by different ground motions is studied. The effect of SSI on edge deformations of the building is obtained by considering two kinds of soil, namely, stiff soil and soft soil having different shear wave velocities. The detrimental effects of SSI are reduced by making use of semi-active MR dampers with friction type damping scheme. The optimum parameter of the friction type damping scheme considering four different ground motions is obtained. The response of the controlled asymmetric structure with three different structural eccentricities is compared with that of the corresponding uncontrolled symmetric structure to assess the effects of lateral torsional coupling. From the results presented above the following conclusions are drawn.

1. The displacement response of the top storey increases from fixed base case, hard soil case ($v_s = 300$ m/s) and soft soil case ($v_s = 100$ m/s) ,respectively. But the torsional rotation of the top storey decreases from fixed base case, hard soil case ($v_s = 300$ m/s), soft soil case ($v_s = 100$ m/s), respectively.
2. The highest reduction in response for friction type damping scheme was found at a transition velocity of 0.25 m/s with dampers placed on first storey.
3. The MR damper with friction type damping scheme is able to achieve significant reduction in displacement of flexible edge, stiff edge and dissipate considerable amount of mechanical energy of the building. Further this response reduction varies widely from ground motion to ground motion. Hence the efficiency of the damper is ground motion dependent.
4. As the eccentricity of the structure increases the flexible edge displacement increases and the stiff edge displacement decreases in comparison to the symmetric structure. The friction type damping scheme could decrease the response of the asymmetric structure to even lower than that of the symmetric structure. So, the detrimental effects of structural eccentricity were mitigated.

References

1. Amini F, Razeghi HR, Shadlou M (2008) New evaluation of soil structure interaction effects on optimal control of structures. In: Proceedings of 14th World conference on earthquake engineering, Beijing, China, pp 46–57
2. Balendra T, Tat CW, Lee SL (1982) Modal damping for torsionally coupled buildings on elastic foundation. J Earthq Eng Struct Dyn 10(5):735–756
3. Braz-Cesar M, Barros R (2017) Optimal control of plan asymmetric structure using magnetorheological dampers. A App Mech 87:893–904
4. Chandra R, Masand M, Nandi SK, Tripathi CP, Pall R, Pall A (2008) Friction dampers for seismic control of La Gardenia Towers South City, Gurgaon, India. In: Proceedings of 12th World conference on earthquake engineering, Auckland, New Zealand, pp 108–120
5. Dutta SC, Bhattacharya K, Roy R (2004) Response of low rise buildings under seismic ground excitation incorporating soil structure interaction. J Soil Dyn Earthq Eng 24:893–914
6. FEMA 450-2003 (2003) NEHRP recommended provisions for seismic regulations for new buildings and other structures. Federal Emergency Management Agency, Washington DC, USA
7. Gazetas G (1991) Formulas and charts for impedances of surface and embedded foundations. J Geotech Eng 117(9):1363–1381
8. Goel RK (1998) Effect of supplemental viscous damping on seismic response of asymmetric plan systems. J Earthq Eng Struct Dyn 27:125–141
9. Indian Standard of Plain and Reinforced Concrete IS456 (2000)
10. Jayalekshmi BR, Chinmayi HK (2014) Effect of soil flexibility on seismic force evaluation of RC framed buildings with shear wall: a comparative study of IS 1893 and EUROCODE 8. J Struct 1–15
11. Kareem A, Kijewski T, Tamura Y (1999) Mitigation of motions of tall buildings with specific examples of recent applications. J Wind Struct 2(3):201–251
12. Lin J, Tsai K, Miranda E (2009) Seismic history analysis of asymmetric buildings with soil structure interaction. J Struct Eng 135(2):101–112
13. Madhekar SN, Jangid RS (2009) Variable dampers for earthquake protection of benchmark highway bridges. Smart Mater Struct 18:1–18
14. Mariani V, Tanganelli M, Viti S, De Stefano M (2016) Combined effects of axial load and concrete strength variation on the seismic performance of existing RC buildings. Bull Earthq Eng 14:805–819
15. Mevada SV, Jangid RS (2012) Seismic response of torsionally coupled system with semi-active variable dampers. J Earthq Eng 16(7):1043–1054
16. Mylonakis G, Gazetas G (2000) Seismic soil structure interaction: beneficial or detrimental? J Earthq Eng 4(3):277–301
17. Nazarimofrad E, Zahrai SM (2018) Fuzzy control of asymmetric plan buildings with active tuned mass damper considering soil structure interaction. J Soil dyn Earthq Eng 115.838–852
18. Raychowdhury P, Singh P (2012) Effect of non-linear soil structure interaction on seismic response of low rise SMRF buildings. J Earthq Eng Vib 11(4):541–551
19. Sivakumaran KS, Balendra T (1994) Seismic analysis of asymmetric multi storey buildings including foundation interaction and P- Δ effects. J Struct Eng 16(8), 609–624
20. Wong HL, Luco JE (1992) Effects of soil structure interaction on the seismic response of structures subjected to active control. In: Proceedings of 10th World conference in earthquake engineering, Balkema, Rotterdam, pp 126–138
21. Wang JF, Lin CC (2005) Seismic performance of multiple tuned mass dampers for soil-irregular building interaction systems. Int J Solids Struct 42(20):5536–5554
22. Yoshida O, Dyke SJ, Truman KZ (2002) Experimental verification of torsional response control of asymmetric buildings using MR damper. J Earthq Eng Struct Dyn, 1–20
23. Zhao X, Wang S, Dongsheng D, Weiquing L (2017) Optimal design of viscoelastic dampers in frame structures considering soil structure interaction effect. Shock Vibr, 1–16

Modelling and Simulation in Blast Resistant Design

A Critical Review of TNT Equivalence Factors for Various Explosives

P. A. Shirbhate and M. D. Goel

Abstract In the present scenario, it has been observed that terrorist attack became global issue and due to its increased frequency, it has gained attention of the researchers all over the world. Thus, it became the need of today's era to understand the basics of explosion and factors affecting the explosion. Types of explosives are one of the important parameters which directly influences the impact of blast on any structural element. The term "TNT equivalence" is considered as the benchmark and is most commonly used to compare the performance of explosives with respect to reference explosive, i.e. trinitrotoluene (TNT). The main purpose of this study is to review TNT equivalence and the factors associated with it. The effect of charge geometry (i.e. spherical, flat, cylindrical, square etc.), confinement (close-in, intermediate or far ranges, etc.) and standoff distance on the peak overpressure and impulse are discussed in the present study as explosive energy and charge mass of the detonating material is related to equivalent weight of TNT. Moreover, effect of scaled distance of high explosives on TNT equivalent is also presented. Further, various numerical methods used to compute the TNT equivalents (TNTe) are reported for improved understanding of the TNT equivalence.

Keywords TNT · Blast · Shape charge · Impulse · Peak overpressure

1 Introduction

These days due to easy production and storage of explosive materials and availability of explosive devices to deliver the explosives, threats due to terrorist attacks have been increased. To evaluate the effect of explosion, it is required to understand different blast wave parameters such as standoff distance, charge mass, detonation velocity, heat of detonation and TNT equivalency. Explosive weight is one of the

P. A. Shirbhate (✉) · M. D. Goel
Department of Applied Mechanics, Visvesvaraya National Institute of Technology, Nagpur 440 010, India
e-mail: payalshirbhate@students.vnit.ac.in

parameter which is needed from safety point of view, during manufacturing and storage of explosive materials, and also to analyse the damaging effects due to the explosion. TNT equivalency is the basis to calculate the weight of explosive material involved during the explosion. The concept of TNT equivalency and the parameters influencing its computation, discussed by various researchers has been extensively reviewed in the present paper.

In the year 1983, Held proposed the concept of effective explosive weight (EEW) which needed the TNT equivalent values, and hence used peak overpressure, P_{max} as well as impulses for equivalency calculation [1]. Esparza also obtained equivalent values of TNT and five other high explosives based on overpressures and impulses [2]. After 4 years, Yong proposed information showing relationship of TNT and range of explosives [3]. Later on, Cooper carried out experimental tests to evaluate TNT equivalence, such as ballistic mortar, plate dent, trauzl, air blast and sand crush due to inaccuracy in prediction of blast wave using available computer code [4]. Afterwards, Formby and Wharton performed experiments on commercial explosives at ground level, and reached to the conclusion that TNT equivalence of different materials depends on whether, the overpressure or impulse data is used for evaluation. The data reported by Formby and Wharton had some deviation, as the hardness of ground was one of the variable that could affect the evaluation of TNT equivalences [5]. In year 2002, Maienschein used thermo-chemical approach to calculate TNT equivalency of any type of explosive. Cheetah code based on thermo-chemical approach used physical and chemical properties to calculate mechanical and thermal energy of detonation. The equivalency obtained from the code was found to be consistent with accuracy of about 3% [6]. Later on, Jeremić and Bajić also proposed the most accurate, reliable and easiest way among other approaches to determine TNT equivalency based on thermo-chemical calculations [7]. After this, Hargather and Settles performed laboratory scale experiments to provide easy alternative to full scale tests. The scaling analysis was done with high speed digital imaging and optical shadowgraphy to find the position of shock wave [8]. In year 2011, Sochet discussed different analytical approaches to determine TNT equivalent based on impulse, pressure, conventional methods and Chapman-Jouguet state equation [9].

Initially, Grisaro and Edri numerically investigated equivalent weight of explosives for impulse and overpressure [10]. Based on this numerical investigation, the concept of equivalency was further extended for internal explosion event [11]. Recently, Panowicz thoroughly discussed six analytical methods to understand the parameters associated with the properties of different charge shape and material of explosive [12].

Some authors [7, 9, 12] discussed analytical approaches used for the evaluation of equivalence for different explosive materials, whereas some authors [2–5, 7, 13, 14] preferred experimental approaches for predication of equivalency by comparing the results with analytically or numerically obtained values, and are discussed briefly in the present work.

2 TNT Equivalence

The blast pressure created due to detonation of high explosive depends on the explosive material, and its composition. Explosive materials with same shape and mass detonate differently due to material properties such as density, energy release rate and mass specific energy [4, 15]. Some common reference should be available so as to predict the effect of different types of explosive materials. Therefore, Trinitrotoluene (TNT) is used as reference explosive due to the availability of large set of experimental data [9]. TNT equivalency is basically a normalization technique that relates the weight of the explosive under consideration with weight of TNT producing equal amount of energy. The ratio of equivalent charge to the actual TNT reference charge is used to get the TNT equivalent factor. For example, 1 kg of Pentolite produces mass equivalent to 1.129 kg of TNT, so the equivalent value of Pentolite is 1.129.

The damaging effects of explosion on the structure are evaluated based on TNT equivalence [3]. The value of TNT equivalent is not constant, and therefore it has been observed that it changes with scaled distance also. The reason for getting different TNT equivalent values are geometry of explosive, experimental condition and the distance from explosive charge to the observation point [7]. Explosives are mainly composed of two components fuel and oxidants, and the energy release mechanism depends on the rate of chemical reactions. Some methods measure total energy released during explosion, whereas some measures blast overpressure without including the release of energy, and therefore TNT equivalent value obtained is not constant for different techniques/method adopted.

3 TNT Equivalent Calculation

TNT equivalent value varies as they are predicted on the basis of blast overpressure, impulse and detonation heat [15]. For evaluation of TNT equivalent, there are few equations which depend on material parameters and some advanced methods are developed on the basis of experimental data, or on trial and error basis. All such methods available for evaluation of equivalency are elaborated in the subsequent section.

In year 1985, Esparza tested six explosives, and at each scaled distance average pressures were measured. Based on scaled distances and corresponding pressures, TNT equivalent ratios were obtained. From the reference TNT curve, incident pressures corresponding to scaled distances were observed to get TNT equivalent value. The relation in Eq. (1) was used for evaluation of equivalent based on scaled distances [2]. Where, Z_x and Z_{TNT} are the scaled distances of reference explosive and TNT explosive, respectively.

$$TNT_{eq} = \left(\frac{Z_x}{Z_{TNT}}\right)^3 \quad (1)$$

Cooper proposed Eq. (2) which depends on the detonation velocities and takes the following form, having D_{exp} and D_{TNT} as the detonation velocities of studied explosive and TNT explosive, respectively [4]. This method calculates square of detonation velocity ratio to predict TNT equivalent.

$$TNT_{eq} = \frac{D_{exp}^2}{D_{TNT}^2} \qquad (2)$$

Another Eq. (3), which is based on the recorded blast pressure and the calculations were carried out with reference to the methods discussed by Maserjian and Fisher [16]. Considering all these parameters Formby and Wharton in 1996 formulated the equation as

$$TNT_{eq} = \frac{W_x}{W_{TNT}} \qquad (3)$$

where W is the effective charge weight of TNT explosive; W_x is weight of explosive under consideration.

Bajić carried out the study of TNT equivalent determination for various explosive materials [17] and proposed one more equation based on detonation pressure P_{CJ} and it is reported as Eq. (4):

$$TNT_{eq} = \frac{P_{CJ} + 20.9}{40} \qquad (4)$$

Basic relation (5) presented in Unified Facilities Criteria (UFC) manual [18] to calculate effective mass of given explosive with respect to standard explosive TNT in case of unconfined explosions is represented as

$$W = \frac{Q_x}{Q_{TNT}} W_x \qquad (5)$$

where W is the effective charge weight of TNT explosive; W_x is weight of explosive; Q_x is the mass specific energy of the explosive; Q_{TNT} is the mass specific energy of TNT.

In above equation, the ratio Q_x/Q_{TNT} represents TNT equivalent which is based on the heat of detonation and is most commonly used expression in the calculation of equivalency.

Later on, in year 2011, Locking further modified Eq. (5) presented in UFC manual [18] based on heat of detonation and reported as Eq. (6) and is based on "Power Index" which considers power of explosive. In the same work Locking presented Eq. (7) based on the heat of detonation (Q) [19]:

$$TNT_{eq} = \frac{H_{EXP}}{H_{TNT}} \times \frac{V_{EXP}}{V_{TNT}} \qquad (6)$$

V_{EXP} and V_{TNT} are the volume of gases per unit mass of explosive and volume of TNT, respectively, at "standard temperature and pressure".

$$TNT_{eq} = \frac{Q_{exp}}{(1-d) \times Q_{TNT} + (m \times Q_{exp})} \quad (7)$$

Here, d represents Y intercept and m represents angular coefficient of straight line curve of heat of detonation ($d = 0.76862$; $m = 0.7341$).

Locking further worked on TNT equivalent evaluation and proposed the Equation based on the method of "Hydrodynamic work", and the equivalency is predicted as per Eq. (8) [20, 21].

$$TNT_{eq} = \frac{P_{CJ, EXP}}{P_{CJ, TNT}} \times \left(\frac{\rho_{0, EXP}}{\rho_{0, TNT}}\right)^{0.96} \quad (8)$$

where $P_{CJ,EXP}$ and $P_{CJ,TNT}$ are the detonation pressure of given explosive and TNT explosive, respectively. $\rho_{0,EXP}$ is the density of unreacted explosive and $\rho_{0,TNT}$ is the density of TNT explosive.

The TNT equivalent values mass specific energy obtained from various researches [7, 14] have been represented in Table 1, these values can be utilized to get the peak overpressure of different explosive materials.

Table 1 TNT equivalent values of different explosives

Explosive type	Heat of explosive Qx(kJ/kg)	TNT equivalent Qx/QTNT
Compound B (60% RDX 40% TNT)	5190	1.148
RDX (Cyclonite)	5360	1.185
HMX	5680	1.256
Nitroglycerien (liquid)	6700	1.481
TNT	4520	1.000
Blasting Gelatin (91% nitroglycerin, 7.9% nitrocellulose, 0.9% antacid, 0.2% water)	4520	1.000
60% Nitroglycerine dynamite	2710	0.000
Semtex	5660	1.250
Hexogen	6334	1.38
Octogen	6538	1.46
Tetryl	5920	1.13
Pentolite 50/50	5860	1.129
PETN (90/10)	6406	1.23
Pentrite	6400	1.13

4 Effective Parameters Affecting TNT Equivalent

The TNT equivalent value has significant effect of certain parameters such as nature of explosive, effect, distance, initiation. Among these parameters, explosive effect is maximum and effect, distance, position of initiation has effect in decreasing order. Several researchers discussed the influence of these parameters [1, 4, 5]. These parameters have been investigated with their respective effects in this section.

4.1 Explosive Nature

Each explosive has certain physical, chemical and nuclear characteristics with reference to standard TNT. Improvised explosive device (IED) may contain homemade explosive or military explosives, with military explosives having TNT equivalent more than 1 [14]. In the similar manner, nature of explosive has significant effect on the TNT equivalent value.

4.2 Effect

Effect of any explosive material also plays major role in equivalency evaluation. Blast overpressure and impulse of the blast wave governs the TNT equivalent values [5, 9, 14]. For example, C-4 has TNT equivalent 1.37 obtained on the basis of blast overpressure, and has the value of 1.19 calculated based on impulse [10]. Also, TNT equivalence obtained from overpressure data is more than that obtained from impulse data [5].

4.3 Shape of the Charge

Most of the study exists on the effect of spherical shape explosive, but it has been observed that other non-spherical shapes affect the TNT equivalency and the effect needs to be studied. Study on cylindrical charge shape on blast wave characteristics has been analysed in earlier research works [1, 13]. In cylindrical shaped explosives there exists gap between incident and reflected shockwave known as the bridge wave. The time elapsed in between primary and secondary shockwave make it respond differently from spherical shape of the explosive. Length to diameter ratios of the cylindrical charge is more than the spherical charge and it has been observed that single value of TNT equivalent is not sufficient and depends on the observation location [14].

4.4 Location of Initiation

In case of the cylindrical explosive, another factor that potentially affects the TNT equivalent is the location at which explosive initiates, i.e. whether the explosive initiates at the centre or at the end. Additionally, the pressure distribution is also significantly affected by the side where explosive initiated [14, 22]. Simoens and Lifebvre [14] carried out experiment and observed that position of initiation has no effect on impulse, however, the overpressure depends on the position of initiation.

4.5 Location of Initiation

Distance of the explosive material influences TNT equivalence and it varies with scaled distance [3]. Cooper carried out experiments on different explosive material and observed from the plot of TNT equivalence versus scaled distance that the value of TNT equivalence changes with scaled distance [4].

In [14] all these factors when considered individually are represented with the relationship in Eq. (9):

$$TNT_e = f_1(\text{nature}) \times f_2(\text{effect}) \times f_3(\text{distance}) \times f_4(\text{shape}) \times f_5(\text{position of initiation}) \quad (9)$$

5 Summary

The main purpose of TNT equivalence is to provide reference for evaluation of effect of explosive under consideration. The paper summarizes the review of evaluation of TNT equivalence and the parameters affecting it. The analytical methods available in various researches have been thoroughly overviewed in the present work. It is observed from the review that TNT equivalent value is not constant and it varies with the method used for calculation and the parameter considered during experimentation. Further, the parameters such as standoff distance, charge shape and weight, material of explosive and heat of detonation influences the TNT equivalence. Study shows that TNT equivalence evaluation is not simple and needs careful understanding of each parameter involved in the calculation. Therefore, to understand the basics involved in TNT equivalence and to predict the damaging effects during explosion this study has been carried out.

References

1. Held M (1983) TNT Equivalent. Propellants Explos Pyrotech 8:158–167
2. Esparza ED (1985) Blast measurement and equivalency for spherical charges at small scaled distances. Int J Impact Eng 4(1):23–40
3. Yong DLV (1989) A study of blast characteristics of several primary explosives and pyrotechnic compositions. J Hazard Mater 21:125–183
4. Cooper PW (1994) Comments on TNT equivalence. In: 20th International pyrotechnics seminar, Colorado
5. Formby SA, Wharton RK (1996) Blast characteristics and TNT equivalence values for some commercial explosives detonated at ground level. J Hazard Mater 50:183–198
6. Maienschein JL (2002) Estimating equivalency of explosives through a thermochemical approach. In: 12th International conference symposium, San Diego
7. Jeremić R, Bajić Z (2006) An approach to determining the TNT equivalent of high explosives. Sci Techn Rev 1:58–62
8. Hargather MJ, Settles GS (2007) Optical measurement and scaling of blasts from gram-range explosive charges. Shock Waves 17:215–223
9. Sochet I (2010) Blast effects of external explosions. In: Eighth international symposium on hazards, prevention, and mitigation of industrial explosions, Japan
10. Grisaro HY, Edri IE (2017) Numerical investigation of explosive bare charge equivalent weight. Int J Prot struct, 1–22
11. Edri IE, Grisaro HY, Yankelevsky DZ (2019) TNT equivalency in an internal explosion event. J Hazard Mater 374:248–257
12. Panowicz R, Konarzewski M, Trypolin M (2017) Analysis of criteria for determining a TNT equivalent. J Mech Eng 63(11), 666–672
13. Ismail MM, Murray SG (1993) Study of the blast waves from the explosion of non-spherical charges. Propellants Explos Pyrotech 18:132–138
14. Simoens B, Lifebvre MH (2011) Influence of different parameters on the TNT-equivalent of an explosion. Cent Eur J Ener Mater 8(1)
15. King KW (2008) Determining TNT equivalency for confined detonations. In: Proceedings PVP 2008 ASME pressure vessels and piping division conference, pp 1–12, Chicago, Illinois, USA
16. Maserjian J, Fisher EM (1951) Determination of average equivalent weight and average equivalent volume and their precision indices for comparison of explosives in air. NAVORD Report 2264, US Naval Ordnance Laboratory, White Oak, Maryland, USA
17. Bajić Z (2007) Determination of TNT equivalent for various explosives (in Serbian). Magister/Master degree thesis, University of Belgrade, Faculty Technology and Metallurgy Belgrade
18. UFC 3-340-02 (2008) Structures to resist the effects of accidental explosions. Unified Facilities Criteria
19. Locking PM (2011) The trouble with TNT equivalence. In: 26th International symposium on ballistics
20. Locking PM (2013) TNT equivalence—experimental comparison against prediction. In: 27th International symposium on ballistics, Freiburg, 22–26 Apr 2013
21. Locking PM (2014) Explosive materials characterisation by TNT equivalence. In: 28th International symposium on ballistics, Atlanta, GA, 22–26 Sept 2014
22. Held M (1999) Impulse method for the blast contour of cylindrical high explosive charges. Propellants Explos Pyrotech 24:17–26

Limitations of Simplified Analysis Procedures Used for Calculation of Blast Response of Structures

K. K. Anjani and Manish Kumar

Abstract The current state of practice in the blast-resistant design of structural members relies on simplified single-degree-of-freedom (SDOF) to calculate the response quantities. An equivalent SDOF representation of a structural member is obtained by assuming a deflected shape with equal deflection at the assumed degree of freedom. The blast load is approximated as a triangular pulse load. The response of an elastic and elasto-plastic SDOF system to triangular pulse is widely available in the literature. The utility of simplified SDOF procedures is mostly limited to calculating displacement response of regular-shaped members subject to far-field detonations. This paper investigates the limitations of the simplified SDOF method and role of various parameters (e.g., shape, standoff distance, boundary conditions, positive phase) on response quantities of interests (e.g., displacement, shear force). The simplified SDOF results are verified using advanced finite element model of the steel column in LS-DYNA. The findings of the study are summarized, and recommendations are provided for usage of simplified SDOF procedures for the blast-resistant design of structures.

Keywords Simplified SDOF analysis · Finite element analysis · Shape functions

1 Introduction

As the threat perception to the risks associated with accidental and malevolent blast loading to critical government and private facilities grows, the elements of blast-resistant design would need to be incorporated for these structures. The research in this area has mostly been confined to military structures, but recently there has been focus on civil structures due to the rise in terrorist threats. The challenges in the blast-resistant design of structures include accurate characterization of blast loads and reliable response estimation of structures. There are limited codes and standards

---

K. K. Anjani · M. Kumar (✉)
Indian Institute of Technology, Bombay 400 076, India
e-mail: 164040015@iitb.ac.in

that provide information on analysis of structures subject to blast loads. ASCE 59-11 [1] provides guidelines for planning, analysis, and design of new and existing structures to resist blast loads. The Indian standard IS 4991 [2] presents the general criteria for blast-resistant design.

The current state of practice in blast-resistant design of structural components relies on simplified single-degree-of-freedom (SDOF) analysis. The accuracy of the analysis depends on the effectiveness in converting the real structure to an equivalent SDOF system. A detailed presentation on simplified SDOF analysis is provided in Biggs [3] and UFC-3-340-02 [4]. The response of elastic and elasto-plastic SDOF systems can be obtained using the shock spectra developed by Biggs [3] and reproduced in DoD [4].

The simplified SDOF procedures for blast-response analysis are mostly appropriate for calculating displacement response of regular-shaped members subject to far-field detonations characterized by uniform blast pressure. There are several limitations of the simplified SDOF analysis that must be addressed for reliable and safe design of structures against blast loading.

Li and Hao [5] proposed a two-step method that combines the traditional SDOF method with FEA to ensure computational efficiency and better accuracy. Al-Thairy [6] performed analytical study to obtain the blast response of steel column using simplified SDOF methods including the effects of axial compressive loading. A new resistance function was developed for beam-columns that consider the effect of axial loads. The accuracy of the developed resistance functions was verified using ABAQUS and validated with experimental results. Lee and Shin [7] extended the empirical chart of elasto-plastic material models for near field explosions. The developed charts were verified using the finite element code LS-DYNA.

The simplified SDOF procedures are used for blast analysis due to its simplicity, but the use of finite element analysis (FEA) is gaining popularity with increase in computational capabilities. However, there is significant effort involved with the development of a verified and validated FEA model, and the numerical precision of computation is much greater than input parameters of blast analysis to justify nonlinear FEA analysis. Hence SDOF method of blast analysis, with awareness of its limitations, would still be the way forward for blast-resistant design of regular structural members adopted by design practitioners. The goal of this paper is to highlight limitations of simplified SDOF procedures for response quantities of interest and provide recommendations on judicious use of response parameters. The rate dependent nature of the material models is neglected for the FEA in this study to allow comparison with the standard results available using SDOF analysis, which do not include the rate effects.

2 Methodology

2.1 Blast Loads

Blast is a rapid and violent chemical reaction that converts the solid or liquid explosive materials into a very hot, dense, and high-pressure gas. The rapid generation of the dense gas compresses the atmosphere ahead of it and generates a high-pressure blast wave. The blast wave propagates into the ambient atmosphere causing a pressure rise that comprises static overpressure and dynamic pressure. The static pressure is due to the compression of the atmosphere by shock front, while the dynamic pressure is caused by energy imparted to the air molecules as the blast wave propagates.

The pressure history due to blast wave at a point located outside the fire ball is shown in Fig. 1. Blast wave reaches at a point after arrival time t_A, and the pressure rapidly increases to peak overpressure (P_{so}) at the point. As the blast wave travels, the pressure decreases gradually and reaches the ambient pressure after a time $t_A + t_0$, where t_0 is referred as positive phase duration. After reaching the ambient pressure, the pressure decreases up to the negative peak P_{so}^- due to overexpansion of gases. The time duration for the negative phase is represented as t_0^-. The blast pressure history is represented using the modified Friedlander equation as

$$p(t) = P_0 + P_s^+ \left(1 - \frac{t}{t_0}\right) e^{-bt/t_0} \qquad (1)$$

The blast parameters (e.g., peak incident and reflected pressure, positive phase duration, impulse) are obtained using the Kingrey-Bulmash charts available in the technical manual UFC-3-340-02 [4]. The charts provide blast parameters for far-field explosions as a function of scaled distance Z, which is defined as

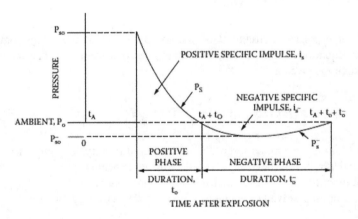

Fig. 1 Pressure history of a blast wave [4]

Fig. 2 Simplified representation of blast pressure history

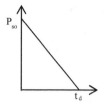

$$Z = \frac{R}{m^{1/3}} \qquad (2)$$

where m is the charge mass and R is the standoff distance.

The effect of negative phase is insignificant on blast response and often neglected so that a simplified triangular pulse representation of blast load can be considered for analysis. The response of a SDOF system to triangular pulse is readily available in literature as standard charts. The simplified representation of the blast load is shown in Fig. 2.

The blast waves undergo reflection and diffraction upon striking a solid structure. The reflected over pressure is developed during the reflection of blast wave from a surface. The reflected pressure depends on the incident over pressure and the angle at which the blast wave strikes the surface. The effective pressure at an angle is calculated based on the reflected pressure (P_r), incident overpressure (P_{so}), and the angle of incidence (θ). The expression for the effective reflected pressure is given as

$$P_{eff} = P_r \cos^2 \theta + P_{so}(1 + \cos \theta - 2\cos^2 \theta) \qquad (3)$$

2.2 Simplified SDOF Analysis

The critical response of a structural member to blast load is obtained at the component level. The equation of motion of the forced vibration of a beam using distributed mass and elasticity is given as

$$m(x)\frac{\partial^2 u}{\partial^2 t^2} + \frac{\partial^2}{\partial^2 x^2}\left[EI(x)\frac{\partial^2 u}{\partial x^2}\right] = p(x,t) \qquad (4)$$

where $p(x, t)$ is the applied force, EI is the flexural rigidity, m is the mass per unit length, and $u(x, t) = \psi(x)z(t)$ is the deflection of the beam expressed as the product of shape function $\psi(x)$ and normalized time coordinate $z(t)$.

The equation is solved to obtain the deflected shape of the beam as

$$\psi(x) = C_1 \sin \beta x + C_2 \cos \beta x + C_3 \sinh \beta x + C_4 \cosh \beta x \qquad (5)$$

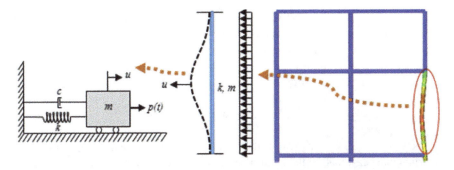

Fig. 3 SDOF analysis of a structural components subject to blast load (Adapted from Hai, 2007)

$$\beta^4 = \frac{\omega_n^2 m}{EI} \tag{6}$$

where C_1, C_2, C_3 and C_4 are the unknown parameters, β is the eigen value parameter, and ω_n is the angular frequency of the column. These parameters are obtained enforcing the boundary conditions. The deflection $u(x, t)$ is obtained by substituting the initial conditions.

It is not always possible to obtain a closed form analytical solution for $u(x, t)$ by solving the partial differential equation. Hence, an alternative approach is adopted where the shape function $\psi(x)$ is assumed based on expected deflected shape, and $z(t)$ is obtained by solving the equation of motion of an equivalent SDOF system. Figure 3 summarizes the equivalent SDOF analysis of a structure subject to blast loads.

The parameters for the equivalent SDOF system (e.g., mass, stiffness and load) are derived using a shape function with equal deflection to the real system at the assumed DOF. The expressions for equivalent mass, stiffness, and load for the SDOF system are

$$\tilde{m} = \int_0^L m(x)[\psi(x)]^2 dx, \quad \tilde{k} = \int_0^L EI(x)[\psi''(x)]^2 dx, \quad \tilde{P} = \int_0^L P(x)\psi(x)dx \tag{7}$$

Transformation factors are calculated as the ratio of properties of the equivalent SDOF to the real system:

$$mass\ factor\ K_M = \frac{\tilde{m}}{\int_0^L m(x)dx}; \quad load\ factor\ K_L = \frac{\tilde{P}}{\int_0^L P(x)dx}; \quad K_{LM} = \frac{K_M}{K_L} a \tag{8}$$

The shape function is selected to have some physical significance for the given boundary condition at different stages of response (e.g., elastic, elasto-plastic). The transformation factors for the one-way elements and two-way elements for different boundary conditions are presented as standard tables in Biggs [3].

The equation of motion of the equivalent elastic SDOF system is

$$\tilde{m}\ddot{x} + \tilde{k}x = \tilde{P} \tag{9}$$

The peak response depends on the ratio of the positive phase (t_d) duration of the triangular blast load to the natural period of the equivalent SDOF system ($T_n = 2\pi/\omega_n$). The loading is considered as impulsive for $t_d/T_N < 0.2$, for which the elastic response is obtained as

$$u(t) = u(0)\cos\omega_n t + \frac{\dot{u}(0)}{\omega_n}\sin\omega_n t \tag{10}$$

where $u(0)$ and $\dot{u}(0)$ are the initial displacement and velocity, respectively, and $\omega_n = \sqrt{\tilde{k}/\tilde{m}}$ is the natural frequency of the system. For impulsive loading, $u(0) = 0$ and $\dot{u}(0) = I_r/\tilde{m}$, where I_r is the reflected impulse. The peak displacement is

$$u_{\max} = \frac{I_r}{\tilde{m}\omega_n} \tag{11}$$

For non-impulsive loading, Biggs [3] has developed charts (shock spectra) of peak displacements vs t_d/T_N (reproduced in DoD [4]) for elastic and elasto-plastic systems.

The reactions of the real system cannot be obtained using the SDOF system as the equivalent system is selected to match the peak deflection and not the force. The dynamic reactions are obtained using the dynamic equilibrium of the real system. Figure 4 shows the free-body diagram of simply supported beam under uniform load for calculation of dynamic reaction.

The dynamic reactions for the fixed–fixed support beam with elastic material properties for point load and uniformly distributed load are [3]

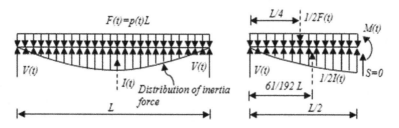

Fig. 4 Calculation of dynamic reaction (adapted from [3])

$$V(t) = 0.71R(t) - 0.21F(t) \tag{12}$$

$$V(t) = 0.36R(t) + 0.14F(t) \tag{13}$$

The internal forces (shear force and bending moments) are also obtained using the static analysis of the real structure subject to an equivalent static force, which is obtained for the given displacement $u(x, t)$ is given as

$$f_s(x, t) = [EI(x)\psi''(x)]'' z(t) \tag{14}$$

where $z(t)$ is the displacement of the equivalent SDOF system. The internal forces obtained using this static force are less accurate than displacements, because it depends on the differential of the shape function which is less accurate than the shape function.

2.3 Finite Element Analysis

Finite element analysis is popularly used to solve complex nonlinear dynamic problems. LS-DYNA [8] is a popular commercial FEA package that is used to obtain blast response of structures. Its strength lies with a robust explicit solver, vast library of materials and inbuilt blast loading functions. The challenge is to define the boundary conditions and specifies material parameters accurately. The response of structural members can be obtained using FEA to gain insight into the dynamic behavior and verify the response obtained using the simplified SDOF procedure. Verified FE models are used to highlight the limitations of the simplified SDOF procedure and provide recommendations on using the analysis results for blast-resistant design.

3 Analysis

3.1 Analysis Model

A series of numerical studies are performed using steel columns of an I and square cross-Sect. (0.406 m × 0.406 m) subject to blast loading are obtained. Geometric and material properties of the columns are presented in Table 1. Fixed boundary conditions with elastic and elasto-plastic material properties are used in the analysis.

Table 1 Geometric and material properties of the steel columns

Properties	Notation	Value	
		I section	Square section
Length (m)	L	5	5
Cross-sectional area (m^2)	A	0.049	0.165
Moment of inertia-strong axis (m^4)	I_{xx}	1.42×10^{-3}	2.26×10^{-3}
Mass density (kg/m^3)	ρ	7850	7850
Young's modulus of elasticity (N/m^2)	E	2×10^{11}	2×10^{11}
Mass per unit length (Kg/m)	m	385	1294
Total mass (Kg)	M	1923	6470
Plastic moment capacity (Nm)	M_p	2.7×10^6	5.9×10^6

3.2 Load Cases

There are several ways in which blast load can be applied to a structural member. The blast load due to a far-field explosion is almost uniform over the height of the member but a close-in detonation results in a concentrated load. The variation of blast load along the length of the member in an actual blast depends on the standoff distance.

A blast load of 10 MN is considered for the present study, which simulates a detonation at the mid-height of the column. Three spatial distributions of the blast load are considered: (1) point load, (2) uniformly distributed load, and (3) spatially varying load. The column response to the first two load distributions are obtained using SDOF and FE methods, but the spatially varying blast load could only be applied in FE methods.

Total blast load of 10 MN is distributed to the nodes at the mid-section and over the height of the FE model for the point and distributed load as shown in Figs. 5 and 6, respectively. A triangular time variation of blast load is considered for the point and uniformly distributed load, but an exponential (Friedlander) is considered for the spatially varying blast load. The spatially varying blast load along the height is applied in LS-DYNA using the *LOAD_BLAST_ENHANCED keyword

Fig. 5 Point load

Fig. 6 Uniform load

Fig. 7 Pressure history

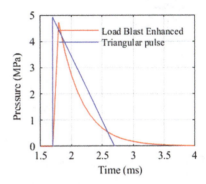

Table 2 Load cases considered for the analysis

S. No:	Load	Spatial variation	Time variation	Load durations	
				I section	Square section
1	Point	Concentrated	Triangular	1, 5, 10 ms	5, 15 ms
2	Uniform	Uniform	Triangular	1, 10 ms	5, 15 ms
3	LSDYNA	Variable	Exponential	1, 10 ms	NA

option, which takes charge mass and standoff distance as input parameters. These two parameters are obtained by equating the peak reflected pressure and impulse with the triangular pulse load of 10 MN (for $t_d = 1$ ms). This gives a charge mass of 6.01 kg of equivalent TNT at a standoff distance of 1.81 m. Figure 7 shows the pressure history for uniformly distributed load and *LOAD_BLAST_ENHANCED keyword option. Different load cases considered for the analysis are summarized in Table 2.

3.3 Simplified SDOF Model

The elastic shape functions for a fixed-end column under point load (ψ_1) at mid-height and uniformly distributed load (ψ_2) are assumed to be their deflected shape

Table 3 Transformation factors for SDOF analysis

Material model	Transformation factors	ψ_1 (point load)	ψ_2 (uniform load)
Elastic	K_L	1	0.53
	K_M	0.37	0.41
Elasto-plastic	K_L	1	0.64
	K_M	0.33	0.50

Table 4 Parameters of resistance curve of the steel columns

Properties	Notation	Point load		Distributed	
		I section	Square section	I section	Square section
Equivalent stiffness (N/m)	k	4.35×10^8	6.96×10^8	6.95×10^8	1.11×10^9
Ultimate resistance (MN)	R_u	4.44	9.36	8.64	18.72
Yield displacement (mm)	y_{el}	10.2	13.4	12.4	16.9

under static loading, and are given as

$$\psi_1(x) = 4\left[\frac{3x^2}{L^2} - \frac{4x^3}{L^3}\right] \tag{15}$$

$$\psi_2(x) = 16\left[\frac{x^2}{L^2} - \frac{2x^3}{L^3} + \frac{x^4}{L^4}\right] \tag{16}$$

Values of the transformation factors for fixed–fixed boundary condition for point load and distributed load conditions are presented in Table 3. The parameters of the resistance curve of the columns are presented in Table 4.

3.4 Finite Element Model

The FE model of the column is developed in LS-PrePost. A uniform mesh size of 15 mm is used for the analysis. Figure 8a and b presents the cross section of the I and square section developed in LS-PrePost. The column is modeled using eight noded hexahedron solid elements with constant stress formulation. Analysis is performed using elastic and plastic kinematic material properties. The stress–strain plot for the elasto-plastic (*PLASTIC_KINEMATIC) material model is shown in Fig. 8c.

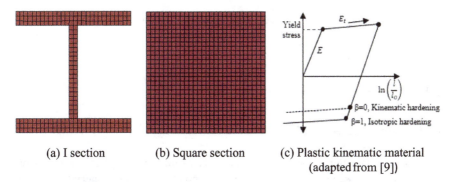

(a) I section (b) Square section (c) Plastic kinematic material (adapted from [9])

Fig. 8 Details of FE model in LS-DYNA

4 Results and Discussion

4.1 *Modal Analysis*

The modal frequencies of the equivalent SDOF system are calculated for the assumed shape functions and compared with the frequencies of the finite element models. The frequencies can also be obtained analytically by solving the beam vibration problem in Eq. (4). The modal frequencies and time-periods obtained using these methods are presented in Table 5. The frequencies obtained using the simplified SDOF and analytical methods are greater than frequencies of the FE model for I column. Simplified procedure and analytical procedures are based on flexural response that neglects in-plane shear deformation in the columns and hence results in a stiffer model. In addition, the boundary condition in LS-DYNA is simulated restraining the nodes at the end section of the I column. This results in a model that is less stiff than the analytical model for which a perfect fixed boundary condition is assumed. The frequencies obtained using different procedures are much closer for the square column due to smaller contribution of shear deformation and better simulation of boundary conditions of the FE model. The modal results provide insight into the dynamic behavior of the columns. The simplified SDOF procedure is more accurate for blast analysis

Table 5 Modal properties of the column

Modal property	I section				Square section			
	Analytical	SDOF		LS-DYNA	Analytical	SDOF		LS-DYNA
		ψ_1	ψ_2			ψ_1	ψ_2	
Frequency (*Hz*)	123	125	122	97	87	90	83	83
Time period (*ms*)	8.2	8.0	8.2	10.3	11.5	11	12	12

of regular-shaped member (e.g., square), and provide results that are closer to FEA results than irregular-shaped members (e.g., I shape).

4.2 Response of I Column

4.2.1 Concentrated Load

The peak displacements and dynamic reactions of the I column subject to concentrated load are presented in Table 6 for elastic and elasto-plastic material model.

For the elastic material, comparable peak displacements are obtained using simplified SDOF analysis and LS-DYNA analysis for the three load durations. For impulsive loading ($t_d = 1$ ms), the SDOF analysis underpredicts the dynamic reactions. The variation can be attributed to the difference between the actual and assumed deflected shape of the element when subjected to blast loading. The static deflected shape and analytical shape obtained by solving Eq. (5) for the fixed boundary conditions are shown in Fig. 9. Figure 10 shows the actual and assumed deflected shape of the column for blast loading.

During initial response, the actual deflected shape has much higher gradient than the shape function assumed for the SDOF analysis. The dynamic reactions and internal forces in the SDOF analysis depend on higher order derivates for the shape function. The arrival time of peak reactions is smaller than the peak displacements. The actual deflection does not initially resemble the assumed shape function

Table 6 Response of the I column subject to blast loading

t_d (ms)	Elastic				Elasto-plastic			
	Disp. (mm)		Reactions (MN)		Disp. (mm)		Reactions (MN)	
	SDOF	LS-DYNA	SDOF	LS-DYNA	SDOF	LS-DYNA	SDOF	LS-DYNA
1	9.2	10.6	2.9	8.5	11.3	16.5	3.4	2.9
5	31.1	41.6	9.0	11.3	102.2	147.2	3.4	2.9
10	45.0	55.4	10.2	11.8	357.9	383.6	3.4	3.6

Fig. 9 Shape functions for fixed-end beam or column

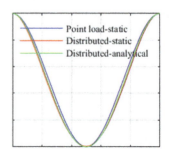

Fig. 10 Assumed and actual (FE) deflected shape of column

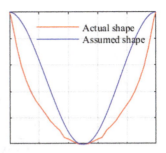

but converges to it at higher deflections. This leads to greater differences for reactions than displacements as forces are obtained as multiple derivatives of the shape function. Therefore dynamic reaction is generally less accurate during early time response for short duration loadings [10]. For higher t_d value, shape becomes close to the assumed shape function. Morison [11] provides coefficients for resistance of the column and applied force (Eq. (12)) derived from dynamic force equilibrium equations using Mathcad calculations. The difference between Biggs [3] and Morison [11] coefficients is less than 20% for most cases. The dynamic reactions obtained for 5 ms and 10 ms show better agreement between simplified SDOF and LS-DYNA results.

The response of the I column is also obtained with the plastic kinematic material model. The equivalent SDOF responses are obtained assuming perfect hinge formation and corresponding suitable shape function at different stages of response. However, the column might undergo partial yielding at the assumed hinge location for small duration loads. The stress vs strain plot at the mid-span cross section of the column for the three load durations are plotted. Figure 11 shows the element locations at the cross section used for plotting stress vs strain graph. The stress vs strain plots for 1 ms and 10 ms load durations are shown in Figs. 12 and 13, respectively.

LS-DYNA provides higher value of peak deflection compared to simplified SDOF analysis using plastic kinematic material model. The dynamic reactions in the column are obtained assuming complete yielding at the ultimate resistance capacity of the

Fig. 11 Elements monitored at mid-height

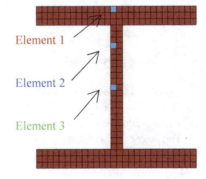

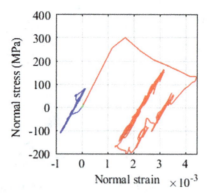

Fig. 12 Element response ($t_d = 1$ ms)

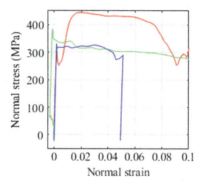

Fig. 13 Element response ($t_d = 10$ ms)

cros-section. The reactions obtained using SDOF analysis is higher compared to LS-DYNA due to partial cross-sectional yielding for 1 and 5 ms load durations. Good agreement is obtained for dynamic reactions for 10 ms load duration with complete yielding of the cross section. Figure 14 shows displacement histories obtained using SDOF and LS-DYNA analysis for different durations of blast loading for elastic material.

4.2.2 Distributed Load

Responses of the I column subject to uniformly distributed point loads and spatially varying *LOAD_BLAST_ENHANCED options are obtained here using SDOF and FE models for blast load durations 1 and 10 ms.

The peak displacements and dynamic reactions are presented in Table 7 for elastic and kinematic plastic material models. Comparable values of the peak displacement and dynamic reactions are obtained using simplified SDOF and LS-DYNA analysis

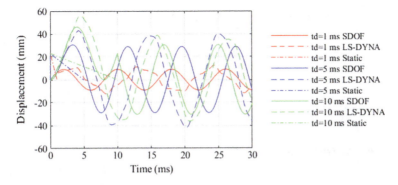

Fig. 14 Displacement history at mid-height of the column (elastic material)

Table 7 Peak response of the I column subject to distributed blast loads

Material model	Displacement (mm)			Dynamic reactions (MN)		
	SDOF	LS-DYNA		SDOF	LS-DYNA	
		Uniform	Variable		Uniform	Variable
Elastic (1 ms)	4.9	5.9	5.0	1.18	1.42	1.79
Plastic (10 ms)	49.7	59.9	45.4	4.1	3.24	3.15

for elastic material. Higher dynamic reaction is obtained using simplified SDOF method for elasto-plastic material. As the section is partially yielded, the reaction obtained assuming hinge formation is higher than the LS-DYNA result. Figure 15 shows displacement history for distributed blast load for elastic material.

Fig. 15 Displacement history at mid-height of the I column subject to distributed blast load (elastic material, $t_d = 1$ ms)

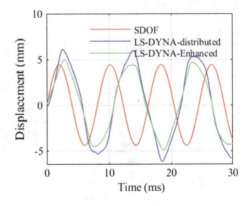

Table 8 Peak response of the square column subject to blast load

t_d (sec) (ms)	Point load				Uniformly distributed			
	Disp (mm)		Reactions (MN)		Disp (mm)		Reactions (MN)	
	SDOF	LS-DYNA	SDOF	LS-DYNA	SDOF	LS-DYNA	SDOF	LS-DYNA
5	19.1	19.8	7.6	10.1	7.5	7.6	4.2	4.77
15	23.7	27.3	9.7	8.5	11.9	11.5	7.0	7.24

4.3 Response of Square Column

The effect of structural shape on accuracy of response calculation using simplified SDOF procedure is investigated. Kinematic plastic material models are used for the analysis. The peak displacements and dynamic reactions of the square column obtained using simplified SDOF analysis and LS-DYNA are presented in Table 8.

A good comparison is obtained for peak displacement using LS-DYNA and simplified SDOF methods. The dynamic reactions for point load shows variation for elastic and elasto-plastic material properties. Good agreement is obtained for dynamic reactions for uniformly distributed load. The column behavior is elastic for kinematic plastic material for 5 and 15 ms load durations. Displacement histories for point load and uniformly distributed load for 5 ms load duration are plotted in Fig. 16. Comparing the response of I section to the square section, it is clear that the shape of the structural member significantly affects the accuracy of simplified SDOF procedure, especially the dynamic reactions. The role of shape of the structure member on blast response becomes less significant when load is applied as uniformly distributed instead of point load.

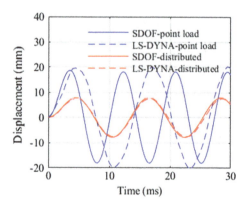

Fig. 16 Displacement history at mid-height of the square column subject to blast load (elastic material, $t_d = 5$ ms)

5 Summary and Conclusions

The limitations of the simplified SDOF analysis used to estimate blast response of structures are investigated by comparing results with finite element analysis in LS-DYNA. The effect of modeling assumptions and parameters on the accuracy of the simplified SDOF procedure is also discussed.

The key conclusions of this study are

1. The simplified SDOF procedure for blast analysis provides a reasonable estimate of the peak displacement but underpredicts the dynamic reactions.
2. The SDOF calculations for inelastic stage of response are inaccurate when there is incomplete hinge formation due to partial yielding of the section for short duration blast loads.
3. Greater difference is observed between the SDOF and LS-DYNA responses when blast load is modeled as concentrated point load than uniform load, because an idealized point load used for simplified SDOF method cannot exactly be simulated in FE models, especially for I section.
4. A reasonable agreement between the SDOF and LSDYNA analysis is obtained for peak response, but differences are observed between response histories. The difference in response histories is greatly diminished for square column with uniformly distributed load.
5. The shape (e.g., I vs. square) of a structural member has a significant effect on accuracy of its blast response obtained using simplified SDOF method, which is more suited to members dominated by flexural response.

Acknowledgements The financial support for this research work was provided through an early career award by the Science and Engineering Research Board of Department of Science and Technology. The authors gratefully acknowledge the support. The authors are also thankful to the Department of Civil Engineering at IIT Bombay who provided travel support to the lead author to attend the conference.

References

1. American Society of Civil Engineers (ASCE) (2011) Blast protection of buildings. In: ASCE/SEI 59-11, Reston, VA
2. Bureau of Indian Standards (BIS) (1968) Criteria for blast resistant design of structures for explosions above ground (Reaffirmed 2008). In: Standard IS 4991, New Delhi, India
3. Biggs JM (1964) Introduction to structural dynamics. McGraw Hill, New York, NY
4. US Department of Defense (DoD) (2008) Structures to resist the effects of accidental explosions. *Report No UFC-3-340-02*, Washington, DC
5. Li J, Hao H (2014) A simplified numerical method for blast induced structural response analysis. Int J Prot Struct 5(3):323–348
6. Al-Thairy H (2016) A modified single degree of freedom method for the analysis of building steel columns subjected to explosion induced blast load. Int J Impact Eng 94:120–133

7. Lee K, Shin J (2016) Equivalent single-degree-of-freedom analysis for blast-resistant design. Int J Steel Struct 16(4):1263–1271
8. LSTC (2012) Computer program LS-DYNA. Livermore Software Technology Corporation, Livermore, CA
9. LSTC (2015) Ls-dyna keyword user's manual, volume II-material models. Livermore Software Technology Corporation, Livermore, California, USA
10. U.S. Army Corps of Engineers, P. D. C. (2008) Methodology manual for the single-degree-of-freedom blast effects design spreadsheets (SBEDS)
11. Morison CM (2006) Dynamic response of walls and slabs by single-degree-of-freedom analysis—a critical review and revision. Int J Impact Eng 32(8):1214–1247

A Comparative Analysis of Computation of Peak Overpressure in Near Field Scenario During High Energetic Material's Detonation

Praveen K. Verma, Rohit Sankrityayan, Devendra K. Dubey, and Anoop Chawla

Abstract A polynomial based empirical function and chart-based techniques are available for computation of peak overpressure in the event of detonation of high energetic material. Such computations are based on experiments carried out using high mass detonative and extrapolation using scaling laws. There are mathematical tools available based on these empirical relationships but have inherent numerical approximations. These tools are observed to be insufficient for accurately predicting peak overpressure in the near field scenarios, i.e., at scaled distance lower than $0.2 \text{ m/kg}^{1/3}$. Furthermore, owing to very high temperature and pressure conditions in such scenarios only limited information is captured. However, simulation techniques using MM-ALE (Multi-Material Arbitrary Lagrangian Eulerian) approach are taxing on computation resources. A comparative study is presented to compare the results obtained using commercially available simulation techniques with results from empirical relations from experiments in near field scenarios (scale distance < $0.2 \text{ m/kg}^{1/3}$). The study highlights merits and challenges of using the available pre- and post-processing techniques. A comparison of results obtained by changing the mesh size and meshing techniques is also discussed.

Keywords Near field detonation · Overpressure · High energetic material · MM-ALE · Scaled distance

1 Introduction

In the case of placement of detonative like Anti-Personnel Mine, a high energetic explosive material is planted to target the personnel, either military or civilian. These explosives are either planted above the ground (hanging in free air) or buried under soil. To represent such conditions in a test environment some accredited laboratories have been conducting experiments, both in free air and charge buried under soil. The

P. K. Verma · R. Sankrityayan · D. K. Dubey (✉) · A. Chawla
Department of Mechanical Engineering, Indian Institute of Technology, New Delhi, India
e-mail: dkdubey@mech.iitd.ac.in

test conditions for very near field experiments are achieved by inclusions of explosive material very close to target, typically 20–100 mm distance. A typical equivalent weight of Tri Nitro Toluene (TNT) is taken to have a scale distance representing anti-personnel mine buried under the soil. For simulation of such conditions of closely placed explosive or a contact explosive in Finite Element (FE) environment (e.g., LS-DYNA), a simplified explosive simulation needs to be conducted.

The most common simulation in FE environment is by placing an explosive charge in air and computing the peak overpressure. Peak overpressure could be computed for an increasing scale distance by post-processing the result at constantly increasing standoff distance. This calculated peak overpressure in the FE environment needs to be validated by an experiment or by some empirical relation driven from experiments. The empirical relation for peak overpressure vs scaled distance during the event of high-energy explosive is available using hemispherical charge. These empirical relations are derived from the experiments conducted in Canada, United Kingdom, and United States. An empirical relationship for peak overpressure measurement was made available in [1]. The hemispherical charges used to carry out surface detonation were 5, 20, and 100 ton of TNT. The above results were further simplified and presented in terms of polynomial equations with an accuracy of 1% in [2]. Since the above equations were both empirical and based on the height quantity of detonative, the polynomial equations might not represent the exact peak overpressure.

It is observed that the above equations were valid for scaled distance of the range of minimum 0.18 m/kg$^{1/3}$. Also, in the actual event of high-energy explosive or in representative experiment, the scaled distance is less than 0.2 m/kg$^{1/3}$, and the Peak overpressure measurement is limited by the availability of the measurement gauges and instrumentation requirement. For such a close field of a very high pressure and temperature environment, measurement of pressure could be erroneous due to damage caused to measurement gauges, instrumentations, and equipments. Therefore, very limited information is available in the near field scenario. Because of this limitation no validation is available for such a close distance. Hence both the FE environments: ConWep and MM-ALE are based on the [1] experiments, where there is a need to develop a simulation framework to validate such near field scenario.

2 Empirical Calculations Available

The empirical calculations for the peak overpressure as discussed above have been the bases for the validation for the simulations. Different available techniques in the form of empirical equations and charts are discussed by Goel and Matsagar [3]. The two different sources readily available are "Kingery- Bulmash Calculator" [4] in the form of MATLAB code and United Nations calculator [5] in the form of html code.

A comparison of the above MATLAB code calculations (both spherical and hemispherical detonatives) with calculation of UN-based code is presented in Fig. 1. The comparison is plotted between peak overpressure vs closed field scaled distance (maximum scaled distance 0.28 m/kg$^{1/3}$). The calculations from UN website and

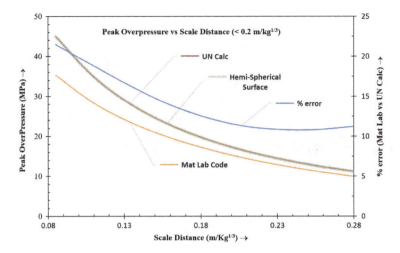

Fig. 1 Comparison of MATLAB calculator with UN code and percentage error between spherical and hemispherical detonatives

MATLAB's hemispherical detonation is over-lapping each other, the orange and gray curves shown in Fig. 1. Therefore, both are predicting same peak overpressure for hemispherical detonation in the scaled distance range being studied (scaled distance 0.28 m/kg$^{1/3}$).

In Fig. 1, the orange curve represents the prediction of the overpressure for a spherical detonative. For an event of high-energy detonation with spherical charge in the air domain is slightly less from a hemispherical charge in term of peak overpressure. Also, the blue curve in Fig. 1 is the representation of the % error between hemispherical and spherical calculations. In the range of 0.20–0.28 m/kg$^{1/3}$ scaled distance of the % error is almost constant and as we reduce the scaled distance the % error increases close to till 21%. It can be concluded that the variation increases as the scaled distance reduces to less than 0.2 m/kg$^{1/3}$.

3 Model for Free Air Detonation

3.1 Free Air Detonation Using MM-ALE

Under high energy material detonation, a shock wave profile is generated and gets propagated as a wave front. For the free air case, the wave profile generates the peak overpressure with respect to time. Free air detonation FE simulation is performed to estimating the wave peak overpressures at a specified distance from the charge center. This is associated with the change of Equation of State (EOS) like pressure, density, and temperature of the ALE group (e.g., Air and TNT). Detonative material is modeled to represent high energy material loading in the air domain with flow out condition. The results in the free air were compared with empirical data obtained from experiment by US Army Research [6].

To assign a volume fraction, LS-DYNA contains a keyword *INITIAL_VOLUME_FRACTION_GEOMETRY which needs coordinates of center, shape, and size of the charge, and the part to be filled with. The material properties for air and TNT were defined using keyword *MAT_NULL and *MAT_HIGH_EXPLOSIVE_BURN, respectively. The equation of state for air was provided using keyword *EOS_LINEAR_POLYNOMIAL and for that of TNT was defined using *EOS_JWL.

3.2 Finite Element Model

Free air high energy detonation is simulated in LS-DYNA using MM-ALE method. A cubical air domain of side 500 mm is used as ALE meshed box with specified weight of TNT at one of its corners filled using initial volume fraction as shown in Fig. 2.

The volume fraction of the TNT was defined as a sphere of radius 18.97 mm (calculated for specified weight of TNT), with its center at one corner of meshed box.

A symmetry is modeled to reduce the number of elements and computation time. Hence, the system used was corner symmetric (1/8th model) of a 500 mm-cube with a spherical charge at center (Fig. 2).

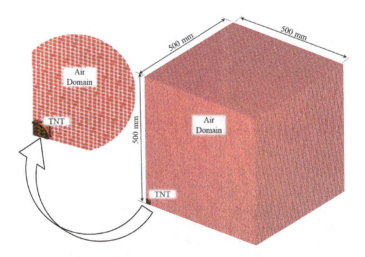

Fig. 2 Corner-Symmetric FE Model for free air in LS-DYNA, mesh size 5 mm

For performing a mesh sensitivity analysis, the element size is varied and results are compared. In this study, three different schemes of element sizes for air domain are chosen. One with a 10 mm element size throughout. Second with more refined 5 mm element size throughout. The third model for air domain is generated with a variation of element size from 2 mm close to the charge (in the center) to 10 mm at the ends of the meshed box. The variation of the mesh size is linear.

3.3 Pressure Measurement in LS-DYNA

In MM-ALE approach in LS-DYNA the post-processing is done either by capturing the pressure at tracer's location or by measuring pressure in the Eulerian elements in the air domain. The tracer measurement history in post-processing is available only at a predefined location in the FE model on the point of interest. Measurement of pressure in the treasure is based on the ALE material (Air and TNT in the model discussed above) flowing through the tracer point. In the MM-ALE model, the flow in the Eulerian elements can contain multi-materials referred as an ALE multi-material group (AMMG). In the case of detonation both Air and TNT could be present simultaneously, especially in the close field measurement. The pressure measured at the tracer location in this case is based on the volume fraction of the ALE material present. This corresponds to the default setting in LS-DYNA card (setting AMMGID to zero; AMGID = 0). Measurement of pressure by a tracer can also be enforced based on single material by setting up AMMGID to a specific multi-material number. Pressure measurement by the tracer for the above-discussed model based on different AMMGID setting is given in Table 1.

The other method of measuring the pressure in the Eulerian elements is an indirect method. In this method, stress in the elements are calculated internally by the software and later the hydrostatic stress component of the element is calculated. The pressure is defined as the negative of average of three principal stresses.

$$Pressure = -\frac{(\sigma_1 + \sigma_1 + \sigma_1)}{3} \quad (1)$$

Since the pressure is compressive in nature negative sign, multiply it. Such a computation is highly dependent on mesh size as there would be a penalty of flow of different materials.

Table 1 Pressure measurement by the tracer at different AMMGID setting

AMMGID	Pressure measurement type	Pressure measurement specific to model
0	Averaging the all ALE material	Average volume fraction of Air and TNT
1	ALE material 1	Only Air
2	ALE material 2	Only TNT

4 Result and Discussions

4.1 Results of Mesh Sensitivity Analysis

A comparative study in terms of peak overpressure and scale distance for varying element size is conducted. In this the element size is chosen as 10, 5 mm and a linear variation of element size of 2 to 10 mm for the air domain.

Figures 3 and 4 represent the peak overpressure vs scaled distance for three different models studied and are compared with Kingry-Bulmash calculations. Figure 4 is the zoomed view of Fig. 3 for scaled distance less than 0.2 m/kg$^{1/3}$. It is observed that the empirical relation (Kingry-Bulmash) is a smooth curve while other curves are not smooth. This is attributed due to the measurement of pressure at the different scaled distance using element.

From both Figs. 3 and 4, it can be seen that element length 5 mm and transition from 2 to 10 mm results almost the same for the closed field (less than 0.2 m/kg$^{1/3}$). Hence, the element size of 5 mm is used for further analysis. Also, it could also be seen that prediction from 10 mm element size is having lot of dispersion especially scale distance below 0.14 m/kg$^{1/3}$.

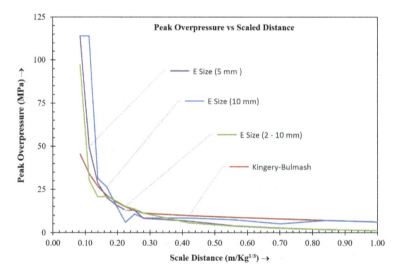

Fig. 3 Comparison of peak overpressure vs scaled distance with empirical Kingery-Bulmash (Brown) and air domain element size variation, 5 mm (purple), 10 mm (blue), and 2–10 mm (green)

A Comparative Analysis of Computation of Peak ... 503

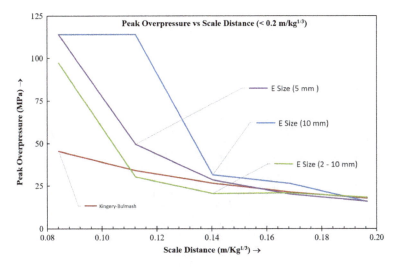

Fig. 4 Zoomed view for scaled distance below 0.2 m/kg$^{1/3}$ of comparison of peak overpressure vs scaled distance with empirical Kingery-Bulmash (Brown) and element size variation

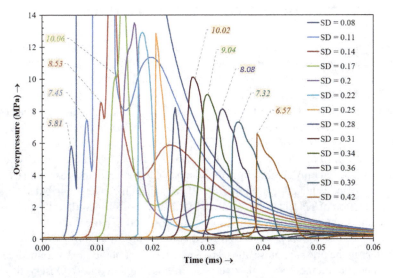

Fig. 5 Tracer output of volume fraction of Air (AMMGID = 1). Dark blue curve is peak overpressure closest to detonative (SD 0.08 m/kg$^{1/3}$). SD = Scale distance

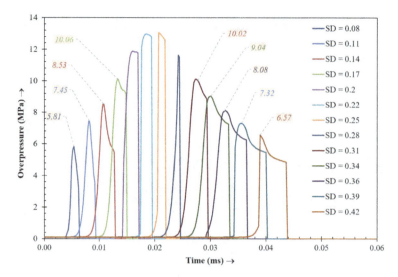

Fig. 6 Tracer output of volume fraction of Air (AMMGID = 0). Dark blue curve is peak overpressure closest to detonative (SD 0.08 m/kg$^{1/3}$). SD = Scale distance

4.2 Peak Overpressure Measurement Using Tracer

Figures 5 and 6 represent the peak overpressure vs time recorded by tracers placed in the increasing distance from the center of the charge. The corresponding scaled distance starting from 0.08 to 0.42 m/kg$^{1/3}$ is plotted in different colors.

By quickly observing the dark blue curve which corresponds to the tracer pickup closest to detonative at a scaled distance of 0.08 m/kg$^{1/3}$. As indicated the peak overpressure recorded by the tracers in Fig. 5 is pressure due to air only, and Fig. 6 is pressure due to volume fraction of Air & TNT. Observation is that the first peak (5.81 MPa) occurring at approximately 5.21 microsecond (0.00521 ms) is peak overpressure due to the contribution of ALE air. While the second peak suddenly increases and extends out of Fig. 5 and is not appearing in Fig. 6. This sudden increase is attributed to large volume fraction of TNT in the close filed. Similar trend is observed for the light blue, brown, and light green curves corresponding to scaled distance of 0.11, 0.14, and 0.17 m/kg$^{1/3}$, respectively, when compared between Figs. 5 and 6.

Further in the pressure pulse after 0.025 ms, the effect of the TNT volume fraction is nullified, hence there are only one peak in both Figs. 5 and 6. Observed in the right most five curves corresponding to scaled distance 0.31, 0.34, 0.36, 0.39, and 0.42 m/kg$^{1/3}$ in both the figures have single peak. Form above, this could be concluded that at closed field the peak overpressure recorded by the tracers are more error prone due to presence of TNT volume fraction. Hence measuring the pressure in the element in the close range of detonation is more accurate for prediction of variable.

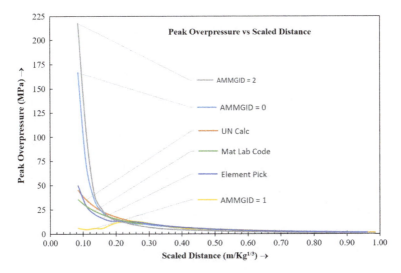

Fig. 7 Peak overpressure vs scaled distance. Comparison of peak overpressure picked up by tracer. Average (AMMGID = 0), ALE air (AMMGID = 1), and ALE TNT (AMMGID = 2) with UN/MATLAB tool. Also pressure measured in ALE element

4.3 Peak Overpressure Using Diagonal Elements

From the conclusion above, we can further capture the peak overpressure by tracers in three different ways, as shown in Fig. 7.

Blue curve (AMMGID = 0) in Fig. 7 represents the pressure captured by average of volume fraction in the location. Violet (AMMGID = 1) curve represents the pressure captured due to ALE air only, and light green (AMMGID = 2) curve represents the pressure captured due to ALE TNT only. Also light blue (Element Pick) curve represents the pressure measured by measurement done on the elements diagonally placed from the center of charge. For comparison, both UN (Red curve) and MATLAB (orange curve) prediction are produced in Fig. 7. Figure 8 is a zoomed view of Fig. 7 for scaled distance less than 0.2 m/kg$^{1/3}$.

For both the figures, it can be re-emphasized that peak overpressure as predicted by tracers in the close field is more erroneous based on volume fraction of ALE element. Also, measuring peak overpressure in the element gives more closed to predictions by Kingery-Bulmash [4]. It can also be seen from Fig. 8 (by comparing green curve with dark blue curve) that with 5 mm element size the peak overpressure measured by elements closely matches with "Kingery-Bulmash Explosive Calculator" in the form of MATLAB code made available by [4]. Hence, a model in LS-DYNA with 5 mm element size and with measurement done on diagonal element gives a close enough result for running simulation in a free air detonation case.

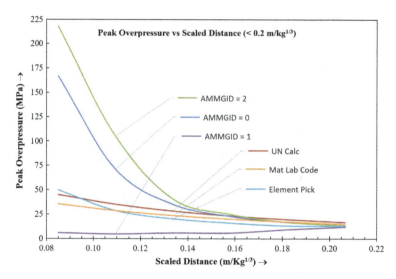

Fig. 8 Peak overpressure vs scaled distance (scaled distance < 0.2 m/kg$^{1/3}$). Zoomed view of earlier figure to capture the scaled distance less than 0.2 m/kg$^{1/3}$

5 Conclusion

Using MM-ALE approach in FE environment, a high energetic detonation in free air is modeled with air and explosive as ALE element. The mathematical tools available based on empirical relationships predict the peak overpressure behavior during the detonation. A comparison of the result is mandatory because of inherent numerical approximation and efficiency to predict peak overpressure. There are limitations in experiments, like measurement of pressure under the very high temperature and pressure environment, hence very limited information is available in the near field scenario. A comparative study is planned to compare the results obtained using commercially available simulation technique, empirical relations, and experiments available, in the near field scenario (scale distance < 0.2 m/kg$^{1/3}$).

For the closed field analysis discussed in the paper, it is concluded that both MATLAB codes based on "Kingery-Bulmash Calculator" and "United Nations calculator" available in the form of html code predict same result for hemispherical detonatives. Also, for the same range of scaled distance the difference between the peak overpressure in case of spherical vs hemispherical detonatives error is 11% close to 0.2 m/kg$^{1/3}$. Further the percentage error increases to 21% in case of reducing scaled distance below 0.2 m/kg$^{1/3}$. For the closed field, the element size of 5 mm predicts sufficiently good result and a linear variation of element size has no significant effect on peak overpressure.

It is also concluded that the peak overpressure captured by tracers in the close field does not predict the behavior as calculated in Kingery-Bulmash mathematical relations. To predict more accurately (close to Kingery-Bulmash calculations) picking up the pressure profile from ALE elements is more accurate. Although the FE simulation is an approximate and needs validation, this is to be kept in mind that pressure profile with the empirical relations and polynomial equations may not be accurately predicting the peak overpressure near close field scenario. Hence with the current advancement of instrumentation and high reliable gauges under very high temperature and pressure environment closer explosive experiments could be attempted.

References

1. Kingery C, Pannill B (1964) Peak overpressure vs scaled distance for TNT surface bursts (hemispherical charges), Army Ballistic Research Lab Aberdeen Proving Ground Md
2. Swisdak Jr MM (1964) Simplified kingery airblast calculations, Naval Surface Warfare Center Indian Head Div Md
3. Goel MD, Matsagar VA, Gupta AK, Marburg S (2012) An abridged review of blast wave parameters. Defence Sci J 62(5), 300–306
4. Rigby S, Tyas A Blast.m - a simple tool for predicting blast pressure parameters, http://cmd.sheffield.ac.uk/?q=software/blastm-simple-tool-predicting-blast-pressure-parameters
5. Kingery-Bulmash Blast Parameter Calculator, https://www.un.org/disarmament/un-saferguard/kingery-bulmash/
6. Dalton JC, Gott JP (2014) Unified Facilities Criteria (UFC) Structures to resist the effects of accidental explosions, Department of Defence, USA

Development of Performance-Based Design Guidelines for Reinforced Concrete Columns Subject to Blast Loads

Vishal Kochar, K. K. Anjani, and Manish Kumar

Abstract Critical civilian facilities and military infrastructure in India have been a target of several terrorist attacks in the last few decades. In order to address the risk of explosion threats, it has become critical to design and construct blast-resistant structures to save lives and protect infrastructure. The focus of this paper is to develop performance-based guidelines for the blast-resistant design of reinforced concrete (RC) columns. Columns are the most vulnerable component of a structure and are at a high risk under a detonation scenario. The paper presents a comprehensive review of the blast-resistant design procedures followed by different design documents. The gaps and challenges in the blast resistance of RC columns were identified and addressed. Specific issues in the axial, flexure, and shear design of columns are highlighted. A performance-based analysis and design approach for blast-resistant design of RC column were developed and is presented in the paper. The design approach was verified using high-fidelity numerical simulations. The performances of the design columns was assessed using nonlinear FE models developed in LS-DYNA. The ultimate capacities of the columns subject to blast loads were obtained. The numerical simulations assist in verifying the proposed design procedures and provide critical information on the failure modes, safety margin, and other parameters for blast-resistant design of RC columns. Recommendations on safe and efficient design of RC members are also provided.

Keywords Reinforced concrete · Column · Blast · FEA · LS-DYNA

1 Introduction

The demand for designing important structures to resist blast load is growing with increase in terrorist threats and for defense applications. The performance goals of a structure are defined based on its function and importance, which determines its

---

V. Kochar · K. K. Anjani · M. Kumar (✉)
Indian Institute of Technology Bombay, Mumbai 400076, India
e-mail: mkumar@iitb.ac.in

failure mode. Columns are the most vulnerable and critical component of a structure that must be designed against blast to achieve the defined performance goals.

Very little literature is available on prescriptive blast-resistant design guidelines of RC columns. The state of practice for the blast-resistant design of RC members mostly follow the procedures suggested in the document UFC-3-340-02 [1]. The preliminary design use simplified SDOF analysis of structural components, which have several limitations [2]. Other guidelines available in ASCE 59-11 [3] and IS 4991-1968 [4] are mostly descriptive and do not provide design and detailing guidelines for RC members. This paper presents a review of the blast-resistant design procedures followed by different design documents. The gaps and challenges in the blast resistance of RC columns are identified and addressed. Specific issues in the axial, flexure, and shear design of columns are highlighted. A unified procedure for blast-resistant design and detailing of RC columns is presented in this paper. Performance of a column designed as per the proposed procedure is compared against an ordinary designed column using nonlinear Finite Element Analysis to verify the effectiveness of the developed guidelines.

2 Background

2.1 Design Objective

The primary objective of the blast-resistant design of a structural element is to provide adequate ductility to achieve deflection consistent with a permitted damage state. This is ensured by appropriate design and ductile detailing so that elements have enough plastic deformation capacity in flexure without premature failure due to other loads (e.g., shear, axial). A concept of resistance-deflection function, $R(x)$, is used for simplified analysis procedures to represent the capacity of the member as a function of deflection and for a given load distribution. Resistance curve is the plot of quasi-static load of same distribution as the applied load that produces deflection is equal to its transient displacement. The resistance curve is used to specify performance goals as function of ductility. The resistance curve of a laterally restrained RC member is shown in Fig. 2. The deflection (for a support rotation) corresponds to a damage state that can be linked to specific performance goals for the structural member.

2.2 Response Characterization

Response of a structure depends on the magnitude and duration of blast. Some literatures provide different guidelines for design based on the response characteristics of the structure. Cormie et al. [5] recommends that response can be classified as impulsive if time to peak response of column (t_m) is greater than three times the duration of

Table 1 Dynamic Increase Factors (DIF) (adapted from [1] and [5])

Action	Concrete	Rebar	
	f'_{dc}/f_{ck}	f_{ds}/f_y	f_{du}/f_u
Flexure	1.25	1.20	1.05
Diagonal tension	1.00	1.10	1.00
Direct shear	1.10	1.10	1.00
Compression	1.15	1.10	NA

applied load (t_d), and dynamic for values smaller than 3. A unified design philosophy is presented here that can be adopted for both dynamic and impulsive response of columns. These design provisions are not applicable for close-in or contact detonation where the failure mode of column is by sudden crushing of concrete or brisance failure [6].

2.3 Material

The blast response of a structure is characterized by very high strain rate, which influences the mechanical characteristics of concrete and steel. The increase in stiffness is negligible but the strength enhancement (dynamic material strength) is calculated using dynamic increase factors (DIF), which are presented in Table 1.

2.4 Performance Goals

The goal of the blast-resistant design of structure is to provide the predetermined level of protection against the estimated hazard. Structures are designed for protection categories 1 and 2 based on performance requirement, which depends on the importance of the structure. The response is measured in terms of support rotation as shown in Fig. 1. The performance goals and associated response limits are summarized in Table 2.

The resistance-deflection curve for a RC column is presented in Fig. 2. The damage states corresponding to different response limits of the RC column are shown in Fig. 3. The resistance increases linearly with deflection until the yielding of tensile reinforcements, subsequent to which it remains constant up to deflection corresponding

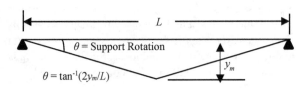

Fig. 1 Support rotation of a structural member

Table 2 Performance goals for a RC column

	Protection category	
	1	2
Rotation (θ)	2°	4°
Importance	Critical	Medium
Performance goals	Limited damage, operational after repairs and retrofit	Extensive damage, Safe replacement or demolition
Damage state	Cracking on the tensile face, cover remains intact	Crushing on compression face and spalling on the tensile face

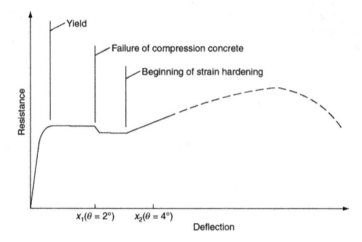

Fig. 2 Resistance-deflection curve [5]

to angular rotation of 2°. The concrete cover is crushed in compression and the loads are transferred to the compression reinforcement. The concrete core remains intact and the strain hardening is observed due to reinforcements. The concrete core becomes ineffective at support rotation of 4° and would lead to failure of the column in absence of any other blast-resistant detailing measures (e.g., lacing) [5].

3 Design Methodology

A unified blast-resistant design methodology for RC column is presented in this paper. Columns behave as a flexural member under lateral blast-pressure loads and are designed to resist movements with adequate capacity in shear and axial compression or tension. This ensures ductile failure of columns while avoiding premature failure in shear or due to axial loads. Special detailing is required near the support to prevent direct shear failure. The connections to adjoining beams or slabs must also

Fig. 3 RC damage state [1]

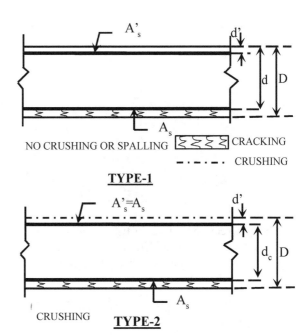

be appropriately detailed to allow development of the full capacity of the column member. Each of these design aspects are discussed and detailing guidelines are provided in the following sections.

3.1 Design for Flexure

The section geometry and material properties of the column are selected to satisfy the design requirements. Equal reinforcements are provided on tension (A_{st}) and compression (A_{sc}) faces to resist the rebound effects and develop enough ductility. The section parameters are defined in Fig. 4. The effective depth of column is calculated as:

$$d = D - c_c - c_r/2 \tag{1}$$

where D is the total depth, c_c is the clear cover, and c_r is the thickness of rebar layer.

The ultimate moment capacity of the column is calculated as

$$M_u = A_{st} \times f_{ds} \times z \tag{2}$$

where f_{ds} is the dynamic strength of steel rebar in flexure and z is the lever arm.

Fig. 4 Column section

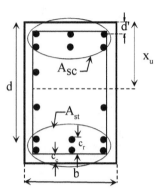

The concrete cover fails in type 2 category and do not contribute in moment capacity. Lever arm is calculated as $z = d - 0.416x_u$ for type I and $z = D - 2c_c - c_r$ type II section. The equivalent depth (x_u) of rectangular stress block is calculated as

$$x_u = \frac{A_s \times f_{ds}}{0.81 \times b \times f'_{dc}} \quad (3)$$

where f'_{dc} are dynamic strength of concrete cylinder.

The ultimate moment capacity of the column is calculated, and the column design is revised if the moment due to blast loads are higher than ultimate capacity. The column is designed as under reinforced section to ensure ductile failure. The following checks are performed for neutral axis depth and percentage of tension reinforcement:

$$x_u < x_{u,balanced} \quad (4)$$

where $x_{u,balanced}$ is the depth of rectangular stress block of balanced section.

$$x_{u,balanced} = \frac{0.0035 \times d}{0.0055 + f_{ds}/E_s} \quad (5)$$

where E_s is the Young's modulus of elasticity of steel.

$$\rho < \left(\frac{A_s}{bd}\right)_{balanced} \quad (6)$$

$$\left(\frac{A_s}{bd}\right)_{balanced} = 100 \times \left(\frac{0.81 \times f'_{dc} \times x_{u,balanced}}{f_{ds} \times d}\right) \quad (7)$$

The resistance curve parameters (stiffness, ultimate resistance, elastic deflection) for the given loading and boundary conditions for the column is obtained using the standard tables provided in standard texts (e.g., [1, 3, 7]) and reported in Table 3.

Table 3 Properties of structural members for different loads (adapted from [7])

Support condition and loading diagram	Elastic resistance (R_e)	Ultimate flexural resistance (R_m)	Stiffness Bending[1,2]	Shear[3]	Ultimate shear
	$\frac{12M_n}{L}$	$\frac{8(M_n+M_p)}{L}$	$K_e = \frac{384EI}{L^3}$ $K_{ep} = \frac{384EI}{5L^3}$ $K_E = \frac{384EI}{5L^3}$	$K_e = K_{ep} = \frac{DE}{0.35L}$	$\frac{R_m\left(\frac{L}{2}-d_e\right)}{L}$
	$\frac{8M_p}{L}$	$\frac{8M_p}{L}$	$K_e = K_E = \frac{384EI}{5L^3}$	$K_e = \frac{DE}{0.35L}$	$\frac{R_m\left(\frac{L}{2}-d_e\right)}{L}$
	$\frac{8M_n}{L}$	$\frac{4(M_n+2M_p)}{L}$	$K_e = \frac{185EI}{L^3}$ $K_{ep} = \frac{384EI}{5L^3}$ $K_E = \frac{160EI}{L^3}$	$K_e = K_{ep} = \frac{DE}{0.35L}$	$LS: \frac{R_m\left(\frac{5L}{8}-d_e\right)}{L}$ $RS: \frac{R_m\left(\frac{3L}{8}-d_e\right)}{L}$
	$\frac{2M_n}{L}$	$\frac{2M_n}{L}$	$K_e = K_E = \frac{8EI}{L^3}$	$K_e = \frac{DE}{1.4L}$	$\frac{R_m(L-d_e)}{L}$

(continued)

Table 3 (continued)

Support condition and loading diagram	Elastic resistance (R_e)	Ultimate flexural resistance (R_m)	Stiffness Bending[1,2]	Shear[3]	Ultimate shear
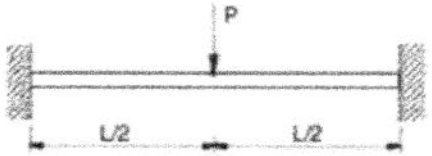	$\dfrac{8M_n}{L}$	$\dfrac{4(M_n+M_p)}{L}$	$K_e = \dfrac{192EI}{L^3}$ $K_{ep} = \dfrac{48EI}{L^3}$ $K_E = \dfrac{192EI}{L^3}$	$K_e = K_{ep}$ $= \dfrac{DE}{0.7L}$	$\dfrac{R_m}{2}$
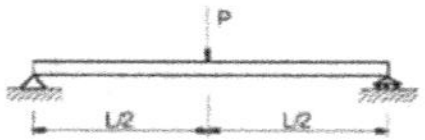	$\dfrac{4M_p}{L}$	$\dfrac{4M_p}{L}$	$K_e = K_E$ $= \dfrac{48EI}{L^3}$	$K_e = \dfrac{DE}{0.7L}$	$\dfrac{R_m}{2}$
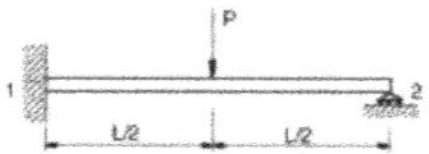	$\dfrac{16M_n}{3L}$	$\dfrac{2(M_n+2M_p)}{L}$	$K_e = \dfrac{107EI}{L^3}$ $K_{ep} = \dfrac{48EI}{L^3}$ $K_E = \dfrac{106EI}{L^3}$	$K_e = K_{ep}$ $= \dfrac{DE}{0.7L}$	$LS: \dfrac{11R_m}{16}$ $RS: \dfrac{5R_m}{16}$

(continued)

Table 3 (continued)

Support condition and loading diagram	Elastic resistance (R_e)	Ultimate flexural resistance (R_m)	Stiffness		Ultimate shear
			Bending[1,2]	Shear[3]	
	$\frac{M_n}{L}$	$\frac{M_n}{L}$	$K_e = K_E$ $= \frac{3EI}{L^3}$	$K_e = \frac{DE}{0.28L}$	R_m

[1] Elasto-Plastic stiffness, K_{ep}, based on first yield occuring at fixed support
[2] Equivalent stiffness, K_E, based on $M_n = M_p$
[3] For concrete members where shear deformation is significant, ($L/D < 5$), it can be accounted for by combining bending stiffness (K_b) and shear stiffness (K_s) in the manner such that $K = 1/(1/K_b + 1/K_s)$
L = span length of member D = Concrete member thickness
K_e = Elatic stiffness K_{ep} = Elasto-plastic stiffness
E = Modulus of Elaticity
K_E = Equivalent elastic stiffness
M_p = Ultimate positive moment capacity (interior) M_n = Ultimate negative moment capacity (support)

The column stiffness is obtained using an average moment of inertia of the column:

$$I = (I_{gross} + I_{cr})/2 \tag{8}$$

where I_{gross} is the gross moment of inertia of the uncracked section, and I_{cr} is the moment of inertia of the cracked section, which is obtained as

$$I_{cr} = F \times b \times d^3 \tag{9}$$

where F is the coefficient of moment of inertia that is obtained using the standard chart provided in Figs. 4.11 and 4.12 of UFC-3-340-02 [1].

An equivalent SDOF system of the column is created and properties are obtained as per the recommended procedures in different codes (e.g., [1, 3]). The transformation factors relate properties of the real system to the equivalent SDOF system (mass, stiffness, loads) and depend on the boundary conditions and load distribution. The natural time period of the equivalent system is obtained as

$$T = 2\pi \times \sqrt{\frac{K_{LM} \times M}{K}} \tag{10}$$

where K_{LM} is the load-mass factor, M and K are the total mass and the stiffness of the column (real system).

Peak transient deflection, y_m, is obtained using the charts (shown in Fig. 5) provided by Biggs [8] as a function of t_d/T and R_m/F_1 ratios, where t_d is the positive phase duration and F_1 is the total magnitude of the applied blast load.

3.1.1 Deflection Check

The deflection check is to ensure that peak deflection of the equivalent SDOF system (y_m), obtained using Fig. 5, is smaller than the deflection corresponding to the support rotation ($y_{lim} = 0.5L\tan\theta$) for the assumed protection category (e.g., $\theta = 2°$ for Type 1). The yield displacement of the real system is calculated as $y_{el} = R_m/K$, where K is the stiffness of the real system. The column needs to be redesigned if y_m is greater than y_{lim}.

3.2 Design for Diagonal Shear

Design for diagonal shear is done to limit tension cracks, which develop at an angle of about 45° from the edge. Shear stirrups are provided to the columns when diagonal shear is more than the shear strength of the concrete. The principle of capacity design is followed to ensure ductile failure in shear. The design shear capacity at the support is determined from the ultimate moment capacity of the section presented in Table 3.

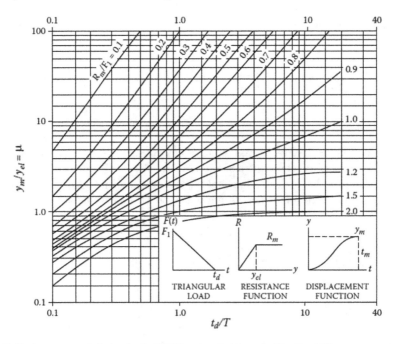

Fig. 5 Peak response of elasto-plastic SDOF system to triangular blast load [9]

The critical distance (d_e) for the location of diagonal shear is assumed as d and z for Type 1 and Type 2 section, respectively. The shear strength contribution for the concrete of the RC section is calculated as

$$\tau_c = \frac{0.85 \times \sqrt{f'_{dc}}\left(\sqrt{1+5\beta} - 1\right)}{6\beta} \tag{11}$$

where f'_{dc} is the characteristic (dynamic) strength of the cylinder which is related to characteristic (dynamic) strength of cube as $f'_{dc} = 0.8 f_{c,dyn}$, and β is given as

$$\beta = \max\left\{\frac{f'_{dc}}{6.89 p_t}, 1\right\} \tag{12}$$

where p_t is the percentage of tensile reinforcement.

Shear capacity of concrete reduces due to axial tension on the column, which is a likely scenario in case of confined blast (e.g., underground structures). The modified design shear capacity of concrete under axial tension is given as [10]

$$\tau_{c,design} = \tau_c\left(1 - \frac{P_u}{3.45 A_g}\right) \tag{13}$$

where A_g is the gross area of the concrete section and P_u is the tensile load. Any shear strength enhancement of concrete due to axial compression should be neglected.

To ensure the ductile shear failure of the RC section through yielding of shear reinforcement and not crushing of concrete in tension, the design strength ($\tau_{c,design}$) of concrete is limited by a maximum value ($\tau_{c,max}$) given by [8]

$$\tau_{c,max} = 0.62 \times \sqrt{f_{c,dyn}} \tag{14}$$

The column needs to be redesigned if it does not satisfy the above condition.

The shear capacity requirement excess of concrete contribution ($V_{uc} = \tau_{c,design} \times b \times d_e$) should be taken by the diagonal shear reinforcement (V_{us}).

If V_u is total shear force, then design shear force for steel can be calculated as

$$V_{us} = V_u - V_{uc} \tag{15}$$

Area of stirrups crossing the critical shear section is given by

$$A_{sv} = n_s \times (\pi/4) \times d_s^2 \tag{16}$$

where d_s is diameter of the rebar and n_s is number of stirrup legs crossing the section.

The spacing between stirrup is calculated using the expression:

$$s_v = \frac{f_{ds} \times A_{sv} \times d_e}{V_{us}} (sin(\theta_s) + cos(\theta_s)) \tag{17}$$

where f_{ds} is the dynamic strength of stirrups, and θ_s is the angle at which stirrup is provided. As a minimum, the detailing of shear stirrups must satisfy the criteria given in clause 8 of IS 13920 [11].

3.3 Design for Direct Shear

The columns are vulnerable to direct shear failure near supports for near-field detonations. The direct shear failure happens instantaneously before any flexural response and mobilization of flexural and diagonal shear capacity. Diagonal bars are provided at the supports to prevent direct shear failure. The deformation in direct shear and the diagonal reinforcement are shown in Figs. 6 and 7, respectively.

For limited damage in type 1 section, concrete is expected to participate in direct shear resistance. However. for type 2 section, the concrete contribution to direct shear resistance is neglected and all the capacity needs are to be provided by the diagonal rebars.

The concrete contribution in direction shear (V_{cd}) is given as following:

For type 1 section with $\tau_{c,design} > 0$:

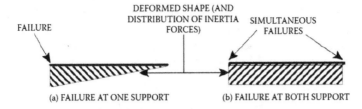

Fig. 6 Deformed shape for direct shear [9]

Fig. 7 Detailing [1]

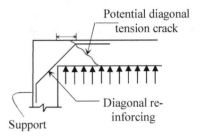

$$V_{cd} = 0.1 \times f'_{dc} \times b \times d_e \qquad (18)$$

$V_{cd} = 0$ for all other cases.

The design direct shear force is calculated as $V_s = 1.5 \times R_s$, where R_s is the support reaction at the column. The required area of diagonal bars is calculated as

$$A_d = \frac{V_s - V_{cd}}{f_{ds} \times \sin \theta_d} \qquad (19)$$

where f_{ds} is the dynamic strength of diagonal reinforcement, and θ_d is the angle between diagonal bars and longitudinal reinforcement.

Number of bars to be provided in one layer can be calculated as

$$n_{rd} = \frac{4 \times A_d}{n_d \times \pi \times d_d^2} \qquad (20)$$

where n_d is the number of layers to be provided and d_d is the diameter of diagonal bars.

3.4 Design for Joint Opening

The connections of columns to beams or slabs can be subjected to joint opening moment in case of a confined blast. The connection must be appropriately detailed to prevent premature failure and ensure ductile response in flexure. A free body diagram (FBD) of forces at the joint connection and formation of cracks are shown in Fig. 8. Radial hoops (a_{sj}) are provided at the joints to prevent connection failure as shown in Fig. 9. Area of a single leg of the radial hoops can be calculated as

$$a_{sj} = \frac{f_{ds}}{f_{yi}} \times \frac{A_{s1}}{n_1 \times n_2} \times \sqrt{1 + \left(\frac{h_1}{h_2}\right)^2} \qquad (21)$$

where f_{yi} is the yield strength of radial rebars in hoops, n_1 is the number of radial hoops, n_2 is the number of hoops per unit width (into the plane), h_1 is the width of colbeam/slab, h_2 is the depth of the column, and A_{s1} is area of longitudinal steel of beam/slab connected with column. The diameter of radial bar is calculated from a_{sj}.

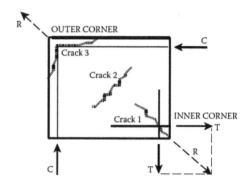

Fig. 8 FBD for joint opening [9]

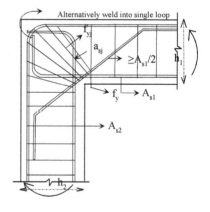

Fig. 9 Connection detailing [9]

3.5 Design Flowchart

A flowchart summary of the blast-resistant design methodology is presented in Fig. 10.

4 Performance Verification

The efficacy of the proposed blast-resistant design methodology is verified using FEA in LS-DYNA. The performance of a column designed using the blast-resistant

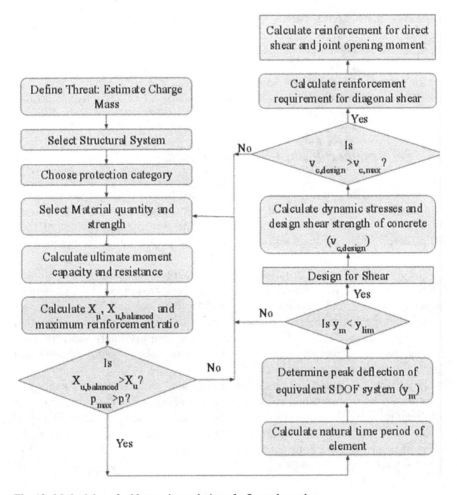

Fig. 10 Methodology for blast-resistant design of a flexural member

design provisions is compared with an ordinary column designed as per IS 456-2000 [12]. The columns were designed for the same moment (300 kN-m) and axial force (1500 kN). Both columns have same geometry but different longitudinal and shear reinforcements, and the blast-resistant column is provided additional diagonal bars to resist direct shear failure. No ductile design provisions were followed for the ordinary column. A design and detailing summary of both columns is presented in Table 4.

Finite element models of both columns are created in LS-DYNA. The CSCM material model was for concrete and kinematic plastic material for steel reinforcements. A uniform mesh size of 10 mm was used for solid elements for concrete and 10 mm beam elements for steel reinforcements. Lagrangian constrain technique was used to simulate the bonding between reinforcements and concrete. The column nodes are restrained in all directions at both support for fixed boundary conditions.

A TNT hemispherical burst at the mid-height of the column and at a standoff distance of 5 m from the column is considered for this study. The charge mass is increased till the failure is obtained for each column. The blast load is applied using LOAD_BLAST_ENHANCED option in LSDYNA.

The failure of the column is defined as the state at which the residual capacity in any of the moment, axial, or shear becomes zero under the action of applied loads.

4.1 Performance of Blast-Resistant Design

The axial capacities of the columns were obtained by applying monotonic axial displacement (no moment) till failure. Similarly, the flexural capacity was obtained by applying monotonic lateral load till failure under a constant axial force of 1500 kN. The FE model and the final failure patterns of both columns are presented in Figs. 11 and 12, respectively.

The plots of the flexural response at the mid-height of both columns subject blast loads at design and the failure charge masses are presented in Fig. 13. Comparable values of moment and ductility are observed for both columns at charge masses smaller than the failure charge mass due to same initial stiffness and extreme duration of loading. The arrival time for the peak moments are extremely short for blast loading. At failure charge, the blast-resistant column shows higher strength and ductility than the ordinary column.

The theoretical design parameters and the summary of ultimate capacities and blast response are presented in Tables 5 and 6, respectively. The blast-resistant rebar detailing leads to enhancement in strength and dutility of the column and increase in the charge mass at failure. Further studies are required to better characterize the effect of boundary conditions on the moment capacities of the columns subject to blast loads.

Table 4 Design and detailing summary of the columns

	Longitudinal rebars	Shear rebars		
		Stirrup		Diagonal rebars at supports
		Interior	Support	
Ordinary Column	16 × 20 mm bars distributed uniformly on all faces	4-legged 8 mm stirrups at 300 mm c/c	4-legged 8 mm stirrups at 300 mm c/c	Not provided
Blast-resistant Column	12 × 25 mm bars distributed uniformly on all the faces	6-legged 8 mm stirrups provided at 300 mm c/c	6-legged 8 mm stirrups at 100 mm c/c till 600 mm from support	4 × 12 mm diagonal bars at the spacing of 100 mm c/c

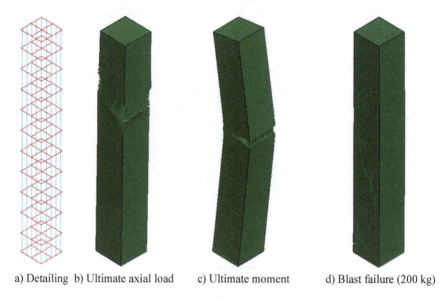

Fig. 11 Performance of ordinary column under quasi-static and blast loads

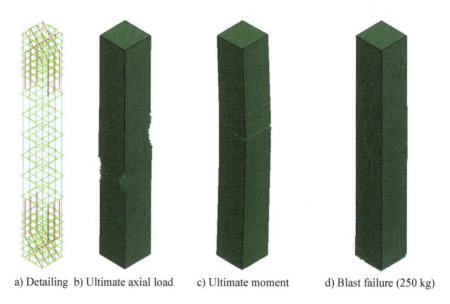

Fig. 12 Performance of blast-resistant column under quasi-static and blast loads

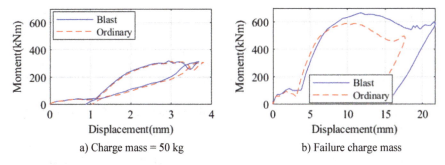

Fig. 13 Flexural response at the mid-height of the columns

Table 5 The theoretical design parameters of the columns

Design type	Axial load capacity (kN)[1]	Blast	
		Design charge (kg)	Allowable deflection (mm)
Ordinary	8437	50	58
Blast	8844	50	58

1. Obtained assuming $P_u = f_{ck}(A_g - A_{st}) + f_y A_{st}$

Table 6 The capacity and blast response of columns obtained using LS-DYNA

Design type	Peak static response		Peak blast response			
	Axial load capacity (kN)	Moment capacity (P = 1500kN)	Failure charge (kg)	Axial force (kN)	Moment (kN-m)	Deflection (mm)
Ordinary	8953	652	200	2509	766	16
Blast	9535	881	250	2711	839	21

5 Summary and Conclusions

A unified methodology for blast-resistant design of columns is presented that addresses different design aspects (flexure, diagonal shear, direct shear, joint connection). The efficacy of this methodology is verified using FEA in LS-DYNA where performances of compliant and non-compliant blast-resistant columns were compared. The key conclusions of this study are

1. The performance goals of RC columns can be linked to the flexure ductility.
2. RC columns need to be designed as flexure member under the action of large lateral loads due to blast.
3. Columns are vulnerable to direct shear failure in case of near-field detonations and additional diagonal reinforcements must be provided to avoid the failure.

4. The connection of columns to beams or slabs must be appropriately detailed with radial hoops to prevent premature failure and allow mobilization of the full capacity of the adjoining RC column and beams.
5. The blast-resistant design provisions enhance the axial and moment capacities by 6.5% and 35%, respectively, while the failure charge mass is increased by 25%.
6. A significant increase in flexural ductility is obtained due to blast-resistant detailing.
7. FEA results are useful in assessing the relative ultimate capacities but the accuracy of the ultimate performance can only be ascertained after experimental validation.

Acknowledgements The financial support for this research work was provided through an early career award by the Science and Engineering Research Board of Department of Science and Technology. The authors gratefully acknowledge the support.

References

1. US Department of Defense (DoD) (2008) Structures to resist the effects of accidental explosions. Report No UFC-3-340-02, Washington, DC
2. Anjani K, Kumar M (2019) Limitations of simplified analysis procedures used for calculation of blast response of structures. In: 7th International congress on computational mechanics and simulation (ICCMS 2019), IIT Mandi, India
3. American Society of Civil Engineers (ASCE) (2011) Blast protection of buildings. In: ASCE/SEI 59-11, Reston, VA
4. Bureau of Indian Standards (BIS) (1968) Criteria for blast-resistant design of structures for explosions above ground (Reaffirmed 2008). In: Standard IS 4991, New Delhi, India
5. Cormie D, Mays G, Smith P (2009) Blast effects on buildings. Thomas Telford, ICE Publishing, London, UK
6. Dua A, Braimah A, Kumar M (2019) Contact explosion response of RC columns: experimental and numerical investigation. Proc Instit Civil Eng Struct Build. https://doi.org/10.1680/jstbu.18.00223
7. Protective Design Centre (PDC) (2008) Methodology manual for the single-degree-of-freedom blast effects design spreadsheets (SBEDS). In: PDC TR-06-01, Rev 1, US Army Corps of Engineers
8. Biggs JM (1964) Introduction to structural dynamics. McGraw Hill, New York, NY
9. Krauthammer T (2008) Modern protective structures. CRC Press
10. American Concrete Institute (ACI) (2011) Building code requirements for structural concrete and commentary. In: ACI 318-11, Farmington Hills, MI
11. Bureau of Indian Standards (BIS) (2016) Ductile design and detailing of reinforced concrete structures subjected to seismic forces -code of practice. In: Standard IS 13920, New Delhi, India
12. Bureau of Indian Standards (BIS) (2000) Plain and reinforced concrete -code of practice. In: Standard IS 456, Rev 4, New Delhi, India

Numerical Damage Modelling of RC Slabs Under Blast Loading Using K&C Concrete Model

K. Akshaya Gomathi and A. Rajagopal

Abstract An explosion near a building can cause catastrophic damage to the building's external and internal structural frames, causing the collapse of the walls and even loss of life. Due to the threat from such extreme loading conditions, efforts have been made during the past three decades to study the behaviour of structural concrete subjected to blast loads and to develop methods of structural analysis to resist blast loads. This paper describes the Finite Element Analysis of RC slabs under blast loading using the predictive engineering software LS-DYNA. The prediction of damage characteristics and mechanisms of reinforced concrete slab exposed to blast loading is done by taking the experimental data from the literature. The reliability of the numerical analysis was modelled using K&C concrete model and are validated by comparing with the experimental results. Further, the parametric study is carried out with K&C concrete model by varying the thickness of the slab, scaled distance and concrete strength. It has been seen that the slab thickness and scaled distance play a major role in blast loading. The maximum deflection has been decreased by increasing the slab thickness and by increasing the scaled distance. But in case of increasing the strength of concrete, the slab shows a very small reduction in maximum deflection. Thus the variation of concrete strength is not an efficient way in case of blast loading. Further, the blast pressure interaction with the reinforced concrete (RC) slabs is investigated by understanding the multi-material Arbitrary Lagrangian Eulerian (ALE) formulation with the ConWep formulation. The above parametric study was carried out with ALE. The advantage and limitation of K&C concrete model are studied in case of concrete structural components subjected to blast loading.

Keywords Blast loading · ConWep · ALE · K&C concrete model

K. A. Gomathi (✉) · A. Rajagopal
Indian Institute of Technology Hyderabad, Hyderabad, India
e-mail: ce19mtech01001@iith.ac.in

1 Introduction

Concrete consist of cement, fine aggregate, coarse aggregate and water. It is a non-homogeneous composite material. The behaviour of concrete under dynamic loading depends on the strength, mix ratio, shape of aggregate and many other factors. The behaviour of concrete under dynamic loading is so different when compared to the behaviour of concrete under static loading.

Blast loads and its effects are considered as one of the severe load or dynamic condition acting on reinforced concrete (RC) structures. A small amount of blast can cause extreme damage because of the sudden release of energy in the event of a blast. Explosions can be categorized as physical, nuclear or chemical events. There exists a need for understanding the behaviour of structures and also to vary the design of the structures in order to improve the strength of structures under extreme loading condition. The understanding of the damage behaviour of components or elements of RC structures under blast loading is essential to understand the damage evaluation in such elements.

There are different methods available to understand the response of structural elements of the building subjected to blast loading. Experimental based testing has been commonly used as a means to analyze particular components. Though experimental evaluation produces real-time response, it consumes time and is expensive. Analytical and numerical methods of finding the consequences of blast loading acting on the structure are more preferable. The actual structure is replaced by an equivalent system of one concentrated mass and one weightless spring which represent the resistance of the structural components against deformation as a single-degree-of-freedom (SDOF) system. The use of finite element analysis (FEA) to analyze these structural components is more suitable since analysis is much faster, and it is an extremely cost-effective analysis. The focus of present work is on numerical analysis of RC slabs subjected to blast loading. Various parametric studies are performed to understand blast response of RC slabs.

2 Analysis Model

2.1 Slab Detailing

The dimension of the concrete slab taken for the experiment was 850 × 750 × 30 mm and 6 mm diameter reinforcement is used in both directions. Three types of reinforcement spacing are considered in both the plane and the results are analyzed. The centre to centre spacings of the reinforcement are taken as 700 mm, 75 mm and 50 mm, respectively. The details of the slab are taken from the experimental work carried out by Yao et al. [1]. The material properties of concrete and steel and details of explosive are given in Tables 1, 2 and 3, respectively.

Table 1 Property of concrete

S. No	Parameter	Value
1	Uniaxial compressive	39.5 MPa
2	Tensile strength	4.2 MPa
3	Young's modulus	28.3 GPa

Table 2 Property of reinforcement

S. No	Parameter	Value
1	Young's modulus	200 GPa
2	Yield strength	395 MPa
3	Ultimate strength	501 MPa

Table 3 Property of explosive

S. No	Parameter	Value
1	Type of explosive	TNT
2	Mass of TNT	0.13 and 0.19 kg
3	Standoff distance	300 mm

The specimen was placed firmly on the steel frame fixed at both end in longitudinal direction. TNT explosives of 0.13 and 0.19 kg were used for the test in order to find the effect of different mass of explosives in terms of damage. In the present study, the mass of 0.13 and 0.19 kg are taken in order to validate the numerical results obtained from LS-DYNA with the experimental results taken from the paper. As shown in Fig. 1, the charge is suspended above the slab at the standoff distance of 300 mm.

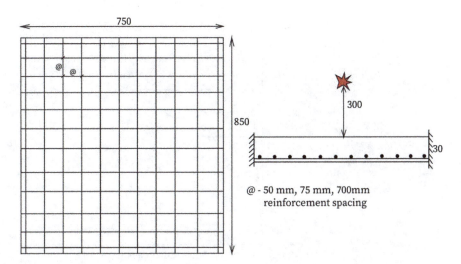

Fig. 1 Detailing of slab [1]

3 Numerical Modelling

Non-linear finite element analysis is carried out using LS-DYNA to understand the dynamic behaviour of the RC slabs which are subjected to extreme loading associated with the detonation of the explosive mass. Under dynamic loading, both tensile and compressive strengths of concrete and steel are increased significantly [2], which are shown in Fig. 2a and b, respectively. Thus, the design allows nominal component strengths to be increased by dynamic increase factors (DIF) to account for rate effects. DIF is the ratio of the dynamic-to-static strength versus strain rate, which is used to study the effect of strain rate in RC structures subjected to the blast loads, the

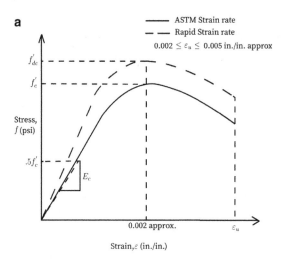

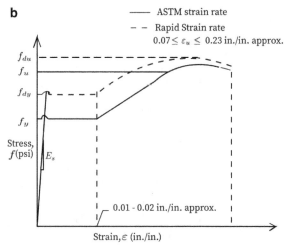

Fig. 2 **a** Stress–strain curve for concrete. **b** Stress–strain curve for steel

strength of the concrete goes on increasing. The effect of strain rate is considered, to investigate the RC structural components under blast loading.

3.1 Blast Load Model

Generally, the blast wave consists of two phases, i.e. positive and negative phase. When an explosion occurs, a sudden release of accumulated energy produces peak overpressure. When the blast wave encounters a structure, there will be sudden decrease of the shock wave and the reflected pressure gets to increase. The reflected and incident pressure wave have a similar form but the reflected wave has a different decay rate. By using the ConWep method, we can generate the appropriate value of pressure history based on the type of blast, location of blast and equivalent mass of TNT. The load segment is defined in LS-DYNA. Thus, the blast load is applied to the structural surface by means of pressure surface identification.

There are two classical algorithms that are used for modelling with finite meshes such as Eulerian and Lagrangian model. In case of Lagrangian model, the meshes of the element moves along with material points. In case of Eulerian model, the meshes of the element are fixed in space, and the material move in pre-planned meshes.

3.1.1 Pure Lagrangian (ConWep) Algorithm

The modelling of components in pure Lagrangian formulation is the simplest way of numerical simulation. The blast pressure time histories are calculated using the ConWep code in LS-DYNA, thus the empirical function of blast loading is formed from the vast amount of experimental data [3]. When a shock wave strikes a target surface at an oblique angle of incidence the effective pressure is given by

$$P_{eff} = P_{ref} \cos^2 \theta + P_{inc}(1 + \cos^2 \theta - 2\cos \theta) \tag{1}$$

where P_{ref} is the reflected pressure, P_{inc} is the incident pressure, θ is the angle of incidence, and it is defined by the tangent to the wave front and the target's surface. Assuming that the decay of explosive occurs in the exponential manner, the pressure history for the free-air explosive is given as

$$P(t) = P_{\max}\left(1 - \frac{t}{t_+}\right) exp\left(-\frac{bt}{t}\right) \tag{2}$$

where $P_{\max}$ is the maximum incident over pressure, t_+ is the positive time duration, b is the decay form factor of the curve. The termination time is given to achieve the complete blast response of the structure. The main disadvantage is that using ConWep method is unable to find the interaction of blast wave with the components of the structure. Another limitation is that in case of spherical air blast, this method is valid over the range of scaled distance $0.147 \text{ m/kg}^{1/3} < Z < 40 \text{ m/kg}^{1/3}$.

3.1.2 Arbitrary Lagrangian Eulerian (ALE) Algorithm

Air is modelled using hydrodynamic material model with 8 noded brick elements. The modelling requires [3] Jones–Wilkins–Lee (JWL) equation of state is given for detonation of gaseous products. In addition to equation of state pressure cut-off, viscosity coefficient and density are to be given as input variables. The structural damage in RC structures is caused mainly due to positive phase of the blast wave and the negative phase does not contribute much under small scaled distance [4]. Thus the pressure cut-off is set to zero. The internal energy of the linear polynomial is linear. The pressure is defined as [4]

$$P = C_0 + C_1\mu + C_2\mu^2 + C_3\mu^3 + (C_4 + C_5\mu + C_6\mu^2)E_0 \qquad (3)$$

The ideal gas law, in linear polynomial equation, the coefficients ($C_0 = C_1 = C_2 = C_3 = C_6 = 0$, $C_4 = C_5 = \gamma - 1$). For perfect gas, pressure is defined as

$$P = (\gamma - 1)\frac{\rho}{\rho_0}E \qquad (4)$$

where $\mu = \frac{\rho}{\rho_0} - 1$, ρ_0 and ρ are the initial and current densities of air, γ is the rate of change to the specific heat of air and E is the specific internal energy (units of pressure). For the initial internal energy, under standard atmospheric pressure $P_0 = 1$ bar, according to the Gamma law calculation at $\gamma = 1.4$, its internal energy is $E_0 = 2.5$ Bar.

JWL equation of state are used for modelling high explosives and can be expressed in the form,

$$P = A\left(1 - \frac{\omega}{R_1 V}\right)e^{(-R_1 V)} + B\left(1 - \frac{\omega}{R_2 V}\right)e^{(-R_2 V)} + \frac{\omega}{V}E \qquad (5)$$

where A, B, R_1, R_2, ω are constants, P is the pressure and V is the relative volume. The parameters for trinitrotoluene (TNT) are given by Dobratz [5] and are listed in Table 4.

4 Karagozian & Case (K&C) Concrete Model (MAT72R3)

The tabulated compaction is used as the EOS for this material model. The tabulated compaction is linear in internal energy. In loading phase, pressure is defined as

Table 4 Material parameters of TNT

S. No	Parameter	Value
1	A (GPa)	371.20
2	B (GPa)	3.231
3	R_1	4.15
4	R_2	0.95
5	ω	0.3
6	E (GPa)	7.0
7	D (m/s)	6930
8	$\rho\ kg/m^3$	1590

$$P = C(\varepsilon_v) + \gamma T(\varepsilon_v)E \qquad (6)$$

where ε_v is the volumetric strain given by the natural logarithm of the relative volume V. Unloading occurs along the unloading bulk modulus to the pressure cut-off and the reloading occurs along the unloading path to the point where unloading began, and it is continuous along the loading path as shown in Fig. 3.

This model uses three shear failure surfaces; they are initial yield surface, maximum failure surface and residual surface, which is expressed as

$$F_i(p) = a_0 + \frac{P}{a_{1i} + a_{2i}.P} \qquad (7)$$

where F_i is the i_{th} failure surface, a_{0i}, a_{1i} and a_{2i} are the parameters of the three-parameter of the failure surfaces.

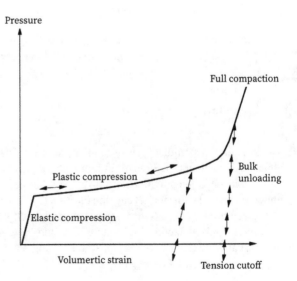

Fig. 3 Pressure versus volumetric strain curve [3]

This models behaviour is such that it is elastic up to the initial yield surface and then increases in an elastoplastic manner up to the maximum yield surface. Then softening occurs until residual surface is reached. Set of three different parameters are used to define each of the three surfaces. The maximum failure surface for concrete is given as

$$\frac{\Delta\sigma}{f'_c} = a_0 + \frac{p/f'_c}{a_1 + a_2 p/f'_c} \tag{8}$$

Thus the expression for DIF of compressive strength of concrete is given by Malver and Ross as

$$\text{CDIF} = \frac{f'_{cd}}{f'_{cs}} = \begin{cases} \left(\frac{\dot{\varepsilon}_d}{\dot{\varepsilon}_{ds}}\right)^{1.026\alpha}, & \dot{\varepsilon}_d \leq 30s^{-1} \\ \gamma(\dot{\varepsilon}_d)^{\frac{1}{3}}, & \dot{\varepsilon}_d > 30s^{-1} \end{cases} \tag{9}$$

where f'_{cd} is the dynamic compressive strength, f'_{cs} is the static compressive strength $\dot{\varepsilon}_d$ = strain rate range of 30 ×10^{-6} to 300 s^{-1}, $\dot{\varepsilon}_{ds}$ = Static strain rate which has a value of 30 ×10^{-6} s^{-1}, γ can be calculated as log $\gamma = 6.156\alpha - 0.49$ where $\alpha = 1/(5 + 3\ f_{cu}/4)$ where f_{cu} = Static Concrete Compressive strength in MPa.

The DIF for tensile strength of concrete is given as

$$\text{TDIF} = \frac{f'_{td}}{f'_{ts}} = \begin{cases} \left(\frac{\dot{\varepsilon}}{\varepsilon_s}\right)^{\delta} \dot{\varepsilon} < 1s^{-1} \\ \beta\left(\frac{\dot{\varepsilon}}{\varepsilon_s}\right)^{\frac{1}{3}} \dot{\varepsilon} > 1s^{-1} \end{cases} \tag{10}$$

where f'_{td} is the Dynamic tensile strength, f'_{ts} is the Static tensile strength, $\dot{\varepsilon}$ is the strain rate, and has a range of 10^{-6}s^{-1}, to 160 s^{-1}, ε_s is the Static strain rate, and has a value of 10^{-6}s^{-1} and β can be calculated using log $\beta = (6\delta - 2), \delta = 1/(1 + 8f_c/f_{c0})$ where $f_{c0} = 10 MPa$, f_c = Static Compressive strength of concrete in MPa.

5 Steel Material Model

The Lagrangian formulation is used in this analysis. In this, the nodes of the mesh are attached to the material. Steel reinforcement is represented as the beam element in the model. Plastic kinematic model is chosen in order to find the response of the element under loading, and this model can find the effects of isotropic and kinematic hardening of the beam elements. Reinforcement is modelled as a separate element since the load understudy is having high rate effect, and bond-slip can be neglected both in the blast and impact analysis. Using the Lagrangian solid formulation, concrete and beam element are tied together by coupling of these elements.

The steel's stress–strain property is particularly sensitive to blast loading. Strain effect has a major effect as that of inertia of the material. Therefore care has to be taken about the stress–strain behaviour of the material. Malver and Crawford [6] has proposed DIF Equation for steel under ASTM standards.

$$\text{DIF} = \frac{(\dot{\varepsilon})^\alpha}{10^{-4}}$$

where $\alpha = 0.074 - 0.040 \frac{f_y}{414}$, f_y = Yeild stress of steel in MPa and $\alpha = 0.019 - 0.009 \frac{f_y}{414}$, f_y = Ultimate stress of steel in MPa.

6 Validation Using Numerical Simulation

The RC slab is modelled and the simulation is carried out in LS-DYNA. It is seen that ALE based modelling is able to give the maximum central deflection value approximately equal to experimental result as shown in Fig. 4a and b. Therefore, further parametric study is carried out by varying the scaled distance, thickness of the slab and concrete compressive strength.

7 Parametric Model

Parametric study is carried out by considering the effect of scaled distance, thickness of the slab and concrete compressive strength using ConWep and ALE method. The numerical calculation is done for 50 mm reinforcement spacing and is given in Fig. 5. It is seen that as the scaled distance and thickness of the slab increases the maximum central deflection is reduced to a greater extend. But increasing the concrete strength does not influence the blast loading effect on the structure shown in Fig. 4. The interaction of blast load on the structure is understood by modelling using ALE formulation. It is shown in Fig. 6.

8 Results and Discussion

Blast wave interaction on the RC slab will cause larger node displacement in short duration, which leads to distortion of mesh elements. This might cause premature stopping of the calculation, when negative element volume is obtained. Thus, high strain problem with larger node displacement cannot predict the accurate result when modelled with Lagrangian algorithm. Thus, the ALE algorithm considering FSI effect is used for modelling explosive and the fluid domain. The residual deflection is shown

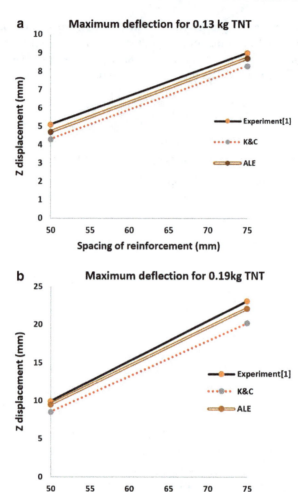

Fig. 4 a Validation of maximum central deflection by varying the reinforcement spacing for 0.13 kg TNT. **b** Validation of maximum central deflection by varying the reinforcement spacing for 0.19 kg TNT

in Fig. 7. It is seen that the ALE was able to predict the maximum central deflection similar to the experiments.

Parametric study shows that the thickness of the slab and the scaled distance play a major role in reducing the maximum central deflection of the slab. But the parametric study carried out by varying the concrete strength does not show much reduction of maximum central deflection. It shows that the concrete strength does not have much influence in case of blast loading.

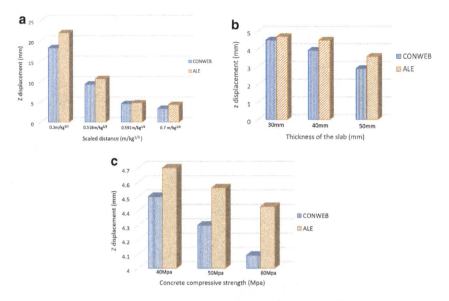

Fig. 5 **a** Influence of maximum central deflection on scaled distance; **b** Influence of maximum central deflection on thickness of the slab; **c** Influence of maximum central deflection on concrete compressive strength

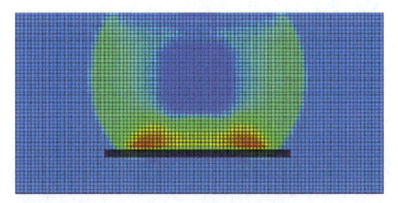

Fig. 6 Blast wave propagation in ALE

The damage characteristic of the slab for three different reinforcement spacings of 50, 75, 700 mm is shown in Fig. 8. As the spacing of reinforcement increases the damage cracks are increasing. For 700 mm reinforcement spacing, the complete spallation of concrete occurs at the centre.

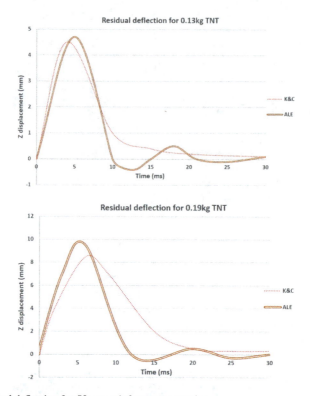

Fig. 7 Residual deflection for 50 mm reinforcement spacing

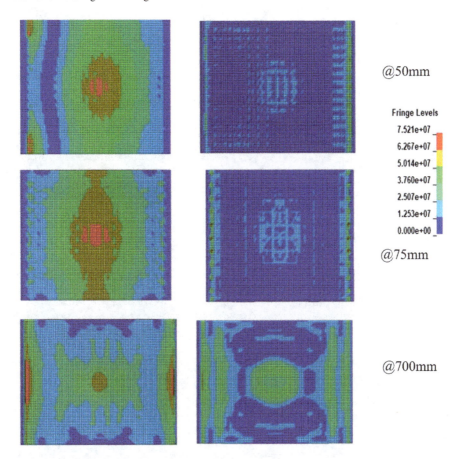

Fig. 8 Von Mises Stress fringe components for reinforcement spacing @50 mm, @75 mm and @700 mm

References

1. Yao S, Zhang D, Chen X, Lu F, Wang W (2016) Experimental and numerical study on the dynamic response of RC slabs under blast loading. Eng Fail Anal 66:120–129
2. UFC 3-340-02 (2008) Structures to resist the effects of accidental explosions. Structures congress 2011
3. Hyde DW (1998) User's guide for microcomputer program CONWEP and FUNPRO, applications of TM 5-855-1. In: Fundamentals of protective design for conventional weapons. US Army Corps of Engineers
4. Livermore Software Technology Corporation (LSTC) (2014) LS DYNA keyword user's manual, vol. II
5. Dobratz BM, Crawford PC (1985) LLNL explosives handbook properties of chemical explosives and explosive simulants. Lawrence Livermore National Laboratory
6. Malver LJ, Crawford JE (1998) Dynamic increase factors for concrete. In: 28th DDESB seminar, pp 1–17, Orlando

Numerical Modeling of Tunnel Subjected to Surface Blast Loading

Jagriti Mandal, M. D. Goel, and Ajay Kumar Agarwal

Abstract Occurrences of terrorist attacks using explosive have frequented in the last two decades. These attacks inflict damage to the structure and ultimately pose threat to the lives of civilians. Public transit systems have been a target of terrorism in the past resulting in casualties and property loss. Indian transit systems consist of underground tunnels mainly to connect locations which otherwise are not easily accessible. Attack on such structures can lead to inaccessibility to isolated places for indefinite period of time. Prevention of such incidents cannot be assured but measures can be adopted to mitigate the damage to these structures. This involves understanding the response of underground structures under various blast scenarios. Thereby, this work presents the response of underground tunnel subjected to surface explosion using multi-material arbitrary Lagrangian Eulerian (MM-ALE) method. The model has been validated by comparing the size of crater formed by explosion with the size calculated using empirical formulae based on the theory of model similarity. Herein, a parametric study has been carried out by varying the TNT charge weight and tunnel lining thickness. Maximum damage is observed in the tunnel with increase in charge weight and decrease in lining thickness. An intervening layer of foam is fitted directly over the tunnel to absorb shock energy generated from detonation. Optimum thickness of foam at which blast damage can effectively be mitigated has been determined by performing numerical simulations.

Keywords Underground tunnel · Surface explosion · Crater size · MM-ALE · JWL · Mitigation

---

J. Mandal (✉)
Department of Mining Engineering, Visvesvaraya National Institute of Technology, Nagpur 440 010, India
e-mail: jagritimandal@students.vnit.ac.in

M. D. Goel · A. K. Agarwal
Department of Applied Mechanics, Visvesvaraya National Institute of Technology, Nagpur 440 010, India

© The Editor(s) (if applicable) and The Author(s), under exclusive license to Springer Nature Singapore Pte Ltd. 2021
S. K. Saha and M. Mukherjee (eds.), *Recent Advances in Computational Mechanics and Simulations*, Lecture Notes in Civil Engineering 103,
https://doi.org/10.1007/978-981-15-8138-0_41

1 Introduction

Tunnels act as a convenient mode of transportation especially in hilly regions where the local terrain proves to be a huge hindrance to construction and maintenance of road due to extreme climatic conditions. Studies have been done on tunnels to understand their behavior under various loading conditions but not many studies have been conducted taking into consideration extreme loading such as blast. In recent past, terrorist attacks using explosives have frequented and in a few of the incidents transportation mediums have been the target of such heinous acts. Blast loading, both internal and external, can cause severe damage to the structural integrity of tunnel and ultimately lead to fatal injuries to civilians. Most of the tunnels are constructed to connect remote areas which otherwise are inaccessible. Any damage to tunnel can cut off the accessibility resulting in isolation of the region.

Internal explosion can cause a lot of damage to tunnel but it can be prevented by enforcing stringent control and security along with active blast mitigation measures. However, such preventive measures cannot protect the tunnel from surface explosions. In case of surface blast, effective mitigation measures should be adopted to minimize the damage to structure which requires an understanding response of tunnel under such loading.

Hence, in this paper, a numerical study has been conducted using finite element (FE) method to determine the response of tunnel in terms of displacement subjected to surface explosion. A parametric analysis has been conducted to study the influence of explosive weight and thickness of tunnel lining on blast damage. Finally, a layer of closed cell foam is fitted over the tunnel in between the soil and the lining. Two types of foams, closed cell aluminum foam (ACCF) and polymeric syntactic foam, have been considered in the study. Numerical simulations are carried out with varying thickness of foam to arrive at the optimum results that will effectively mitigate the blast damage.

2 Geometry and Finite Element Model

The geometry of tunnel considered in this study was 8×10 m with a soil cover of 3.75 m. One-fourth of the whole field was modeled based on Alekseenko test [1] to save computational time as shown in Fig. 1. Considering the most critical scenario, the explosive was centered above the tunnel. Convergence test was carried out with three mesh sizes of 250, 200, and 150 mm to ensure the accuracy of the FE model. Mesh size of 150 mm produced sufficiently accurate results and hence was chosen.

Non-reflecting boundary conditions were applied on infinite boundaries to avoid reflection of blast waves from the sides. Further, translation of the nodes on other two lateral sides perpendicular to the symmetrical planes was restricted. The explosive is modeled using INTERIOR_VOLUME_FRACTION_GEOMETRY option provided by LS-DYNA® in spherical shape.

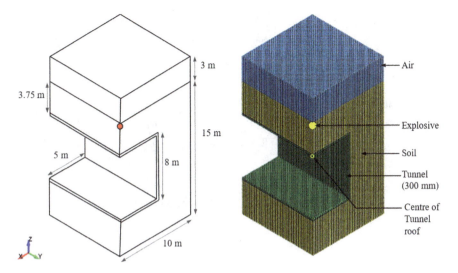

Fig. 1 One-fourth geometry and FE model of tunnel

3 Materials Model

3.1 Air

MAT_NULL material model was used to model the air domain while linear polynomial equation of state was used to define the pressure in terms of volume and internal energy which is expressed as

$$P = C_0 + C_1\mu + C_2\mu^2 + C_3\mu^3 + (C_4 + C_5\mu + C_6\mu^2)E \qquad (1)$$

The constants C_0, C_1, C_2, C_3, and C_6 were set to zero, and C_4 and C_0 were equated to $(\gamma-1)$ which modified Eq. (1) to ideal gas equation, where

$$\gamma = \frac{C_p}{C_v}. \qquad (2)$$

Equation (1) then becomes

$$P = (\gamma - 1)\frac{\rho}{\rho_0}E \qquad (3)$$

where
P = Pressure
γ = ratio of specific heat
ρ_0 = initial density of the air

Table 1 Material properties for air

P (kg/m³)	C_0	C_1	C_2	C_3	C_4	C_5	C_6	E_0 (MPa)	ρ_0 (MPa)
1.29	0	0	0	0	0.4	0.4	0	0.25	1

Table 2 Material properties for TNT

P (kg/m³)	υ_D (m/s)	P_{cut} (MPa)	A (MPa)	B (MPa)	R_1	R_2	ω	V_0	E_0 (MPa)
1630	6930	2.1×10^4	3.738×10^5	3.747×10^3	4.15	0.9	0.35	1	6000

E = internal energy per initial volume.

The material parameters necessary to model air domain is given in Table 1 [2].

3.2 Explosive

High explosive burn material model has been used to model the explosive which utilizes parameters such as density (P), velocity of detonation (υ_D), and Chapman–Jouguet pressure (P_{cut}). Whereas Jones-Wilkins-Lee equation of state was used to define the pressure in terms of relative velocity (V) and internal energy per initial volume (E) and the expression is as follows:

$$p = A\left(1 - \frac{\omega}{R_1 V}\right)e^{-R_1 V} + B\left(1 - \frac{\omega}{R_2 V}\right)e^{-R_2 V} + \frac{\omega E}{V} \quad (4)$$

where ω, A, B, R_1, and R_2 are JWL parameters and they depend of the type of explosive. V is defined as ratio of the density of explosive to the density of the detonating materials. Table 2 enlists the material properties of TNT and its JWL parameters [2].

3.3 Soil

Sandy loam soil is found commonly in most part of the world and hence for the current numerical study it has been assumed that the tunnel is surrounded by sandy loam. MAT_SOIL_AND_FOAM material model has been used to model the soil. Material properties of soil and its tri-axial hydrostatic compression data are given in Tables 3 and 4, respectively [3]. The soil properties used in the simulation are density (ρ), shear modulus (G), bulk modulus (K_u), yield function constants a_0, a_1, and a_2, and pressure cutoff for tensile fracture (P_{cut}).

Table 3 Material properties for soil

ρ (kg/m^3)	G (MPa)	K_u (MPa)	a_0	a_1	a_2	P_{cut} (MPa)
1255	1.724	5.516	0	0	0.8702	0

Table 4 Tri-axial hydrostatic compression data for soil

True volumetric strain	0.05	0.1	0.15	0.2	0.25	0.3	0.33
Pressure (MPa)	0.02	0.05	0.07	0.12	0.2	0.34	0.5

Table 5 Material properties for tunnel lining (Reinforced concrete)

P (kg/m^3)	E (GPa)	υ	σ_y (MPa)	E_{tan} (MPa)	β	ε_f
2650	39.1	0.25	100	4000	0.5	0.8

3.4 Tunnel

Reinforced concrete is generally used to make tunnel lining. They exhibit non-linearity along with strain rate dependent behavior under dynamic loading such as blast. However, to keep the numerical simulation simple and less time consuming the strain rate dependent behavior of reinforced concrete has been neglected in the study. Both concrete and reinforcement were assumed and modeled as one whole part using principle of equivalent [2, 4, 5]. Plastic Kinematic material model was used to define the behavior of reinforced concrete which includes parameters such as density (ρ), Young's modulus (E), Poisson's ratio (υ), yield stress (σ_y), tangent modulus (E_{tan}), hardening coefficient (β), and effective plastic strain (ε_f) for eroding material (Table 5).

4 Validation of FE Model

For validation, a soil domain of 20 × 20 × 20 m and an air domain with a height of 5 m was considered. The TNT of weight 250 kg was buried in the soil in such a way that its upper side coincides with the ground surface as shown in Fig. 2. The soil and air were created using eight-node solid element of size 150 mm.

Craters: When a shallow buried explosive detonates it causes two combined effects: the material below the explosive gets crushed and compacted while the material above the explosive gets blown away. This leads to crater formation. The shape of the crater depends on the shape of the explosive while its size depends on the charge weight and the soil properties such as compressive strength, density, mineral composition, and voids [1].

The size of the crater for a certain type of soil and given charge weight can be predicted using an empirical formula which is based on the theory of model similarity:

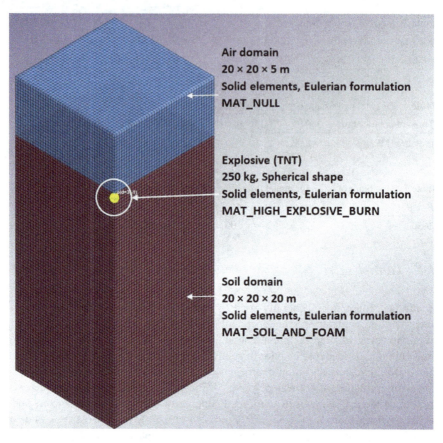

Fig. 2 One-fourth symmetrical FE model of the soil and air domains with explosive using MM-ALE method

$$R_{vd} = k_{vd}^* \sqrt[3]{W} \qquad (5)$$

where R_{vd} is the radius of the crater formed in m, k_{vd} is the coefficient of proportionality, and W is the charge weight in kg. For sandy loam, the value for k_{vd} is 0.694 (Henrych 1979). Substituting the values of k_{vd} and W with 0.694 and 250 kg, we get

$$R_{vd} = 0.694 \times \sqrt[3]{250} = 4.37 m$$

The result from the simulation showed that a crater of a radius of 4.4 m (approx.) was formed in the soil due to the explosion which can be seen in Fig. 3.

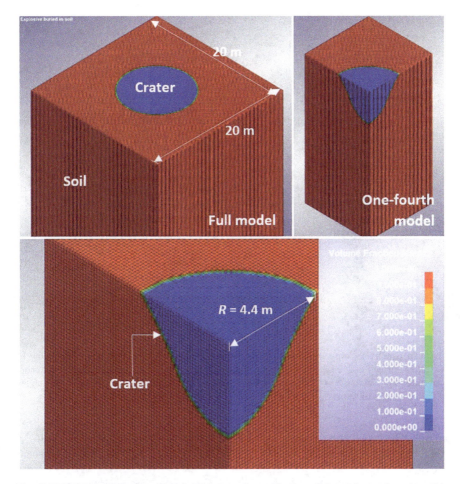

Fig. 3 Volume fraction contour showing the crater formed in the soil due to detonation where R is the radius of the crater

5 Results and Discussion

The effect of blast loading due to detonation of 250 kg TNT on tunnel of lining thickness 300 mm is shown in Fig. 4a. The figure also shows the variation of pressure on tunnel roof with respect to time. At peak A, pressure is the maximum which signifies incident pressure, whereas the second peak B signifies reflected pressure. On comparing pressure versus time curve with displacement versus time curve it can be observed that a significant correlation can be established between pressure variation and resultant deformation. As the incident pressure increases the deformation at the center of tunnel roof increases and when the pressure reaches peak A the deformation hits the first dip C. With decrease in incident pressure the tunnel roof tries to rebound back to its original position; however, at approximately 0.105 s the reflected pressure

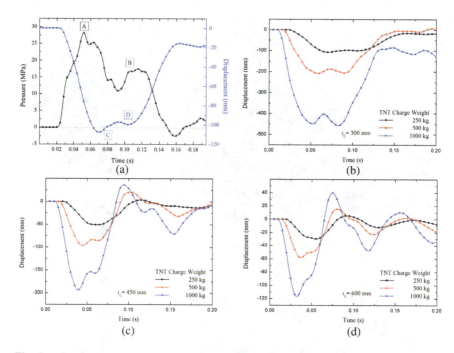

Fig. 4 a Graphs showing variation in pressure and deformation with time for tunnel lining of 300 mm subjected to 250 kg TNT explosion. Time histories of displacement at the center of tunnel roof for lining thickness of **b** 300 mm, **c** 450 mm, and **d** 600 mm for three different charge weights of 250, 500, and 1000 kg

arrives which makes the center of roof dip again resulting in the second dip D. Therefore, dip D can be attributed to the combined effect of incident and reflected pressure.

Further, the sensitivity of structural response of tunnel was investigated by conducting a parametric analysis with lining thickness and explosive charge weight as two variables as shown in Fig. 4b, c and d. Decrease in deformation can be observed as the lining thickness is increased. However, increase in weight of TNT leads to increase in deformation hence proving detrimental to the structural stability. It can further be noticed that the shape of deformation versus time curve for each tunnel lining thickness differs for given TNT weight. This is essentially due to the fact that mode of vibration of any structure depends on its natural vibration which is greatly influenced by stiffness and mass of the said structure.

6 Mitigation

The simplest technique to reduce the deformation caused due to blast loading is by increasing inertial resistance of the structure which can be done by increasing its mass. In this case, the tunnel lining thickness was increased to achieve the same. For further mitigation of blast damage two energy absorbing materials, closed cell aluminum foam (ACCF) and syntactic foam, were chosen and the resultant deformation after their application were compared to arrive at the most efficient solution. Foam of thickness 100 mm was introduced in between the soil and tunnel lining as a protective barrier against the blast energy (Fig. 5). A thin sheet of steel of 5 mm thickness was fitted over the foam to take care of the initial bending stresses and to transfer the load to foam. It was assumed to have elastic plastic nature with strain hardening. Piecewise Linear Plasticity material model was used to define its behavior which includes parameters such as density, $\rho = 7800\,\text{kg/m}^3$, Young's modulus, $E = 210\,\text{GPa}$ and Poisson's ratio, $\upsilon = 0.3$ [6]. Deshpande Fleck foam and crushable foam material models were used to model ACCF and syntactic foams, respectively. The former material model includes parameters such as density, Young's modulus, plateau stress (σ_p), and other material parameters like γ, α_2, and β. Whereas parameters required to define the material model of syntactic foam are density, Young's modulus, tension cutoff (σ_t), Poisson's ratio (υ), damping coefficient (η), and yield stress (σ_y). Their respective material parameters are given in Tables 6 and 7 [6, 7]. Stress–strain graph for syntactic foam of 460 kg/m^3 density is given in Fig. 6.

Figure 7 shows comparison of time history curves of deformation at roof center of bare tunnel ($t_l = 300$ mm) with foam fitted tunnel of two different lining thicknesses

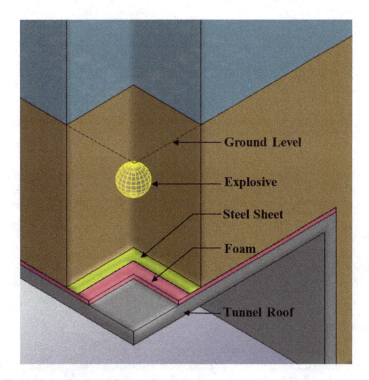

Fig. 5 Tunnel lining fitted with foam and steel sheet

Table 6 Material properties for ACCF foam

Density (kg/m³)	E (GPa)	γ (MPa)	α₂ (MPa)	β	σ_p (MPa)
270	1.1	0.0253	1.605	2.17	0.77

Table 7 Material properties for syntactic foam

Density (kg/m³)	E (GPa)	σ_t (MPa)	υ	η	σ_y (MPa)
460	1.5	8	0.3	0.3	60

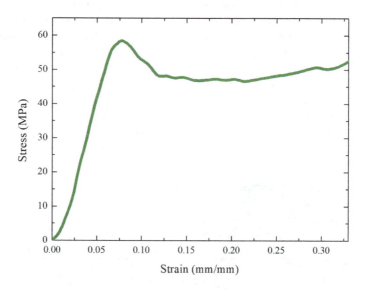

Fig. 6 Stress–strain curve of syntactic foam

($t_1 = 300$ and 450 mm). From the figure, it can be observed that performance of polymeric syntactic foam (PSF) is better than ACCF foam for $t_1 = 300$ mm. However, the difference in deformation in the case of $t_1 = 450$ mm is not significant. Table 8 shows precise maximum deformation at the center of tunnel roof after fitting ACCF and PSF foam over tunnel for two lining thicknesses (t_1) of 300 and 450 mm. It also shows percentage difference in deformation between bare tunnel ($t_1 = 300$ mm) and foam fitted tunnel ($t_1 = 300$ and 450 mm).

7 Conclusions

Herein, the behavior of shallow tunnel subjected to surface blast loading was investigated using Multi-Material Arbitrary Lagrangian Eulerian technique (MM-ALE). The response was measured in terms of deformation at the roof center of tunnel

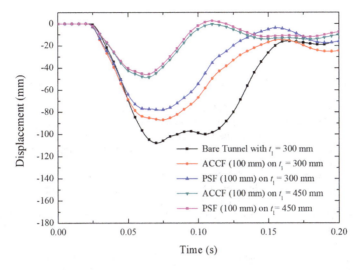

Fig. 7 Comparison of deformation time histories between bare tunnel ($t_1 = 300$ mm) and foam fitted tunnel ($t_1 = 300$ and 450 mm)

Table 8 Comparison of maximum displacements at roof center for tunnels of thickness 300 and 450 mm with two different foam types

t_1 (mm)	ACCF		PSF	
	100 mm		100 mm	
	Displacement (mm)	Percentage Difference (%)	Displacement (mm)	Percentage Difference (%)
300	87	19.25	78	27.6
450	48.6	54.9	45.9	57.4

and its sensitivity to variation in tunnel lining thickness and explosive charge weight was determined. Displacement at roof center of tunnel with thickness, $t_1 = 300$ mm, increased by 94% and 324% when the explosive charge weight was increased from 250 kg to 500 and 1000 kg, respectively. Significant reduction in displacement was noted when the tunnel lining thickness was increased. An increase of 52.85% and 72.43% was observed when the thickness was increased to 450 and 600 mm, respectively.

Deformation was further reduced when a layer of energy absorbing material was fitted over the tunnel. On comparison of performance of the two foams used in the study, polymeric syntactic foam showed better results than ACCF foam for lining thickness of 300 mm. However, slight difference was observed in the case of lining thickness of 450 mm. Thus, in conclusion the damage caused by blast loading can be mitigated by increasing the thickness of tunnel lining along with providing a layer of polymeric syntactic foam or ACCF foam over the tunnel.

References

1. Henrych J (1979) The dynamics of explosion and its use. Elsevier Scientific Publishing Company
2. Yang Y, Xie X, Wang R (2010) Numerical simulation of dynamic response of operating metro tunnel induced by ground explosion. J Rock Mech Geotech Eng 2:373–384. https://doi.org/10.3724/SP.J.1235.2010.00373
3. Kulak RF, Bojanowski C (2011) Modeling of cone penetration test using SPH and MM-ALE approaches. In: 8th european LS-DYNA users conference. pp 1–10, Strasbourg
4. Mobaraki B, Vaghefi M (2015) Numerical study of the depth and cross-sectional shape of tunnel under surface explosion. Tunn Undergr Space Technol 47:114–122. https://doi.org/10.1016/j.tust.2015.01.003
5. Mussa MH, Mutalib AA, Hamid R, Naidu SR, Radzi NAM, Abedini M (2017) Assessment of damage to an underground box tunnel by a surface explosion. Tunn Undergr Space Technol 66:64–76. https://doi.org/10.1016/j.tust.2017.04.001
6. Goel MD, Matsagar VA, Gupta AK (2014) Blast resistance of stiffened sandwich panels with closed-cell aluminum foam. Lat Am J Solids Struct 11:2497–2515. https://doi.org/10.1590/S1679782520140013000010
7. Gupta N, Pinisetty D, Shunmugasamy VC (2013) Reinforced polymer matrix syntactic foams: effect of nano and micro-scale reinforcement. Springer International Publishing

Dynamic Response of Tunnel Under Blast Loading and Its Blast Mitigation Using CFRP as Protective Barrier

V. S. Phulari and M. D. Goel

Abstract The terrorist activities throughout the world increasing from last decades that leads to loss of life and property. Tunnels are underground structures used for many purposes and their collapse will lead to complete hectic situation for any nation. Hence, it becomes greatest importance to safeguard such structures under extreme loading conditions such as resulting from explosion. Not as much of research had been reported on underground tunnel subjected to blast loading as compared to other structures. To minimize the loss of human life and property, it is very important to understand the response of underground tunnel under explosion. Herein, numerical investigation of underground tunnel is carried out using FE package ABAQUS/Explicit[®] (Dassault Systèmes Simulia Corporation. France, 2014 [1]). Complete structure is modeled using CEL (Coupled Eulerian and Lagrangian) volume fraction method as per ABAQUS/Explicit[®] (Dassault Systèmes Simulia Corporation. France, 2014 [1]). First of all, FE analysis is validated with the available experimental results and then parametric investigation is carried out. Herein, tunnel structure is investigated under varying charge weight for the better understand of dynamic response structure. Further, CFRP (Carbon Fiber Reinforced Polymer) is used as a protective barrier between blast waves and structure to mitigate the structure damage against blast energy. Based on this investigation, it is observed that tunnel structural damage is significantly reduced by employing CFRP as protective barrier.

Keywords Underground tunnel · Blast load · ABAQUS/Explicit[®] · Numerical simulation and CFRP

V. S. Phulari · M. D. Goel (✉)
Department of Applied Mechanics, Visvesvaraya National Institute of Technology, Nagpur 440 010, India
e-mail: mdgoel@apm.vnit.ac.in

© The Editor(s) (if applicable) and The Author(s), under exclusive license to Springer Nature Singapore Pte Ltd. 2021
S. K. Saha and M. Mukherjee (eds.), *Recent Advances in Computational Mechanics and Simulations*, Lecture Notes in Civil Engineering 103,
https://doi.org/10.1007/978-981-15-8138-0_42

1 Introduction

Terrorist activities affect the economic and social aspect of a nation to great extent. Public structures are mainly at the targeted end of these acts. Tunnel is one of the most important underground structures used for public transportation. However, due to increased frequency of terrorist activities, the threat to such structures has increased. Most of the research work available on underground tunnel is with general loading conditions, and less research work is accessible with regard to underground tunnels subjected to blast loading. Hence, it is important for designers and structural engineers to work together to understand the effects of such extreme loading on tunnels and thereby, bring down the extent of damage.

In this paper, the response of underground tunnel is analyzed by using FE package ABAQUS/Explicit® [1]. The obtained results from ABAQUS/Explicit® are validated with data presented by Soheyli et al. [2]. However, mere determining the response of the structure under blast loading is not sufficient, there is also a great need for research on the damage mitigation of the structure against blast loading.

In the present investigation after finding the response of the tunnel structure under blast loading, the research is continued further toward the blast mitigation of the same. The response of the front face of the tunnel structure under blast loading is tensile in nature. To reduce the extent of these tensile stresses under the required limits, herein, Carbon Fiber Reinforced Polymer (CFRP) material is used in between the tunnel and the blast waves. Carbon Fiber Reinforced Polymers (CFRP) has been gaining popularity over years as a repair and retrofitting construction material. It is used to enhance the shear and flexural strength of reinforced concrete under various kinds of loading conditions. Having the advantage of light weight and high strength CFRP is best suited for retrofitting of RC structures in the case of blast.

2 Structural Geometry and FE Modeling

The study was performed on 4 m long box-shaped tunnel, which was buried in soil at 1.5 m below the ground surface with 10 cm thick walls separated by 80 cm. The cross section of the bare tunnel and CFRP fitted tunnel is shown in Fig. 1a and b.

All the concrete components are reinforced with a steel mesh of 8 mm diameter bars at 100 mm c/c spacing. The detailed steel mesh is shown in Fig. 2a. For loading, TNT (Tri-nitro-toluene) explosive is applied at a depth of 2 m inside the soil from the ground surface and at 4 m away from the surface of tunnel lining, as shown in Fig. 2b.

The CEL volume fraction method, which represents the ratio by which each element is filled with Eulerian material, is used to find the dynamic response of reinforced concrete tunnel. ABAQUS/Explicit® is used for modeling the whole structure [1]. For 1.69 kg TNT the sphere radius is calculated to be 0.063 m. Similarly, for 3 kg and 5 kg TNT, the sphere radius is found to be 0.076 m and 0.09 m, respectively. The

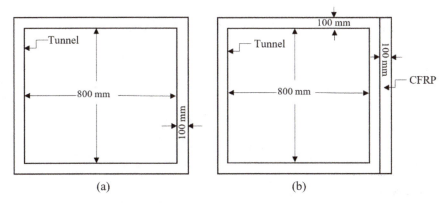

Fig. 1 **a** Cross section of the tunnel without any protective barrier, **b** Cross section of the tunnel with protective barrier

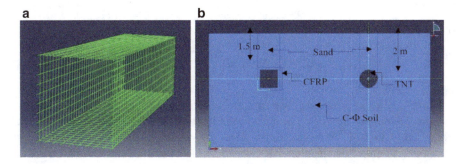

Fig. 2 **a** Mesh network inside the RC tunnel, **b** Cross section of the assembled structure inside the soil medium

explosive is modeled separately using Jones-Wilkins-Lee (JWL) equation of state with Eulerian formulation. EC3D8R elements are used for the Eulerian elements. Validation of the FE model has been carried out by comparing the numerically obtained results with the results from experimental tests [2].

3 Material Properties

In this section, the materials properties used for the simulation are discussed. The reinforcement is made up of mild steel with Young's modulus, $E = 210$ GPa, density, $\rho = 7800$ kg/m^3, Poisson's ratio, $\mu = 0.3$ and tensile strength of steel is 300 MPa [3]. For explosive material, JWL (Jones-Wilkns–Lee) equation of state (EOS) is used as reported in Eq. 1.

Table 1 JWL material properties for TNT explosive [4]

Detonation wave speed	6930 m/s
Density (ρ)	1630 kg/m^3
A	3.738×10^{11} Pa
B	3.747×10^{9} Pa
Ω	0.35
R_1	4.15
R_2	0.9
Detonation energy density	3.68×10^{6} J/kg

Table 2 Material properties for Ideal gas [4]

Density (ρ)	1.225 kg/m^3
Gas constant	287.058
Ambient pressure	101325 Pa
Specific heat	718 J/kg
Dynamic viscosity	1.7894×10^{-5} N-s/m^2

$$P = A\left(1 - \frac{\omega}{R_1 \rho}\right) e^{-R_1 \rho} + B\left(1 - \frac{\omega}{R_1 \rho}\right) e^{-R_2 \rho} + \omega \rho e_{\text{int}} \qquad (1)$$

where, A, B, R_1, R_2 and ω are the user defined material constants, and ρ is the density of the detonation products. The material properties for JWL equation of state are reported in Table 1. To model the air, ideal gas equation of state is used where its behavior is defined using parameters such as ambient pressure, density, gas constant, specific heat and dynamic viscosity (Refer Table 2).

Two different types of soil materials are used. The tunnel is placed inside the cohesive and frictional (C-Φ) soil and only frictional soil is used for filling. Both the types of soil are modeled using Mohr–Coulomb failure criterion. The material properties for C-Φ soil and frictional soil are reported in Tables 3 and 4, respectively.

Table 3 Material properties for C-Φ soil [2]

Density (ρ)	1850 kg/m^3
Cohesion	24 kPa
Friction angle	31°
Young's modulus (E)	29 MPa

Table 4 Material properties for friction soil [2]

Density (ρ)	1650 kg/m^3
Cohesion	0 kPa
Friction angle	20°
Young's modulus (E)	15 MPa

Table 5 Material properties for concrete [2]

Density (ρ)	2400 kg/m^3
Compressive strength	23 MPa
Tensile strength	2.3 MPa
Young's modulus	30 GPa

Table 6 Material properties for CFRP [5]

Density (ρ)	1800 kg/m^3
Poison's ratio	0.37
Young's modulus (E)	176.515 GPa
Ultimate tensile strength	2380.95 MPa

Concrete material is used for the tunnel lining, and its behavior is modeled using concrete damaged plasticity model available in ABAQUS® [1]. This material model incorporates both the tensile and compressive plasticity to define the inelastic nature of concrete. The yield surface is thus defined by two hardening variables, each corresponding to failure mechanism under compressive and tensile loading. The material properties for concrete are reported in Table 5.

In order to reduce the displacements of the structure, CFRP is used. The CFRP is assumed to exhibit linear elastic behavior followed by sudden rupture and its material properties are reported in Table 6. The failure mode of CFRP is defined using strain cut-off failure criterion.

4 Results and Discussion

The deformation of structure is studied carefully to understand its behavior under blast load application. The results are computed at 20 cm away from the central axe of the tunnel and 40 cm below the roof. The complete process is carried out using ABAQUS®/Explicit and the results are discussed in this section.

The results of acceleration time history, the peak acceleration from the front wall to the explosion and the distribution of displacement on the front wall caused by the explosion is obtained by using FE software ABAQUS® CEL method and these results are validated with the experimentally obtained results from Soheyli et al. [2]. The structure is analyzed for a time period of 0.12 s. According to the results obtained from Soheyli et al. [2], peak acceleration of 22.7 m/s^2 is reached at 0.0077 s. From CEL method, peak acceleration of 21.3 m/s^2 is reached at 0.006 s. The two acceleration-time histories are compared below, as shown in Fig. 3a.

The maximum displacement is 1.1 mm at the center of the front wall of tunnel as obtained in Soheyli et al. [2], whereas, from CEL method, the maximum displacement is obtained to be 1.09 mm at the center of the front wall of the tunnel. The distribution of displacement on front wall of the tunnel is compared as shown in Fig. 3b.

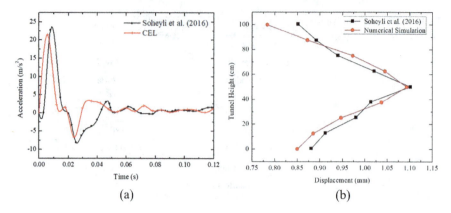

Fig. 3 **a** Acceleration-time history at the center of the tunnel wall, **b** Distribution of displacement on the front wall of the tunnel

After validation, two more explosions are applied by increasing the charge weight and the distribution of displacement on the front wall of the tunnel is compared.

Initially, 1.69 kg mass TNT is applied according to Soheyli et al. [2]. For better understanding of the effect of TNT charge weight on tunnel response the TNT weight is increased to 3 and 5 kg. The graphical comparison of distribution of the displacement by changing the charge weights is shown in Fig. 4.

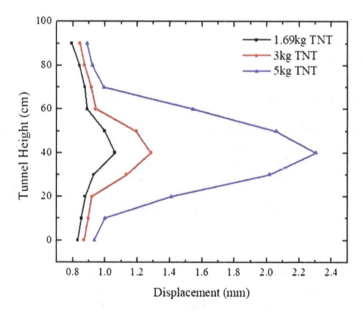

Fig. 4 Distribution of displacement on the front wall for different charge weight

After finding the response of the tunnel structure, further investigation is carried out to reduce the displacement of tunnel under same charge weights. In this investigation, for the reduction of the displacement, CFRP material is used as a protective barrier between tunnel surface and blast waves. Initially, 100 mm thick CFRP material is placed along the surface of the tunnel. General contact is applied between Eulerian and Lagrangian elements, surface to surface contact is applied between the CFRP and the tunnel elements, which is available in ABAQUS®/Explicit [1].

After the application of the CFRP as a protective barrier, the obtained displacement of the tunnel structure is 0.32 mm. Similarly, two more charge weights, i.e., 3 and 5 kg of TNT is replaced one after the other and obtained displacement of the tunnel structure without any protective barrier is 1.3 mm and 2.3 mm, respectively. After the application of the protective barrier, decreased displacements of the tunnel under 3 kg and 5 kg of TNT by using CFRP material are 0.38 mm and 0.46 mm, respectively. The graphical representation of distribution of displacements along the length of the tunnel with and without protective barrier for 1.69, 3, and 5 kg of TNT is shown in Fig. 5a, b, and c, respectively.

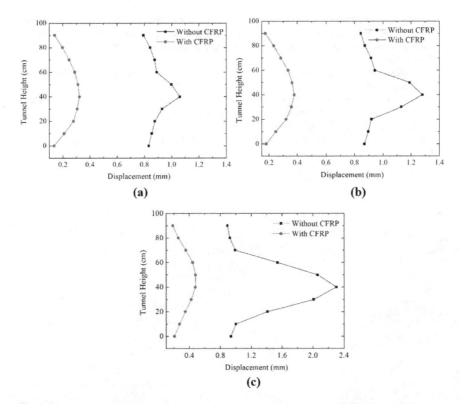

Fig. 5 Comparison of distributed displacements for **a** 1.69 kg, **b** 3 kg and **c** 5 kg TNT on front wall of bare tunnel and CFRP fitted tunnel

5 Conclusions

Dynamic analysis of tunnel was performed to understand its response to blast loading. The sensitivity of the structure was determined by varying the explosive weight in terms of displacement variation along the length of the tunnel wall. Further, to reduce the blast damage a protective material was used. Displacement of the bare tunnel structure for 1.69 kg TNT is 1.06 mm, which reduced to 0.32 mm by placement of CFRP material as protective barrier. Similarly, for 3 kg and 5 kg of TNT displacements of the bare tunnel structure are 1.3 mm and 2.3 mm, displacements of CFRP fitted tunnel are 0.38 mm and 0.49 mm respectively. It can be observed from the obtained results that 70–75% displacement can be reduced by using CFRP material as a protective barrier.

References

1. ABAQUS/Explicit® user's manual, V6.14 (2014) Dassault systèmes simulia corporation. France
2. Soheyli MR, Akhaveissy AH, Mirhosseini SM (2016) Large-scale experimental and numerical study of blast acceleration created by close-in buried explosion on underground tunnel lining. Shock Vibr
3. Goel MD, Matsagar VA, Gupta AK (2011) Dynamic response of stiffened plates under air blast. Int J Prot Struct 2(1):139–155
4. Tiwari R, Chakraborty T, Matsagar V (2016) Dynamic analysis of a twin tunnel in soil subjected to internal blast loading. Indian Geotechn J 46(4):369–380
5. Rathod B, Goel MD, Chakraborty T, Matsagar VA, Guégan P, Binetruy C (2019) Experimental and numerical investigation of high strain rate behavior of CFRP laminates. SN Appl Sci 1:736, https://doi.org/10.1007/s42452-019-0732-9

Uncertainty Quantification and Reliability Analysis

Uncertainty Quantification of Random Heterogeneous Media Using XFEM

Ashutosh Rawat and Suparno Mukhopadhyay

Abstract The macroscopic mechanical properties of two-phase heterogeneous materials, consisting of random inclusions in a solid medium, are governed by the individual material properties of the inclusions and the medium, as well as the volume fraction and spatial distribution of the inclusions. In design and analysis using such materials, the macroscopic material properties are often used. The characterization of the macroscopic properties based on the accurate micro-structure modeling may often be computationally expensive. Hence, homogenization of representative micro-structures is often used to get these macroscopic properties. In this paper too we adopt this approach for a medium with random elliptical inclusions. Monte Carlo (MC) simulations are used to obtain different micro-structure realizations, enabling the statistical modeling of macroscopic material properties. The material is modeled using the extended finite element method (XFEM), where the inclusions are modeled independent of the finite element mesh using the level set method, thereby reducing the computational cost involved in the MC simulations. The effective elastic modulus and Poisson's ratio of the heterogeneous material are obtained using Hill's averaging theorem, following which the failure stress is obtained using the computed homogenized elastic properties. The excellent synergy between XFEM and MC simulations gives the statistical characteristics of these effective material properties, including the variation in their estimates induced by the random micro-structure. It is shown that the uncertainty in the homogenized failure stress is much higher than the uncertainty in the elastic properties.

Keywords XFEM · Two-phase heterogeneous material · Monte Carlo simulation · Homogenization · Elastic properties · Failure stress

---

A. Rawat · S. Mukhopadhyay (✉)
Indian Institute of Technology Kanpur, Kanpur, India
e-mail: suparno@iitk.ac.in

© The Editor(s) (if applicable) and The Author(s), under exclusive license to Springer Nature Singapore Pte Ltd. 2021
S. K. Saha and M. Mukherjee (eds.), *Recent Advances in Computational Mechanics and Simulations*, Lecture Notes in Civil Engineering 103,
https://doi.org/10.1007/978-981-15-8138-0_43

1 Introduction

Heterogeneous materials are widely used in present day engineering applications, from composites in aerospace structures to concrete in civil engineering structures. The effective designing of a structure constructed using such materials is dependent on the properties of these materials, like elastic modulus, Poisson's ratio, and failure stress. These properties are governed by the microscopic structure and properties, like the properties of the individual material constituents, their distribution and orientation. Hence, it is important to understand the micro to macro relation of the material properties in such heterogeneous materials. However, the characterization of macroscopic properties based on accurate micro-structure modeling may be computationally expensive and often not easily tractable. Hence, the homogenization of the complex micro-structure to obtain the macroscopic properties have become popular in the recent years [1–3].

In the earlier days, the classical Finite Element Method (FEM) was commonly used for micro-structure modeling, with the FE mesh conforming to every material interface boundary in the heterogeneous medium [4]. While the FEM would give an accurate micro-structure model, it is inefficient for the homogenization process, where averaging is used over a large number of simulations of sample micro-structures. While studies have introduced fast meshing techniques [5], the repeated generation of different FE meshes, conforming to different realizations of the micro-structure, would still be computationally expensive. To overcome the problem of re-meshing, some studies introduced finite elements methods which required minimum or no re-meshing [6, 7]. The development of FEM without re-meshing led to an efficient technique called the eXtended Finite Element Method (XFEM). In XFEM a regular mesh is used to model the medium, while the discontinuities, strong (cracks), or weak (inclusions), are modeled using local enrichment functions. In homogenization, XFEM coupled with Monte Carlo (MC) simulations may be used to obtain the macroscopic properties of a heterogeneous material with randomly distributed inclusions in a medium. This is computationally very efficient, as only the enrichment function changes in each MC simulation, with no need of re-meshing.

This paper deals with the homogenization of two-phase heterogeneous material, consisting of elliptical inclusions uniformly distributed in a medium. The material is modeled using XFEM, with the inclusions modeled using level sets as enrichment functions [8]. The uncertainties are quantified using MC simulations of the random micro-structure. The homogenization is carried out for the linear elastic properties, namely the elastic modulus and Poisson's ratio, as well as for the failure stress marking the start of nonlinear behavior. The distribution of the homogenized properties obtained through the MC simulation is used to quantify the uncertainties induced by the random micro-structure.

2 Modeling Elliptical Inclusions Using XFEM

Consider a heterogeneous medium ($\Omega \subset \mathbb{R}^2$) with two phases, the base material (Ω^+) and the random inclusion (Ω^-), as shown in Fig. 1. The medium consists of outer boundary (Γ_t and Γ_u), where either traction or displacement is given and internal boundary (Γ_{inc}) where there are inclusions. The governing equations for the medium are

$$\begin{aligned} \nabla \sigma + b &= 0 \; in \; \Omega \\ u &= \bar{u} \; in \; \Gamma_u \\ \sigma.n &= t \; in \; \Gamma_t \\ \sigma.n_{inc} &= 0 \; in \; \Gamma_{inc} \end{aligned} \quad (1)$$

where σ is the Cauchy stress, b is the body force, $\bar{u}$ and t are the specified displacement and traction at the boundaries Γ_u and Γ_t, respectively, n and n_{inc} are the unit normal to the outer boundary and inclusion boundary, respectively, and $u(x)$ is the displacement field which satisfies the weak form of Eq. (1).

In XFEM, local enrichment functions $\psi(x)$, satisfying the local displacement field, are used to model the inclusions, resulting in virtual DOFs (α_i) introduced at the nodes. The displacement field under the influence of multiple inclusions, with an appropriate shift to satisfy the Kronecker-δ property, is given as

$$u(x) = \sum_{i \in I} N_i(x) \left[u_i + \sum_{j \in J} (\psi_j(x) - \psi_j(x_i)) \alpha_{ij} \right] \quad (2)$$

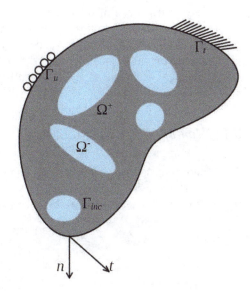

Fig. 1 Two-phase heterogeneous medium

where I is the set of element nodes and J is the set of inclusions within the element domain. With this displacement field, using the Galerkin finite element method we get

$$\begin{bmatrix} K_{uu} & K_{ua} \\ K_{au} & K_{aa} \end{bmatrix} \begin{Bmatrix} u \\ \alpha \end{Bmatrix} = \begin{Bmatrix} f_u \\ f_\alpha \end{Bmatrix} \qquad (3)$$

where K_{uu}, K_{aa} and K_{ua} are the blocks of the stiffness matrix associated with the standard FE approximation, the enriched approximation and the coupling between the standard and enriched approximations, respectively. For modeling the weak discontinuity due to the inclusions at the material interface, absolute value of signed distance function is used as the enrichment function:

$$\psi(x) = |\xi(x)| \qquad (4)$$

where $\xi(x)$ represents the equation of the ellipse with major axis α and minor axis β in the transformed $(\bar{x}, \bar{y})$ coordinate system:

$$\xi(x) = \sqrt{\frac{\bar{x}^2}{\alpha^2} + \frac{\bar{y}^2}{\beta^2} - 1} \qquad (5)$$

3 Homogenization

For the homogenization using MC simulations, a large number of sample microstructures are generated, following the approach given by Hiriyur et al. [2]. Each sample is described by the volume fraction (ϕ) and number (n_{incl}), along with the probability distributions of the center (f_c), aspect ratio (f_r), major axis (f_α) and orientation (f_θ) of the elliptical inclusions. To obtain the homogeneous material properties, each sample is analyzed under an applied displacement boundary condition. Considering all the samples in MC simulation, the distributions and statistics (mean, standard deviation) of the homogenized properties are obtained.

3.1 Homogenized E_{eff} and v_{eff}

The homogenized elastic modulus (E_{eff}) and Poisson's ratio (v_{eff}) are obtained based on Hill's averaging theorem [9], whereby the energy of the homogenized macro-structure is equal to that of the micro-structure in the average sense:

$$\bar{\sigma} : \bar{\varepsilon} = \frac{1}{|V|} \int_\Omega \sigma : \varepsilon \, dA \qquad (6)$$

where A is the coordinate system of the sample. The macroscopic stress and tangent moduli of the sample can be computed from the forces and stiffness properties at boundary nodes [10]. Using a macroscopic strain tensor $\bar{\varepsilon}$, nodal displacements are set on the surface of the micro-structure as

$$u_b = D_b^T \bar{\varepsilon} \qquad (7)$$

where D_b depends on the nodes (b) on the boundary of the sample:

$$D_b = \frac{1}{2}\begin{bmatrix} 2x & 0 \\ 0 & 2y \\ x & y \end{bmatrix} \qquad (8)$$

where $(x, y) \in A$. Using the nodal force vector at the boundary f_b, the macroscopic stress is calculated as

$$\bar{\sigma} = \frac{1}{|V|} \sum_{b \in B} D_b f_b \qquad (9)$$

where B is the set of all boundary nodes. Then by the stress–strain relationship:

$$\bar{\sigma} = C\bar{\varepsilon} \qquad (10)$$

Applying the macroscopic strain $\bar{\varepsilon} = [1\ 0\ 0]^T$, we can then evaluate the elements of the linear elastic isotropic constitutive matrix C, obtaining E_{eff} and v_{eff}.

3.2 Homogenized Failure Stress

In this paper, failure of the heterogeneous material is assumed to coincide with the yielding of either the base medium and/or the inclusion. The homogenized failure stress (Y_{eff}) is thus obtained by comparing the yielding of the micro-structure and the homogenized macro-structure. For the yielding of materials constituting the micro-structure, von Mises yield criterion is used. Using the nodal displacements from the XFEM solution, the stress field in each element is obtained as

$$\sigma(x) = CB(x)u \qquad (11)$$

This stress field is used to find the von Mises stress, $\sigma_v(x)$, which is maximized over the heterogeneous medium Ω to find the potential failure point:

$$x^* = \mathrm{argmax}_\Omega\ \sigma_v(x) \qquad (12)$$

The macroscopic strain $\bar{\varepsilon}$ is then scaled so as to make the σ_v at x^* equal to the yield stress Y of the material (inclusion or the base medium) at x^*:

$$\bar{\varepsilon}^* = \frac{Y}{\sigma_v(x^*)}\bar{\varepsilon} \tag{13}$$

where $\bar{\varepsilon}^*$ is the strain at which the micro-structure fails due to material (inclusion or base medium) yielding. It is assumed that the homogenized macro-structure also fails under an applied strain of $\bar{\varepsilon}^*$. The stress at which the homogenized macro-structure fails is then obtained as

$$\bar{\sigma}^* = C\bar{\varepsilon}^* \tag{14}$$

with the corresponding von Mises stress taken as the homogenized failure stress.

4 Numerical Example

In this section, a numerical example of the homogenization of a unit cell is presented. As shown in Fig. 2, the unit cell consists of elliptical inclusions, of random sizes and orientations, randomly distributed in a base medium. The inclusions and the base medium have elastic moduli E_{inc} and E_m, respectively, Poisson's ratio ν_{inc} and ν_m, respectively, and yield stress Y_{inc} and Y_m, respectively. In this example, $E_{inc}/E_m = 41/7$, $\nu_{inc} = \nu_m = 0.3$, and $Y_{inc}/Y_m = 27/5$. For the homogenization, the approach described in Sect. 3 is followed. Two different volume fractions are considered: $\phi = 0.2$ and 0.4. To get the statistics of the macroscopic material properties, 500 MC simulations are performed for each volume fraction, with $n_{inc} = 15$ elliptical inclusions in each sample.

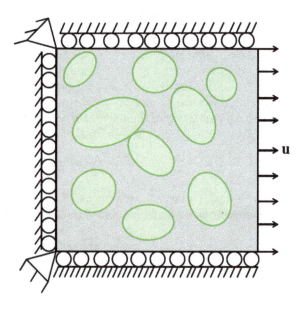

Fig. 2 Sample unit cell with elliptical inclusions

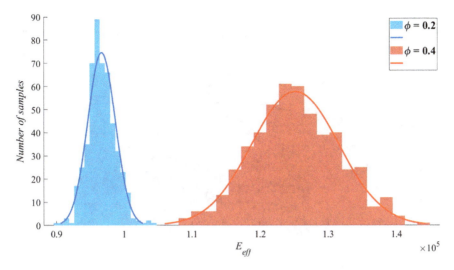

Fig. 3 Distribution of homogenized elastic modulus (E_{eff})

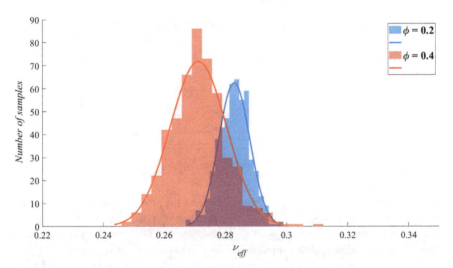

Fig. 4 Distribution of homogenized Poisson's ratio (ν_{eff})

Figures 3, 4 and 5 show the distributions (histograms) of the homogenized elastic modulus, Poisson's ratio and failure stress, respectively, for the two volume fractions of 0.2 and 0.4. It is evident that the distributions of all three properties can be well approximated via normal distributions. The mean, standard deviation and coefficient of variation (COV) of these homogenized properties are listed in Table 1. The mean elastic modulus is found to increase with the volume fraction, as also observed by Hiriyur et al. [2]. The mean Poisson's ratio is found to be slightly underestimated in both cases (noting that $\nu_{inc} = \nu_m = 0.3$) and shows a slightly decreasing trend

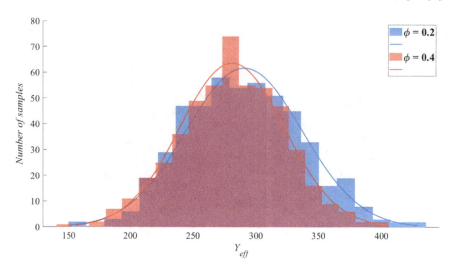

Fig. 5 Distribution of homogenized failure stress (Y_{eff})

Table 1 Statistics of homogenized properties for different volume fractions (ϕ)

Property	ϕ	Mean	Std. devn.	COV
E_{eff} (MPa)	0.2	9.67×10^4	2.03×10^3	0.02
	0.4	1.25×10^5	6.39×10^3	0.05
ν_{eff}	0.2	0.28	0.005	0.018
	0.4	0.27	0.009	0.033
Y_{eff} (GPa)	0.2	290.52	46.25	0.16
	0.4	280.88	41.69	0.15

with volume fraction. The uncertainty in the elastic modulus and Poisson's ratio increases with the volume fraction. This may be expected as in the limiting case of a homogeneous medium ($\phi = 0$) the uncertainty (COV) should also reduce to zero. Both the mean and the uncertainty in the failure stress is found to be less affected by the volume fraction. The uncertainty in the failure stress is found to be significantly higher (one order of magnitude higher COV) than the uncertainty in the elastic modulus and Poisson's ratio, indicating that the randomness of the microstructure induces a higher uncertainty near failure (yielding), compared to estimates of homogenized elastic constants.

5 Conclusion

This paper used homogenization using XFEM and MC simulations to obtain the macroscopic material properties of two-phase heterogeneous materials. The homogenization was illustrated using an unit cell with random elliptical inclusions in a base medium. It is shown that this is a good technique to obtain the homogenized elastic modulus and Poisson's ratio for lower volume fraction of the inclusions, while for higher volume fraction the uncertainty in these homogenized elastic properties increase. Further, the paper showed that the uncertainty induced by the random micro-structure is higher near failure, defined in this paper as the yielding of either the base medium and/or the inclusion. This indicates a higher uncertainty in the nonlinear (post-yield) regime as opposed to the linear elastic regime of material behavior, suggesting the cautious use of homogenization at nonlinear regimes.

References

1. Baxter SC, Hossain MI, Graham L (2001) Micromechanics based random material property fields for particulate reinforced composites. Int J Solids Struct 38(50):9209–9220
2. Hiriyur B, Waisman H, Deodatis G (2011) Uncertainty quantification in homogenization of heterogeneous microstructures modeled by XFEM. Int J Numer Methods Eng 88(3):257–278
3. Savvas D, Stefanou G, Papadrakakis M, Deodatis G (2014) Homogenization of random heterogeneous media with inclusions of arbitrary shape modeled by XFEM. Comput Mech 54(5):1221–1235
4. Zohdi TI, Wriggers P (2008) An introduction to computational micromechanics, 2nd edn. Lecture notes in applied and computational mechanics, vol 20. Springer, Heidelberg
5. Shan Z, Gokhale A (2002) Representative volume element for non-uniform micro-structure. Comput Mater Sci 24(3):361–379
6. Moës N, Dolbow J, Belytschko T (1999) A finite element method for crack growth without remeshing. Int J Numer Methods Eng 46(1):131–150
7. Belytschko T, Black T (1999) Elastic crack growth in finite elements with minimal remeshing. Int J Numer Methods Eng 45(5):601–620
8. Sukumar N, Chopp DL, Moës N, Belytschko T (2001) Modeling holes and inclusions by level sets in the extended finite-element method. Comput Methods Appl Mech Eng 190(46):6183–6200
9. Hill R (1963) Elastic properties of reinforced solids: some theoretical principles. J Mech Phys Solids 11(5):357–372
10. Miehe C, Koch A (2002) Computational micro-to-macro transitions of discretized microstructures undergoing small strains. Arch Appl Mech 72(4):300–317

Uncertainty Propagation in Estimated Structural Parameters Owing to Univariate Uncertain Parameter Using RSM and PDEM

Kumar Anjneya, Divya Grover, and Koushik Roy

Abstract The existence of uncertainty even in a single structural parameter may lead to random responses from the structure. This results in propagation of uncertainty in some other parameters estimated using these random responses. Therefore, to understand the behavior of any structure, investigation of the stochastic system is essential in structural health monitoring (SHM). For this purpose, response surface methodology (RSM) is applied for stiffness calculation. Probability density evolution method (PDEM) is employed for quick and efficient generation of the probability density function (PDF). In this study, the top floor mass is considered as random input parameter for PDEM. PDF of the random parameter is discretized into representative points. RSM is used to carry out inverse optimization for finding the structural properties (i.e., stiffness). PDEM is then employed for generating the PDF of stiffness. From the PDF of stiffness, it can be seen how randomness propagates from the system uncertainty into the estimated parameters.

Keywords Response surface methodology (RSM) · Probability density evolution method (PDEM) · Finite-element(FE) model · Uncertainty propagation

---

K. Anjneya
Department of Civil Engineering, IIT Kharagpur, Kharagpur 721302, India
e-mail: Kanjneya@gmail.com

D. Grover · K. Roy (✉)
Department of Civil and Environmental Engineering, IIT Patna, Patna 801103, India
e-mail: koushik@iitp.ac.in

D. Grover
e-mail: divyagrover016@gmail.com

© The Editor(s) (if applicable) and The Author(s), under exclusive license to Springer Nature Singapore Pte Ltd. 2021
S. K. Saha and M. Mukherjee (eds.), *Recent Advances in Computational Mechanics and Simulations*, Lecture Notes in Civil Engineering 103,
https://doi.org/10.1007/978-981-15-8138-0_44

1 Introduction

In the field of structural health monitoring, non-destructive damage detection techniques have been emerging since the last couple of decades. Vibration-based damage detection techniques are most common among them because of their global damage identification capabilities. These methods are based on the fact that due to change in the structural properties like stiffness, caused due to damage, the dynamic response of the structure changes. For the success of any vibration-based technique, it is important that the developed methodology should also perform well under uncertain situations.

Uncertainty in the structural parameters is inevitable [1]. There are number of factors that introduce uncertainty into a system which can be modeling errors as well as measurement errors. All these uncertainties make the deterministic approach a difficult task to carry out. If the level of uncertainty is high, then actual information may get concealed. Therefore, there is a need to study the developed methodology under the effect of uncertainty. In the structural parameter identification process, the uncertainty in one of the parameters produces randomness in other derived parameters, i.e., randomness propagates in the system. Hence, for checking the applicability of any methodology, uncertainty in the structural parameters should be considered.

When uncertainty is considered as a random variable then Monte-Carlo simulation (MCS) can be used which involves generating huge data sets for each random parameter. However, carrying out the analysis using each of these random data is practically inefficient in order to obtain the PDF of the desired parameters. Also, the uncertainty can be considered as a random process varying with time. Therefore, in order to solve all these problems and also to deal with the random process, PDEM can be used [2].

PDEM was developed in the past decade by Li et al [3]. The greatest advantage of using PDEM is to reduce the computational time required for analysis. In PDEM, the domain of the random parameter is discretized into smaller number of domains and representative points are selected in each domain. Corresponding to these representative points analysis is carried out and probability density equation (PDE) is solved to obtain the PDF of the desired structural parameters.

In the present study, a six-storey shear building has been considered, where randomness has been introduced in the mass of the top floor. The parameter is discretized into domains and representative point in each such domain is selected. The responses corresponding to each of these representative points are obtained. In this study, RSM [4] based structural parameter identification process has been used, where a regression equation is developed between the frequencies and the stiffness at different stories. These response surface (RS) equations are then used for predicting the stiffness at each storey through inverse optimization technique. Thereafter, PDE equation is solved to obtain the PDFs of the stiffness of each storey.

Thus, with lesser computations corresponding to representative points, PDFs of the actual stiffness can be derived.

2 Methodology

2.1 Response Surface Methodology

The RSM is a statistical regression technique which generates regression equations relating the output responses with that of the input parameters [5]. For every regression technique, a number of design points relating output to the inputs are required to fit a regression equation between them. Selection of these design points in RSM is done through design of experiments (DoE) [6]. There are various types of DoE which can be used for this purpose, namely, central composite design (CCD), Box Behnken design (BBD), D-optimal, etc [7]. The first step toward the development of RS model is to carry out a screening experiment so as to sort out the insignificant parameters. After doing this, a low order regression equation is developed to move closer to the region of optimum. Closer to the region of optimum, a more detailed higher order regression equation is generated to have a better relationship between the parameters and the responses. RS equations are used to optimize a process, however it can also to used obtain the responses corresponding to a particular set of the variables. This can further be extended to obtain the parameters that would cause these responses using inverse optimization.

Response Surface (RS) Model Generation. The screening experiment is carried out [8] using a low order model to find which of the factors are contributing significantly toward the total variance of the model. The equation used for this purpose can be expressed as follows:

$$Y = \beta_0 + \beta_1 X_1 + \beta_2 X_2 + \beta_3 X_1 X_2 + e \quad (1)$$

where β_o, β_1, are regression coefficients calculated using the method of least squares and e is the modeling error.

After carrying out the screening experiment, next step is the generation of a second-order regression equation for the numerical modeling of the structure. This equation is used to find the structural parameters through inverse optimization using the responses obtained from the structure. A second-order RS equation can be expressed as shown in the following Equation:

$$Y = \beta_o + \beta_1 X_1 + \beta_2 X_2 + \beta_3 X_1 X_2 + \beta_4 X_1^2 + \beta_5 X_2^2 + e \quad (2)$$

where β is the regression coefficient and e is the modeling error. The regression coefficients are obtained through the method of least square. Generally, variable X is written in the normalized form so that the maximum and the minimum value corresponds to +1 and −1, respectively. The final regression equation can be written as follows:

Fig. 1 Two factorial two level CCD

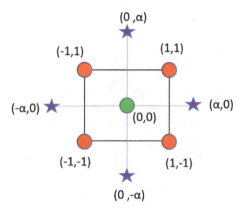

$$\widehat{Y} = \beta_0 + \sum_{i=1}^{k} \beta_i + X_i + \sum_{i=1}^{k} \beta_{ii} X_i^2 + \sum_{i<j} \sum_{j=2}^{k} \beta_{ij} X_i X_j \qquad (3)$$

Central Composite Design. There are varieties of DoE that can be used, however CCD [6] has been used in this study because of its wide popularity. CCD can either be factorial or fractional factorial with center points augmented with the star points. A full factorial CCD with two factors have been shown in Fig. 1. Figure 1 shows four corner (± 1) factorial design points and four axial (± 1) α star points. Depending upon the values of the α the CCD are of various types, namely, face-centered, rotatable and practical [6]. The α for practical is calculated as per following Eq. (4):

$$\alpha = \left(k^{1/4} \right) \qquad (4)$$

where k = number of factors or variables considered in the design, This is used if the number of the factors are six or more.

Model Building. First of all, design points are generated using DoE. Design bounds for each of the parameter are fixed around 20–30% about their mean value. Corresponding to each of the set of input parameters, responses are obtained using the eigenvalue analysis. The method of adjusted and the sequential sum of squares are used for building up the model. For screening out the parameters as well as the selection of independent or interaction terms in RS model of the responses, a analysis of variance(ANOVA) test is carried out. In the ANOVA test, F value is calculated to find which of the parameters are contributing significantly towards the total variance of the model.

$$F_A = \frac{SSR_A/k}{SSE/(n-k-1)} \sim F_{0.05,k,n-k-1} \qquad (5)$$

where n is the number of data points and k is the number of β parameters minus one. F_A and $F_{0.05,k,n-k-1}$ denotes the F-test value of an input parameter A and the chosen F-test criterion value, respectively. SSR and SSE are the sums of squares due to the regression and residual, respectively. $F_A > F_{0.05,k,n-k-1}$ indicates that A contributes significantly to the total variance of the model. Corresponding to each of the responses a separate RS equation is generated.

$$\widehat{\lambda_n} = \beta_0 + \sum_{i=1}^{k} \beta_i + X_i + \sum_{i=1}^{k} \beta_{ii} X_i^2 + \sum_{i<j}^{k} \sum_{j=2} \beta_{ij} X_i X_j \quad (6)$$

where n represents the nth mode. Once the model is ready, it needs to be checked for its adequacy as shown.

$$R^2 = \left(\frac{SSR}{SST}\right) = 1 - \left(\frac{SSE}{SST}\right), 0 \leq R^2 \leq 1 \quad (7)$$

Values of R^2 should be closer to 1. There are other criteria also that needs to be checked because the addition of non-significant parameter to the model increases the value of R^2. So other two criteria are also must.

$$R_{adj}^2 = 1 - \frac{SSE/(n-p)}{SST/(n-1)} = 1 - \frac{n-1}{n-p}(1 - R^2) \quad (8)$$

$$R_{pred}^2 = 1 - \frac{PRESS}{SST} \quad (9)$$

where $p = k + 1$, and *PRESS* is the predicted residual error sum of squares. Their values should be closer to 1 and both values should have a maximum difference of 0.2. R_{adj}^2 decreases with addition of the insignificant terms to the model, and R_{pred}^2 decreases if model has too many insignificant terms. SST is the total sum of squares which is calculated as the summation of error sum of squares and the residual sum of squares.

Multi-objective Inverse Optimization. After obtaining the responses from the actual structure through experiment or generated through simulation, inverse optimization algorithms are used to obtain the corresponding parameters. In the present study, a six-storey shear building has been considered to carry out the study in which the stiffness at different stories has been taken as the input variables. In order to carry out the inverse optimization, the difference between the first three frequencies predicted through RS equations and one obtained form the actual structure are minimized. Thus, there are three objectives that need to be satisfied simultaneously. For this purpose, MATLAB-based "fgoalattain" multi-objective optimization tool-box is used [7]. The fgoalattain tries to minimize the difference between the frequencies by looking for a set of stiffness values within the upper and lower bound provided, that will make it possible, thus finding out the actual stiffness of each storey. This goal attainment problem tends to minimize the slack variable y as follows:

$$\underset{x,y}{\text{Minimize } y} \quad \begin{cases} F(x) - wy \leq goal \\ lb \leq x \leq ub \end{cases} \quad (10)$$

where

$$F(x) = \left\{ abs\left(\frac{f_{rsm} - f_{exp}}{f_{exp}}\right) \right\} \quad (11)$$

Here, $F(x)$ is an dimensionless objective function. Here, goal is set to zero. Weight w controls over and under attainment of the objective. "y" is the slack element used as dummy, lb and ub are the lower and upper bounds of the parameters, respectively. The "fgoalattain" finds the value of stiffness in this range only.

2.2 Uncertainty Analysis Using Probability Density Evolution Method

Due to propagation of uncertainty in the system, stiffness of the structure becomes dependent on the random input parameters for PDEM, i.e., $k(\theta)$ where $\theta = (\theta_1, \theta_2, ..., \theta_j)$ are mutually independent random variables and j is the number of random parameters. The joint probability density function $P(\theta) = \prod_1^j P(\theta_j)$ where $j = 1$ in this case. For plotting the PDF, a virtual random process is defined for each storey stiffness parameter, A virtual random process is generated as [9]

$$z_q(\theta, t) = k_q(\theta) \sin\frac{\pi}{2}t \quad (12)$$

where $k_q(\theta)$ is the qth storey stiffness parameter. On differentiating above equation,

$$\dot{z}_q(\theta, t) = \frac{\pi}{2}k_q(\theta)\cos\frac{\pi}{2}t \quad (13)$$

z(t) is unique and continuously dependent on θ which is given by

$$z_q(\theta, t) = H_q(\theta, t) \quad \text{and} \quad \dot{z}_q(\theta, t) = h_q(\theta, t) \quad (14)$$

The joint probability density function of (z_q, θ) is denoted as $P(z_q, \theta, t)$. According to the principle of preservation of probability, the joint PDF of (z_q, θ) should satisfy the generalized probability density evolution equation (GPDEE) as

$$\frac{\partial P(z_q, \theta, t)}{\partial t} + \dot{z}_q(\theta, t)\frac{\partial P(z_q, \theta, t)}{\partial z} = 0 \quad (15)$$

Substituting Eq. (13) into Eq. (15),

$$\frac{\partial P(z_q, \theta, t)}{\partial t} + \frac{\pi}{2}k_q(\theta)\cos\frac{\pi}{2}t\frac{\partial P(z_q, \theta, t)}{\partial z} = 0 \quad (16)$$

With initial condition,

$$P_{z_q}(z, \theta, t_0) = \delta(z - z_0)P(\theta) \tag{17}$$

$$P_{z_q}(z, t) = \int P_{z_q}(z, \theta, t_0)d\theta \tag{18}$$

Note that, $k_q(\theta) = z_q(t)|_{t=1}$.

Solution of GPDEE via a Family of δ-Sequences. One of the way of representing Dirac delta function is to approach it as the limit of different families of functions known as Dirac's sequence [10]. It means, if $\lim_{\lambda \to 0} \delta_\lambda(z - z_0) = \delta(z - z_0)$, $\delta_\lambda(z - z_0)$, is a Dirac sequence for $\lambda \in R$. There are many types of one-dimensional δ sequences like Legendre polynomials, harmonic functions, Hermite polynomials and many probability density functions [10]. Out of the options available, one based on the normal distribution is more preferred as it guarantees the non-negative nature and satisfies the consistent condition of probability [10]. The normal distribution of δ-sequence is shown as follows [10]:

$$\delta_\lambda(z - H_q(\theta, t)) = \frac{1}{\sqrt{2\pi}\lambda} exp\left[\frac{-1}{2\lambda^2}(z - H_q(\theta, t))^2\right] \tag{19}$$

Substituting Eq. (19) in Eq. (18) yields,

$$P(z, t) = \int \delta(\frac{z_q - H_q(\theta, t)}{H_{q,0}}) \frac{1}{H_{q,0}} P_\theta(\theta)d\theta \tag{20}$$

where $H_{q,0} = max[H_q(\theta, t)]$

Error Minimization for PDF. The first step for application of PDEM is to choose representative points from the sample space of the random parameter [11]. There are different types of methods available for this purpose. These methods are based on the number of uncertain parameters considered for the analysis. In the present study, grid-type point sets method [12] has been used for selecting these representative points. As the number of the representative points increases and tends to infinity, the value of λ decreases toward zero. This value turns impractical for the analysis of the stochastic system due to heavy computational effort involved. Hence, the value of λ should not be very small because of the condensation effect in the neighborhood of $z = H_q(\theta, t)$ which distorts the real PDF [10]. The value of λ is essential and is determined as per the computational experiences.

Now, as the approximate PDF $P_z(z, t)$ is a function of λ, the value of λ should be suitably chosen so that the PDF matches with the actual one. For minimizing the error, first, and second-order error estimation, mostly used in practice, are given as follows:

$$E_{1rr}(\lambda) = \frac{|E(t) - E(t,\lambda)|}{|E(t)|} \qquad (21)$$

$$E_{2rr}(\lambda) = \frac{|V(t) - V(t,\lambda)|}{|V(t)|} \qquad (22)$$

where $E(\cdot)$ and $V(\cdot)$ are the first and second-order moments, respectively, and $E_{1rr}(\lambda)$ and $E_{2rr}(\lambda)$ are essentially the errors of the mean and variance, respectively.

Intuitively, it can be thought that optimal λ will make both the error terms minimum, however, it has been found that λ reaches optimal value when $E_{1rr}(\lambda)$ reaches its minimum [10]. Also as stated earlier, λ depends upon the number of representative points, therefore the optimal λ should also depend upon them. As λ increases, the smearing effect becomes less but the high-frequency oscillation becomes more obvious. If λ is too large, the effect of dissipation is too large, on the contrary if λ is too small, the effect of high-frequency oscillation is too high [10].

3 Simulation Model Study

3.1 Model Description

For the present study, a six-storey shear building has been considered as shown in Fig. 2. The stiffness value for each storey is assumed to be 2×10^8 N/m while the mass of each floor is assumed as 1×10^5 kg. The mass of top floor (i.e., M_6) is assumed to follow a normal distribution with coefficient of variation as 5 percent, i.e., standard deviation (σ) = 5×10^3 kg. For M_6, random numbers are generated in the range of $[\mu - 3\sigma, \mu + 3\sigma]$ corresponding to their mean values. In this study, 1000 realizations of stiffness matrix are generated considering the uncertainty in the top floor mass. Corresponding to uncertain data points, frequencies are obtained using eigenvalue analysis. Here, first three frequencies are used.

For the model considered in the present study, PDF for the input random mass parameters has been shown in Fig. 3. These PDFs are discretized into 100 domains each. Following the methodology step-by-step, the PDF of the output, i.e., stiffness values at each storey is obtained.

3.2 Methodology Step-by-Step

Flowchart for overall process is shown in Fig. 4. The flowchart has been explained in detail as follows:

(1) *Introducing randomness into the system*- Randomness is introduced in mass matrix of the top floor with a mean value of 1×10^5 kg and coefficient of variation as 5 percent.

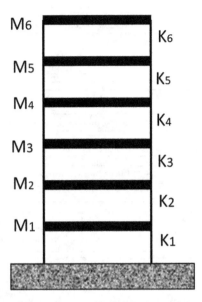

Fig. 2 Shear building model

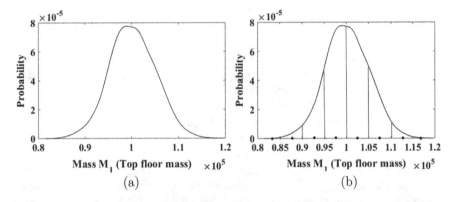

Fig. 3 PDF for **a** M_1 (Top floor mass), **b** Discretization of M_1 into domains

(2) *Discretizing the random parameter-* The random mass parameter is discretized into domains and representative points are selected [12]. Probability in their respective domains is calculated by taking the area under the curve in each domain. Let the discretized points in domain of M_1 is represented by α_q, where $q = 1,2,3,.....N_{sel}$.

(3) *Modal response extraction-* Corresponding to each of the random data set, the responses from the structure are taken. In the present study, first three modal frequencies are taken for stiffness estimation.

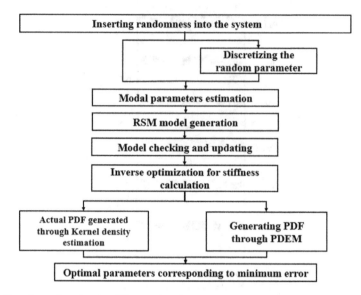

Fig. 4 Flowchart of stochastic stiffness estimation

(4) *Response surface model generation-* For generating the RS model, design bounds for the stiffness are fixed as 1×10^8 kN/m as lower bound and 4×10^8 kN/m as upper bound.
(5) *Model checking and updating-* Once the model is ready, it needs to be checked and updated, if required. However, updating step is not required in this case as design bounds have been taken large enough.
(6) *Inverse optimization for stiffness calculation-* In order to carry out the inverse optimization for stiffness calculation, the frequencies are obtained corresponding to each representative point of discretized mass parameter.
(7) *PDF generation using PDEM-* A virtual random process is generated (Eq. 12). Corresponding to each representative point, stiffness values are calculated and are substituted in Eq. 15. GPDEE is then solved using the method of characteristics to obtain $P_z(z, \theta, t)$, and then with the help family of delta sequence $P_z(z, t)$ corresponding to each representative point is obtained. At t=1, PDF of z will be equal to PDF of stiffness.
(8) *Error minimization-* A suitable value of λ for each PDF is obtained to minimize the error between the actual PDF and the one obtained from the numerical solution to the PDE equation. For actual PDF generation of the stiffness at each floor, inverse optimization is carried out for each realization.

4 Results and Discussions

The PDFs of the stiffness are shown in Figs. 5, 6 and 7. The corresponding optimal λ values required for solving the PDE equation numerically have been shown in Table 1. From the PDFs, it can be seen that the uncertainty introduced in the mass parameter is propagated to stiffness at each floor. From Fig. 5, 6 and 7, it can be seen that PDF generated through PDEM is overlapping the actual one satisfactorily. The predicted stiffness values and corresponding standard deviation and Coefficient

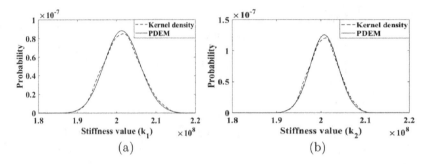

Fig. 5 PDF for **a** K_1 (Bottom storey), **b** K_2 (Second storey)

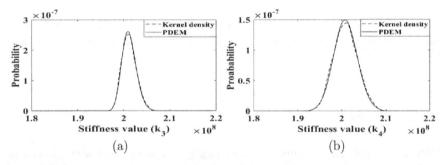

Fig. 6 PDF for **a** K_3 (Third storey), **b** K_4 (Fourth storey)

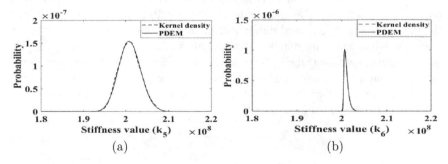

Fig. 7 PDF for **a** K_5 (Fifth storey), **b** K_6 (Top storey)

Table 1 Optimal λ for each storey

S.No.	Storey stiffness	Optimal λ
1.	K_1	0.004
2.	K_2	0.0025
3.	K_3	0.0016
4.	K_4	0.0025
5.	K_5	0.0036
6.	K_6	0.00042

Table 2 Estimated stiffness values

S.No.	Storey stiffness	Mean value	Standard deviation	Coefficient of variation
1.	K_1	2.02×10^8	5.5281×10^6	2.74
2.	K_2	2.01×10^8	4.2002×10^6	2.09
3.	K_3	2.0165×10^8	2.3238×10^6	1.14
4.	K_4	2.0115×10^8	3.6517×10^6	1.82
5.	K_5	2.0115×10^8	3.4785×10^6	1.73
6.	K_6	2.017×10^8	0.85147×10^6	0.422

of variation can be seen from Table 2. It can be observed from Table 2 that due to randomness in the mass of the top floor, bottom floor stiffness is more uncertain and has highest coefficient of variation among all.

5 Conclusion

RSM is very efficient in dealing with uncertainty. The inverse optimization process takes very short time to predict the stiffness in each realization. The PDF for stiffness at each storey is obtained using PDEM. PDEM reduces the number of realizations required to obtain the PDFs, thus reducing the overall time required to generate the PDFs. The results obtained through RSM coupled with PDEM can further be improved with increasing number of realizations. The proposed methodology is quite efficient in obtaining the stiffness at each storey which can be employed for the damage identification purpose. This study is done by considering randomness in top storey mass however the algorithm will be more helpful when it will be independent of physical parameters of the system.

References

1. Xu LY, Zhang J, Li J, Wang XM (2011) Stochastic damage detection method for building structures with parametric uncertainties. J Sound Vibr 330:4725–4737. https://doi.org/10.1016/j.jsv.2011.03.026
2. Li J (2016) Probability density evolution method: background, significance and recent developments. Prob Eng Mech 44:111–117. https://doi.org/10.1016/j.probengmech.2015.09.013
3. Li J (2004) Probability density evolution method for dynamic response analysis of structures with uncertain parameters. Comput Mech 34:400–409. https://doi.org/10.1007/s00466-004-0583-8
4. Rutherford CA, Inmanb JD, Parka G, Hemeza MF (2005) Use of response surface metamodels for identification of stiffness and damping coefficients in a simple dynamic system. Shock Vibr 12:317–331. https://doi.org/10.1155/2005/484283
5. Box GEP, Wilson KB (1951) On the experimental attainment of optimum conditions. J R Stat Soc, Ser B (Methodol) (1951)
6. Raymond MH, Douglas MC, Christine AM (2009) Response surface methodology. Wiley Inc, Hoboken
7. Umar S, Bakharya N, Abidin ZRA (2018) Response surface methodology for damage detection using frequency and mode shape. Measurement 115:258–268. https://doi.org/10.1016/j.measurement.2017.10.047
8. Fang SE, Perera R (2009) TA response surface methodology based damage identification technique. Smart Mater Struct 18:065009 (14pp). https://doi.org/10.1088/0964-1726/18/6/065009
9. Yang Y, Li LJ, Zhou HC, Law SS, Lv L (2019) Damage detection of structures with parametric uncertainties based on fusion of statistical moments. J Sound Vibr 442:200–219. https://doi.org/10.1016/j.jsv.2018.10.005
10. Fan W, Chen J, Li J (2009) Solution of generalized density evolution equation via a family of δ sequences. Comput Mech 43:781–796. https://doi.org/10.1007/s00466-008-0345-4
11. Li J, Chen J (2009) Stochastic dynamics of structures. Wiley (Asia) Pte Ltd (2009). https://doi.org/10.1002/9780470824269
12. Li J, Chen BJ (2007) The number theoretical method in response analysis of nonlinear stochastic structures. Comput Mech 39:693–708. https://doi.org/10.1007/s00466-006-0054-9

Seismic Response of Liquid Storage Tank Considering Uncertain Soil Parameters

Hitesh Kumar and Sandip Kumar Saha

Abstract Properties of soil and structure govern the effects of soil-structure interaction on the seismic response of any structure. However, it is well known that there involves high uncertainty in the engineering properties of the soil in its natural state. The present study attempts to investigate the effects of the uncertain soil properties on the seismic responses of liquid storage tank. Critical soil parameters, such as shear wave velocity and mass density, are represented by suitable probability distribution functions and used in the analyses. The liquid storage tank is modeled using the lumped mass idealization and the soil domain is modeled utilizing the finite element approach. A sufficiently large number of coupled soil-tank models are generated with each model having a unique variation of soil parameters along the depth. Utilizing Monte Carlo simulation technique, it is observed that the uncertain soil parameters have a significant effect on the peak response quantities of the liquid storage tank.

Keywords Liquid storage tank · Seismic response · Soil-structure interaction · Uncertainty

1 Introduction

Soil-structure interaction (SSI) is a well-known phenomenon that influences the seismic behavior of a structure, and the degree of influence depends on the properties of structure and soil. An earthquake excitation causes a permanent deformation in the soil which influences the wave propagation from soil to structure and vice versa. However, the dynamic properties of a structure are generally considered deterministic since they are factory made or manufactured/constructed ensuring quality control. Nevertheless, there involves high level of uncertainty in the engineering properties of the soil due to our limited knowledge about the natural process of soil formation. Geotechnical testing at a site can give a sound understanding of the properties of the

---

H. Kumar (✉) · S. K. Saha
Indian Institute of Technology Mandi, Mandi 175075, India
e-mail: s17002@students.iitmandi.ac.in

© The Editor(s) (if applicable) and The Author(s), under exclusive license to Springer Nature Singapore Pte Ltd. 2021
S. K. Saha and M. Mukherjee (eds.), *Recent Advances in Computational Mechanics and Simulations*, Lecture Notes in Civil Engineering 103,
https://doi.org/10.1007/978-981-15-8138-0_45

local soil. However, the correctness of the information depends upon the depth up to which the soil is explored, randomness in the spatial distribution around the testing location, sampling method (size and total number of soil specimens), accuracy of instruments, and human errors. Treating soil as a homogeneous medium does not necessarily put the design on the conservative side especially for a liquid storage tank structure that has a huge capacity (mass) and founded on relatively soft soil. Moreover, due to its enormous importance in industrial and civilian applications more care should be taken for the design of tank structures. Popescu et al. [1] reported that deterministic analysis underestimates the bearing capacity of soil as compared to the analysis of soil with spatial variability. Kim et al. [2] reported that the spatial variability of elastic modulus affects the wave propagation and results in a complex phenomenon of wave reflection and mode conversion. Lizarraga and Lai [3] from the study of spatial variability of soil reported that the deterministic analysis underestimates the crest settlement of the earthen dam as compared to the stochastic analysis. Since the Indian subcontinent have more than half of its geographical area categorized as earthquake prone [4] and have the second largest population in the world, the associated seismic risks are severely high. However, it is less in practice to perform a detailed geotechnical exploration of the construction site, thereby making it crucial to consider the uncertainty associated with the soil parameters in the framework of soil-structure interaction.

The present study attempts to investigate the effect of uncertainty of critical soil parameters by assigning appropriate probability distribution functions and consider them in the analyses. Uncertainty of the soil parameters has been studied in past by several researchers. Paice et al. [5] have studied the settlement of foundation on spatially varied soil by considering Young's modulus of the soil as uncertain while fixing values of all other parameters. Fenton and Griffiths [6] also considered Young's modulus of the soil as uncertain. Wang and Hao [7] investigated a layered soil model and reported that variation of mass density and shear modulus have significant effects on the ground motion as compared to the variation of damping ratio of the soil. Nour et al. [8] reported that the uncertainty of Poisson's ratio has no significant effect, therefore the variability associated with it can be neglected. However, the variation of shear modulus of the soil is critically important. Popescu et al. [1] have considered the undrained shear strength and Young's modulus of the soil as uncertain. Hamidpour and Soltani [9] have considered the shear modulus of soil as uncertain.

Hydrodynamic pressure exerted by the fluid content of the tank on its walls (fluid-tank interaction) under random loading (e.g., earthquake) makes its dynamic behavior unique from other structures, such as buildings or bridges. Further, limited research has been reported on the effect of soil-structure interaction on the seismic responses of liquid storage tanks [10–15]. Therefore, it is important to consider soil parameters as uncertain in the framework of SSI, for seismic response analyses of liquid storage tanks. Herein, ground supported cylindrical liquid storage tank with flexible (steel) walls is considered for the seismic performance evaluation.

The present study considers uncertainty in two soil parameters, namely, shear wave velocity and mass density. Herein, the layered soil model is considered, i.e., the soil parameters can vary along the vertical direction and the spatial variation is not considered along horizontal directions. To investigate such problems with layered soil, Popescu et al. [1] have emphasized the use of Monte Carlo simulation (MCS). The present study utilizes the MCS on a set of 150 coupled soil-tank models, with each model having a unique profile of soil parameters, to investigate the effect of uncertain soil parameters on the seismic response of tank structure. The primary objective of this study is to investigate the variation in the seismic response of liquid storage tank when the soil properties are considered random in nature along the depth.

2 Modeling of Liquid Storage Tank

The liquid content of the cylindrical tank is modeled as three uncoupled masses [16], namely, sloshing mass, impulsive mass, and rigid mass. The sloshing mass (m_c) is attached to the rigid tank wall at a height of H_c from the tank base with spring (stiffness expressed by k_c) and dashpot (damping ratio expressed by ξ_c). The impulsive mass (m_i) is attached to the rigid tank wall at a height of H_i from the tank base with spring (stiffness expressed by k_i) and dashpot (damping ratio expressed by ξ_i). The rigid mass (m_r) is rigidly connected to the tank base at a height of H_r. The geometry of the tank (i.e., slenderness ratio) is generally expressed in terms of the ratio of height (H) and radius (R) of the liquid column. Various parameters of the lumped mass idealization of the tank are a function of its slenderness ratio ($S = H/R$), mass density of the liquid content (ρ_w), and the mass density (ρ_s) and Young's modulus (E_s) of the tank wall material. Details about the tank parameters and their calculation procedure can be found elsewhere [11, 15–19].

3 Modeling of Soil Domain

Layered soil model is considered for modeling the soil domain. The bedrock is considered at a depth of 30 m and the soil domain is assumed to be divided into 30 layers each having one meter of depth. Shear wave velocity (V_s) and mass density (ρ_s) are the basic parameters for defining the computational soil domain, and the shear modulus is a function of these two parameters. Poisson's ratio of the soil is considered to be 0.33.

Soil variability is defined by the mean value and the coefficient of variation (CV) of the parameter. For the layered model, it is assumed that as the depth increases the coefficient of variation decreases leading to less randomness and relatively more uniformity near to the bedrock. The bedrock is assumed as a homogenous linear-elastic semi-infinite half-space and modeled using LK (Lysmer-Kuhlemeyer) [20] boundary to avoid any spurious wave reflections [21].

Any effect of foundation embedment is kept outside the scope of this study as the structure is supposed to be founded on a shallow foundation. The top 3 m of the soil is

Table 1 Shear wave velocity distribution corresponding to different soil depths

Soil depth (m)	V_s
	Mean (m/s); Lognormal(μ, σ); CV(%)
3–10	300; LogN(5.59,0.22); 50
11–20	500; LogN(6.13,0.18); 45
21–30	800; LogN(6.61,0.15); 40

assigned the parameters such that shallow foundation would be a natural choice thus, excluding any case of soft soil in the top 3 layers (3 m). Use of a deep foundation for the tank structure is not considered in this study. The computational soil domain is divided into three regions each with 10 m depth. Further, the top 10 m is subdivided into two regions with 3 m and 7 m depths, respectively. For top 3 m, the shear wave velocity can vary uniformly between 180 and 300 m/s. For the underlying 7 m depth (after first 3 m from top surface), the minimum value of shear wave velocity is limited to 90 m/s. For rest of the soil domain, probability distribution parameters are given in Table 1.

In the present study, both shear wave velocity and mass density of the soil are assumed to follow lognormal distribution. This assumption is based on the fact that properties of the soil being a physical parameter cannot take a negative value. Similar assumptions may be found in earlier studies [1–3, 5–9]. The lognormal probability distribution functions [$LogN(\mu, \sigma)$] for the random soil parameters are defined by their mean [$e^{\mu+0.5\sigma^2}$] and standard deviation [$e^{\mu+0.5\sigma^2}\sqrt{(e^{\sigma^2}-1)}$]. In order to assign parameters to the distribution function, it is assumed that, as the depth (from the ground surface) increases, the coefficient of variation ($CV = \sqrt{e^{\sigma^2}-1}$) decreases. In other words, it is assumed that high variation in the soil parameters at greater depths is less probable and the soil is more uniform near the bedrock as compared to the soil near ground surface.

The soil parameters used herein for the classification of soil types (refer to Table 2), and the possible range of these parameters for a specific soil type (which is an important factor to define the distribution parameters) are assumed taking in consideration the existing literature [22–25] and open access test results of various sites available on the internet. Further, the distribution parameters (μ and CV) are such that the

Table 2 Soil classification based on its shear wave velocity, and mass density distribution corresponding to various soil types

Soil ID	Soil type	V_s (m/s)	ρ_s
			Mean (kg/m³); Lognormal(μ, σ); CV(%)
S1	Firm to hard rock	700–1450	2400; LogN(0.876,0.00001); 0
S2	Soft to firm rock	375–700	2150; LogN(0.74,0.05); 22
S3	Stiff clay	200–375	1900; LogN(0.60,0.09); 30
S4	Soft soil	90–200	1750; LogN(0.52,0.09); 30

subsequent soil profile affects the seismic responses of the tank, as well as cause a noticeable difference in the peak responses with a change in the soil profile.

Each layer of the soil domain is assigned the value of shear wave velocity using the data given in Table 1. The shear wave velocity of each layer is compared with the values given in Table 2 (Column 3) to find the soil type for that layer. Corresponding to the soil type, mass density is assigned to each layer using the distribution parameters given in Table 2 (Column 4). Subsequently, the number of elements in the vertical (or horizontal) direction is calculated by dividing the depth (or width) of soil domain by the height (or width) of soil element. The width of the soil element is assumed same for all layers and set equal to the minimum height of soil element in the entire domain. The number of elements (n) per wavelength and the size of soil element needs to be considered in a way that the important aspects of the shear wave propagation can be captured well within a cutoff frequency ($f = V_s/\lambda$, where λ is the wavelength of shear wave). Herein, n and f are assumed as 8 [26] and 15 Hz [27], respectively, as suggested in literature. The height of soil element (Δ_i) in the ith layer can be expressed as

$$\Delta_i = \frac{\lambda_i}{n} = \frac{(V_s)_i/f}{n} \text{ ; here } i = 1 \text{ to } 30. \quad (1)$$

From the above equation, it can be noted that the size of soil elements within each layer depends upon the value of shear wave velocity in that layer, therefore each layer can have different number of elements based on its shear wave velocity. However, the mesh of finite soil elements is such that it strictly refines as the depth decreases. In case of sandwiching of soil layers of different types, the total numbers of elements in a layer are such that the underlying layer will always have a lesser number of elements. The width of the soil domain is set as 250 m ensuring that further increase in the width does not cause a significant variation in the response quantities.

4 Details and Validation of Numerical Model

OpenSees [28], a finite element based seismic analysis tool, is used for modeling the tank-foundation system along with the soil domain and performing seismic analyses. The liquid storage tank is modeled in two dimensions with three degrees of freedom. The sloshing mass (m_c) and impulsive mass (m_i) have translational degrees of freedom whereas the rigid mass (m_r) is rigidly connected to the tank base which has rotational degree of freedom. The spring and dashpot are assigned at their respective locations (refer to Fig. 1) using link elements having uniaxial material properties. Flexibility of the foundation is not considered, and it is modeled using beam-column element having very high stiffness. The nodes that link the tank foundation to the soil domain are assigned equal degrees of freedom along both the translational directions. The soil is modeled in two dimensions with two translational degrees of freedom at each node. Four noded quadrilateral elements with plane strain material formulation

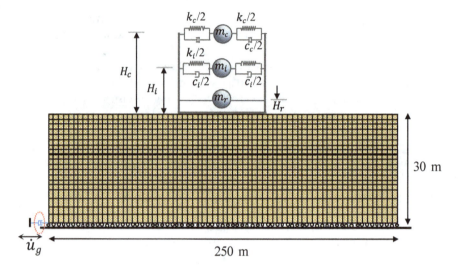

Fig. 1 The mechanical analog of liquid storage tank resting on the soil domain

are assigned to the soil domain since a properly defined two-dimensional plane strain soil model can approximate the responses of a three-dimensional soil domain [29]. However, the radiation damping associated with the low-frequency modes may be overestimated [29]. Nevertheless, the focus of the present study is to evaluate the possible range of responses from a large number of samples (soil-tank models) and not the specific value of peak response from the seismic analysis. Therefore, it is feasible as well as computationally efficient to use a two-dimensional plane strain model for the soil.

Groundwater is not considered within the computational soil domain, and it is assumed that the shear behavior of the soil element is not affected by the confinement changes. Therefore, a pressure independent elastic–plastic material having multi-yield surfaces (Von Mises type), is assigned to the soil element, that possesses a linear-elastic behavior under gravity loading and elastic–plastic behavior under dynamic loading. Soil domain is laterally excited, and a shear deformation behavior is considered. Nodes at the horizontal boundaries of the soil domain, sharing the same ordinate, are constrained to have the same response along both degrees of freedom. The nodes at the bottom of the soil domain are restrained along vertical degree of freedom and constrained to have the same response along horizontal degree of freedom. Two additional nodes are added near the corner of the soil domain, one is fixed along vertical degree of freedom and the other is fixed along both degrees of freedom. These nodes are assigned LK dashpot [20] having damping coefficient $C = V_r \times \rho_r \times A_s$, where V_r is the shear wave velocity of the bedrock which is assumed to be 1400 m/s; ρ_r is the mass density of the bedrock which is assumed to be 2700 kg/m^3; and A_s is the base area of the soil domain. The out-of-plane thickness of soil domain is considered as thrice the width of the foundation. Equal degree of freedom along horizontal direction is assigned to the base nodes of the soil domain and the LK dashpot node. The input excitation is given at LK dashpot node in terms of velocity time history [30, 31].

The acceleration responses of the three lumped masses are recorded to evaluate the resulting inertial forces. Base shear can be calculated by taking the summation of inertial forces at the tank base, and inertial forces multiplied by their respective lever arms about the tank base gives the base overturning moment (or simply base moment). Base shear (V_b) and base moment (M_b) can be expressed as

$$V_b = m_c(\ddot{u}_c + H_c\ddot{u}_\theta) + m_i(\ddot{u}_i + H_i\ddot{u}_\theta) + m_r(\ddot{u}_r + H_r\ddot{u}_\theta) \qquad (2)$$

$$M_b = m_c H_c(\ddot{u}_c + H_c\ddot{u}_\theta) + m_i H_i(\ddot{u}_i + H_i\ddot{u}_\theta) + m_r H_r(\ddot{u}_r + H_r\ddot{u}_\theta) + (I_r + I_f)\ddot{u}_\theta \qquad (3)$$

where $\ddot{u}_c$, $\ddot{u}_i$ and $\ddot{u}_r$ denotes the absolute acceleration of the three lumped masses, respectively; I_r and I_f are the moment of inertia of the mass which is rigidly connected at the tank base [32] and the foundation, respectively; $\ddot{u}_\theta$ is the rotational acceleration of the tank base.

4.1 Validation of Liquid Storage Tank Model

For the validation of the liquid storage tank model, the soil domain is assigned the parameter such that it behaves as rigid ($V_s = 5000$ m/s). The model of ground supported tank is validated by comparing the seismic responses of the present model with published results. Saha et al. [33] has investigated the effectiveness of different models of base isolation system for liquid storage tanks by comparing the peak responses of base-isolated tanks with corresponding fixed base tanks. They studied the effects of various parameters (e.g., slenderness ratio of the tank, and natural period of the base isolation system) on the seismic responses of base-isolated liquid storage tanks and compared the results with the response of fixed base liquid storage tanks. Nevertheless, the focus of present study is to evaluate the peak seismic responses of a fixed base liquid storage tank. Therefore, time history responses of fixed base liquid storage tank (slender geometry and without base isolation) from the published study [33] are reproduced using the model of the present study for numerical verification. The tank was subjected to the N00E component of Loma Prieta (1989) earthquake, recorded at LGPC, having a peak ground acceleration (PGA) of 0.57 g; where g is the acceleration due to gravity. Table 3 presents the various tank parameters considered in the published study.

Table 3 The tank parameters considered in the published study [33]

S	H (m)	ρ_w (kg/m^3)	ρ_s (kg/m^3)	E_s (GPa)	ξ_c	ξ_i
1.85	11.3	1000	7900	200	0.005	0.02

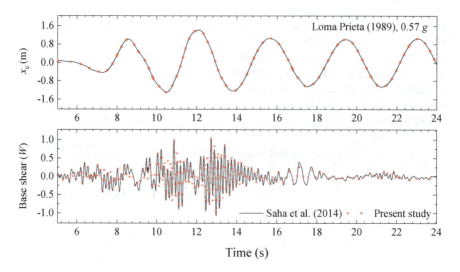

Fig. 2 Comparison of results from the tank model of the present study and Saha et al. [33]

With tank modeled for identical parameters (Table 3) and subjected to the same input motion, the sloshing displacement (x_c, i.e. relative to the tank base) and the base shear normalized with respect to the seismic weight of the tank ($W = Mg$; where $M = m_c + m_i + m_r$) are obtained using the current numerical model, and compared with the published results. Figure 2 shows a close match among the response time histories obtained from the current analysis and that presented in Saha et al. [33].

4.2 Validation of Soil Model

For the validation of soil model, the tank is uncoupled from the soil domain. The soil is assumed as homogeneous half-space and subjected to two outcrop motions, namely, (i) Loma Prieta (1989) earthquake, recorded at Diamond Heights, having PGA of 0.113 g, and (ii) Loma Prieta (1989) earthquake, recorded at Yerba Buena Island, having PGA of 0.067 g.

The acceleration obtained at the ground surface from the two-dimensional model of the soil domain in OpenSees is compared with the corresponding response obtained from the one-dimensional shear soil column with identical soil parameters (refer to Table 4). A commercially available program ProShake 2.0 [34] is used to evaluate

Table 4 Various parameters of homogeneous soil domain considered for validation

V_s (m/s)	ρ_s (kg/m^3)	Bulk modulus (kPa)	Soil cohesion (kPa)	Peak shear strain	Soil friction angle
250	1700	7.1×10^4	95	0.1	0.0

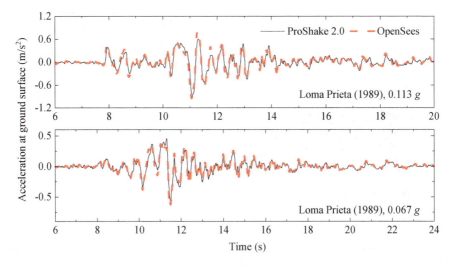

Fig. 3 Comparison of ground surface acceleration time histories from OpenSees and ProShake 2.0

the ground surface response of shear soil column. Figure 3 shows close match of both the acceleration time histories.

5 Numerical Analysis

To study the effect of uncertainty in the soil parameters on the seismic responses of liquid storage tanks, a set of 150 coupled soil-tank models are generated. The random soil parameters for each of these models are generated through MCS using the distributions of the parameters as presented in Tables 1 and 2. For the selection of input motions, 40 ground motions are selected from PEER strong motion database [35] such that these motions have a magnitude greater than 4.0 and recorded at a site having shear wave velocity greater than 800 m/s. All 40 ground motions are normalized for a PGA of 0.35 g and applied as bedrock motion at shear soil column to evaluate the corresponding free field motion at the ground surface. Acceleration response spectra evaluated for all free field motions are plotted in Fig. 4 and out of these motions two ground motions are selected whose response spectrum lies closest to the average spectrum of 40 free field motions. The selected input motions are, (i) the Whittier Narrows (E1) earthquake (10/4/1987), recorded at LA—Wonderland Ave and having a PGA of 0.0165 g, and (ii) the Whittier Narrows (E2) earthquake (10/1/1987), recorded at LA—Wonderland Ave and having a PGA of 0.0414 g. Time histories of both the input motions are normalized at the peak values of 0.25 g and 0.35 g which closely corresponds to the seismic zone factor of Zone IV and V, respectively, as per IS 1893:2016 [36].

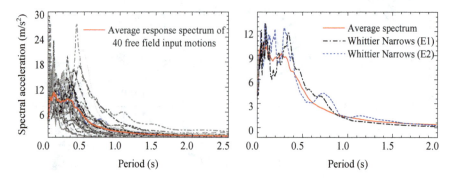

Fig. 4 Acceleration response spectra of 40 free field motions and average spectrum compared with the response spectra for Whittier Narrows (E1 and E2) input motions

The properties of the slender ($S = 1.85$) liquid storage tank are considered same as presented in Table 3. Further, it is assumed that the roof of the tank does not interfere with the sloshing mass of the liquid content, and the tank is properly anchored at its base. The base shear and overturning moment at the base of the liquid storage tank are evaluated and normalized with respect to its seismic weight (W).

5.1 Effect of Uncertainty in Soil Parameters

The peak base shear and base moment observed from the complete set of 150 soil-tank models are expressed in terms of frequency distribution plots (histogram). The normalized response quantities are evaluated for both the input motions (E1 and E2) with peak intensities of 0.25 and 0.35 g. Figures 5 and 6 show the frequency distributions of the responses under the earthquake E1 and E2, respectively. Lognormal fit for the frequency distribution is also presented for all the response quantities. Using the fitted lognormal distribution, the cumulative distributions of both the response quantities are evaluated and presented in Fig. 7.

From the frequency distribution plots, it is observed that the resulting base shear (or base moment) lies within a lower bound and upper bound, e.g., for E1 (normalized PGA = 0.25 g) the base shear ranges between 0.160 and 0.205 W with response quantities in-between showing a skewed normal distribution. Further, it can be assumed that such distributions will also exist for PGA values in-between 0.25 and 0.35 g. Therefore, for a particular input motion, say E1, the value of base shear can be expected to fall between 0.16 and 0.28 W for the PGA of input motion ranging from 0.25 to 0.35 g. Similarly, the base moment is expected to range between 0.88 and 1.46 Wm under the same earthquake and PGA range.

However, the range of peak response quantities is not the same for both the input motions and a significant difference between these ranges can be observed from Figs. 5 and 6. Under the earthquake E2, the expected range of base shear lies between

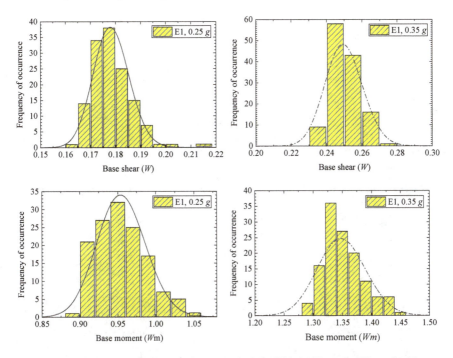

Fig. 5 Frequency distribution of the peak responses under Whittier Narrows (E1) earthquake

0.18 and 0.38 W for the considered range of PGA (Fig. 6). Whereas the expected range of base moment lies between 1.11 and 1.80 Wm. Further, the distribution characteristics of the response quantities are different for different ground motions. This implies that, for the same set of soil-tank models, the characteristics of the foundation motion or more generally the characteristics of input motion (frequency content) affect the expected range of peak seismic response quantities of the tank. Nevertheless, as evident from Figs. 5 and 6, considerably large ranges of the response quantities show the significant effects of uncertain soil parameters on the peak seismic responses of liquid storage tank, regardless of the characteristics of input motion. Therefore, it is crucial to consider the uncertainty related to soil characteristics in the seismic analyses of the tank structure.

It can also be noted that noticeable amplification of peak responses (refer Fig. 5 and 6) is observed with increase in PGA, regardless of the earthquake. Nevertheless, the maximum frequencies of the peak responses are not same for the considered earthquakes. The distributions of peak responses vary considerably when the earthquake changes even with the same PGA. Therefore, it can be concluded that characteristics of the input motion (both PGA and frequency content) have a significant effect on the peak seismic responses of liquid storage tank resting on soil with uncertain parameters.

The limiting values of base shear and overturning moment are obtained as 0.27 and 8.64 Wm [19, 37]. From Fig. 7, for PGA = 0.25 g it can be observed that the

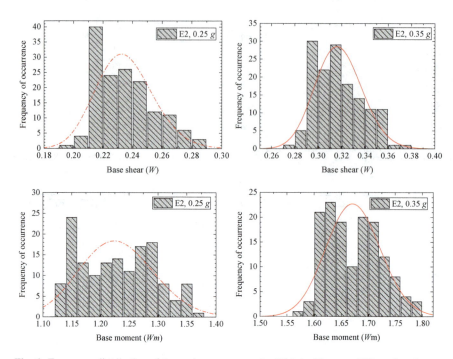

Fig. 6 Frequency distribution of the peak responses under Whittier Narrows (E2) earthquake

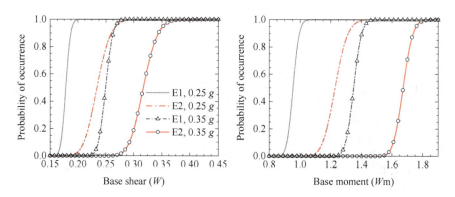

Fig. 7 Probability of occurrence of the peak responses under different earthquakes (E1 and E2)

probability of occurrence of a value greater than or equal to the limiting value of base shear is negligible for earthquake E1 and 0.03 for earthquake E2. For PGA = 0.35 g, the probability of occurrence of the same under the earthquakes E1 and E2 is 0.02 and 0.99, respectively. It is further observed that the probability of occurrence of a value greater than or equal to the peak base shear for E2 (normalized at 0.25 g) is near the corresponding curve for E1 (normalized at 0.35 g). This clearly shows the critical influence of the characteristics of input motion (e.g., frequency content), apart from its PGA, on the resulting peak responses of the liquid storage tank. Nevertheless, the realizations of peak base moment are much lesser than the limiting value, irrespective of the characteristics of the input motion. Therefore, it can be concluded that the liquid storage tank resting on soil with uncertain parameters is more likely to fail due to the exceedance of limiting value of base shear as compared to the base moment.

6 Conclusion

The effect of uncertainty of soil parameters on the seismic responses of a liquid storage tank is evaluated. Soil shear wave velocity and mass density are considered as uncertain parameters. A simplified approach is followed to define the probability distribution functions to the considered soil (uncertain) parameters. A large number of coupled soil-tank models are created with each model having a unique profile of soil parameters, developed through MCS, with respect to the depth. The input motions are normalized to have PGA values closely corresponding to the seismic zone factors of Zones IV and V as specified in IS 1893:2016 [36]. The peak response quantities of the tank are evaluated, namely, base shear and base moment. Further, the probability of occurrence of critical base shear and overturning moment are also presented. Following conclusion are drawn from the present study.

1. The uncertain soil parameters significantly affect the amplitude and probability distribution of peak responses of ground supported liquid storage tank.
2. The amplified peak intensity of input motion results in an overall amplification of peak response quantities of the liquid storage tank.
3. The characteristics of input motion significantly affect the range as well as the overall distribution of the peak response quantities of liquid storage tank.
4. Liquid storage tank resting on soil with uncertain parameters is much likely to undergo a failure mode triggered by the exceedance of limiting base shear as compared to base moment.

References

1. Popescu R, Deodatis G, Nobahar A (2005) Effects of random heterogeneity of soil properties on bearing capacity. Probab Eng Mech 20(4):324–341
2. Kim HK, Narsilio GA, Santamarina JC (2007) Emergent phenomena in spatially varying soils. In: Probabilistic applications in geotechnical engineering, pp 1–10
3. Lizarraga HS, Lai CG (2014) Effects of spatial variability of soil properties on the seismic response of an embankment dam. Soil Dynam Earthq Eng 64:113–128
4. BMTPC (2006) Vulnerability Atlas of India: earthquake, windstorm and flood hazard maps and damage risk to housing. Published by Building Materials and Technology Promotion Council. Ministry of Housing and Urban Poverty Alleviation, Government of India
5. Paice GM, Griffiths DV, Fenton GA (1996) Finite element modeling of settlements on spatially random soil. J Geotech Eng 122(9):777–779

6. Fenton GA, Griffiths DV (2002) Probabilistic foundation settlement on spatially random soil. J Geotech Geoenviron Eng 128(5):381–390
7. Wang S, Hao H (2002) Effects of random variations of soil properties on site amplification of seismic ground motions. Soil Dynam Earthq Eng 22(7):551–564
8. Nour A, Slimani A, Laouami N, Afra H (2003) Finite element model for the probabilistic seismic response of heterogeneous soil profile. Soil Dynam Earthq Eng 23(5):331–348
9. Hamidpour S, Soltani M (2016) Probabilistic assessment of ground motions intensity considering soil properties uncertainty. Soil Dynam Earthq Eng 90:158–168
10. Veletsos AS, Tang Y (1987) Rocking response of liquid storage tanks. J Eng Mech 113(11):1774–1792
11. Haroun MA, Abou-Izzeddine W (1992) Parametric study of seismic soil-tank interaction I: horizontal excitation. J Struct Eng 118(3), 783–797
12. Kim JM, Chang SH, Yun CB (2002) Fluid-structure-soil interaction analysis of cylindrical liquid storage tanks subjected to horizontal earthquake loading. Struct Eng Mech 13(6):615–638
13. Kim MK, Lim YM, Cho SY, Cho KH, Lee KW (2002) Seismic analysis of base-isolated liquid storage tanks using the BE–FE–BE coupling technique. Soil Dynam Earthq Eng 22(9–12):1151–1158
14. Ormeño M, Larkin T, Chouw N (2019) Experimental study of the effect of a flexible base on the seismic response of a liquid storage tank. Thin Walled Struct 139:334–346
15. Meng X, Li X, Xu X, Zhang J, Zhou W, Zhou D (2019) Earthquake response of cylindrical storage tanks on an elastic soil. J Vibr Eng Technol 7(5):433–444
16. Haroun MA, Housner GW (1981) Earthquake response of deformable liquid storage tanks. ASME J Appl Mech 48(2):411–418
17. Haroun MA (1983) Vibration studies and tests of liquid storage tanks. Earthq Eng Struct Dynam 11(2):179–206
18. Shrimali MK, Jangid RS (2002) Seismic response of liquid storage tanks isolated by sliding bearings. Eng Struct 24(7):909–921
19. Saha SK, Sepahvand K, Matsagar VA, Jain AK, Marburg S (2016) Fragility analysis of base-isolated liquid storage tanks under random sinusoidal base excitation using generalized polynomial chaos expansion–based simulation. J Struct Eng 142(10):04016059
20. Lysmer J, Kuhlemeyer RL (1969) Finite dynamic model for infinite media. J Eng Mech Div 95(4):859–878
21. Zhang Y, Yang Z, Bielak J, Conte JP, Elgamal A (2003) Treatment of seismic input and boundary conditions in nonlinear seismic analysis of a bridge ground system. In: 16th ASCE engineering mechanics conference. University of Washington, Seattle, USA
22. NAVFAC (1986) Soil mechanics, design manual 7.01. Naval Facilities Engineering Command
23. Kelly D (2006) Seismic site classification for structural engineers. Structure 21:1–4
24. Subramanian N (2011) Steel structures-design and practice. Oxford University Press
25. Coduto DP, Kitch WA, Yeung MCR (2016) Foundation design: principles and practices, vol. 3. Upper Saddle River, Prentice Hall
26. Bao H, Bielak J, Ghattas O, Kallivokas LF, O'Hallaron DR, Shewchuk JR, Xu J (1998) Large-scale simulation of elastic wave propagation in heterogeneous media on parallel computers. Comput Methods Appl Mech Eng 152(1–2):85–102
27. Zhang Y, Conte JP, Yang Z, Elgamal A, Bielak J, Acero G (2008) Two-dimensional nonlinear earthquake response analysis of a bridge-foundation-ground system. Earthq Spectra 24(2):343–386
28. McKenna F, Fenves GL, Scott MH (2000) Open system for earthquake engineering simulation. Pacific Earthq Eng Res Center (PEER)
29. Luco JE, Hadjian AH (1974) Two-dimensional approximations to the three-dimensional soil-structure interaction problem. Nucl Eng Des 31(2):195–203
30. Joyner WB, Chen AT (1975) Calculation of nonlinear ground response in earthquakes. Bull Seismol Soc Am 65(5):1315–1336

31. Lysmer J (1978) Analytical procedures in soil dynamics. NASA STI/Recon Technical Report No. 80
32. Haroun MA, Ellaithy HM (1985) Model for flexible tanks undergoing rocking. J Eng Mech 111(2):143–157
33. Saha SK, Matsagar VA, Jain AK (2014) Earthquake response of base-isolated liquid storage tanks for different isolator models. J Earthq Tsunami 8(05):1450013
34. ProShake. 2.0 (Ground response analysis program), Edupro Civil System Inc. Retrieved from http://www.ProShake.com
35. Pacific Earthquake Engineering Research Center (PEER) (2012) NGA database. Retrieved from http://peer.berkeley.edu/nga/. Accessed on 05 Jan 2019
36. IS1893: 2016 Indian Standard (2016) Criteria for earthquake resistant design of structures. Bureau of Indian Standards, New Delhi, India
37. Okada J, Iwata K, Tsukimori K, Nagata T (1995) An evaluation method for elastic-plastic buckling of cylindrical shell under shear forces. Nucl Eng Des 157(1–2):65–79

Reliability Assessment of CFRP Composite Laminate Subjected to Low Velocity Impact Damage

Shivdayal Patel, Akshay Sontakke, and Suhail Ahmad

Abstract The stochastic behaviors of the carbon fiber reinforced plastic (CFRP) materials are investigated under the low velocity impact to consider the scatters of the material properties to predict the safety and the reliability of the structure. The probabilistic and the deterministic responses of the CFRP materials for different ply orientations are considered to obtain the optimum design for the structures. Composites have numerous applications in modern industries for the low velocity impact to predict the failure behavior of the structure. Failure of composite laminate is catastrophic in nature due to brittle nature of matrix, fiber, uncertainty in volume of constituent material, anisotropic characteristics, and in homogeneity. The stochastic continuum damage model is used for the composite materials to predict the different modes of failure such as fiber, matrix, and delamination. Modern industries prefer a material having more likelihood to perform a required function with desire life span. The stochastic finite element method is performed to account the scatter in the random fields to determine the stochastic response and reliability of the structure. The probabilistic response of the CFRP structure is determined using Gaussian process response surface method to investigate the probability of failure or reliability of composite structures.

Keywords Composites · Low velocity impact · Safety and reliability

1 Introduction

The carbon fiber reinforced plastic (CFRP) composite materials mostly used for the automobile industry, an aerospace industry, and the offshore structure design due to the light weight and higher specific strength and stiffness of the composite materials.

S. Patel (✉) · A. Sontakke
Indian Institute of Information Technology, Design and Manufacturing, Jabalpur 482005, India
e-mail: shivdayal@iiitdmj.ac.in

S. Ahmad
Indian Institute of Technology, Delhi 110016, India

© The Editor(s) (if applicable) and The Author(s), under exclusive license to Springer Nature Singapore Pte Ltd. 2021
S. K. Saha and M. Mukherjee (eds.), *Recent Advances in Computational Mechanics and Simulations*, Lecture Notes in Civil Engineering 103,
https://doi.org/10.1007/978-981-15-8138-0_46

Structures made up of CFRP material due to Low Velocity Impact (LVI) to predict the damage behavior [1–4]. This damage is mostly occurred from the drop of object or impact from exploded particles. Impact energy dissipation occurred through the different failure modes in structures, i.e., fiber failure, matrix failure, and normal and shear failure in delamination [2–4]. For better designing of composite materials and structure a designer/engineer should have a clear understanding of impact dynamic behavior of composite materials, as these materials will be subjected to the function in industrial sectors hence, one can not underestimate the probability of catastrophic failure of these materials. Most of the time impact damage is too small that it may go ignored reducing the design capability and performance of structure, leads to devastating damage to the system, wealth, and life. The CFRP composite laminate subjected to the LVI and the scatter arises in damage due to the geometric uncertainty, loading uncertainty, and uncertainty in material properties. The reliability methods provided an approximate solution for accurate prediction for the probability of failure and reliability of structure for a given loading conditions [5, 6].

The LVI on CFRP material is extensively studied by Abrate [1], Jagtap et al. [2], Shi et al. [3], and Salvetti et al. [4] to determine the inter and intra-laminate damage in CFRP composite using the continuum damage approach. Finite element analysis-based numerical techniques are used to evaluate the damage behavior and the residual mechanical characteristics of impact. Abrate [1], Jagtap et al. [2], Shi et al. [3] and Salvetti et al. [4], Liao et al. [5], Metouiet al. [6], and Li et al. [7] explored the effects of different failure criteria including Hashin damage criteria, Chang–Chang criteria and Puck criteria for the prediction of the progressive damage behaviors and the different stacking sequence on T700GC/M21 carbon/epoxy composite laminate subjected to the LVI using the user defined subroutine VUMAT code in ABAQUS. Patel et al. [8] presented the probabilistic and sensitivity analysis of the composite laminate under LVI. Gaussian Response Surface Method (GRSM) and Monte Carlo simulation (MCS) techniques were performed to predict the probability of failure and also found that the GRSM showed the great reduction in computation cost and time to achieve the same level of accuracy. Randomness behavior was applied to the elastic modulus, the strength properties, and impactor velocity to predict the realistic failure of the composite structure. Failure probability of each plate was determined using Hashin damage failure criteria and delamination failure criteria as a limiting state function. Randomness in material properties was discretize and model using Karhunen-Loeve (KL) decomposition technique. Gayathari et al. [9] build a reliability model for composite laminate subjected to uniformly distributed load. Probabilistic study is performed to detect the variation in maximum deflection in damaged and undamaged plies by introducing matrix crack. Patel et al. [10–13] determined the failure of probability of each plie of composite laminate subjected to LVI using damage initiation and propagation failure due to delamination and matrix cracking as a limiting state function. Scatter in material properties and variability in impactor velocity is used as randomize variable. GRSM was performed to determine the stochastic behavior of the contact force and the stochastic displacement of the composite plate. Stochastic finite element method (SFEM) was presented for analysis of Offshore composite structure subjected to LVI.

The focus of this studied to determine the probabilistic response of CFRP composite plate subjected to low velocity impact to consider the scatter in the material properties. The stochastic finite element analysis is performed to determine the stochastic contact force of the composite structures.

2 Composite Damage Modeling

2.1 Progressive Damage Modeling of the Composite Plate

The finite element modeling is an essential tool to determine the approximate solution for the engineering problem. In the present study finite element code ABAQUS software is used for the dynamic explicit analysis of CFRP composite laminates.

The governing equation for analysis of explicit dynamics is explained in below [10].

$$[M]\{\ddot{u}\} + [K]\{u\} = \{(F)(t)\} \tag{1}$$

Here, [M] is mass matrix, [K] is stiffness matrix, {F (t)} is instantaneous external forces, and t, $\{\ddot{u}\}$ and $\{u\}$ is time, acceleration and displacement. ω_{max} is maximum frequency step time is given by

$$\Delta t = \frac{2}{\omega_{max}}$$

In acceleration in terms of mass matrix is given by

$$\{\ddot{u}\} = [M]^{-1}(\{(F)(t)\} - [K]\{u\}) \tag{2}$$

In this work, Hashin damage criterion is used to define the damage initiation and damage propagation in material subjected to loading conditions. Relation between stress and strain is given by,

$$\sigma = C_0 \varepsilon \tag{3}$$

where C_0 = stiffness matrix

σ, ε = Cauchy stress and strain.

The effective stress $\{\bar{\sigma}\}$ in the equivalent undamaged continuum related to Cauchy stress $\{\sigma\}$ in the damaged continuum as

$$\{\bar{\sigma}\} = [\Omega]\{\sigma\}$$

$$\Omega_{11} = (1 - d_f^t)(1 - d_f^c)$$

$$\Omega_{22} = (1 - d_m^t)(1 - d_m^c)$$

$$\Omega_{33} = (1 - d_m^t)(1 - d_m^c)$$

$$d_s = (1 - (1 - d_f^t)(1 - d_f^c)(1 - d_m^t)(1 - d_m^c))$$

where $[\Omega]$ = effective damage tensor, Ω_{11}, Ω_{22}, and Ω_{33} represent thediagonal damage values of damage tensor.

d_f, d_m and d_s = Fiber, matrix and shear damage variable and superscript t- tension, c- compression.

2.2 Stochastic Finite Element Method

Stochastic modeling of the CFRP composite material is considered for the uncertainty in material properties to accurately predict the probability of failure for the composite structure under the low velocity impact. The K-L expansion method is used for the discretization of the random fields. The stochastic finite element analysis is performed using the scatters in the materials properties to run the different sets of FEA. The Gaussian response surface method is investigated to accurately predict the stochastic behavior of the CFRP composite structures.

3 Numerical Modeling CFRP Materials

The CFRP composite structure dimension for the plate and impactor are showed in Fig. 1 and the thickness of the composite laminates is 4 mm with a stacking sequence of [02/452/902/−452]2 s. A 3D FE model was designed with a full size of 125 × 75 × 4 mm^3 representing the composite laminate. Two-ply having a same material orientation is treated as a single unit this improved computational efficiency. For closed impact zone 72 × 36 mm meshing with an element size of 1.2 × 0.9 × 0.5 mm was selected. Element SC8R (8-node quadrilateral continuum shell element with reduced integration and hourglass control) is used in meshing of composite lamina. Element R3D4 (4-node 3D bilinear rigid quadrilateral) were used to mesh the rigid impactor body. Composite laminate was constituent of Toray T700 fiber and M21 epoxy matrix material combined together to form a laminate T700CG/M21 [7]. Numerical properties of the composite laminate are used from the literature [7, 10]. The stainless steel rigid projectile was de-signed with a hemispherical head of 16 mm diameter. A clamping steel plate with hole of area 125 × 75 mm in the middle was

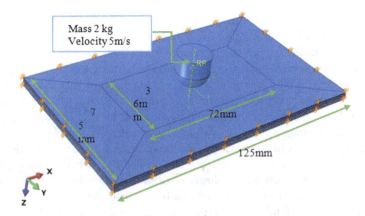

Fig. 1 Assembly of the composite plate and impactor

used to hold the laminate samples during the impact test. A drop velocity of 5 m/s and the mass of the impactor was equal to 2 kg resulting in the impact energies of 25 J, respectively. The LVI impact behavior of the CFRP composite laminate considered the different stacking orientations such as $[0_2/90_2]_{2S}$ and $[0_2/45_2/-45_2/90_2]_S$ is introduced in the FE model. Deterministic responses such as the peak contact force and the peak displacement of the CFRP composite laminate are determined using the different stacking sequences of the structure. All impact characteristics of numerical modeling and material properties of composite laminate kept same for all three models.

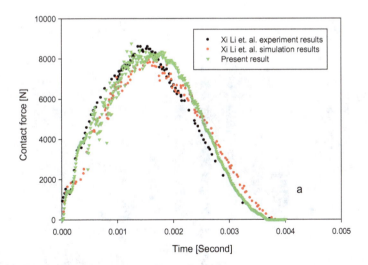

Fig. 2 Validation study of the CFRP material contact force time history

The contact force time history is shown in Fig. 2 and it is observed that the contact force slightly declines at time of 0.18 and 0.35 ms it is because of the matrix cracking and delamination initiations failure started for the composite structure. At 1.6 ms maximum contact force of the peak value 8.6 kN was seen in this study which is still 0.58% less than the experimental results. A maximum displacement 5.381 mm was observed in the study which is still 5.08% more than the experiment results but 2% better than the numerical simulation of the Xi et al. [7]. The damage area of 5720 mm^2 and damage volume of 2449 mm^3 was seen.

Cracking of matrix in the transverse direction was mostly seen in the simulation for the top two plies and bottom two plies of the laminate. Matrix cracking failure started for the top plies and the last plies of composite laminate because of the contact stresses and the shear stresses as well as the bending stresses and the shear stresses respectively most predominant on the top and bottom plies. The delamination failure is also started from the top and the bottom plies because the interlaminar shear stresses and the normal stress are mostly predominant for the composite laminate. The damage area is observed maximum in between 45/90 and 90/−45 interfaces because of the dissimilar ply orientation of the plies. The direction of delamination damage propagation area was in the direction of fiber orientation in the interface. In deterministic analysis fiber failure in tensile mode is observed in almost all the plies while only first and last ply showed a sign of damage compression. The contact force and the displacement responses of the both ply orientation $[0_2/45_2/-45_2/90_2]_S$ and $[0_2/45_2/90_2/-45_2]_S$ are very similar but the peak contact force and the peak displacement of the $[0_2/90_2]_{2S}$ composite laminate is showed that 26% lesser than and 14% greater than of the peak force and the displacement $[0_2/45_2/-45_2/90_2]_S$ composite laminate, respectively.

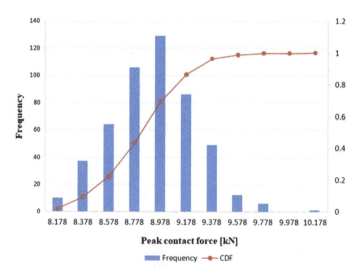

Fig. 3 PDF and CDF of distribution of the peak contact force for $[0_2/45_2/90_2/-45_2]_S$

4 Probabilistic Numerical Modeling

The probabilistic peak contact force of CFRP composite laminate showed the different stacking sequence of the composite laminate such as $[0_2/45_2/90_2/-45_2]_S$, $[0_2/90_2]_{2S}$, and $[0_2/45_2/-45_2/90_2]_S$. Stochastic finite element method performed to determine the probability distribution function (PDF) and Cumulative Distribution Function (CDF) of CFRP structure. The maximum, minimum, and mean values showed the probabilistic analysis which is compared with the mean value response of deterministic study.

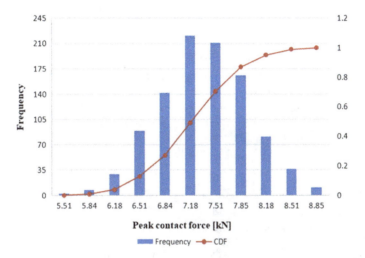

Fig. 4 PDF and CDF of distribution of the peak contact force for $[0_2/90_2]_{2S}$

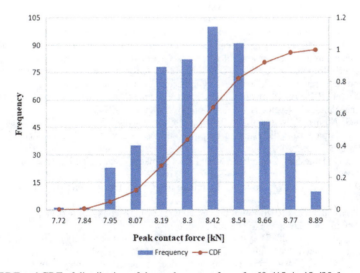

Fig. 5 PDF and CDF of distribution of the peak contact force for $[0_2/45_2/-45_2/90_2]_S$

Probabilistic finite element analysis is performed to estimate the probabilistic response of the peak contact force of the CFRP composite structures. The PDF and CDF plots of the CFRP composite laminate showed in Figs. 3, 4, and 5 for the stacking sequence $[0_2/45_2/90_2/-45_2]_S$, $[0_2/90_2]_{2S}$ and $[0_2/45_2/-45_2/90_2]_S$, respectively. After comparing the stochastic response obtained from the maximum and minimum of the peak contact force with the mean value response of deterministic analysis. It has been found that for $[0_2/45_2/90_2/-45_2]_S$ laminate, there is underestimation of 17.6% in analysis of peak contact force for $[0_2/90_2]_{2S}$ and $[0_2/45_2/-45_2/90_2]_S$ laminate there is underestimation of 36 and 9% is noted.

5 Conclusions

SFEM methodology is developed to calculate the probabilistic response of composite laminate subjected to the low velocity impact damage. Comparative study of stochastic response of the peak contact force using PDF and CDF with the mean value response of deterministic analysis shows the 17% underestimation in $[0_2/45_2/90_2/-45_2]_S$ laminate. For $[0_2/90_2]_{2S}$ and $[0_2/45_2/-45_2/90_2]_S$ laminate shows the underestimation of 36% and 5% in peak contact force, respectively.

References

1. Abrate S (1998) Impact on laminated composite materials. ASME J Appl Mech Rev 44(4):155–1991
2. Jagtap R, Ghorpade Y, Lal A, Singh BN (2017) Finite element simulation of low velocity impact damage in composite laminates. Mater Today Proc 4(2):2464–2469
3. Shi Y, Swait T, Soutis C (2012) Modeling damage evolution in composite laminates subjected to low velocity impact. Compos Struct 94(9):2902–2913
4. Salvetti M, Gilioli A, Sbarufatti C, Manes A, Giglio M (2018) Analytical model of the dynamic behaviour of CFRP plates subjected to low-velocity impacts. Compos Part B Eng 142:47–55
5. Liao BB, Liu PF (2017) Finite element analysis of dynamic progressive failure of plastic composite laminates under low velocity impact. Compos Struct 159:567–578
6. Metoui S, Pruliere E, Ammar A, Dau F (2018) A reduced model to simulate the damage in composite laminates under low velocity impact. Comput Struct 199:34–45
7. Xi Li et al (2019) Assessment of failure criteria and damage evolution methods for composite laminates under low-velocity impact. Compos Struct 207, 727–739
8. Patel S, Ahmad S, Mahajan P (2014) Reliability analysis of a composite plate under low-velocity impact using the Gaussian response surface method. Int J Comput Methods Eng Sci Mech 15(3), 218–226
9. Gayathri P, Umesh K, Ganguli R (2010) Effect of matrix cracking and material uncertainty on composite plates. Reliab Eng Syst Saf 95(7):716–728
10. Patel S, Ahmad S (2017) Probabilistic failure of graphite epoxy composite plates due to low velocity impact. ASME J Mech Des 139(4), 44501
11. Patel S, Soares CG (2018) Reliability assessment of glass epoxy composite plates due to low velocity impact. Compos Struct 200:659–668
12. Patel S, Soares CG (2017) System probability of failure and sensitivity analyses of composite plates under low velocity impact. Compos Struct 180:1022–1031
13. Patel S, Ahmad S, Mahajan P (2016) Probabilistic finite element analysis of S2–glass epoxy composite beams for damage initiation due to high velocity impact. ASME J Risk Uncertainty Part B Mech Eng 2(4), 044504-1–044504-3

Passive Vibration Control of Tall Structures with Uncertain Parameters—A Reliability Analysis

Said Elias, Deepika Gill, Rajesh Rupakhety, and Simon Olafsson

Abstract The paper studies the reliability analysis and passive vibration control (PVC) of tall structures with uncertain parameters. A tall structure is modeled using certain parameters, Monte Carlo method is employed to create the uncertain parameters for the selected tall structure. The robust tune-able PVC (TPVC) schemes are installed to mitigate the response of the structure under earthquake excitations. Different tuned mass damper (TMD) schemes are used to mitigate the response of the structure. These schemes are single TMD (STMD), multiple TMDs (MTMDs), and distributed MTMDs (d-MTMDs). Newmark's integration method is used to solve the equation of motion for the coupled system. It is found that ignoring the uncertainties in parameters of structure cause reduction in performances of the TPVC. Different probability distribution functions (PDF) are compared to select the most suitable one.

Keywords Earthquake · Passive vibration control · Reliability analysis

1 Introduction

Buildings are often subjected to earthquake excitations. Therefore, during last many decades progress has been made to improve the performances of the buildings while subjected to earthquakes. Seismic risk, safety, reliability analysis (RA), and fragility analysis (FA) of buildings has been widely investigated by researchers. Datta [1] reported that the most crucial characteristic of RA was considering the uncertainties that make buildings vulnerable to collapse for a specified limit state. Based on RA the optimum stiffness of various floors, which was used to simplify the analysis of a

---

S. Elias (✉) · R. Rupakhety · S. Olafsson
Earthquake Engineering Research Centre, University of Iceland, Selfoss 800, Iceland
e-mail: said@hi.is

D. Gill
Indian Institute of Technology (IIT) Delhi, New Delhi 110016, India

ten-story reinforced concrete (RC) building [2]. A detailed review of structural reliability methods for seismic safety assessment is provided in reference [3]. Based on the information reported [1–3], three forms of uncertainties exist, which are dominant in seismic RA, specifically: (a) unpredictability and inconsistency of excitation; (b) statistical uncertainty, which ascends because of approximation of parameters relating statistical models; and (c) model uncertainty, which ascends because of inadequacy of structural modeling. Not only these uncertainties can cause failure to conventional type structures, but also can reduce the performances of the controlled structures [4]. Gill et al. [4] proposed a robust control scheme that worked in existence of uncertainty in stiffness or damping of the building. Later, the application of that scheme was verified for response mitigation of non-linear structure while subjected to number of ground excitations [5]. Earlier, the robust system was examined for response mitigation of different structures subjected to dynamical loadings [6–15]. Saha et al. [16, 17] studied the uncertainty analysis of base-isolated (BI) structures under random excitations. Earlier a detailed literature survey about research progress of passive vibration control (PVC) such as tuned mass dampers (TMDs) was presented by Elias and Matsagar [18]. The detailed reliability analysis (RA) of controlled structures is missing in the literature. Hence, the intention of this study is to conduct the reliability analysis of tall structures controlled with different tunable PVC schemes. These schemes are single TMD (STMD), multiple TMDs (MTMDs), and distributed MTMDs (d-MTMDs). Also, a suitable distribution is identified to choose suitable probability density function (PDF). Uncertainties in structures' stiffness and damping are determined by employing Monte Carlo method. The performances of the controller schemes are checked in presence of structural uncertainties.

2 Mathematical Modeling

A 80.77 m tall building having plan dimensions of 30.48 m × 36.58 m is considered for the present study. The detail of the building can be found in earlier studies [4, 5]. The technique employed in this study gives design parameters such as mass (m_i), damping (c_i), and stiffness (k_i) of the ith TPVC (Fig. 1) which are not excessively vulnerable to the frequency detuning due to the uncertainties in damping matrix (DM) [C_N] and stiffness matrix (SM) [K_N] of the structure. Figure 1 also illustrates a flowchart for the reliability analysis which is implemented to design a robust TPVC scheme. To be noted that mass [M_N] of the structure is unchanged in this procedure. Firstly, it is important to select proper probability density function (PDF) of the SM, [K_N] of the building. Then, generation of several (P) SM [K_N], j ($j = 1, 2, 3, ..., P$) employing the MC method is required. The frequencies of the building are premediated for [$K_{N,j}$], and DM [$C_{N,i}$] are estimated based on the reduced system and adopting the assumption of modal damping. The cumulative distribution function (CDF) is defined based on selected PDF ones. The CDF is denoted by p_f and can be calculated by

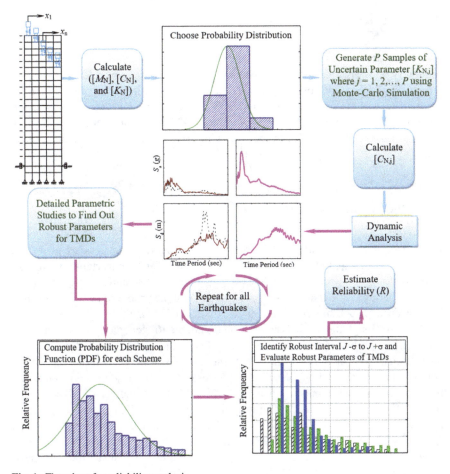

Fig. 1 Flowchart for reliability analysis

$$p_f = p[g(x) \leq 0] = \int f(x)dx \quad (1)$$

here $f_{(x)}$ is the joint probability density function of $g_{(x)} = 0$ and the integration is applied over the domain D where $g_{(x)} < 0$. The reliability is defined as

$$R = 1 - p_f \quad (2)$$

The change in the SM and DM of the main building is ±15% [4]. The reliability of the controlled structure is assed by considering the maximum displacement, drift, and acceleration response of the controlled structure normalized by the uncontrolled one while subjected to earthquake excitations. Maximum peak displacement is denoted by (J_1), maximum peak drift is denoted by J_2, and maximum peak acceleration is denoted by (J_3).

3 Numerical Study

A building having 20 floors as explained in References [4, 5] is considered for reliability analysis presented in this study. In first trial 500 new stiffness matrices are developed and controlled by the TPVC schemes while subjected to different earthquake excitations. In second trail 500 new damping matrices are developed and controlled using the TPVC schemes. These TMD schemes denoted as STMD, MTMDs, and d-MTMDs and the real earthquake excitations are explained in Gill et al. [4].

3.1 Uncertainty in Stiffness

By considering uncertainty in stiffness of the building, new matrices are generated with the help of Monte Carlo simulation. While considering uncertainty in the SM of the building, $P = 500$ random SM are generated. In this study $\pm 15\%$ change in the SM is assumed, so that to assess the efficiency of the proposed technique even in the existence of such superior uncertainties than indicated in the benchmark problem ($\pm 10\%$). Although the normal distribution is suitable for generating the matrices, in this study other distributions such as Inverse Gaussian, Weibull, and Lognormal are also verified. To be noted for each new stiffness matrix, Raleigh method is applied to compute the corresponding DM. The responses of the new structures are controlled by the three controller schemes, namely, STMD, MTMDs, and 5d-MTMDs (Figs. 2 through Figs. 3, 4, 5, 6, and 7). The response is normalized by the peak response of the NC ones (donated by $J: J_1, J_2,$ and J_3). The distribution of the objective function (peak displacement, drift, and acceleration) obtained with respect to uncertainties is shown in Figs. 2 through Figs. 3, 4, 5, 6, and 7, in order to evaluate the robustness and reliability of the STMD, MTMDs, and d-MTMDs.

To be concerned about variation of displacement response (J_1) as shown in Figs. 2 and 3. It is found that using any of the distribution techniques will not affect the design performance significantly. Since number of earthquakes are used therefore, Lognormal distribution is recommended to the most suitable one. It is also observed that generally the d-MTMDs are robust and reliable than STMD and MTMDs while subjected to earthquakes.

The variation in acceleration (J_3) control of the building equipped with three TMD schemes subjected to earthquakes is shown in Fig. 6. The variation in the J_3 for the STMD is the least under all the earthquakes. Hereafter, it is apparent that the STMD is truly robust but least effective in acceleration response reduction. However, it is demonstrated in Fig. 6 that in case where the building is installed with STMD, objective function is distributed around 1, indicating that marginally reduced response is achieved. In addition, it is observed that the variation of the relative frequency of objective function is higher in case of the MTMDs and d-MTMDs. This shows higher

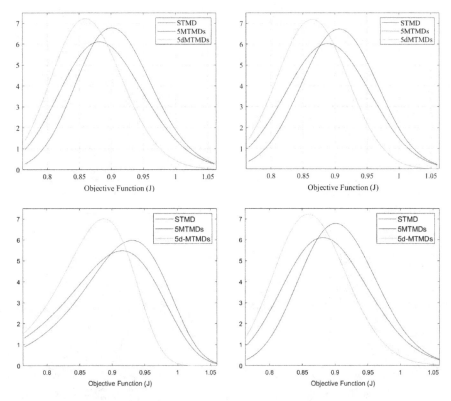

Fig. 2 Effect of stiffness uncertainties on the peak displacement response using MC simulation based on a different probability density function (PDF): Inverse Gaussian (top left), Normal (top right), Weibull (bottom left), and Lognormal (bottom right)

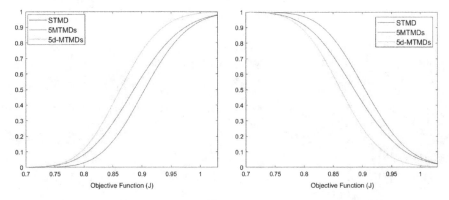

Fig. 3 Robustness (left) and reliability (right) curves based on stiffness uncertainties for the peak displacement response using MC simulation (cumulative density function, CDF based on Lognormal)

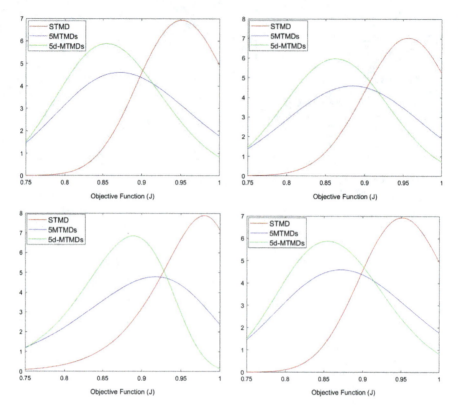

Fig. 4 Effect of stiffness uncertainties on the peak drift response using MC simulation based on different a probability density function (PDF): Inverse Gaussian (top left), Normal (top right), Weibull (bottom left), and Lognormal (bottom right)

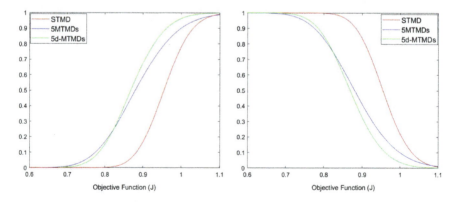

Fig. 5 Robustness (left) and reliability (right) curves based on stiffness uncertainties for the peak drift response using MC simulation (cumulative density function, CDF based on Lognormal)

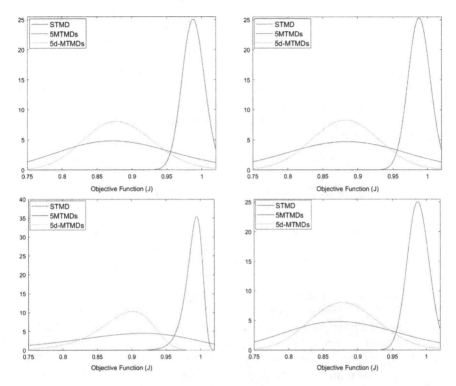

Fig. 6 Effect of stiffness uncertainties on the peak acceleration response using MC simulation based on a different probability density function (PDF): Inverse Gaussian (top left), Normal (top right), Weibull (bottom left), and Lognormal (bottom right)

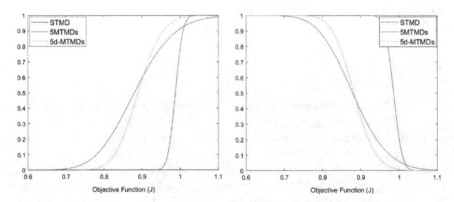

Fig. 7 Robustness (left) and reliability (right) curves based on stiffness uncertainties for the peak acceleration response using MC simulation (cumulative density function, CDF based on Lognormal)

response reduction and relatively low robustness. Yet, under the earthquakes considered in this study, MTMDs and d-MTMDs are successfully reducing the acceleration response of the steel building (see Fig. 7). Nevertheless, d-MTMDs have advantage over MTMDs as the masses are distributed on different floors. In general, the figures confirm the reliability of multi-modal response mitigation of 20 story steel building under earthquakes. Hence, it is suggested to go for d-MTMDs as the heavy mass is divided and placed on different floors. Also, fitting is simpler as matched by a heavy STMD added on top floor.

3.2 Uncertainty in Damping

By considering uncertainty in damping of the main structure, 500 new matrices are produced. Performance of the STMD, MTMDs, and 5d-MTMDs in presence of uncertainty in damping of the parent structure under seismic excitations is plotted in Fig. 8.

It is evident from the figure that the uncertainty in DM of the building has lesser effect on performance of the TMDs. Mostly, it is found that STMD is not reliable scheme in controlling the drift and acceleration response. However, the peak displacement is somehow controlled by STMD. The concentration of the objective function around 1 shows the robustness of the STMD but not the effectiveness. MTMDs are less reliable in mitigating the displacement but reliable in controlling the drift and acceleration response. It is found that d-MTMDs are marginally more reliably scheme for seismic response control of structures in existence of uncertainty in the damping of the building. Therefore, this study recommends use of d-MTMDs for seismic response mitigation of structures.

4 Conclusions

The reliability and robustness of three tuned mass damper (TMD) schemes such as single TMD (STMD), multiple TMDs (MTMDs), and distributed MTMDs (d-MTMDs) in the existence of uncertainty in stiffness and in damping of parent structure while subjected to seismic excitations is investigated. It is concluded that use of the d-MTMDs has advantages over MTMDs and STMD schemes because the total mass is divided and they are positioned on different floors. Also, fitting them based on architecture needs is simpler than a heavy STMD on top floor. In addition, d-MTMDs are more reliable in earthquake response control of structures in presence of uncertainty in stiffness or damping than STMD and in some cases of MTMDs.

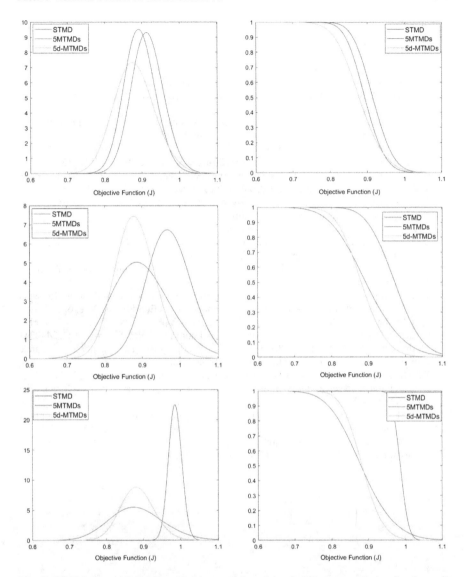

Fig. 8 PDF (left) and reliability (right) curves based on damping uncertainties for the peak displacement, drift, and acceleration response using MC simulation (Lognormal distribution)

References

1. Datta TK (2010) Seismic analysis of structures. John Wiley & Sons
2. Beck AT, dos Kougioumtzoglou IA, Santos KR (2014) Optimal performance-based design of non-linear stochastic dynamical RC structures subject to stationary wind excitation. Eng Struct 78:145–153
3. Der Kiureghian A (1996) Structural reliability methods for seismic safety assessment: a review. Eng Struct 18(6):412–424

4. Gill D, Elias S, Steinbrecher A, Schröder C, Matsagar V (2017) Robustness of multi-mode control using tuned mass dampers for seismically excited structures. Bull Earthq Eng 15(12):5579–5603
5. Elias S, Matsagar V (2019) Seismic vulnerability of a non-linear building with distributed multiple tuned vibration absorbers. Struct Infrastruct Eng 16:1–6
6. Elias S (2018)Seismic energy assessment of buildings with tuned vibration absorbers. Shock Vibr 2018
7. Stanikzai MH, Elias S, Matsagar V, Jain AK (2019) Seismic response control of base-isolated buildings using tuned mass damper. Aust J Struct Eng 1–2
8. Stanikzai MH, Elias S, Matsagar V, Jain AK (2019) Seismic response control of base-isolated buildings using multiple tuned mass dampers. Struct Des Tall Spec Build 28(3):e1576
9. Elias S, Matsagar V, Datta TK (2019) Dynamic response control of a wind-excited tall building with distributed multiple tuned mass dampers. Int J Struct Stab Dyn 19(06):1950059
10. Elias S, Matsagar V, Datta TK (2019) Along-wind response control of chimneys with distributed multiple tuned mass dampers. Struct Control Health Monit 26(1):e2275
11. Elias S, Matsagar V (2018) Wind response control of tall buildings with flexible foundation using tuned mass dampers. In: Wind engineering for natural hazards: modeling, simulation, and mitigation of windstorm impact on critical infrastructure, pp 55–78
12. Elias S (2019) Effect of SSI on vibration control of structures with tuned vibration absorbers. Shock Vibr
13. Elias S, Matsagar V, Datta TK (2017) Distributed multiple tuned mass dampers for wind response control of chimney with flexible foundation. Proced Eng 199:1641–1646
14. Elias S, Matsagar V (2017) Effectiveness of tuned mass dampers in seismic response control of isolated bridges including soil-structure interaction. Latin Am J Solids Struct 4(13):2324–2341
15. Matin A, Elias S, Matsagar V (2019) Distributed multiple tuned mass dampers for seismic response control in bridges. Proc Inst Civil Eng Struct Build 1–8
16. Saha SK, Sepahvand K, Matsagar VA, Jain AK, Marburg S (2013) Stochastic analysis of base-isolated liquid storage tanks with uncertain isolator parameters under random excitation. Eng Struct 57:465–474
17. Saha SK, Matsagar V, Chakraborty S (2016) Uncertainty quantification and seismic fragility of base-isolated liquid storage tanks using response surface models. Probab Eng Mech 43:20–35
18. Elias S, Matsagar V (2017) Research developments in vibration control of structures using passive tuned mass dampers. Annu Rev Control 44:129–156

Dual Polynomial Response Surface-Based Robust Design Optimization of Structure Under Stochastic Blast Load

Gaurav Datta and Soumya Bhattacharjya

Abstract The present study deals with robust design optimization (RDO) of an underground bunker under stochastic blast-induced ground motion. Since the direct Monte Carlo Simulation (MCS) requires extensive computational time to solve an RDO problem, the polynomial response surface method (RSM) is highly appreciated as an alternative to reduce the computational burden. Thus, the present study attempts to explore the advantage of the moving least-squares method (MLSM) based dual RSM to take into account the record-to-record variation in the blast load in place of the conventional least-squares method (LSM). The blast-induced ground shock has been artificially generated by considering the Tajimi-Kanai power spectral density function (PSDF) and incorporating uncertainty in the blast parameters, viz. explosion distance and explosive charge weight. The application of the proposed dual RSM in RDO not only evades several finite element analyses runs in the simulation loop during the optimization process, but also improve the computational efficiency significantly. The RDO is formulated by simultaneously optimizing the expected value and variance of the performance function by using the weighted sum approach. The results show that the present MLSM-based RDO strategy yields more accurate and robust solutions than the conventional LSM-based approach when compared with the direct MCS results.

Keywords Dual response surface method · Robust design optimization · Stochastic blast load · Underground bunker

---

G. Datta · S. Bhattacharjya (✉)
Indian Institute of Engineering Science and Technology, Shibpur, Howrah 711103, India
e-mail: soumya@civil.iiests.ac.in

G. Datta
e-mail: gaurav.rs2015@civil.iiests.ac.in

© The Editor(s) (if applicable) and The Author(s), under exclusive license to Springer Nature Singapore Pte Ltd. 2021
S. K. Saha and M. Mukherjee (eds.), *Recent Advances in Computational Mechanics and Simulations*, Lecture Notes in Civil Engineering 103,
https://doi.org/10.1007/978-981-15-8138-0_48

1 Introduction

The prevalent philosophy in practice in India is the deterministic approach to blast-resistant design as outlined by the Indian Standard codal provisions [1]. However, it is well accepted that even in the case of a clearly defined blast loading condition, the load-related parameters may show a remarkably high degree of uncertainty [2]. Moreover, the optimal design of structure under blast load disregarding uncertainty may lead to unsafe design, inviting failure consequences [3, 4]. Hence, the paradigm is gradually shifting from deterministic design optimization (DDO) of structure to robust design optimization (RDO).

The RDO is an attractive option since it is capable of assuring an improved performance of structure by minimizing the variation of performance and ensuring necessary safety. The study on RDO of reinforced concrete (RC) structures subjected to stochastic seismic load is notable in this regard [5]. It has been generally observed in the literature that the studies on RDO of structures under blast loading are scarce [6–9].

Israel and Tovar [6] presented a robust and reliable design of blast mitigating shell structure under stochastic blast loading by adopting the univariate dimension reduction methodology. However, they mainly dealt with topology optimization of the plate structure of a lightweight ground vehicle. Marjanishvili and Katz [7] outlined a decision-based framework by which the magnitude of the consequences to blast-induced local damage can be calculated and used to assess structural resiliency. Stewart [8] predicted reliability-based design load factors considering air blast variability on a model of RC column in direct Monte Carlo Simulation (MCS) framework. Hadianfard et al. [9] conducted a reliability analysis of steel columns under blast loading due to uncertainties associated with blast loading and material properties. They have also used MCS to obtain the damage probability. The direct MCS approach is the most accurate way to obtain RDO solution under random dynamic loading [4, 6]. However, the main drawback of the MCS approach is the extensive computational time requirement. Hence, in the present RDO study, an adaptive polynomial response surface method (RSM) has been applied to overcome the associated computational burden. The conventional RSM is hinged on the concept of least-squares method (LSM). However, it has been reported in the literature that the LSM is a major source of error in the RSM [10]. Thus, in the present study, a moving least-squares method (MLSM) based RSM is explored for RDO under stochastic blast-induced ground motion (BIGM) considering parameter uncertainty.

The application of RSM to deal complex mechanical models of the structure under dynamic load is quite straightforward. But there remains a critical issue for structural response approximation under stochastic BIGM, as the load time-history varies significantly even with the same value of load-related parameters. In this regard, the concept of dual RSM [11, 12] is felt to be more elegant for efficient sampling encompassing (i) random dynamic loading and (ii) design variable (DV) and design parameter (DP) vector. This procedure helps to lucidly construct the Design of Experiment (DOE) containing stochastic load parameters in only one go.

The concept of dual RSM has been successfully applied for fragility analysis and optimization under stochastic earthquake [12, 13]. However, the dual RSM is not yet applied in RDO under blast load and builds the uniqueness of this study.

2 Robust Design Optimization of Structure

The RDO is fundamentally concerned with minimizing the effect of uncertainty in the DVs, x and DPs, z. Let us consider two designs: the conventional optimal design (x_{opt}) and robust optimal design (x_{rob}). When both the designs are perturbed with the same amount of input deviation Δx due to uncertainty, the output response probability density function (PDF) has considerably less deviation for RDO (Δf_{rob}) than the other case Δf_{opt}, i.e., $\Delta f_{rob} < \Delta f_{opt}$. The RDO captures comparatively a flatter insensitive region of the performance function. The RDO can be expressed as [14],

$$\text{find } \mathbf{x}, \text{ to minimize } \phi(\mathbf{x}, \mathbf{z}) = (1-\alpha)\frac{\mu_f(\mathbf{x},\mathbf{z})}{\mu_f^*} + \alpha\frac{\sigma_f(\mathbf{x},\mathbf{z})}{\sigma_f^*}, 0 \leq \alpha \leq 1$$

$$\text{subjected to } \Gamma_j(\mathbf{x},\mathbf{z}) = F_{g_j(\mathbf{x},\mathbf{z})}^{-1}\left(1 - P_{f,j}^{acc}\right) \leq 0, \forall j \in J, x_i^L \leq x_i \leq x_i^U \quad (1)$$

where $\Gamma_j(\mathbf{x},\mathbf{z})$ is the value of the jth constraint function obtained from the inverse CDF of $g_j(\mathbf{x},\mathbf{z})$, i.e., $F_{g_j(\mathbf{x},\mathbf{z})}^{-1}(.)$ for a user-specified acceptable probability of constraint feasibility $1 - P_{f,j}^{acc}$. The relationship between $\Gamma_j(\mathbf{x},\mathbf{z})$ and $F_{g_j(\mathbf{x},\mathbf{z})}^{-1}(.)$ can be written as,

$$F_{g_j(\mathbf{x},\mathbf{z})}\left(\Gamma_j(\mathbf{x},\mathbf{z})\right) = P\left(g_j(\mathbf{x},\mathbf{z}) \leq \Gamma_j(\mathbf{x},\mathbf{z})\right) = 1 - P_{f,j}^{acc}$$
$$\Rightarrow \Gamma_j(\mathbf{x},\mathbf{z}) = F_{g_j(\mathbf{x},\mathbf{z})}^{-1}\left(1 - P_{f,j}^{acc}\right) \quad (2)$$

3 Dual RSM in RDO

In the dual RSM, the responses are evaluated at each DOE point for all the input blast loads in the suite. Then, the mean Y_μ and the standard deviation (SD), Y_σ of any desired response 'Y' are computed at the considered design parameters. The response surface for mean and SD are obtained for the considered responses, i.e.,

$$Y_\mu = h_1(\mathbf{x},\mathbf{z}) \text{ and } Y_\sigma = h_2(\mathbf{x},\mathbf{z}) \quad (3)$$

Finally, the CDF of the response is constructed by using Y_μ and Y_σ. For example, if the response is observed to follow Extreme Value type I distribution, then CDF of

Y is given as,

$$F_Y(\mathbf{Y}) = \exp\left[-\exp\{-\alpha_n(Y - u_n)\}\right] \qquad (4)$$

where $\alpha_n = 1\big/\sqrt{6}(\pi/Y_\sigma)$ and $u_n = Y_\mu - 0.5772/\alpha_n$.

In Eq. (4), u_n and α_n are the parameters of the distribution. The dual RSM can be implemented by both the LSM and the MLSM-based RSM. As already discussed, noting the drawbacks of the conventional LSM-based RSM for improved response approximation, the MLSM-based dual RSM is explored in the present study.

4 Implementation of the Proposed RDO Approach

The implementation procedure of the proposed RDO procedure in the dual RSM framework using MLSM-based RSM is explained in this section. The Uniform Design (UD) scheme [15] has been adopted in the present study for the construction of DOE. At each DOE point, l numbers of blast load time histories are generated to capture the record-to-record variations of blast-induced ground motion. In the present study, l is taken as 30. Details of the generation of artificial blast load time history are presented in Sect. 5. It may be noted here that if the number of DOE point is P, dynamic responses are evaluated for $P \times l$ times. In this way, one will have a P-dimensional vector for the mean response (Y_μ) and the SD of response (Y_σ). Thereafter, dual response surfaces for Y_μ, i.e., $h_1(\mathbf{x}, \mathbf{z})$ and Y_σ, i.e., $h_2(\mathbf{x}, \mathbf{z})$ are obtained. These RSM expressions are validated for a wide range of data points other than the DOE points applying standard error norms. It may be noted here that the MLSM-based dual RSM generates a new set of expressions for $h_1(\mathbf{x}, \mathbf{z})$ and $h_2(\mathbf{x}, \mathbf{z})$ at each iteration of RDO depending on the value of the DVs corresponding to that optimization iteration. Expressions for $h_1(\mathbf{x}, \mathbf{z})$ and $h_2(\mathbf{x}, \mathbf{z})$ are then used to estimate CDF of constraint function (see Eq. (1)). Since the MCS is not used inside the optimization iterations by the proposed approach, a substantial amount of computational time is saved. The RDO problem is solved by the Sequential Quadratic Programming (SQP).

5 Generation of Stochastic Blast Loading

Artificial BIGM acceleration time histories compatible with prescribed power spectral density function (PSDF) have been generated by using MATLAB script. Based on field measured data and numerical simulation results, the effects of frequency content, amplitude, and duration of BIGM are approximately represented by the Tajimi and Kanai spectrum [16] as shown in Eq. (5).

$$S_g(\omega) = \frac{1 + 4\xi_g^2\omega^2/\omega_0^2}{\left(1 - \omega^2/\omega_0^2\right)^2 + 4\xi_g^2\omega^2/\omega_0^2} S_0 \quad (5)$$

where S_0 is a scaling factor, ω_0 is the principal frequency and ξ_g is a parameter governing the spectral shape. The parameter ξ_g has a constant value of 0.6. The principal frequency can be written as follows [17]:

$$\omega_0 = 465.62\left(r/q^{1/3}\right)^{-0.13}, 0.3 \leq r/q^{1/3} \leq 10 \text{ (Hz)} \quad (6)$$

In Eq. (6), r is the distance in meters measured from the charge center and q is the TNT charge weight in kilograms. The scaling factor of the spectrum is [18],

$$S_0 = 1.49 \times 10^{-4} r^{-2.18} q^{2.89} \left(\text{m}^2/\text{s}^3\right) \quad (7)$$

The stationary random acceleration time history, $\ddot{u}_g(t)$ is generated from the PSDF function (see Eq. (5)) as [19],

$$\ddot{u}_g(t) \cong \sum_{\tilde{n}=1}^{\tilde{N}} \sqrt{(S_g(\omega)\Delta\omega)}\left(A_{\tilde{n}}\cos(\omega_{\tilde{n}}t) + B_{\tilde{n}}\sin(\omega_{\tilde{n}}t)\right) \quad (8)$$

In Eq. (8), $A_{\tilde{n}}$ and $B_{\tilde{n}}$ are independent Gaussian random variables with zero mean and unit random variance. The frequency step, $\Delta\omega$ is given by $(\omega_u/\tilde{N})$, where ω_u is the upper cut-off frequency, above which the values of the frequency spectrum are insignificant. The interval $[0, \omega_u]$ is divided into $\tilde{N}$ equal parts. The non-stationary BIGM $\ddot{x}_g(t)$ is then obtained by multiplying a deterministic modulating function $\phi(t)$ to $\ddot{u}_g(t)$. Thereby, $\ddot{x}_g(t)$ is obtained as [20];

$$\ddot{x}_g(t) = \phi(t)\ddot{u}_g(t) \text{ and } \phi(t) = e^{-\bar{c}(t-t_1)} \quad (9)$$

where $\bar{c}$ is a constant and t_1 is the strong motion duration.

Now, for the nonlinear solution of dynamic equilibrium equation, the matrix equation of motion with nonlinear stiffness under blast excitation for multi-degree of freedom system can be written as:

$$[\mathbf{M}]\{\ddot{\mathbf{u}}(\mathbf{t})\} + [\mathbf{C}]\{\dot{\mathbf{u}}(\mathbf{t})\} + [\mathbf{K}]\{\mathbf{u}(\mathbf{t})\} = -[\mathbf{M}]\ddot{\mathbf{x}}_g(\mathbf{t}) \quad (10)$$

In the above, [**M**], [**C**] and [**K**] are the global mass, damping and stiffness matrices of the structure, respectively. **ü(t)**, **u̇(t)** and **u(t)** are acceleration, velocity, and displacement vector of the structure due to stochastic blast load, $\ddot{\mathbf{x}}_g(\mathbf{t})$. The dynamic response is needed to be obtained by performing time-history analysis of the finite element model of the structure under study.

A MATLAB script was written to simulate blasting wave and when considering the loading model of explosion; the site condition indirectly through the amplitude envelope function was considered. In order to evaluate the effect of the blast-induced ground motion on the nonlinear response of the structure, charge weights with three different charge centers were simulated. The acceleration–time histories of two typical charge intensity chosen as 500 and 1000 kg with distances of 100 m is illustrated in Fig. 1. Figure 2 shows the record-to-record variation in blast-induced ground motion for the same intensity of charge (300 kg) and explosion distance (300 m), thereby indicating the need for dual RSM in response approximation under stochastic blast load.

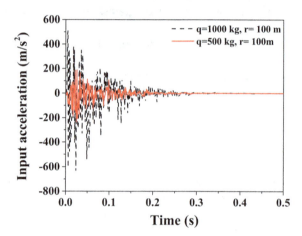

Fig. 1 Two typical blast-induced ground motions: $q = 500$ kg, $r = 100$ m, and $q = 1000$ kg, $r = 100$ m

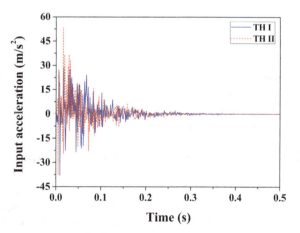

Fig. 2 Time histories showing record-to-record variations for $q = 300$ kg and $r = 300$ m

6 Numerical Study

An underground concrete bunker (Fig. 3) is taken up to elucidate the effectiveness of the present RDO approach. The floor level of the bunker is at a depth of 4 m from the ground level and it spans 15 m and 10 m in the transverse and lateral dimensions, respectively. The finite element modeling is done in SAP2000 software [21] using 3-D frame elements for beams and columns and thin shell element for walls and slabs (Fig. 4). The shell thickness is 300 mm for top and bottom slabs, while the walls are 250 mm thick. The dead load consists of the self-weight of structural and non-structural members as per the Indian Standard code of practice [22]. A uniformly distributed surcharge load of 4 kN/m^2 is applied over the top slab. The concrete grade is considered as M40 (i.e., the compressive strength of concrete is 40 N/mm^2). The reinforcing steel grade is taken as Fe500 (i.e., yield strength of steel is 500 N/mm^2).

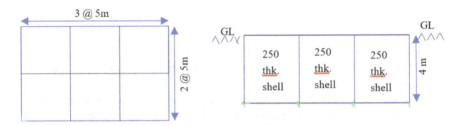

Fig. 3 Plan and elevation of the concrete bunker

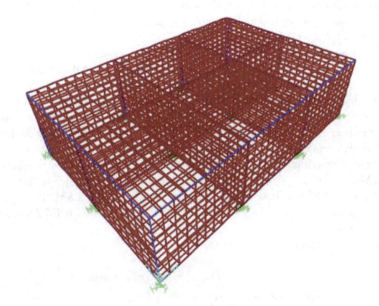

Fig. 4 Finite element model of the bunker

Table 1 Details of the uncertain DVs and DPs

	Variable (and notation)	Unit	Nominal value	COV (%)	pdf
DPs	Explosive charge intensity (q)	kg	100–1000	10–50	Normal
	Explosion distance (r)	m	100–300	10–30	Normal
	Young's Modulus of concrete (E_c)	GPa	29.58	10	Log-Normal
	Unit cost of concrete (C_c)	INR	12000/cu. m.	10	Normal
DVs	Wall thickness (x_1)	mm	To be optimized	5	Normal
	Top slab thickness (x_2)	mm	To be optimized	5	Normal
	Bottom slab thickness (x_3)	mm	To be optimized	5	Normal
	Width of beam (x_4)	mm	To be optimized	5	Normal
	Depth of beam (x_5)	mm	To be optimized	5	Normal
	Width of column (x_6)	mm	To be optimized	5	Normal
	Depth of column (x_7)	mm	To be optimized	5	Normal

The foundation is raft type. The lateral earth pressure is considered on the outer walls of the bunker. The effect of soil-structure interaction is taken by horizontal and vertical springs according to FEMA 356 [23].

The objective function of the DDO problem involves minimization of the material cost of the structure with maximum lateral deflection constraint. The details of the assumed statistical properties of the DVs and DPs are presented in Table 1. The variation in the load-related DPs is considered to be maximum of 50%. The DOE points are obtained for eleven random variables by arranging 30 equidistant levels of each variable according to the UD table. Thereafter, thirty time-histories are generated for each of the DOE points.

The nonlinear dynamic response of the structure is obtained by using SAP2000 [21] software. The response approximation capabilities by the MLSM and the LSM-based metamodels are studied by comparing the response obtained by the metamodels with the FEA-based actual results in Fig. 5 for ten test points at different q. These ten test points are different from the DOE points needed to construct the metamodel. It can be observed that the MLSM-based RSM predictions are in close conformity with the FEA results than by the conventional LSM-based RSM predictions. The coefficient of determination (R^2) values calculated based on these test points is 0.99 for the MLSM and 0.76 for the LSM. It can be observed that R^2 value for the MLSM is nearly 1.0, whereas it is significantly lesser than 1.0 for the LSM. This endorses the superiority of the MLSM-based RSM over the LSM-based RSM for response approximation.

The DDO problem is formulated as,

$$\text{Min.} f(\mathbf{x}, \mathbf{z}) = \text{cost of the structure}$$
$$\text{subjected to: } g(\mathbf{x}, \mathbf{z}) : \delta(\mathbf{x}, \mathbf{z}) \leq \delta_a; \mathbf{x}^L \leq \mathbf{x} \leq \mathbf{x}^U \quad (11)$$

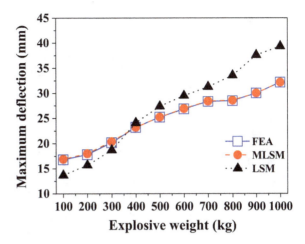

Fig. 5 Comparison of the MLSM and the LSM-based response predictions with the actual response

where δ and δ_a are the maximum lateral deflection and allowable deflection, respectively. δ_a is taken as ($H_s/250$), where H_s is the height of the structure [24].

The CDF of the response generated by the dual RSM is then used to formulate the RDO of Eq. (1). The RDO is performed by the proposed MLSM-based dual RSM and the conventional LSM-based RSM approach. The results are then compared with the most accurate direct MCS-based RDO. The results are presented for varying explosive charge weights. The optimal cost and the SD of optimal cost obtained by the RDO are presented in Figs. 6 and 7, respectively. It can be observed that the optimal cost and its SD by the MLSM-based dual RSM approach is in close conformity with the most accurate direct MCS-based approach. It is also important to note here that the results by the conventional LSM-based RSM approach is significantly deviated

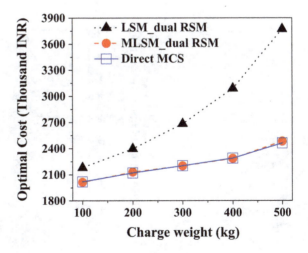

Fig. 6 The robust optimal cost with varying explosive charge weight

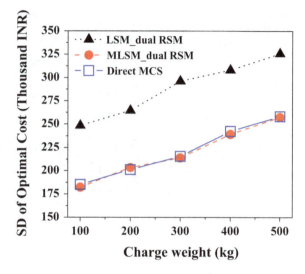

Fig. 7 The SD of the optimal cost with varying explosive charge weight

from the direct MCS results. This warns the application of the LSM-based RSM in the RDO.

The Pareto-fronts [25] obtained by the proposed and the conventional RDO approaches are plotted in Fig. 8. The Pareto-front obtained by the MLSM-based dual RSM is much closer to the direct MCS-based RDO results in comparison to that obtained by the LSM-based dual RSM. The dispersion in optimal cost is maximum of 8.3% (by the proposed approach) for input uncertainty considered as high as 50%, which indicates the robustness of the structure with respect to the considered uncertainties. The computational time by the RDO to generate complete Pareto front when direct MCS is used without taking the help of a metamodel is 8 h (with one 4 GB

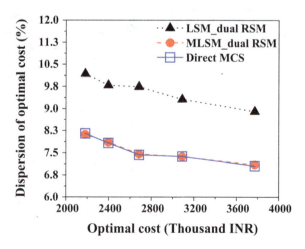

Fig. 8 The Pareto-optimal curve

RAM CPU) approximately. The proposed MLSM-based dual RSM and the conventional LSM-based dual RSM yields the Pareto front in approximately 30 and 26 min, respectively (with one 4 GB RAM CPU). The associated numbers of function calls are 363280 and 16549 by the MCS-based RDO and the MLSM-based dual RSM, respectively. This endorses the computational efficiency of the proposed MLSM-based dual RSM in RDO. It can be further observed that the inaccurate LSM-based approach takes lesser computational time than the proposed MLSM-based approach. This is due to the fact that MLSM-based dual RSM generates one new RSM expression at each iteration of RDO.

7 Summary and Conclusions

An efficient RDO procedure in the MLSM-based dual RSM framework is presented under random blast-induced ground motion. The efficiency of the proposed approach is demonstrated by optimizing an underground bunker made up of concrete. It has been observed that with respect to the conventional deterministic design the RDO yields 9% higher optimal cost considering uncertainty in load and other system parameters. But, at the same time, COV of the objective function is reduced to 8.3% (maximum) by the proposed RDO approach. The present MLSM-based RDO approach is observed to be more accurate than the conventional LSM-based RDO approach when compared with direct MCS as reference. In this regard, the drawback of the conventional LSM-based RSM approach is clearly distinguishable. Also, the proposed RDO approach is computationally efficient as it yields the Pareto-front in substantially lesser time than the direct MCS-based RDO approach. The RDO procedure presented here is a general one and can be easily extended to large complex structures, which is under consideration at this stage.

Acknowledgements The research work presented here was supported by the Institute fellowship to the first author from the Ministry of Human Resource Development, Government of India. The second author gratefully acknowledges the funding received from the CSIR to carry out this research work (Scheme No. 22(0779)/18/EMR-II dated 02/05/2018).

References

1. IS 4991 (1968, Reaffirmed 2003) Criteria for blast resistant design of structures for explosions above ground. Bureau of Indian Standards, New Delhi, India
2. Campidelli M, El-Dakhakhni WW, Tait MJ, Mekky W (2015) Blast design-basis threat uncertainty and its effects on probabilistic risk assessment. ASCE J Risk Uncertainty Eng Syst Part A Civil Eng 1(4). https://doi.org/10.1061/ajrua6.0000823
3. Hao H, Stewart MG, Li ZX, Shi Y (2010) RC column failure probabilities to blast loads. Int J Prot Struct 1(4):571–591. https://doi.org/10.1260/2041-4196.1.4.571

4. Xu J, Spencer BF Jr, Lu X, Chen X, Lu L (2017) Optimization of structures subject to stochastic dynamic loading. Comput Aided Civil Infrastruct Eng 32:657–673. https://doi.org/10.1111/mice.12274
5. Bhattacharjya S, Chakraborty S (2011) Robust optimization of structures subjected to stochastic earthquake with limited information on system parameter uncertainty. Eng Optim 43(12):1311–1330. https://doi.org/10.1080/0305215X.2011.554545
6. Israel JJ, Tovar A (2013) Investigation of plate structure design under stochastic blast loading. In: 10th World congress on structural and multidisciplinary optimization, USA
7. Marjanishvili SM, Katz B (2015) A framework for performance-based optimization of structural robustness. In: 12th International conference on applications of statistics and probability in civil engineering, ICASP12, Canada
8. Stewart MG (2018) Reliability-based load factor design model for explosive blast loading. Struct Saf 71:13–23. https://doi.org/10.1080/15732479.2019.1566389
9. Hadianfard MA, Malekpour S, Momeni M (2018) Reliability analysis of H-section steel columns under blast loading. Struct Saf 75:45–56. https://doi.org/10.1016/j.strusafe.2018.06.001
10. Kim C, Wang S, Choi KK (2005) Efficient response surface modeling by using moving least squares method and sensitivity. AIAA J 43(1):2404–2411. https://doi.org/10.2514/1.12366
11. Lin DKJ, Tu W (1995) Dual response surface optimization. J Qual Technol 27(1):34–39. https://doi.org/10.1080/00224065.1995.11979556
12. Rathi AK, Chakraborty A (2017) Reliability-based performance optimization of TMD for vibration control of structures with uncertainty in parameters and excitation. Struct Control Health Monit 24(1):e1857. https://doi.org/10.1002/stc.1857
13. Ghosh S, Ghosh S, Chakraborty S (2017) Seismic fragility analysis in the probabilistic performance-based earthquake engineering framework: an overview. Int J Adv Eng Sci Appl Math. https://doi.org/10.1007/s12572-017-0200-y
14. Venanzi I, Materazzi AL, Ierimonti L (2015) Robust and reliable optimization of wind-excited cable-stayed masts. J Wind Eng Ind Aerodyn 147:368–379. https://doi.org/10.1016/j.jweia.2015.07.011
15. Fang KT, Lu X, Tang Y, Yin JX (2004) Constructions of uniform designs by using resolvable packings and coverings. Discrete Math 274:25–40. https://doi.org/10.1016/S0012-365X(03)00100-6
16. Kanai K (1961) An empirical formula for the spectrum of strong earthquake motions. Bull Earthq Res Inst, University of Tokyo, 39
17. Wu C, Hao H, Lu Y (2005) Dynamic response and damage analysis of masonry structures and masonry infilled RC frames to blast ground motion. Eng Struct 27:323–333
18. Hao H, Wu C (2005) Numerical study of characteristics of underground blast induced surface ground motion and their effect on above-ground structures part II effects on structural responses. Int J Soil Dyn Earthq Eng 25(1), 39–53. https://doi.org/10.1016/j.engstruct.2004.10.004
19. Rice SO (1944) Mathematical analysis of random noise. Bell Syst Techn J 24(1):282–332
20. Jennings PC, Housner GW (1969) Simulated earthquake motion for design purpose. In: Proceedings of the 4th World conference on earthquake engineering (A-1). Santiago, pp 145–160
21. SAP2000 (2007) Integrated software for structural analysis and design. Computers and Structures Inc., Berkeley, California
22. IS875 (Part 2) (1987) Indian standard code of practice for design loads (other than earthquake) for buildings and structures, part 2 imposed loads (Second Revision). Bureau of Indian Standards, New Delhi
23. FEMA (2000) Prestandard and commentary for the seismic rehabilitation of buildings, FEMA-356. Federal Emergency Management Agency, Washington, DC
24. IS1893 (Part 1) (2002) Indian Standard criteria for earthquake resistant design of structures part 1 general provisions and buildings (Fifth Revision). Bureau of Indian Standards, New Delhi
25. Deb K (2001) Multi-objective optimization using evolutionary algorithms. Wiley, New York

Stochastic Modal Damping Analysis of Stiffened Laminated Composite Plate

Sourav Chandra, Kheirollah Sepahvand, Vasant Matsagar, and Steffen Marburg

Abstract Laminate composite plates have several significant applications in aerospace and automobile industries due to their high strength-to-weight ratio, thermal stability, etc. However, thin laminated composite plate has a tendency of buckling when subjected to some adverse loading conditions. The addition of stiffener can avoid the buckling tendency of the thin laminated composite plate. The effect of stiffener on natural frequency and random modal damping are studied herein. The effect of uncertainty in modal damping is accounted while analyzing the composite plate to evaluate the uncertainty in damped dynamic response of the stiffened composite plate. The modal damping of the composite is determined using viscoelastic damping (VED) model. The randomness in the modal damping is propagated from uncertainty in loss factor of the lamina, and stochastic finite element method (SFEM) based on generalized polynomial chaos (gPC) is applied to evaluate the uncertainty in the modal damping of the stiffened laminated composite plate. First-order shear deformation theory (FSDT), including rotary inertia, is adapted to develop collocation-based stochastic finite element formulation of the composite plate with stiffener. Addition of the stiffener increases the frequency of the composite structure. Uncertainty in the modal damping due to the varying layers of the plates and stiffener directions have been investigated.

Keywords Stiffened laminated composite plate · Visco-elastic damping (VED) · Modal damping · Generalized polynomial chaos (gPC)

S. Chandra (✉) · K. Sepahvand · S. Marburg
Chair of Vibroacoustics of Vehicles and Machines, Technical University of Munich (TUM), 85748 Garching bei Munich, Germany
e-mail: sourav.chandra@tum.de

V. Matsagar
Department of Civil Engineering, Indian Institute of Technology (IIT) Delhi, Hauz Khas 110 016, New Delhi, India

© The Editor(s) (if applicable) and The Author(s), under exclusive license to Springer Nature Singapore Pte Ltd. 2021
S. K. Saha and M. Mukherjee (eds.), *Recent Advances in Computational Mechanics and Simulations*, Lecture Notes in Civil Engineering 103,
https://doi.org/10.1007/978-981-15-8138-0_49

1 Introduction

Nowadays, thin laminated composite plates are widely used in aerospace and automobile industries due to their some specific advantages such as light-weight, superior stiffness and strength, and dimensional stability. For taking advantage of the high stiffness, designer often uses the thin laminated plate to some specific structure for reducing the overall weight of the structure. However, thin laminated composite plate often tends to buckle due to some adverse loading or environmental conditions. The designer address this issue by adding a stiffener with the thin plate. The attachment of the stiffener reduces the buckling tendency of the composite plate without much weight penalty to the overall structure. Moreover, the addition of stiffener increases the frequency of the structure. Research on the stiffened plate has initiated in 70's by Olson and Hazel [1]. They had presented an experimental study and conducted finite element (FE) analysis on the isotropic stiffened plates considering in-plane and bending motions. Samanta and Mukhopadhyay [2] have presented static and dynamic analyses of the folded plate structure based on an orthotropic model. Later, they extended this formulation for the free vibration analysis of the stiffened shell [3]. Halder and Sheikh [4] have presented a free vibration analysis of the folded plate with various boundary conditions using high-precision shear deformable triangular element. Sadek and Tawfik [5] have shown an FE formulation to analyze stiffened laminated plate using nine node and three-node plate elements using higher order shear deformation theory (HSDT). This FE model can efficiently analyze thick and moderately thick stiffened laminated composite plate. Attaf and Hallaway [6] have studied vibration responses of eccentrically stiffened and unstiffened composite plate due to applied in-plane loading. Lee and Lee [7] have investigated the influence of the stiffener size on natural frequency and subsequent mode shape of the composite plate. Niyogi et al. [8] have investigated free and forced vibration responses of the laminated composite folded plate structures using nine node Lagrangian plate bending element with 5 degrees of freedom (DOFs) per node, by incorporating rotary inertia. Pal and Niyogi [9] have reported dynamic analysis of the one- and two-folded stiffened laminated composite and sandwich plates using first-order shear deformation theory. Topal and Uzman [10] have presented a frequency optimization of a folded laminated composite plate with respect to the fiber orientation. However, the effect of stiffener on the modal damping of the composite plate has not received much attention until now, although knowledge of the modal damping is essential while evaluating the damped dynamic response of the stiffened laminated composite plate. Hu and Dokainish [11] have reported a comparative analysis of two damping models such as visco-elastic damping (VED) model and specific damping capacity (SDC) model to quantify the modal damping of the laminated composite plate. They have presented a detailed discussion on the FE modeling of the two damping models and their specific advantages while applying them to the laminated composite plate.

Furthermore, stochastic behavior of the modal damping for stiffened laminated composite plate is remained unnoticed, although it is a well-known fact that elastic moduli and damping parameters are quite uncertain in nature. The uncertainty in these parameters have significant influence on the stochastic damped dynamic response of the composite plate. The effects of the uncertainties in elastic and damping parameters on the dynamic response of the composite plate have been shown by Sepahvand et al. [12, 13] using generalized polynomial chaos (gPC) expansion method. Of late, Chandra et al. [14] have presented a stochastic dynamic analysis of graphite-epoxy composite plate subjected to random thermal environment using gPC expansion method. Sepahvand [15, 16] presented a stochastic harmonic analysis of the composite plate due to uncertain modal damping parameters. However, the effect of uncertainty in principle loss factor on modal damping for stiffened and unstiffened laminated composite plate has not been studied yet. The influence of damping from the visco-elastic layer has been investigated in [17]. The paper aims to develop an FE model for stiffened laminated plate using folded plate formulation as presented in [8] to evaluate the natural frequency and modal damping. Subsequently, this deterministic solver is used to evaluate the uncertainty in modal damping for various stiffener orientations and thicknesses using the gPC expansion method.

2 Mathematical Formulation

A thin laminated composite plate of length L, width W, and uniform thickness h, along with a stiffener in a particular direction is considered for the analysis here. The depth of the stiffener is d_s. The thickness of the stiffener and plate is kept the same and consist of n numbers of multi-directional lamina. Classical laminated plate theory considering first-order shear deformation theory (FSDT) with 5 DOFs is used for analysis of the stiffened plate. To account for three translations and three rotations, a drilling DOF is introduced and followed by proper transformation to obtain global matrices of the stiffened composite structure. The modal damping of the composite plate is evaluated by inserting VED model into the FE formulation of the stiffened laminated composite plate. In this study, the loss factor in the principle directions of the lamina is considered as uncertain, and corresponding uncertainty in the modal damping of the stiffened and unstiffened plates are determined using the gPC expansion method. A 3-dimensional 2nd order Hermite polynomial and deterministic coefficients are used to develop the corresponding expansion of the uncertain parameters.

2.1 FE Formulation of the Stiffened Laminated Composite Plate

The generalized displacement field of the composite plate using the FSDT is described by $\{d\} = \{u \quad v \quad w \quad \theta_x \quad \theta_y\}^T$ where, u, v, and w are linear displacement in x, y, and z directions, respectively and θ_x and θ_y are the rotations about x and y direc-

Fig. 1 Laminate geometry and fiber angle orientations with respect to the global axes

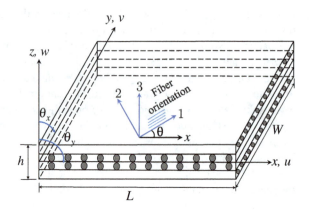

tions, respectively, as shown in Fig. 1. The corresponding displacement relationship is given by

$$u = u_0 + z\theta_y,$$
$$v = v_0 - z\theta_x, \qquad (1)$$
$$w = w_0,$$

where u_0, v_0, and w_0 are the mid-plane displacements along x, y, and z directions, respectively. Shear rotations φ_x and φ_y for the flat plate are stated as

$$\varphi_x = w,_x + \theta_y,$$
$$\varphi_y = w,_y - \theta_x. \qquad (2)$$

In the present formulation, eight-node element is used for conducting the FE analysis. The nodal displacement of the element is given by $\{d_i\}$ and relates with generalized displacement $\{d\}$ using proper interpolation function $[N]$ as

$$\{d\} = [N]\{d_i\}. \qquad (3)$$

The elemental stiffness matrix $[K_e]$ is taken in the following form

$$[K_e] = \int_{A_e} [B]^T [D][B] dA_e, \qquad (4)$$

where $[B]$ is strain-nodal displacement matrix, and written as

$$\{\epsilon\} = [B]\{d_i\}. \qquad (5)$$

Herein, $\{\epsilon\}$ is mid-plane strain vector of the laminated composite plate. The average stiffness matrix $[D]$ is given by

$$[D] = \begin{bmatrix} A_{11} & A_{12} & A_{16} & B_{11} & B_{12} & B_{16} & 0 & 0 \\ A_{12} & A_{22} & A_{26} & B_{12} & B_{22} & B_{26} & 0 & 0 \\ A_{16} & A_{26} & A_{66} & B_{16} & B_{26} & B_{66} & 0 & 0 \\ B_{11} & B_{12} & B_{16} & D_{11} & D_{12} & D_{16} & 0 & 0 \\ B_{12} & B_{22} & B_{26} & D_{12} & D_{22} & D_{26} & 0 & 0 \\ B_{16} & B_{26} & B_{66} & D_{16} & D_{26} & D_{66} & 0 & 0 \\ 0 & 0 & 0 & 0 & 0 & 0 & A_{44} & A_{45} \\ 0 & 0 & 0 & 0 & 0 & 0 & A_{45} & A_{55} \end{bmatrix}, \quad (6)$$

where $A_{ij}, B_{ij}, D_{ij} = \sum_{k=1}^{N} \int_{z(k-1)}^{z_k} (Q_{ij})_k (1, z, z^2) dz$, $i, j = 1, 2, 6$ and $A_{ij} = \sum_{k=1}^{N} \int_{z(k-1)}^{z_k} \kappa(Q_{ij})_k dz$, $i, j = 4, 5$, $\kappa = 5/6$. Here, $(Q_{ij})_k$ is element of off-axis stress–strain relationship matrix $[Q]$ of kth lamina and is obtained from $(C_{ij})_k$ with proper transformation on $[C]$. Here, $(C_{ij})_k$ is element of on-axis stress–strain relationship matrix $[C]$. The elements of the $[C]$ matrix for the kth orthotropic lamina are $C_{11} = E_{11}/(1 - \nu_{12}\nu_{21})$, $C_{12} = \nu_{21} E_{11}/(1 - \nu_{12}\nu_{21})$, $C_{22} = E_{22}/(1 - \nu_{12}\nu_{21})$, $C_{44} = G_{23}$, $C_{55} = G_{13}$, and $C_{66} = G_{12}$. Herein, E_{11} and E_{22} are elastic moduli, G_{12}, G_{13} and G_{23} are shear moduli, and ν_{12} and ν_{21} denote Poisson's ratios of the lamina in the principal lamina direction. The elemental mass matrix is written as

$$[M_e] = \int_{A_e} [N]^{\text{T}} [m] [N] dA_e, \quad (7)$$

where $[m]$ is the inertia matrix of the composite plate.

Transformation matrix for stiffener: The positive directions of the linear displacements u, v, and w, and the rotations θ_x and θ_y of the plate element are shown in Fig. 2. A transformation matrix $[T]$ [8] is developed to relate local displacement $\{d\}$ of the stiffener and global displacement $\{d'\}$ as

$$\{d\} = [T]\{d'\}. \quad (8)$$

Here $[T]$ matrix posses properties of orthogonality. Thereafter, the transformed elemental stiffness $[K'_e]$ and elemental mass $[M'_e]$ matrices are expressed in global coordinate system as

$$[K'_e] = [T]^{\text{T}} [K_e][T],$$
$$[M'_e] = [T]^{\text{T}} [M_e][T]. \quad (9)$$

Prior to applying the transformation, the 40 × 40 elemental stiffness and mass matrices are expanded to 48 × 48 size by inserting θ_z drilling degree of freedom at each node of the element. The off-diagonal terms corresponding to θ_z are considered as zero, whereas a very small positive value is inserted in corresponding leading diagonal terms. Generally, inserted positive value is assumed as 100000 times smaller than the smallest leading diagonal term [18]. This formulation often described as folded plate formulation.

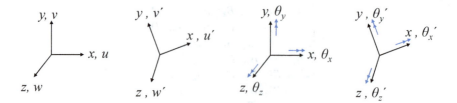

(a) Transformation of translations from local (x, y, z) to global (x', y', z') axes.

(b) Transformation of rotations from local (x, y, z) to global (x', y', z') axes.

Fig. 2 Transformation from local displacement $\{d\}$ to global displacement $\{d'\}$

The global stiffness matrix $[K']$ and the global mass matrix $[M']$ are developed after assembling the elemental stiffness $[K'_e]$ and elemental mass matrices $[M'_e]$, respectively, in global coordinate system. Therefore, undamped free vibration analysis of the stiffened composite plate involves the solution of

$$[M']\{\ddot{d}\} + [K']\{d\} = 0, \tag{10}$$

where $\{\ddot{d}\}$ denotes acceleration corresponds to $\{d\}$.

2.2 Modal Damping Formulation Using VED Model

The dissipation of specific energy in the form of acoustic wave and heat is responsible for passive damping during vibration of the structure. Here, mathematical representation of the damping is made by using the visco-elastic damping (VED) model. Accordingly, energy dissipation from the vibrating structure is represented by complex modulus approach. It is assumed that stress, $\sigma(t)$ and strain, $\varepsilon(t)$ are harmonically time dependent for linear visco-elastic material and are modeled as $\sigma(t) = \sigma_0 e^{i\omega t}$ and $\varepsilon(t) = \varepsilon_0 e^{i\omega t}$. Subsequently, the linear visco-elastic constitutive relationship is expressed as

$$\sigma(t) = \int_{-\infty}^{t} C^*(t - \tau) d\varepsilon(\tau). \tag{11}$$

Here, the complex modulus, C^* is expressed as a combination of real and imaginary components as $C^* = C' + iC''$. Real component, C' is termed as storage modulus, whereas imaginary component, C'' is termed as loss modulus. Accordingly, the complex elastic constants of the composite plate in the principle lamina direction are presented as

$$E_{11}^* = E_{11}(1+i\eta_{11}), \ E_{22}^* = E_{22}(1+i\eta_{22}), \ G_{12}^* = G_{12}(1+i\eta_{12}). \tag{12}$$

Herein, η_{11}, η_{22}, and η_{12} are the damping loss factors along the longitudinal, transverse, and shear directions of the lamina, respectively. The complex elastic moduli and shear modulus are inserted in Eq. (6), and corresponding global complex stiffness matrix, $[K'^*]$ is evaluated for stiffened laminated composite plate as

$$[K'^*] = [K'_R] + i[K'_I], \tag{13}$$

where $[K'_R]$ is storage stiffness matrix, and $[K'_I]$ is loss stiffness matrix. Hence, damped free vibration equation of the stiffened composite plate is written as

$$[M']\{\ddot{d}\} + [K'^*]\{d\} = 0. \tag{14}$$

The general solution of Eq. (14) is stated as $\{d\} = \{\phi_i^*\}e^{i\omega_i^* t}$, where ϕ_i^* and ω_i^* are i^{th} complex eigen mode and complex eigen frequency, respectively. Homogeneous solution of the Eq. (14) is written as

$$\{[K'^*] - \omega_i^{*2}[M']\}\phi_i^* = 0, \quad i = 1, 2, ..., n. \tag{15}$$

The corresponding complex eigen frequency is written as

$$\omega_i^* = \omega_{iR}(1+i\eta_i)^{1/2}, \tag{16}$$

where ω_{iR} is the real part of the complex eigen frequency. Hence, modal loss factor, η_i of the composite plate is determine from

$$\eta_i = \frac{\text{Im}(\omega_i^{*2})}{\text{Re}(\omega_i^{*2})}. \tag{17}$$

The modal loss factor, η_i can be written in terms of specific damping capacity (SDC) Ψ_i for i^{th} mode as $\Psi_i = 2\pi\eta_i$. Henceforth, i^{th} modal damping of the stiffened laminated composite plate is described by SDC, Ψ_i.

2.3 Stochastic Modal Damping Formulation

In the present study, loss factors, η_{11}, η_{22}, and η_{12} in the principle directions of the lamina are considered as uncertain parameters. Hence, the complex stiffness matrix of the composite plate become uncertain. Considering these uncertainties, Eq. (14) can be written as

$$[M']\{\ddot{d}(\xi)\} + [K'^*(\xi)]\{d(\xi)\} = 0, \tag{18}$$

where $d(\boldsymbol{\xi})$ is the unknown displacement vector, which is random in nature. The vector $\boldsymbol{\xi} = \{\xi_1 \; \xi_2 \; \xi_3\}^T$ represents all the random variables involved to generate random system responses. Random modal damping of the composite is calculated as a random variable with reference to Eq. (17) as

$$\eta_i(\boldsymbol{\xi}) = \frac{\text{Im}(\omega_i^{*2}(\boldsymbol{\xi}))}{\text{Re}(\omega_i^{*2}(\boldsymbol{\xi}))}. \tag{19}$$

The random modal damping can be expressed in terms of the SDC as

$$\Psi_i(\boldsymbol{\xi}) = 2\pi \frac{\text{Im}(\omega_i^{*2}(\boldsymbol{\xi}))}{\text{Re}(\omega_i^{*2}(\boldsymbol{\xi}))}. \tag{20}$$

The random modal damping, $\Psi_i(\boldsymbol{\xi})$ of the composite is approximated by deterministic coefficients and random orthogonal function. The random input parameters are assumed to be in normal distribution, thereby Hermite polynomial is used here as orthogonal basis function. The uncertain modal damping is represented by the mth-order truncated gPC expansion as

$$\Psi_i(\boldsymbol{\xi}) = \sum_{m=0}^{N} a_{im} H_{im}(\boldsymbol{\xi}), \tag{21}$$

where a_{im} is the deterministic unknown coefficients and $H_{im}(\boldsymbol{\xi})$ is the orthogonal stochastic basis functions in the form of Hermite polynomial. The unknown deterministic coefficients, a_{im} are calculated from realizations of the system response on a set of collocation points. A 2nd-order gPC expansion with 3-dimensional random vector $\boldsymbol{\xi}$ is used to construct random Hermite polynomial, which leads to 10 unknown deterministic coefficients. The minimum number of collocation points should be at least equal to the numbers of unknown deterministic coefficients. For stochastic collocation-based gPC expansion method, random loss factors are generated from each set of collocation points and realized the corresponding structural response in terms of the modal damping using deterministic FE solver as discussed in Sect. 2.2. The collocation points are generated from zero and roots of higher order Hermit polynomials; the three collocation points are $(0, \pm 1.732)$. However, $3^3 = 27$ set of collocation points are generated, and subsequent structural response are yielded at pre-generated collocation points. Unknown coefficients, a_{im} for each modal frequency are evaluated by employing least squares minimization technique. The numerical algorithm is given in Fig. 3.

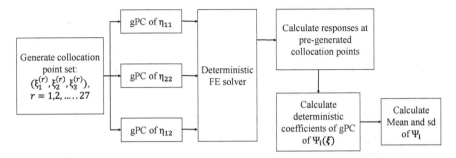

Fig. 3 Numerical algorithm to calculate mean and sd of modal damping of stiffened and unstiffened laminated composite plate

3 Numerical Study

A numerical study is presented here to evaluate the modal frequency and uncertain modal damping of the stiffened and unstiffened laminated composite plates. The modal damping of the composite plate is determined using the VED method. Stochastic study of the modal damping is conducted here considering the uncertainty in the damping loss factor along the principle directions of the lamina. The mean, μ_η and standard deviation (sd) σ_η of the damping loss factor of the lamina are listed in Table 1. The nominal values of the elastic moduli and shear modulus considered in this study are presented in Table 2. As random parameters are in normal distribution, Hermite polynomial is used as an orthogonal basis function.

Geometric dimensions and positioning of the stiffener are described in Table 3, and the geometry of stiffened Plates 4 and 5 is shown in Fig. 4. Plates 1 and 2 are used to validate the in-house MATLab® code developed for evaluating the modal frequency

Table 1 The mean μ_η and standard deviation σ_η of the damping loss factor [11]

Random damping loss factor	Type of distribution	μ_η (1/2π %)	σ_η (1/2π %)
η_{11}	Normal	0.87	0.174
η_{22}	Normal	4.75	0.95
η_{12}	Normal	6.13	1.226

$\Psi_i = 2\pi\eta_i$

Table 2 Nominal values of the elastic parameters and density of the composite laminate [11]

$G_{13} = G_{12}$, $G_{23} = 0.5 G_{12}$ and $\nu_{21} = \nu_{12} \frac{E_{22}}{E_{11}}$			
Elastic constant (GPa)	$E_{11} = 34.49$	$E_{22} = 9.40$	$G_{12} = 4.49$
Poisson's ratio (–)	$\nu_{12} = 0.30$	$\nu_{21} = 0.30$	–
Density (kg/m^3)	1813.9	–	–

Table 3 Geometric dimension of the composite plates with and without stiffener

Plate type	Length L (mm)	Width W (mm)	Thickness h (mm)	Direction of the stiffener	Depth of the stiffener d_s (mm)
Plate 1 [11]	227	227	2.05	–	–
Plate 2	200	200	2	x	25
Plate 3	300	200	2, 6, 12, 20	–	–
Plate 4	300	200	2, 6, 12, 20	x	25
Plate 5	300	200	2, 6, 12, 20	y	25

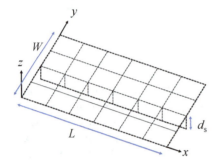
(a) Plate 4: stiffener in x-direction

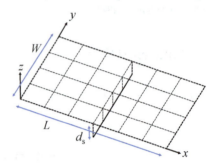
(b) Plate 5: stiffener in y-direction

Fig. 4 Geometry of the stiffened Plates 4 and 5

and modal damping of the unstiffened and stiffened composite plates, respectively. The stochastic analysis of the modal damping is presented with reference to the unstiffened Plate 3, and stiffened Plates 4 and 5 for various thicknesses.

3.1 Validation of the VED Damping Model for Stiffened Composite Plate

The FE model of the stiffened laminated composite plate has been developed using folded plate formulation and is compared with the representative results. The natural frequency and modal damping of the composite plate for free–free boundary condition with and without stiffener are evaluated using the present formulation and are compared with the natural frequency and damping reported by Hu and Dokainish [11] and ANSYS® simulation to establish validity of the present in-house code. Plate 1 is used to calculate the natural frequency and modal damping (Table 4) of the unstiffened composite plate with free–free boundary condition. The Plate 1 is discretized in finite element mesh as 4 × 4. First five natural frequencies and modal

Table 4 Comparison of natural frequency and modal damping for Plate 1 [11]

Mode no	Present		Hu and Dokainish [11]		ANSYS®
	Natural frequency f_i (Hz)	Damping Ψ_i (%)	Natural frequency f_i (Hz)	Damping Ψ_i (%)	Natural frequency f_i (Hz)
1	66.37	5.91	66.43	5.93	66.35
2	126.42	2.34	126.30	2.22	126.40
3	160.14	1.33	160.00	1.49	160.14
4	184.17	4.17	184.00	4.21	184.08
5	206.37	3.20	206.10	3.22	206.32

$f_i = (\omega_i / 2\pi)$

Table 5 Comparison of natural frequency and modal damping for Plate 2 [11]

Mode no.	Present		ANSYS®
	Natural frequency f_i (Hz)	Damping Ψ_i (%)	Natural frequency f_i (Hz)
1	88.21	5.92	88.16
2	148.92	2.34	148.89
3	212.81	4.17	212.74
4	265.16	3.30	265.10
5	300.11	3.39	298.96

damping calculated using the present formulation represent a good agreement with the results reported by Hu and Dokainish [11]; furthermore, the natural frequency is also sufficiently in agreement with the simulation carried out in ANSYS®. It confirms the validity of the present code to calculate the natural frequency and modal damping for unstiffened laminated composite plate. Plate 2 is used to validate the folded plate formulation to analyze stiffened laminated composite plate with free–free boundary condition. For the Plate 2, a stiffener with 25 mm depth is attached in x-direction at the middle of the plate. It is observed from Table 5 that the natural frequency calculated using the present folded plate formulation closely matches with the natural frequency obtained from the corresponding simulation carried out in ANSYS®. It proves, the present formulation can sufficiently evaluate the natural frequency and modal damping for the composite plate with and without stiffener. The modal damping for the Plate 2 is also presented in Table 5. This deterministic FE code is used to generate the modal damping response at pre-generated collocation points for stochastic modal analysis.

3.2 Numerical Results

In the present study, nominal natural frequencies for Plates 3, 4, and 5 for various thickness are shown in Fig. 5a, b, and c, respectively with fixed boundary condition. Here, various thickness of ($0°/90°/0°/90°$) laminated composite plate is used for analysis, and corresponding uncertain damping parameters and nominal elastic properties of the lamina are illustrated in Tables 1 and 2. The first natural frequency for Plate 3, i.e., unstiffened laminated composite plate for corresponding plate thickness is reported as lowest with comparison to the stiffened plates and depicted in Fig. 5a. For Plate 4, i.e., plate with stiffener in x-direction, made-up with 2 mm thick laminate represents higher modal frequencies with comparison to the Plate 5. Whereas first natural frequency for Plate 5 is higher than the Plate 4 while they are made-up using 6, 12, and 20 mm thick laminates. It is observed that addition of the stiffener has increased the first natural frequency of the composite structure. The variation of modal frequencies for the corresponding unstiffened and stiffened composite plates are also shown in Fig. 5. It (Fig. 5) reveals that first natural frequency of the Plate 5 is higher than the Plate 3, made-up using 12 and 20 mm thick laminates, however, higher order modal frequencies of Plate 3 using the corresponding laminates are higher than Plate 5. It indicates that increasing the thickness of the laminate added mass to the stiffened structure thereby, higher order natural frequencies are reducing for Plate 5 with comparison to the unstiffend Plate 3. Therefore, optimum stiffness of the stiffened composite structure can be obtained by suitable combination of the stiffener orientation and laminate thickness.

The mean and sd of the modal damping for the unstiffened and stiffened composite plates are presented in Figs. 6, 7, and 8 for various thicknesses. It is found that Plate 5 with 20 mm thickness gives the highest first mean modal damping whereas, corresponding lowest value is obtained for Plate 3 with 6 mm thick laminate. First mean modal damping value for the Plate 5 remains higher than Plate 4 using 2, 12, and 20 mm thick laminate. Furthermore, Plate 3 gives the lowest first mean modal damping value with comparison to the stiffened plates, i.e., Plate 4 and Plate 5 for 6, 12, and 20 mm thick laminates. However, the magnitudes of the higher damping mode are not in increasing order rather, they varied nonlinearly. It is also noticed that variation in the sd of modal damping exhibits a similar variation trend as that of the corresponding mean values. It implies that the level of uncertainty with respect to the mean value remains constant regardless of the increment in the mode and thickness of the composite plate. It is also observed that randomness in the damping loss factors of the lamina contributed a significant level of uncertainty to the modal damping of the stiffened and unstiffened laminated composite plates. Therefore, effect of uncertainty in the modal damping for unstiffened plate and stiffened plate with various directions of the stiffener should be investigated separately to evaluate random damped dynamic responses of the unstiffened and stiffened composite plates, respectively.

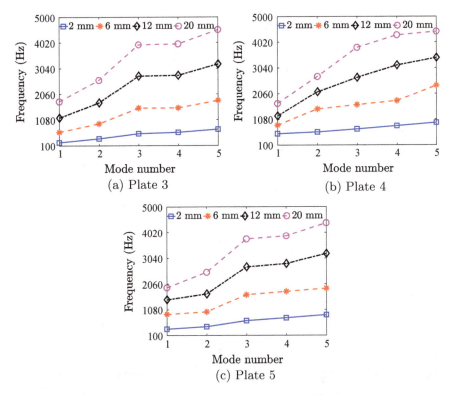

Fig. 5 Natural frequency of Plates 3, 4, and 5 for (0°/90°/0°/90°) laminate with fixed boundary condition

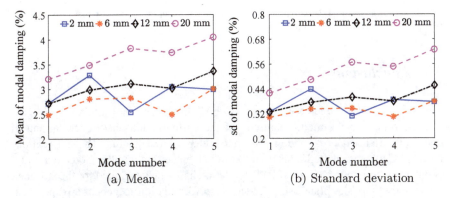

Fig. 6 Mean and sd of modal damping for (0°/90°/0°/90°) laminate (Plate 3) with fixed boundary condition

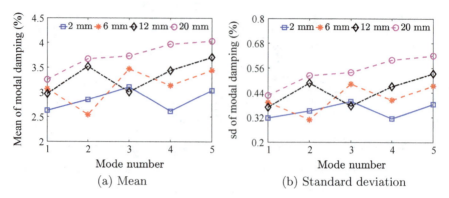

Fig. 7 Mean and sd of modal damping for (0°/90°/0°/90°) laminate with stiffener in *x*-direction (Plate 4) with fixed boundary condition

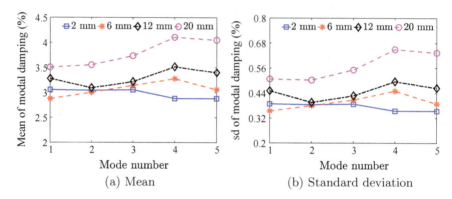

Fig. 8 Mean and sd of modal damping for (0°/90°/0°/90°) laminate with stiffener in *y*-direction (Plate 5) with fixed boundary condition

3.3 Conclusions

The stochastic finite element method (SFEM) has been employed to investigate the uncertainty in the modal damping of the stiffened and unstiffened laminated composite plates due to random loss parameters of the lamina. The collocation-based gPC expansion method is used to evaluate uncertainty in the modal damping. A deterministic finite element (FE) formulation has been developed for the stiffened laminated composite plate using folded plate formulation to generate response at the predefined collocation points. First-order shear deformation theory (FSDT) has been adapted considering rotary inertia into the folded plate formulation. Modal damping of the stiffened laminated composite plate has been determined using visco-elastic damping (VED) formulation. The VED formulation is incorporated into the FE formulation of the stiffened laminated composite plate to calculate the modal damping of the stiffened laminated composite plate. To investigate the effect of stiffener, natural frequency of the composite plate has been determined for various thicknesses and stiffener directions. Plate 5 provides the higher first natural frequency than Plate 3 and Plate 4 for the corresponding laminate thicknesses of 6 mm and more. The mean and standard deviation (sd) of the modal damping is determined for the stiffened and unstiffened composite plates. It is also found that level of uncertainty in the modal damping does not vary much with the variation in the thickness using gPC expansion method. Furthermore, the highest first mean modal damping is reported for 20 mm thick Plate 5. This study indicates that a wise application of stiffener and laminate thickness can increase the stiffness and modal damping of the structure as well. Moreover, this study paved the way for further study on the stiffened laminated composite plate for random damped dynamic responses due to random loss factor.

Acknowledgements The first and third authors gratefully acknowledge the financial support extended by the Deutscher Akademischer Austauschdienst (DAAD) under Research Grants-Doctoral Programmes in Germany and that by the Alexander von Humboldt Foundation under Fellowship for Experienced Researcher.

References

1. Olson MD, Hazell CR (1977) Vibration studies on some integral rib-stiffened plates. J Sound Vibr 50(I)L:43–61
2. Samanta A, Mukhopadhyay M (1999) Finite element static and dynamic analysis of folded plate. J Eng Struct 21:277–287
3. Samanta A, Mukhopadhyay M (2004) Free vibration analysis of stiffened shell by the finite element technique. Eur J Mech A/Solids 23:159–179
4. Haldar S, Sheikh AH (2005) Free vibration analysis of isotropic and composite folded plates using a shear flexible element. Finite Elements Anal Design 42:208–226
5. Sadek EA, Tawfik SA (2000) A finite element model for the analysis of stiffened laminated plates. J Comput Struct 75:369383
6. Attaf B, Hollaway L (1990) Vibrational analyses of stiffened and unstiffened composite plates subjected to in-plane loads. J Compos 21(2):117–126
7. Lee DM, Lee I (1995) Vibration analysis of anisotropic plates with eccentric stiffeners. J Comput Struct 57(1):99–105
8. Niyogi AG, Laha MK, Sinha PK (1999) Finite element vibration analysis of laminated composite folded plate structures. J Shock Vibr 6(5–6):273–283
9. Pal S, Niyogi AG (2008) Application of folded plate formulation in analyzing stiffened laminated composite and sandwich folded plate vibration. J Reinf Compos 27(7):698–710
10. Topal U, Uzman Ü (2009) Frequency optimization of laminated folded composite plates. Mater Design 30:494–501
11. Hu B-G, Dokainish MA (1993) Damped vibrations of laminated composite plates Modeling and finite element analysis. Finite Elements Anal Design 25:103–124
12. Sepahvand K, Scheffler M, Marburg S (2015) Uncertainty quantification in natural frequencies and radiated acoustic power of composite plates: analytical and experimental investigation. Appl Acoust 87:23–29
13. Sepahvand K, Marburg S (2017) Spectral stochastic finite element method in vibroacoustic analysis of fiber-reinforced composites. Proc Eng 199:1134–1139
14. Chandra S, Sepahvand K, Matsagar VA, Marburg S (2019) Stochastic dynamic analysis of composite plate with random temperature increment. Compos Struct 226:111159

15. Sepahvand K, Marburg S (2016) Random and stochastic structural - acoustic analysis. In: Chapter 10 in engineering vibroacoustic analysis: methods and applications. Hambric SA, Sung SH, Nefske DJ (eds). Wiley (2016), pp 305–338
16. Sepahvand K (2017) Stochastic finite element method for random harmonic analysis of composite plates with uncertain modal damping parameters. J Sound Vibr 400:1–12
17. Sepahvand K, Marburg S (2014) Stochastic dynamic analysis of structures with spatially uncertain material parameters. Int J Struct Stab Dyn 14(08):1440029
18. Bathe KJ (2016) Finite element procedures in engineering analysis, 2nd edn. Prentice Hall, Pearson Education, Inc., USA

Support Vector Model Based Thermal Uncertainty on Stochastic Natural Frequency of Functionally Graded Cylindrical Shells

Vaishali and S. Dey

Abstract This paper presents the effect of temperature on stochastic natural frequencies of cylindrical shells, composed of functionally graded materials (FGM) by using machine learning quadratic Support Vector Machine (SVM). An eight noded isoperimetric quadratic element is considered for the finite element formulation. The power law is employed to construct the material modelling of FGM cylindrical shells. Monte Carlo Simulation (MCS) is carried out in conjunction with stochastic eigenvalue solution. In the present study, zirconia (ceramic) and aluminium (metal) are considered to compose the FGM. The machine learning SVM model is constructed to reduce the computational iteration time and cost and validated with the traditional MCS model. The statistical analyses are conducted to portray the first three modes of frequencies. The results show that due to the increase of the temperature, the values of both deterministic as well as the stochastic mean of the first three natural frequencies decreases along with the decrease in sparsity. Sensitivity analysis is also carried out to enumerate the significant important input parameters contributing to influence the output quantity of interest (QoI). The statistical results obtained are the first known results.

Keywords Functionally graded materials · Cylindrical shells · Support vector machine · Monte Carlo simulation and sensitivity

1 Introduction

Functionally graded materials (FGM) are the advanced materials which have gained immense popularity because of its exclusive properties like heat and corrosion resistance in addition to high stiffness and strength. They are extensively used in aerospace, marine, civil construction and mechanical industries. These materials are the example of nonhomogeneous materials which is graded at intervals. Therefore

---

Vaishali (✉) · S. Dey
National Institute of Technology Silchar, Assam 788010, India
e-mail: vaishali765@gmail.com

© The Editor(s) (if applicable) and The Author(s), under exclusive license to Springer Nature Singapore Pte Ltd. 2021
S. K. Saha and M. Mukherjee (eds.), *Recent Advances in Computational Mechanics and Simulations*, Lecture Notes in Civil Engineering 103,
https://doi.org/10.1007/978-981-15-8138-0_50

at different sections, unique properties are present which doesn't resemble the properties of parent materials. By using power law, the gradations of material properties are varied across the thickness.

In general, engineering structures are susceptible to vibration, so dynamic analysis plays an important role. Using a deterministic approach, various authors [1–4] have carried out a vibration analysis of FGM structures. Some researchers worked on the analytical solution [5, 6] of functionally graded plates. In various dimensions, many researchers worked on functionally graded (FG) structures [7–11]. But stochastic responses of FGM shells are yet not intrusively addressed. In the present study, it is aimed to assess the stochastic first three natural frequencies of functionally graded cylindrical shells considering the effect of variation of temperature. In this paper, various sections are presented after Sect. 1 is an introduction, such as Sect. 2: which illustrates the theoretical formulation of FGM cylindrical shells, Sect. 3: depicts the stochastic results obtained followed by discussion while Sect. 4: portrays the concluding remarks.

2 Theoretical Formulation

In FGM, the variability in material properties [12] vary with change in temperature which can be shown by given formulation,

$$Q = Q_0 + Q_{-1}T^{-1} + 1 + Q_1 T + Q_2 T^2 + Q_3 T^3 \tag{1}$$

where the temperature coefficients are represented by $Q_0, Q_{-1}, Q_1, Q_2, Q_3$ while T represents temperature (in Kelvin). The material properties variation across the depth can be expressed by various laws such as sigmoid law, exponential law and power law. In the present study, power law [13] is considered which can be expressed as

$$R(\hat{w}) = R_m(\hat{w}) + [R_c(\hat{w}) - R_m(\hat{w})]\left[{w}/{t(\hat{w})} + 0.5\right]^P \tag{2}$$

where R_m and R_c represent properties of metal and ceramic respectively. Here, $\hat{w}$ represents the degree of stochasticity. The geometry of the cylindrical shell is shown in Fig. 1. Finally, the equation of motion for free vibration is obtained in the global form [7]

$$[M(\hat{w})]\{\ddot{\delta}_p\} + ([K(\hat{w})] + [K_\sigma(\hat{w})])\{\delta_p\} = 0 \tag{3}$$

where $\{\delta_p\}$ is the displacement vector, $[M(\hat{w})]$ is the mass matrix while $[K(\hat{w})]$ and $[K_\sigma(\hat{w})]$ are the random elastic stiffness matrix and the random geometric stiffness matrix, respectively. Considering eigenvalue problem [14], the stochastic natural frequencies ($\omega(\hat{w})$) can be obtained from Eq. (3) as

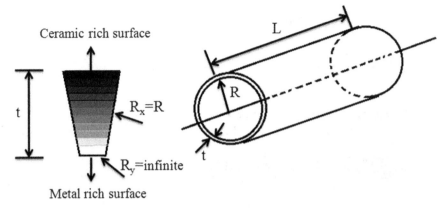

Fig. 1 Functionally graded cylindrical shell geometry

$$[A(\hat{w})]\{\phi\} = \lambda(\hat{w})\{\phi\} \qquad (4)$$

where

$$[A(\hat{w})] = ([K(\hat{w})] + [K_\sigma(\hat{w})])^{-1}[M(\hat{w})]$$
$$\text{and } \lambda(\hat{w}) = 1/[\omega(\hat{w})]^2 \qquad (5)$$

The summary of the entire procedure is represented in the form of flow diagram as shown in Fig. 2.

3 Results and Discussion

In the present study, Monte Carlo simulation is employed for stochastic natural frequency analysis of functionally graded cylindrical shells having Rx = 0.1e20, Ry = 2. The composition of the materials considered for the present analysis is, namely, zirconium (ceramic) and aluminium (metal), whose properties are furnished in Table 1. In the present FE formulation, an eight noded isoperimetric quadratic element is considered. As a surrogate, machine learning quadratic Support Vector Machine (SVM) is used. From the errors (see Table 2) obtained for sample 256, 512 and 1024, the root mean square error (RSME) for sample size 1024 is minimum and the scatter plot (see Fig. 3) for same sample size 1024 shows the least deviation from the perfect prediction, so for further study sample size 1024 is considered. The first three natural frequencies are determined by varying the temperature of functionally graded cylindrical shells i.e. 300, 600, 900 and 1200 K. The present study is validated with the previous research results obtained in past literature [11] as furnished in Table 3. The results show (see Fig. 4) that due to the increase of the temperature, the values of both deterministic, as well as the stochastic mean of the first three natural frequencies

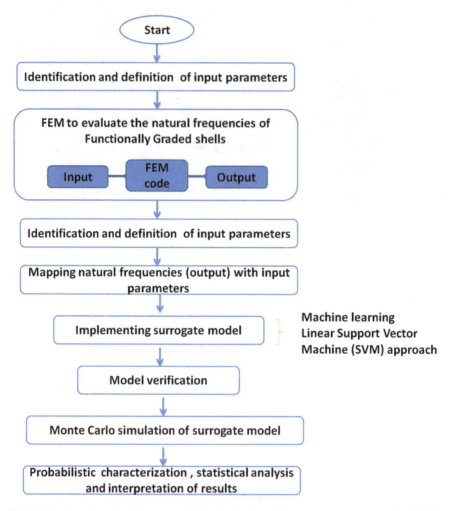

Fig. 2 Flowchart for natural frequency (N_f) analysis using a linear support vector machine (SVM) surrogate model

Table 1 Material properties [10]

	E [GPa]	υ	ρ [kg/m^3]
Metal	700	0.3	2707
Ceramic	168	0.3	5700

decreases along with the decrease in sparsity. Further sensitivity analysis (see Fig. 5) is also carried out which shows the influence of individual material properties. It is observed that Young's Modulus has a maximum effect while the effect of shear modulus and mass density is moderate and Poisson's ratio has the least influence.

Table 2 Root Mean Square Error (RMSE) for first, second and third natural frequency (rad/sec) while using quadratic SVM surrogate on functionally graded cylindrical shells with a sample size of 256, 512 and 1024

	Fundamental natural frequency (rad/s)	Second natural frequency (rad/s)	Third natural frequency (rad/s)
256	0.838	0.2487	0.79457
512	0.14649	0.15137	0.43908
1024	0.088871	0.14839	0.26091

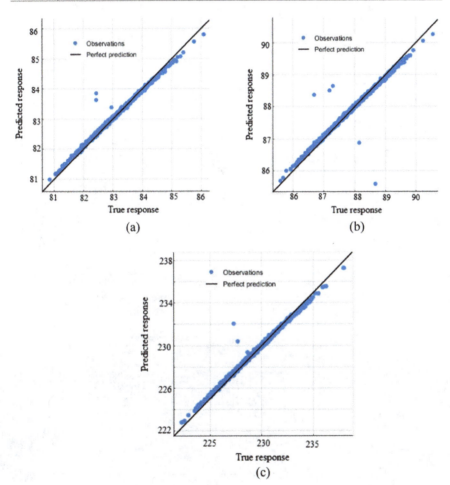

Fig. 3 Scatter plot due to combined variation of random input parameters considering quadratic SVM surrogate model with a sample size of 1024

Table 3 The fundamental natural frequency of FG square plate for symmetric boundary conditions

P	t/L	Baferani et al. [11]	Present Study
1	0.1	0.0891	0.0883
2	0.1	0.0819	0.0797

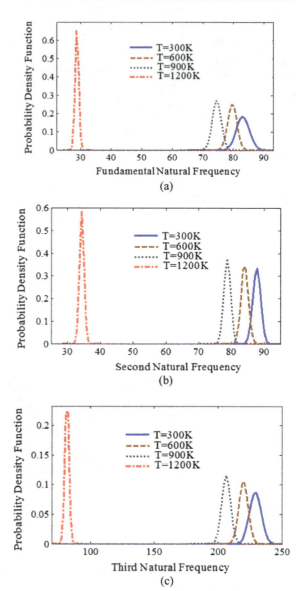

Fig. 4 Stochastic natural frequency (rad/s) of FGM cylindrical shells (**a**) Fundamental (**b**) Second and (**c**) Third modes considering temperature(T) = 300 K, 600 K, 900 K, 1200 K

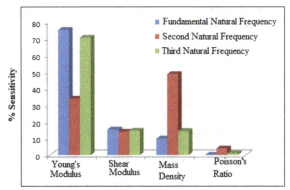

Fig. 5 Sensitivity analysis for various random input parameters for fundamental, second and third natural frequency

4 Conclusions

Based on machine learning quadratic Support Vector Machine (SVM) in combination with finite element formulation, the natural frequency of functionally graded (FG) cylindrical shell is studied. The surrogate used increases the computational efficiency along with the reduction in computational cost. The novelty of present work includes the effect of temperature on stochastic natural frequencies of FGM cylindrical shells. Due to unavoidable inherent randomness present in these structures, a band of deviation in the values of natural frequencies is observed compared to the deterministic mean value. Therefore, to map the degree of influence of sources causing uncertainties, sensitivity analysis is carried out in order to ensure the reliability and safety of such shell structures. Based on these observations, the present work can be further extended to more complex geometries and structures.

Acknowledgements The first author acknowledges Ministry of Human Resource and Development (MHRD) and Technical Education Quality Improvement Programme (TEQIP-III), Govt. of India, for the financial support provided during research work.

References

1. Reddy BS, Kumar JS, Reddy CE, Reddy KVK (2014) Free vibration behaviour of functionally graded plates using higher-order shear deformation theory. J Appl Sci Eng 17(3): 231–241
2. Wang YQ, Jean WZ (2017) Nonlinear dynamics of functionally graded material plates under dynamic liquid load and with longitudinal speed. Int J Appl Mech 9(4):1750054
3. Dey S, Sarkar S, Das A, Karmakar A, Adhikari S (2015) Effect of twist and rotation on the vibration of functionally graded conical shells. Int J Mech Mater Des 11(4):425–437
4. Sharma AK, Chauhan PS, Narwaria M, Pankaj R (2016) Vibration analysis of functionally graded plates with multiple circular cutouts. Int J Curr Eng Sci Res 2(3):15–21
5. Apuzzo A, Raffaele B, Raimondo L (2015) Some analytical solutions of functionally graded Kirchhoff plates. Compos B Eng 68:266–269

6. Barretta R, Raimondo L (2015) Analogies between Kirchhoff plates and functionally graded Saint-Venant beams under torsion. Continuum Mech Thermodyn 27(3):499–505
7. Karsh PK, Mukhopadhyay T, Dey S (2018) Stochastic dynamic analysis of twisted functionally graded plates. Compos B Eng 147:259–278
8. Dey S, Adhikari S, Karmakar A (2015) Impact response of functionally graded conical shells. Lat Am J Solids and Struct 12(1):133–152
9. KarshPK, Mukhopadhyay T, Dey S (2018) Stochastic investigation of natural Frequency for functionally graded plates. IOP Conf Series Mater Sci Eng 326(1)
10. Tornabene F, Erasmo V (2013) Static analysis of functionally graded doubly-curved shells and panels of revolution. Meccanica 48(4):901–930
11. Baferani AH, Saidi AR, Ehteshami H (2011) Accurate solution for free vibration analysis of functionally graded thick rectangular plates resting on elastic foundation. Compos Struct 93(7):1842–1853
12. Touloukian YS (1967) Thermophysical properties of high temperature solid materials. McMillan, New York
13. Loy CT, Lam KY, Reddy JN (1999) Vibration of functionally graded cylindrical shells. Int J Mech Sci 41(3):309–324
14. Bathe K-J, Edward LW (1973) Solution methods for eigenvalue problems in structural mechanics. Int J Numer Meth Eng 6(2):213–226

CPSIA information can be obtained
at www.ICGtesting.com
Printed in the USA
BVHW012353081220
595246BV00001B/1